Teaching and Learning Elementary and Middle School Mathematics

SECOND EDITION

W9-AKV-811

Douglas E. Cruikshank
Linfield College

Linda Jensen Sheffield
Northern Kentucky University

Merrill, an imprint of
Macmillan Publishing Company
New York

Maxwell Macmillan Canada
Toronto

Maxwell Macmillan International
New York   Oxford   Singapore   Sydney

Editor: Linda James Scharp
Developmental Editor: Kevin M. Davis
Production Editor: Rex Davidson
Art Coordinator: Peter A. Robison
Text Designer: Anne Flanagan
Cover Designer: Russ Maselli
Production Buyer: Patricia A. Tonneman

This book was set in Century Old Style by Carlisle Communications, Ltd. and was printed and bound by Semline, Inc., a Quebecor America Book Group Company. The cover was printed by Phoenix Color Corp.

Macmillan Publishing Company
866 Third Avenue
New York, NY 10022

Macmillan Publishing Company is part of the
Maxwell Communication Group of Companies.

Maxwell Macmillan Canada, Inc.
1200 Eglinton Avenue East, Suite 200
Don Mills, Ontario M3C 3N1

Library of Congress Cataloging-in-Publication Data
Cruikshank, Douglas E., 1941–
    Teaching and learning elementary and middle school mathematics/
Douglas E. Cruikshank, Linda Jensen Sheffield.—2nd ed.
        p.   cm.
    Rev ed. of: Teaching mathematics to elementary school children.
c 1988.
    Includes bibliographical references and index.
    ISBN 0-02-326095-5
    1. Mathemathics—Study and teaching (Elementary) I. Jensen, Linda
R., date   . II. Cruikshank, Douglas E., 1941–   Teaching
mathematics to elementary school children. III. Title.
QA135.5.C778 1992
372.7—dc20   91–9670
                                                                  CIP

Printing: 2 3 4 5 6 7 8 9   Year: 3 4 5

Preface

This second edition of *Teaching Mathematics to Elementary School Children: A Foundation for the Future* now titled *Teaching and Learning Elementary and Middle School Mathematics* is designed to assist you as you teach mathematical concepts and skills to elementary and middle school children, grades kindergarten to eight. Our focus is on the future; children with strong mathematical foundations will be better equipped for the world they will face in the twenty-first century. No one can predict the changes that will take place by that time, but we do know that it will be insufficient to teach elementary and middle students merely to compute or to solve the problems of today. The mathematical foundations are laid when children actively construct or invent mathematics, discuss and write about their work, and solve a variety of problems. Students must be able to think mathematically, logically, visually, and creatively. They should be able to use calculators and computers.

We have expanded and strengthened the second edition of this text by:

1. Incorporating both the curriculum and professional standards of the National Council of Teachers of Mathematics in each chapter.
2. Strengthening the presentation of mathematical concepts and skills at the middle school level.
3. Encouraging the use of constructivist approaches to develop mathematical reasoning, including activities and discussions.
4. Adding an introductory chapter and new chapters on technology, problem solving, and algebra.
5. Adding sections in each chapter on grouping students for learning and verbal and written communication.
6. Including expanded sections on assessment alternatives and cooperative learning.

This book will help you understand and carry out the teacher's role in elementary and middle school mathematics instruction. To this end, we have drawn research from a variety of sources, reviewed recommendations from professional organizations, such as the National Council of Teachers of Mathematics (NCTM), and employed experiences gained from children and teachers. All of these have been carefully applied to the teaching of mathematics. Two documents intended to guide reform in school mathematics during the 1990s have been published by the NCTM since the first edition of this text. They are *Curriculum and Evaluation Standards for School Mathematics* (1989) and *Professional Standards for Teaching Mathematics* (1991). We have included many of the recommendations from these two documents at the beginning of each chapter.

Many elementary and middle school teachers rely heavily on mathematics textbooks and teachers' manuals for their lessons. To supplement textbook use, we present numerous activities to illustrate how to teach mathematical concepts and principles more actively and to provide you with models to use in your classroom. Computers and calculators are emphasized in each chapter, for they are integral instructional tools in each mathematics unit. Rote memorization of facts and algorithms is de-emphasized, while higher-level thought processes, mental calculation and estimation, and problem creating and solving are stressed. We have embraced the principles of constructivism and have emphasized the importance of allowing children to construct mathematics and to discuss their work.

We hope you will try many of the activities we have presented. Teachers who discover the joy of actively participating in learning mathematics will be more enthusiastic and confident when teaching. Your interaction with children will be richer and more exciting. You are encouraged to add to the collection of activities in this text, modify others for your particular students, and share activities and experiences with other teachers.

The features that made the first edition of this text distinctive have been maintained:

1. Focusing on the future.
2. Incorporating current research on how children learn specific mathematical topics.
3. Considering children's different learning styles.
4. Integrating computers and calculators as tools in the teaching of each content strand.
5. Focusing on problem creating and solving and using children's own thought processes in teaching all mathematical content.
6. Presenting activities in each chapter as models for concrete concept development and reinforcing multiple embodiments of topics.
7. Displaying pages from elementary mathematics textbooks and discussing their instructional uses.
8. Including probability, statistics, and graphing which has been emphasized recently due to the influence of personal computers.
9. Presenting bibliographies in each chapter that direct the reader to recent research material in each content area.
10. Presenting, in appendixes, blackline masters for frequently used materials and addresses of publishers of software and commercial manipulative materials.
11. Including topics such as evaluating, estimating and mental calculation, and problem creating and solving in each chapter.
12. Separating the teaching of concepts and the teaching of skills, with many suggestions for each.

We hope you are the finest teachers ever to assist children in learning. We need more skillful, knowledgeable, and compassionate teachers. If this textbook can, in any way, help prepare, direct, and encourage you in nurturing the mathematical growth of children, then we will be fulfilled.

We would like to thank the many children we worked with as we were preparing this manuscript for their invaluable feedback on activities and the mathematics they were learning. A special thanks to the children in Carlton and McMinnville, Oregon; Hebron, Kentucky; Findlay, Ohio; and Ft. Thomas, Kentucky, for their help and inspiration. Thanks to the preservice and inservice teachers we have worked with over the last 23 years of college teaching. We could not have produced this book without all your feedback and suggestions.

We would like to express our appreciation to the reviewers, whose suggestions made this a much stronger textbook. The reviewers of the first edition: Lucy Dechene, Fitchburg State College; Jon Engelhardt, Arizona State University; Boyd Holtan, West Virginia University; Charles Lamb, University of Texas at Austin; Walter Secada, University of Wisconsin-Madison; Richard Shumway, Ohio State University; and Alan Zollman, University of Kentucky. The reviewers of the second edition: Louis Fillinger, Fort Hays State University; Helene J. Sherman, University of Missouri-St. Louis; Jerry Becker, Southern Illinois University at Carbondale; William L. Merrill, Central Michigan University; Anne G. Dorsey, University of Cincinnati; Joan C. Carson, University of Mississippi; and Hiram D. Johnston, Georgia State University. The editorial staff at Macmillan has been very supportive. Finally, our families and friends deserve much credit for their understanding and support. We are deeply grateful to Linda, Lori, Julie, Bill, Maureen, and Danny.

D. E. C.
L. J. S.

Contents

3 Children and Mathematics   21

The Children's World   21

Psychological Considerations in Teaching Mathematics   24

Teaching Children Mathematics   29

4 Solving and Creating Problems, Estimating, and Mental Calculating   35

Problem Solving   37

Estimating   42

Mental Calculating   43

Grouping Students for Problem Solving, Estimating, and Mental Calculating   45

Communicating in Problem Solving   48

Evaluating Problem Solving, Estimating, and Mental Calculating   48

Something for Everyone   51

10 Teaching Operations with Rational Numbers   195

Contents

You are about to start on a very important journey. This journey will be an investigation of how you can be successful in your role as an elementary or middle school teacher of mathematics. We hope that by the end of the journey you will have gained confidence as a person and as a teacher, and that you will be ready to share the joy and excitement that emerges from sound mathematics instruction. This is not a journey where you will be asked to sit back and relax; rather, you will be challenged to develop new confidence in your own mathematical ability, to attack intriguing mathematical problems, to look inward at yourself as an individual and as a teacher, and to set personal goals for your elementary or middle school teaching. This textbook is intended to provide assistance and offer direction on your journey. We realize, however, that the most effective test of your mathematics teaching ability will be your own teaching experiences. You will be supported by your teachers, supervisors, fellow students, and, of course, by your students, who will appreciate your methods of teaching.

How do individuals prepare to teach mathematics? At first glance, the natural question, "How will I teach mathematics?" appears to be one that will never be satisfactorily answered. The beginning of the journey to answer this question is found here in Chapter 1. Two important topics are examined to give you a foundation for your investigation of how to teach mathematics:

1. Philosophical considerations.
2. Influences on the school mathematics curriculum.

Philosophical Considerations

You have taken mathematics courses to prepare yourself for teaching. It is common to have taken such courses, perhaps for several years, and never to have been introduced to the nature of mathematics and what it means for you to know mathematics. Investigating these topics can prove helpful for you as you become a teacher of elementary or middle school mathematics. Knowing something about the nature of mathematics and what it means to know mathematics is considered fundamental by the Mathematical Sciences Education Board of the National Research Council (1990, p. 9). They have pointed out that "To realize a new vision of school mathematics will require public acceptance of a realistic philosophy of mathematics that reflects both mathematical practice and pedagogical experience."

The notion of a new vision of school mathematics comes from several sources that will be discussed in this and the next chapter. It refers to a new way to think about how children learn mathematics and how the subject may be effectively taught. You will be a part of an exciting period of new emphasis in school mathematics. Let us turn to two questions that will serve as a philosophical basis for school mathematics in the 1990s and beyond.

What Is Mathematics?

Mathematics is particularly well defined by the National Research Council in its publication *Everybody Counts:*

> As a practical matter, mathematics is a science of pattern and order. Its domain is not molecules or cells, but numbers, chance, form, algorithms, and change. As a science of abstract objects, mathematics relies on logic rather than on observation as its standard of truth, yet employs observation, simulation, and even experimentation as means of discovering truth. (1989, p. 31)

The idea of patterns is key to the nature of mathematics. Children observe patterns long before entering school; they bring with them the ability to discover and recognize patterns. Elementary and middle school children are introduced to patterns in counting, addition tables, geometry, fractions, and decimals. Students understand mathematics because they are able to discern patterns in manipulative materials, pictures, and symbols. Number and geometric patterns exhibit a regularity that is consistent and makes sense to young learners. Thinking of mathematics as the science of patterns and order helps you as a teacher when you begin to teach children mathematics. You will want to assist children in looking for and developing patterns in their study of mathematics.

What Does It Mean to Know Mathematics?

What it means to know mathematics emerges from the nature of mathematics. Thus, to know mathematics means to know patterns and relationships among patterns. The National Research Council expands on this notion. The learner needs

> . . . to be able to discern patterns in complex and obscure contexts; to understand and transform relations among patterns; to classify, encode, and describe patterns;

to read and write in the language of patterns; and to employ knowledge of patterns for various practical purposes. (1990, p. 12)

Children must have the opportunity to study, discover, and invent patterns of many different types while in the process of learning and eventually knowing mathematics. This suggests that the process of knowing mathematics involves active mental activity that is an outgrowth of many types of experiences. As a teacher, you have a responsibility to help children in their quest to know patterns and relationships among patterns.

Implications of These Philosophical Considerations

By developing a philosophical basis for mathematics education we hope to illustrate how the new vision of school mathematics will influence the teaching of mathematics to elementary and middle school students. Classifying mathematics as a science suggests that mathematics is actively explored through experimentation, discovery, manipulation, and discussion, and that calculators and computers can be used as tools of mathematics. This view contrasts with the view that mathematics is only a paper and pencil exercise that relies on rules, formulae, and memory.

Thinking of mathematics as a search for patterns suggests that the content of a mathematics program should include patterns of many types, including those discovered in the study of numbers, geometry, measurement, and algebra, topics usually considered in contemporary mathematics programs. Problem situations, problem solving, and problem posing are provided as ways to search for patterns. This view contrasts with the view that elementary school mathematics programs should focus only on arithmetic—rules and algorithms.

Your philosophical beliefs about mathematics will serve you as you engage in teaching. You will be able to envision a broader mathematics curriculum, one that includes more than just arithmetic. You will feel comfortable about encouraging students to introduce problems from their own experience. You will see merit in the use of modern technology in teaching. You will be convincing as you discuss your beliefs with parents and colleagues. In short, you will have a strong rationale for the new vision of school mathematics.

Influences on the School Mathematics Curriculum

Many forces influence the mathematics content taught in elementary and middle schools. Some have greater influence than others, and the influence shifts as times change. Nevertheless, all of the forces interact with one another. Among the most influential forces are professional organizations, mathematics textbooks, standardized achievement tests, and state governmental bodies.

Professional Organizations

The mathematics education community in this country has strong, active membership in several professional organizations. Among those organizations are the Mathematical Association of America (MAA), National Council of Teachers of Mathematics (NCTM), School Science and Mathematics Association (SSMA), Research Council for Diagnostic and Prescriptive Mathematics (RCDPM), and Psychology of Mathematics Education (PME). As well, individual state mathematics organizations play an important role by supporting sound programs of mathematics in their states.

The National Council of Teachers of Mathematics is perhaps best known to elementary and middle school teachers. Its members assume leadership roles at the national, state, and local levels to provide sound mathematics programs. The NCTM develops position papers based on the work of its many national committees. The NCTM disseminates information to teachers and administrators through its publication house. Pamphlets, yearbooks, small books, and journals provide readers with up-to-date information about teaching mathematics. Elementary and middle school teachers find many teaching ideas for classroom use in the *Arithmetic Teacher,* a monthly journal.

The NCTM and its affiliated state organizations hold outstanding conferences throughout the school year. Each year, there is one national meeting, and there are numerous regional conferences. These conferences are highlighted by hundreds of sessions and workshops that show teachers effective ways to present mathematics.

The leadership provided by the NCTM helps influence the direction of the school mathematics curriculum. Illustrative of this leadership are two recent documents. The documents are *Curriculum and Evaluation Standards for School Mathematics* (1989) and *Professional Standards for Teaching Mathematics* (1991). Each will be discussed below.

Curriculum and Evaluation Standards for School Mathematics.   This document consists of a set of standards for a K–12 mathematics curriculum for schools in North America. As well, it consists of a set of standards for evaluating both the curriculum and students' mathematical achievement. Accordingly, the NCTM (1989, p. v) noted that "The *Standards* is a document designed to establish a broad framework to guide reform in school mathematics in the next decade. In it a vision is given of what the mathematics curriculum should include in terms of content priority and emphasis." The *Standards* is a remarkable document in that it represents the thinking of a broad cross section of the mathematics education community in the United States, including classroom teachers. It is also the first time that the NCTM has recommended a broadly

TABLE 1–1

Curriculum Standards

Grades K–4	Grades 5–8
1. Mathematics as Problem Solving	1. Mathematics as Problem Solving
2. Mathematics as Communication	2. Mathematics as Communication
3. Mathematics as Reasoning	3. Mathematics as Reasoning
4. Mathematical Connections	4. Mathematical Connections
5. Estimation	5. Number and Number Relationships
6. Number Sense and Numeration	6. Number Systems and Number Theory
7. Concepts of Whole Number Operations	7. Computation and Estimation
8. Whole Number Computation	8. Patterns and Functions
9. Geometry and Spatial Sense	9. Algebra
10. Measurement	10. Statistics
11. Statistics and Probability	11. Probability
12. Fractions and Decimals	12. Geometry
13. Patterns and Relationships	13. Measurement

(NCTM, 1989, p. 15, 65. Reprinted by permission.)

stated K–12 mathematics curriculum. It is expected that the direction of mathematics education in the 1990s and beyond will be greatly influenced by the message and recommendations contained in this document.

The *Standards* contains five general goals for all students along with several implications:

> . . . (1) that they learn to value mathematics, (2) that they become confident in their ability to do mathematics, (3) that they become mathematical problem solvers, (4) that they learn to communicate mathematically, and (5) that they learn to reason mathematically. These goals imply that students should be exposed to numerous and varied interrelated experiences that encourage them to value the mathematical enterprise, to develop mathematical habits of mind, and to understand and appreciate the role of mathematics in human affairs; that they should be encouraged to explore, to guess, and even to make and correct errors so that they gain confidence in their ability to solve complex problems; that they should read, write, and discuss mathematics; and that they should conjecture, test, and build arguments about a conjecture's validity. (1989, p. 5)

These goals are appropriate for you as an elementary or middle school teacher. Your enthusiasm for mathematics and your ability to effectively teach mathematics will depend, in part, on your acceptance of these goals and their implications as a personal challenge.

The curriculum standards presented in this document are divided into three levels, grades K–4, 5–8, and 9–12. The evaluation standards are presented in a fourth section. Twenty-six standards are presented in levels K–4 and 5–8 (see Table 1–1). Each standard is then discussed and illustrated with examples of how the standard might be realized in the classroom. You will find the standards included throughout this textbook.

The *Standards* clearly calls for change in mathematics education. It suggests that increased attention be given to topics previously given little or cursory attention, such as developing the meanings of fractions, decimals, numbers, operations, and spatial sense; probability and statistics; and reasoning. It suggests that decreased attention be given to topics that previously dominated instruction, such as performing complex paper and pencil computations, memorizing rules and algorithms, and manipulating symbols. While not covering all of the areas of increased and decreased attention, these examples should give you a sense of the shift in direction recommended by the NCTM. You are encouraged to become familiar with the *Curriculum and Evaluation Standards for School Mathematics* and to use your knowledge of these standards to establish yourself as a forward-looking teacher of mathematics. These standards are presented in the chapters that follow.

Professional Standards for Teaching Mathematics. This document is designed to complement the *Curriculum and Evaluation Standards* discussed above. It is intended to set standards for teachers and administrators so they will be able to provide the kind of instruction recommended by the *Curriculum and Evaluation Standards*. The professional standards for teaching mathematics involve four different aspects of professional responsibility: (1) teaching, (2) evaluation of teaching, (3) professional development of teachers, and (4) support and development of teachers and teaching. There are a total of 24 professional standards (see Table 1–2). Following the presentation of each standard there is an elaboration of the standard followed by a vignette or an example that illustrates how the standard might be met. Professional teaching standards are included in Chapter 15.

The professional standards will serve you as you implement your goals for teaching mathematics based on the *Curriculum and Evaluation Standards*. We are very fortunate to have these two sets of standards. Hopefully, you will be able to take advantage of the wealth of recommendations put forth.

TABLE 1–2

Professional Standards for Teaching Mathematics

Standards for Teaching Mathematics

Tasks
1. Worthwhile Mathematical Tasks
Discourse
2. Teacher's Role in Discourse
3. Students' Role in Discourse
4. Tools for Enhancing Discourse
Environment
5. Learning Environment
Analysis
6. Analysis of Teaching and Learning

Standards for the Professional Development of Teachers of Mathematics

1. Experiencing Good Mathematics Teaching
2. Knowing Mathematics and School Mathematics
3. Knowing Students as Learners of Mathematics
4. Knowing Mathematical Pedagogy
5. Developing as a Teacher of Mathematics
6. The Teacher's Role in Professional Development

Standards for the Evaluation of the Teaching of Mathematics

The Process of Evaluation
1. The Evaluation Cycle
2. Teachers as Participants in Evaluation
3. Sources of Information
The Foci of Evaluation
4. Mathematical Concepts, Procedures, and Connections
5. Mathematics as Problem Solving, Reasoning, and Communication
6. Mathematical Disposition
7. Assessing Students' Mathematical Understanding
8. Learning Environment

Standards for the Support and Development of Mathematics Teachers and Teaching

1. Responsibilities of Policymakers in Government, Business, and Industry
2. Responsibilities of Schools
3. Responsibilities of Colleges and Universities
4. Responsibilities of Professional Organizations

(NCTM, 1991, p. 19, 71, 123, 177. Reprinted by permission.)

Mathematical Textbooks

The mathematics textbook, commonly referred to as the "math text," has historically been very influential in determining the elementary and middle school mathematics curriculum. It establishes an important curricular framework. It provides a continuity from September to June and consistency from one grade level to another. The mathematical topics included in math texts have evolved steadily for over a hundred years of textbook publication. The textbooks change when new trends emerge and are supported by teachers, mathematics leaders, parents, and administrators. Publishers respond to their customers.

Well-designed textbooks in the hands of skillful teachers are powerful educational tools. With texts, teachers can diagnose and evaluate children's mathematical performance. They can assign exercises to reinforce concepts and skills already learned as well as to teach and reteach important mathematical concepts and skills.

Skillful teachers know when to go beyond the pages of the textbook. Activities must often replace pictures and symbols. Teaching styles that differ from the textbook presentation will be needed. Mathematical applications will require children's more active participation. The daily textbook routine should give way to other methods to allow mathematics to come alive in the minds of the children.

The National Research Council expressed the challenge faced by textbook publishers in light of recent research findings:

New textbooks must be designed and written to reflect the important principles of mathematics curricula: genuine problems; calculators and computers; relevant applications; reading and writing about mathematics; and active strategies for learning. (1990, pp. 49–50)

You will need to be a discerning consumer of mathematics textbooks. You must decide if textbooks serve your instructional needs and you must decide how math textbooks best serve the needs of children. This is the art of teaching.

Standardized Achievement Tests

Nearly every school district administers standardized achievement tests at least once each academic year. These tests measure a variety of elementary school skills and knowledge. The results of standardized achievement tests are reported in terms of nationally established norms and reflect a child's general achievement. Some districts test only selected grade levels, while others test all grade levels. Considerable importance is accorded test results. A summary of the results is commonly published in local newspapers. As a result, many teachers set high achievement test goals. When teachers spend large amounts of time preparing their children to take achievement tests, the mathematics curriculum is heavily influenced by the test.

Those who carefully design achievement tests do not intend that the tests be used to determine the curriculum.

The tests are developed to measure mathematical concepts and skills commonly found in elementary mathematics programs throughout the country. They rely heavily on the math texts as sources of test items. They also rely on their customers for suggestions when new editions of tests are prepared. To be of continuing value, these tests must reflect the changes that are recommended by the NCTM *Standards* and other forward-looking documents.

State Governmental Bodies

Influenced by numerous national reports on education and a public concerned with the quality of education, state legislative education committees and legislatures, state departments of education, and teacher certification offices have developed laws, policies, and regulations that have an impact on elementary and middle school mathematics curricula. There has been concern about the overall quality of education in this country. As a result, state and local study committees have made recommendations about how to improve the quality of education. As new statewide goals are established, curricular areas such as mathematics are being reviewed. Thus, revisions are being made to statewide mathematics curricula, in part, to align them with the new standards.

Changes in mathematics curricula in various large states result in changes in textbooks and necessitate changes in standardized testing. The forces that influence the curriculum are interrelated. The mathematics curriculum with which you will be involved has been affected by the forces mentioned above. The response you and your students have toward mathematics and the textbook will likely influence curricular change in the future.

Looking Ahead

Most of the remaining chapters of this book (Chapters 5–14) include the following sections:

A. NCTM *Standards*
B. Developing Concepts
C. Developing Skills
D. The Math Book
E. Estimating and Mental Calculating
F. Problem Creating and Solving
G. Grouping for Instruction
H. Communicating
I. Evaluating Mathematics
J. Something for Everyone

Each chapter includes a brief section, Key Ideas, that summarizes the chapter. We discuss how mathematical concepts are developed in Chapter 3, problem creating and solving and estimating and mental calculating in Chapter 4, and grouping for instruction in Chapter 15. What follows is a brief discussion of the remaining sections.

NCTM Standards

The *Standards* was discussed earlier in this chapter. Each standard is presented in more detail at the beginning of the chapter that contains the content strand mentioned in the standard. As a result, you will be able to see the connection between a particular standard and suggestions for developing that content strand in the elementary or middle school classroom.

Developing Skills

Once mathematical concepts have been introduced, skills associated with those concepts may be taught. Skills include the basic facts and procedures that children apply as they calculate and solve problems. Many of the skills children use are memorized. For example, the basic addition facts, those from $0 + 0$ to $9 + 9$, are important when children perform mental and paper and pencil calculations.

Included along with activities for developing skills are activities that employ calculators and computers. As mentioned above, these tools play an important role in the learning of elementary and middle school mathematics.

The type of understanding that comes from learning skills has been called "procedural understanding" (Hiebert, 1986) or "instrumental understanding" (Skemp, 1987). It is substantively different from conceptual understanding, which is discussed in Chapter 3. Hiebert (1986, p. 8) explained that, "Conceptual knowledge, by our definition, must be learned meaningfully. Procedures, on the other hand, may or may not be learned with meaning. We propose that procedures that are learned with meaning are procedures that are linked to conceptual knowledge." He goes on to explain that procedures that have not been connected with conceptual knowledge are more easily forgotten and more difficult to reconstruct. We concur that when skills or procedures are taught they should be associated whenever possible with their underlying concepts.

The Math Book

Each of Chapters 4–15 contains an illustration of a mathematics textbook page that corresponds to the content emphasis of the chapter. In presenting the textbook examples, we recognize the important role that the math text plays in the instructional program; on the other hand, we have devoted the bulk of the current textbook to presenting activities you can use to help children learn mathematics with or without a textbook. In the math book inserts, we illustrate a variety of textbooks to give you an idea of those that are currently on the market. We also take the opportunity to comment on the textbook page illustrated, to show how the author recommends that it be used, and to offer additional suggestions when we feel that they are appropriate.

Communicating

Throughout the learning process, communication is crucial. The ways that teachers and students communicate affect the quality of learning. New emphasis is being given to the oral and written communication that occurs during the learning of mathematics. As children construct and invent mathematics, their discussions and writing are rich with explanations of their thinking processes. A brief description of ways that children may communicate as they learn mathematics has been included in each of the content chapters.

Evaluating Mathematics

The most commonly used tools for evaluating mathematics learning are the curriculum-embedded tests provided with basal textbook series and standardized achievement tests. The textbook tests provide diagnosis at the beginning of a chapter or section, evaluate the children's progress during and at the end of each chapter, and check their ability to recall concepts and skills from earlier chapters. Standardized tests that include sections designed to test mathematics achievement are commonly administered once a year to a group of children.

There are, however, alternative assessment techniques that overcome the shortcomings of textbook tests and standardized tests. Among them are observation and questioning, performance-based assessment, diagnostic interviews, teacher-designed tests, writing activities, and group problem solving. These techniques have an important role in broad-based evaluation programs. In each chapter, evaluation procedures are discussed with regard to the content emphasis of the chapter. A more detailed description of evaluation procedures is included in Chapter 15.

Something for Everyone

All schools should provide elementary and middle school children with the most effective instruction possible. Teachers should consider the learning styles of their students when planning lessons. While children learn using all of their senses, some children depend more on a particular sense. **Visual learners** learn more readily by looking at pictures, illustrations, and objects; **auditory learners**, by listening and talking; and **kinesthetic learners**, by touching objects and moving about. Visual learners should have plenty of pictures, illustrations, and manipulative materials available. Auditory learners should be given clearly stated directions and explanations and should be encouraged to discuss their own knowledge and understanding. Kinesthetic learners should be provided with opportunities to move about and to touch and manipulate objects. Children with a tendency toward a systematic, logical approach will do well with the textbook. Children with a tendency toward a graphic approach will do well with a more active, pictorial, manipulative approach. Thus, while the daily mathematics program may be textbook based,

you must adjust your teaching to allow for the differences among your students. Each chapter discusses ways that you can help children with different learning styles to learn the content emphasized by the chapter.

There are other children with special needs, those who have unusual difficulty learning mathematics and those who have particular talents when it comes to mathematics. Both groups require the teacher's attention. Chapter 15 further discusses the special needs of various children.

Teaching Considerations

Finally, we need to consider our interaction with the children with whom we work. Teaching is a tremendous challenge, and both teachers and students reap enormous personal rewards. Beliefs, upbringing, prejudices, values, self-concept, and personality affect an individual's teaching style. The approach a teacher takes to children, classroom organization, and disruptions is an extension of who that teacher is. Three aspects of teaching in general, as well as of teaching mathematics in particular, are failure, management, and encouragement. Each is briefly discussed below.

Failure.   Children, especially young children, seem to cope naturally with failure. Gaining an initial understanding of life, which includes learning about the environment and roles within the family and society, involves trial and error. Failure in this context is used as a springboard to future success. The healthy use of mistakes can provide a foundation for growth.

On the other hand, failure may be harmful to the child. When teachers or parents teach that failure should be avoided at all costs, children develop sophisticated mechanisms to avoid failure. Holt (1964) discussed many reactions or strategies that children develop to avoid failing. For instance, when a teacher asks a question, some children excitedly wave their hands even though they do not know the answer because they know that the teacher will call on the children who look apprehensive. Or, for example, a child begins to give a response and then closely observes the teacher for clues, usually nonverbal clues, that the response he or she has begun is correct. Teachers may feel that it is their duty to point out all mistakes. Holt and others appropriately warn of the consequences of such behavior. It is far healthier to look for children's strengths and assets. Teachers should not ignore incorrect mathematical concepts, but they should teach concepts without destroying children's egos.

Children need to feel secure. The following are several suggestions for developing children's abilities so that they may become confident learners. Teachers and parents should:

1. Accept their own mistakes openly in front of children. Show children how failure is useful in learning and unimportant in determining one's overall worth.

2. Allow for failure and treat it as a natural part of the learning process. Let children make mistakes and learn from their mistakes. Avoid embarrassing them for making mistakes.

3. Provide opportunities for children to succeed. Develop their confidence through success to the extent that they will maintain confidence when they do not succeed.

4. Avoid blaming children who do not meet adult expectations. Expect high-quality performance, but keep in mind the capabilities of the youngsters.

5. Accept children as individuals worthy of respect. Believe that children deserve respect. Listen when they wish to share ideas and feelings. Respond in a warm manner.

6. Provide opportunities for children to make decisions that affect their lives within the school setting. Make sure the decisions are real and the children can and will abide by the logical consequences of their decisions.

7. Discuss failure with individuals and with the class group. Come to agreements about how failure will be dealt with. Live up to those agreements.

Management.   Management and discipline are of utmost concern to teachers and parents regardless of the age of the children. Many factors affect children's behaviors. Some of these factors are genetic, but most behaviors are learned in the home environment. Others result from the school experience and interaction with society at large. When children are brought together in a group, individual and group behavior should be channeled constructively. Even free play activities involving several individuals in parallel play should be managed so that children with overlapping interests avoid conflict. Cooperation and compromise should be introduced and nourished to the degree that children understand those concepts. Even though egocenteredness is characteristic, respect for the rights of others should be fostered.

Themes commonly found in textbooks on classroom management include these teacher responsibilities: developing and maintaining a working environment, defining the rules for individual and group work, providing a quality educational program, being the educational leader in the classroom, involving students in decision making, being consistent, providing stability, monitoring and guiding activities, and treating misbehavior. You are urged to select one or more books on classroom management and to incorporate in your teaching style management techniques characterized by an ongoing respect for youngsters and their talents.

Encouragement.   Encouragement has been carefully studied. Self-evident as it may appear, encouragement is seldom recognized as a crucial factor in the growth and development of healthy young minds. Teachers and parents tend to discourage youngsters even while they profess the need for encouraging them. Discouragement

emerges as one of the greatest single causes for failure. Dinkmeyer and Dreikurs comment on the importance of encouragement in working with children:

> We, as educators, as parents, and as teachers, are in charge of the greatest treasure society possesses, the next generation. The urgent question which confronts us today is whether we will be able to guide them into becoming capable and responsible human beings or whether we will have to wait until youth itself claims its right to proper guidance and education. This question will be decided, in our opinion, by our ability to change from a punitive, retaliatory, and mistake-centered educational practice to one of encouragement for all those who have failed to find their way toward fulfillment. (1963, pp. 124–125)

As a result of their extensive work on the encouragement process, Dinkmeyer and Dreikurs (1963, p. 50) have delineated nine methods of encouragement:

1. Place value on children as they are.
2. Show faith in children to enable them to have faith in themselves.
3. Have faith in children's abilities; win their confidence while building their self-respect.
4. Recognize effort and a job well done.
5. Utilize the group to facilitate and enhance the development of the child.
6. Integrate the group so that children can be sure of their place in it.
7. Pace skills sequentially and psychologically to permit success.
8. Recognize and focus on strengths and assets.
9. Utilize the interests of children to energize instruction.

KEY IDEAS

A philosophical foundation incorporating the nature of mathematics and what it means to know mathematics is important for teaching the subject. Mathematics is the science of pattern and order. To know mathematics is to know patterns and relationships among patterns. Thus, the study of mathematics is an active investigation of many types of patterns.

Among the major influences on the mathematics curriculum are professional organizations, mathematics textbooks, standardized achievement tests, and state governmental bodies. The National Council of Teachers of Mathematics has published a set of standards for mathematics curricula along with a set of standards for evaluation. These are looked upon as influential guidelines for mathematics instruction in the 1990s and beyond.

REFERENCES

CARSON, JOAN C., and BOSTICK, RUBY N. *Math Instruction Using Media and Modality Strengths.* Springfield, Ill.: Charles C. Thomas, 1988.

DINKMEYER, DON, and DREIKURS, RUDOLF. *Encouraging Children to Learn: The Encouragement Process,* Englewood Cliffs, N.J.: Prentice-Hall, 1963.

HIEBERT, JAMES, and LEFEVRE, PATRICIA. "Conceptual and Procedural Knowledge in Mathematics: An Introductory Analysis." In Hiebert, James, ed. *Conceptual and Procedural Knowledge: The Case of Mathematics.* Hillsdale, N.J.: Lawrence Erlbaum Associates, 1986.

HOLT, JOHN. *How Children Fail.* New York: Pitman Publishing Corp., 1964.

NATIONAL COUNCIL OF TEACHERS OF MATHEMATICS. *Curriculum and Evaluation Standards for School Mathematics.* Reston, Va.: NCTM, 1989.

———. *Professional Standards for Teaching Mathematics.* Reston, Va.: NCTM, 1991.

NATIONAL RESEARCH COUNCIL. *Everybody Counts: A Report to the Nation on the Future of Mathematics Education.* Washington, D.C.: National Academy Press, 1989.

———. *Reshaping School Mathematics: A Philosophy and Framework for Curriculum.* Washington, D.C.: National Academy Press, 1990.

SKEMP, RICHARD R. *The Psychology of Learning Mathematics.* Hillsdale, N.J.: Lawrence Erlbaum Associates, 1987.

2 Technology in Teaching Mathematics

As we approach the twenty-first century, the differences between the industrial age of the early twentieth century and the information age of today are ever more apparent. We can no longer train children for the future by giving them only the "shopkeeper skills" of computing with whole numbers using paper and pencil. We must prepare children for the future by teaching them to think, reason, and solve problems using as much technology as possible, always keeping in mind that the technology they will need to use in the future will be far more advanced than anything available today. We would be doing students a great disservice, however, if we did not let them use all that is available now. This means that we must encourage students of all ages to take full advantage of calculators, computers, and other technology.

The National Council of Teachers of Mathematics has recommended that

- appropriate calculators should be available to all students at all times;
- a computer should be available in every classroom for demonstration purposes;
- every student should have access to a computer for individual and group work;
- students should learn to use the computer as a tool for processing information and performing calculations to investigate and solve problems. (1989, p. 8)

Having this technology available does not mean, as some have suggested, that students will no longer need to think to do mathematics. In fact, the availability of technology offers us the opportunity to emphasize the thinking, reasoning, and problem solving and to de-emphasize the rote memorization that so often characterized the teaching of basic facts and algorithms. Using a calculator or computer to assist with computation should be for a mathematician what using a word processor is for a writer. The technology does not do the thinking, but it frees the writer, or mathematician, to focus on the work that he or she is creating. Students should be afforded this same opportunity. What a waste of a student's years in elementary school if all we teach students is to perform paper and pencil computations that could be done more accurately and efficiently on a $5 calculator!

It is important for teachers to know what children can do with calculators other than check homework, and to know computer applications other than those that transfer drill sheets from a workbook to a computer screen. The following sections investigate some common misconceptions about technology and suggests some uses for it.

Calculators

When I hear parents today complaining that their children are using calculators, I am reminded of my grandfather complaining that my father was allowed to use paper and pencil when he did his mathematics in school. My grandfather learned to do all his mathematics mentally and scoffed at the idea of having to write down anything to do multiplication or long division. He was quite concerned that my father might need to do some arithmetic on the job and would find that he did not have a pencil or a piece of paper. Today, parents worry that children might not have a calculator with them or that it might be too dark to use a light-activated calculator. Now, mathematics educators are realizing that my grandfather had the more urgent concern. As adults, we are called upon to do mathematics mentally much more frequently than we are to do calculations with paper and pencil. Mental computation and estimation and calculators and computers are used to do much of the computation work on the job. The need to do long paper and pencil computation is minimal and losing

importance every day, while the need to mentally compute and estimate is growing in importance, and the optimal use of calculators and computers is essential.

Research has consistently shown that we do not need to worry that children will become dependent upon calculators and never learn their basic facts once they have a calculator available. From a review of several studies on the effects of calculator usage, Hembree and Dessart (1986) concluded that not only do calculators not harm children's knowledge and use of basic facts and algorithms, but that with proper use, they can improve the average student's basic skills with paper and pencil and can improve problem-solving skills. Further, Wheatley and Clements (1990, p. 22) suggested that learning opportunities can be presented by using calculators in problem situations. These learning opportunities, from a constructivist perspective, focus on meaning, assist students as they solve problems, and allow the learner to consider increasingly complex tasks. In addition, calculators can help improve a student's self-concept in mathematics and lessen math anxiety. It is time to stop worrying and to start using calculators.

How Should Calculators Be Used?

Calculators should be used as an everyday part of the mathematics curriculum. They should be used as instructional tools and as computational tools. Teachers should plan lessons that employ the calculator to assist children in developing their thinking abilities. Children should be encouraged to decide when a calculator would be useful and when it would be quicker and easier to solve problems mentally. Throughout this book we will suggest ways in which to use calculators. The following list will give you just a few ideas:

1. *To develop an understanding of the calculator.* Students need to be taught what the calculator does and what individual keys do. Introductory activities include opportunities for children to explore the calculator as they would any manipulative material, to see what they can discover about pushing the keys. Other activities include recognizing numerals, both from the keypad and from the display, and learning the keyboard numeral sequence. To help with the latter skill, a large keyboard may be placed on the floor for students to walk on, and a student can hop from one key to another as directed by the teacher or another student. Later, the keystroke sequence for keying a problem can be acted out on the large keypad. Teachers or students can use a calculator on the overhead projector that matches the individual calculators used by students at their desks to demonstrate certain procedures. Counting and skip counting on the calculator and other activities follow. As a part of these initial activities, the children should be encouraged to learn the vocabulary of the calculator and to discuss what they find as they work with the calculator and what they think is happening.

2. *To develop number sense and place value concepts.* Students should frequently use calculators in conjunction with concrete materials in order to develop solid concepts of the numbers they are using. Understanding leads to number sense, the ability to recognize number relationships, the ability to determine if operations are reasonable, and the ability to interpret numbers used in daily life. For young children, this may mean using a calculator as they count a pile of pennies they have saved. As one child moves one penny at a time from one pile to another, another child can push +1 on the calculator each time. The first child can then announce the number of pennies in the new pile as the second child pushes the equals sign on the calculator to check the total. Special notice should be taken when the number in the tens place changes as a result of adding 1. Ask the children to predict what number will show in the display if you begin at 19 and add 1.

Applying number sense with the calculator means, in part, being able to determine quickly if a calculator

result is reasonable. For example, when asked to find 6×189 the student might think that 189 is near 200 and 6×200 is 1200. Thus, the result should be near 1200, and because 189 is less than 200, the result will be less than 1200. On the other hand, the student may think that because 189 is 11 less than 200, 6×189 will be 6×11 or 66 less than 6×200, that is, the result will be $1200 - 66$. The final result, 1134, may be computed on a calculator or mentally. This type of procedure and others are presented by Bobis (1991) in a discussion of number sense and calculators.

Older children might use the calculator to help them determine if they are at least one million days old, one million hours old, one million minutes old, or one million seconds old. They could also use it to determine if one million ping pong balls or one million mathematics books would fit in their classroom. These activities help them develop a much better sense of the relative magnitudes of numbers than would filling in a worksheet that asks them which numeral is in the millions place in 23,456,741.

3. *To develop and recognize patterns.* Using a calculator takes much of the drudgery out of computations and allows students to perform computations that might be too time-consuming using only paper and pencil. Third graders might wish to explore what happens when numbers are multiplied by 9. As they multiply 9×1, 9×2, 9×3, etc., on a calculator, many patterns begin to emerge. They should notice that when multiplying by a one-digit number, the sum of the digits in the answer add to 9 and the ones digit decreases by one while the tens digit increases by one. They could probably also discover these patterns without a calculator, but what happens when the multiplier becomes larger than 10? Will the sum of the digits in the answer always be a multiple of 9? Will the pattern in the ones and tens places continue? With a calculator, children can explore patterns involving very large numbers and can focus on finding reasons for the patterns rather than on the computation itself. They can then go on to other patterns, such as those in multiples of 3, 5, 10, or 11. They may wish to record their findings on a hundreds chart and expand their search to include such topics as prime and composite numbers. Many mathematical puzzle books offer other ideas for students to explore, especially in sections on mathematical magic, which give students ideas for fooling their friends and families using some of the structure of our numeration system.

An interesting exploration with the calculator was discussed by Zollman (1990). Zollman noted that subtracting various sequences of digits that follow certain *geometric* patterns results in the difference 198. Figure 2–1 illustrates three of these patterns.

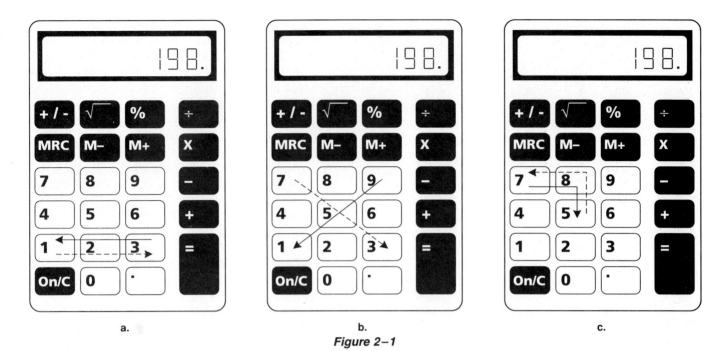

a.   b.   c.

Figure 2–1

Figure 2–1a illustrates the *rows* pattern. For example, 321 − 123 results in 198. Likewise, 654 − 456 and 987 − 789 both result in 198. Figure 2–1b shows the *diagonals* pattern. For example, 951 − 753 results in 198, as does 357 − 159. Figure 2–1c shows the *right angles* pattern. For example, 785 − 587 results in 198 and 624 − 426 also results in 198. There are other combinations that follow these patterns and result in 198. As well, there are other patterns, such as the *crosses, squares, obtuse angles,* and *acute angles* patterns. Can you find them? Encourage the students to explore why these patterns exist.

4. *To develop concepts of operations.* A calculator is of no use if you do not know which operation to use to solve a given problem. Children can explore with a calculator to determine why addition and subtraction are inverse operations or why multiplication may be used instead of repeated addition. Second graders may wish to determine the number of eggs stored in the cafeteria. One student may note that there are 15 cartons of eggs in the school refrigerator, each holding a dozen eggs, and may suggest that they add 12 + 12 + 12. . .15 times to determine the total number of eggs. Another student may suggest simply multiplying 12 × 15. Ask half of the students to find the answer using repeated addition while the other half use multiplication, and discuss the results. Ask the students to suggest several problems of this type and work them using both addition and multiplication to convince the students that the answers will always be the same.

Students in the middle grades might use a calculator that displays and computes with fractions in the $\frac{p}{q}$ format. This can help them see that one does not add numerators and denominators together when adding fractions. They can then use concrete materials to explain what is happening on the calculator.

5. *To develop problem-solving and thinking abilities.* When children are given problem situations involving the calculator they have the chance to increase their skills in reasoning and in constructing mathematical relationships. Wheatley and Clements (1990) describe activities that facilitate these thinking processes. A description of one of these activities follows. This is an activity for two students using a nonscientific calculator. Some calculators cannot be used for this activity and you should check the calculators available before starting the activity. One student picks a mystery number, such as 43, and enters 43 ÷ 43 = . This will result in a *1* in the calculator display. The calculator is given to the second student, who is instructed to "Guess my number." The second student enters his or her guess and pushes the equals key. The second student will have successfully guessed the number when the calculator again displays a *1*. A series of guesses is shown below that eventually result in achieving the goal.

$$8 = 0.1860465$$
$$13 = 0.3023256$$
$$25 = 0.5813953$$
$$50 = 1.1627907$$
$$45 = 1.0465116$$
$$44 = 1.0232558$$
$$43 = 1$$

The mystery number is 43. The process students use to successfully determine the target number may initially involve random guessing, but soon students will recognize the relationship among the strings of digits. Eventually, the activity will take on greater meaning as understanding of decimal numeration is increased. Students who solve this problem are constructing meaning in a way that makes sense to them.

6. *To get a graphic picture of the data.* Middle school students can explore the uses of a graphics calculator. With a prompt of "Y =" on the calculator, the students can discover what happens to the graph of the formula $y = mx + b$ as m or b changes. They can graph formulae that they have seen in other classes, such as force = mass × acceleration ($F = ma$), and study the effects of changing one of the variables, such as the mass or the acceleration. Again, this should accompany a concrete experiment to avoid trying to develop concepts in isolation. Even middle school students are rarely able to develop a new concept on a completely abstract level, although we may be tempted to try to present concepts on an abstract level with the aid of a calculator.

7. *To solve problems where the computation might otherwise be prohibitive.* There are lots of examples of problems in everyday life that we would hesitate to solve if we did not have a calculator available. Young students might wish to plan a budget for their allowances. If they are trying to save enough money to buy a toy for $3.75 on an allowance of $.50 a week, first graders might use a calculator to determine the number of weeks they will have to save. They might suggest subtracting $.50 from $3.75 until no money remains and then counting the number of times they subtracted. This might also be a good opportunity to discuss the use of division as repeated subtraction. Middle school students may be starting to think about saving money for college tuition and could use the calculator to determine which of several banks or savings and loans are offering the best interest rate. Ask students to be on the lookout for problems in everyday life that you would use a calculator to solve. You might set up a store or bank in your classroom that the students would run with the aid of a calculator.

8. *To improve mental computation and estimation skills.* Calculators should be used in conjunction with mental computation and estimation and not as a replacement for these skills. Students should always estimate to determine whether an answer on the calculator makes sense. You might try playing a target game to help the students improve these skills. In this game, the leader puts a number sentence on the board such as 16 × _____ = 4562. Each student then writes down his or her best estimate. After the estimate is recorded, the student is allowed to use the calculator only to multiply that estimate by 16. After this calculation is performed, the students are polled to determine the winner of the round. For example, if the best estimate is 300, the leader would write on the board 16 × 300 = 4800. Students would then use this information in order to come up with a better estimate. Play continues for several rounds or until the answer is within 10 of the correct product. Older students could be encouraged to use decimals until the answer is less than 1 away from the correct product. Games such as this one help students develop their mental computation skills and can be adjusted to the level of the students. Younger children might play the game with addition and subtraction, rather than multiplication and division.

What Types of Calculators Should Be Purchased?

This is a very difficult question and one upon which the "experts" disagree. The answer will depend on the level of the students with whom you are working and your budget. For many computations, a simple four-function calculator will be sufficient. If you choose this type of calculator, be sure to choose a brand that is light-activated; otherwise, the expense of batteries would soon amount to more than the cost of the calculator. You should look for a calculator that has memory and constant functions and a key for square roots. These can usually be found for under $5. The main drawback is that the operations will probably be done in the order in which they are entered into the calculator and not in correct mathematical order. This should be discussed with the students if you opt for this type of calculator. For about $8, you can purchase a four-function calculator from Texas Instruments, which uses correct order of operations. You should also get a calculator for the overhead projector that matches the calculator the students are using so that you can demonstrate its use to all of the students simultaneously.

For middle school students, you might consider getting a calculator that will operate on common fractions in the $\frac{p}{q}$ form and that displays the result of division as an integral answer and a remainder rather than as a decimal. The most popular of these is the Texas Instruments Explorer calculator. There is also an overhead calculator to match this calculator. For older students, you might also want to consider a graphics calculator. They are versatile and provide for a number of applications.

Computers

Like calculators, computers are finding an ever more important place in society and in the schools. In many schools, they are even more commonplace than calculators

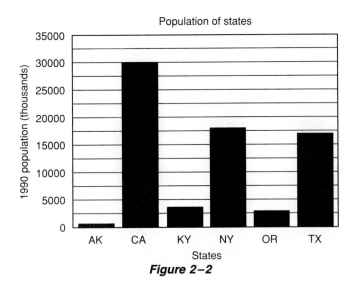

Figure 2-2

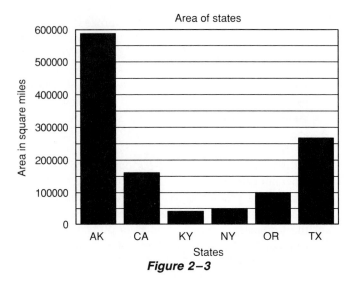

Figure 2-3

in spite of their cost. With the proliferation of computers, we must be careful to ensure that they are used to their advantage and not simply for completing worksheets on a screen instead of on paper. The following are a few of the ways in which computers may be used in a mathematics class:

1. *To teach programming.* Learning to write a computer program teaches children to be very careful and to pay attention to details. Some calculators can be programmed, as can all computers. Two programming languages that are popular in elementary and middle schools are Logo and BASIC. BASIC is a well-known, general purpose programming language and is available on most types of computers. Commands for the computer are written in English and are relatively easy to learn, although students must be very careful to get all spacing and punctuation correct in all commands. Logo is a computer language that was developed at the Massachusetts Institute of Technology by Seymour Papert specifically for children. With it, children can write graphics programs as well as do word-processing and other types of programming. More will be said about Logo in the chapter on geometry.

2. *As a tool for either student or teacher use.* The computer can help the teacher and the student with a variety of tasks. These include word-processing, record keeping, constructing graphs and charts, and telecommunications. Teachers may use the tools to keep track of student progress and manage individualized assignments or to help with assigning grades or recording attendance. With the new emphasis on communication in mathematics, word-processing makes it easier for students to write problems for other students to solve, to record how they solved a problem, or to keep a daily log of what they are learning in mathematics. Graphics and music programs even allow students to illustrate their work

and set it to music. Students and teachers alike may use a data base to collect information to solve real life problems. For example, students could collect census information about the class, and they could use the data to create graphs and other displays to answer questions of interest. For example, students could compare information such as the population and the area of selected states. Figure 2-2 graphically shows the population of six states.

The data are entered into a computer graphing application and the type of graph is selected. The computer generates the appropriate graph.

If another graph such as that shown in Figure 2-3 is produced, then a graphic comparison can be made between the two types of information displayed, population and area.

Most graphing applications give the operator a choice between different types of graphs: bar, column, line, pie, scatter, and so forth. A variety of interesting comparisons can be made using graphs.

Spreadsheets may be used to collect data that can also be displayed graphically, such as data that provide information related to the characteristics of various sports balls. These data, presented in the *Arithmetic Teacher* (November, 1990), lend themselves to some interesting explorations. For example, Figure 2-4 illustrates a spreadsheet that shows the characteristics of ten sports balls and extends this information to a cubic meter of these balls.

Data are entered in columns A–E and simple formulae are entered in columns F, G, and H. Thus, the cost of filling a cubic meter with baseballs (column F) is determined by multiplying the figure in column D by the figure in column E in row 4. The formula is quickly *pasted* into the spreadsheet and applied to all appropriate data in the spreadsheet. Do you see what formulae were used for columns G and H?

An extension of the sports-balls data is shown in Figure 2-5. Here, using the same data, we are able

	A	B	C	D	E	F	G	H
1		Approx.	Average	Approx.	Approx.	Cost of balls	Length of 1 cubic	Weight of
2		diameter	weight	cost per	number of balls	to fill	m of balls put	1 cubic m
3	Type of Ball	(cm)	(g)	ball	in 1 cubic m	1 cubic m	end to end (m)	of balls (kg)
4	Baseball	7.6	145	$ 5.00	2200	$11,000	167.2	319
5	Basketball	24	596	$50.00	70	$ 3,500	16.8	41.72
6	Croquet	8.6	340	$ 5.80	1500	$ 8,700	129	510
7	Golf	4.3	46	$ 2.30	12200	$28,060	524.6	561.2
8	Handball	4.8	65	$ 2.95	8800	$25,960	422.4	572
9	Racquetball	5.7	40	$ 1.50	5200	$ 7,800	296.4	208
10	Soccer	22	425	$ 9.95	80	$ 796	17.6	34
11	Softball	9.8	187	$ 5.95	1000	$ 5,950	98	187
12	Table tennis	3.7	2	$ 0.92	19700	$18,124	728.9	39.4
13	Tennis	6.5	57	$ 1.00	3800	$ 3,800	247	216.6
14								

Figure 2–4
Sports-Balls Information

	A	B	C	D	E	F	G	H	I
1		Approx.	Approx.	Average	Approx.	Filling our classroom with balls			
2		no. of balls	diameter	weight	cost per	No. of balls	Cost to fill	Length end	Wt. of balls in
3	Type of Ball	in 1 cubic m	(cm)	(g)	ball	in classroom	classroom	to end (km)	classroom (kg)
4	Baseball	2200	7.6	145	$ 5.00	897600	$ 4,488,000	68.2	130152
5	Basketball	70	24	596	$ 50.00	28560	$ 1,428,000	6.9	17021.76
6	Croquet	1500	8.6	340	$ 5.80	612000	$ 3,549,600	52.6	208080
7	Golf	12200	4.3	46	$ 2.30	4977600	$11,448,480	214.0	228969.6
8	Handball	8800	4.8	65	$ 2.95	3590400	$10,591,680	172.3	233376
9	Racquetball	5200	5.7	40	$ 1.50	2121600	$ 3,182,400	120.9	84864
10	Soccer	80	22	425	$ 9.95	32640	$ 324,768	7.2	13872
11	Softball	1000	9.8	187	$ 5.95	408000	$ 2,427,600	40.0	76296
12	Table tennis	19700	3.7	2	$ 0.92	8037600	$ 7,394,592	297.4	16075.2
13	Tennis	3800	6.5	57	$ 1.00	1550400	$ 1,550,400	100.8	88372.8
14									
15	Size of our classroom in cubic m =				408				
16									

Figure 2–5
Sports-Balls Classroom Data

to calculate information about filling our classroom with various balls.

Key information is put into cell E15 in the spreadsheet. Once the size of the classroom is entered, the information in columns F–I is calculated quickly. Again, it was easy to enter formulae into columns F–I to perform this task. For example, the weight of baseballs to fill the classroom in kilograms is calculated by the formula D4*F4/1000. D4 and F4 are the cells where the data are located, "*" means to multiply, and "/" means to divide. It is easy to see from the spreadsheet which of the balls would take a million to fill the classroom. Questions and discussions emerge when using spreadsheets. They should be pursued.

In telecommunication, students and teachers use a modem and a telephone line to access data bases across the country. This may include finding information from a free source such as NASA (this does require a long-distance phone charge) or using any of a multitude of bulletin boards or other information sources.

3. *For drill and practice.* Many programs are written to give children practice in various skills, especially computation. These are not designed to take the place of teaching, but are to be used after the children have already learned the concepts, preferably using concrete materials. These programs are often more fun than the practice exercises in a textbook,

but it should be kept in mind that their main purpose is still for memorization of facts and skills.

4. *To simulate real life.* Some programs are designed to give students practice with problems that they might encounter in everyday life that are too expensive, time-consuming, or difficult to recreate in a classroom. These include programs that set up a shop, bank, or stock market and give students a chance to explore the effects of different types of management. They also include programs that give students a chance to create different machines on the screen and run a variety of tests on them.

5. *To give students experience in problem solving, reasoning, and logic.* Many programs include a wide range of problems for students to solve, from saving the princess from the evil wizard to solving difficult logic problems. Many of these programs are so involving and challenging for the students that they will pay to use them in video arcades and never realize that they are using problem-solving skills.

6. *As a tutorial.* These programs are designed to teach the students concepts. They are frequently used in conjunction with manipulative materials that the students use simultaneously with the programs. The computer is an excellent way to individualize instruction. It can determine which concepts a student understands and which need to be repeated or presented in another form.

7. *To aid in spatial visualization.* Computer graphics are a good way to present geometric concepts that are difficult to illustrate in a textbook. Again, many of the graphic geometric programs are best used along with manipulative materials. Computers can show three-dimensional objects from a variety of different viewpoints and ask children to predict what an object would look like from another viewpoint or after it has been rotated or flipped. As with other types of programs, it is a good idea to use the spatial programs either along with concrete materials or after the students have had experience rotating and flipping paper models or concrete models.

Other Technology

Overhead Projectors

The overhead projector is not an example of new technology, but it is very useful in the mathematics classroom. Most commercial manipulative mathematics materials that are sold for students also have a matching transparent material that students and teachers can manipulate on the overhead projector. This is very useful when a teacher is demonstrating a concept for the students to follow with their own materials or when a student wishes to explain to other students a new idea that he or she has discovered. Teachers can also make their own transparent materials by cutting them out of colored transparency film or colored plastic report covers. The different colors of plastic are very useful for making attribute materials or other materials where the color of the material is an integral part of the concept being taught. Teachers and students alike will find that it is much easier to manipulate materials on the overhead projector than to draw models on the chalkboard.

Lettering Machines

The lettering machine is a device that will cut out letters, numerals, and other figures for the teacher to use in the classroom. Figures may be cut out of paper, cardboard, felt, railroad board, transparency film, plastic, sandpaper, and a number of other materials. For the mathematics teacher, some of the most useful dies include attribute blocks, pattern blocks, two- and three-dimensional geometric models, base ten blocks, tangrams, pentominoes, fraction pieces, and small figures such as teddy bears and smiley faces that can be used as counters. Teachers find this machine useful for making these manipulatives of laminated railroad board or felt rather than purchasing commercial materials.

Videotapes and Cameras

Many good videotapes, such as *Donald Duck in Mathemagic Land* and *Powers of Ten,* are available that demonstrate mathematical concepts whose presentation is enhanced by the movement and sound available on video. When looking for videotapes or films to use in the classroom, do not forget that the presentation of mathematics can also be enhanced by this medium. Students may want to try their hand at making their own videotapes to illustrate mathematical concepts that they have discovered.

Slides and Cameras

Still photos and slides are also good for demonstrating mathematical concepts, especially in geometry. The NCTM has an excellent set of geometry slides that includes resources for the teacher to use in teaching a number of geometric concepts such as symmetry, tessellations, transformations, and other concepts of both two- and three-dimensional geometry. Students may also use a camera to illustrate their own geometry discoveries.

Interactive Video, Telecommunications, and Other Newer Technologies

The field of technology is growing rapidly and educational applications are constantly being developed. Interactive videos are frequently used in tutorials to respond to individual students. They are generally programmed to give a response based on the answer of the user. Teachers should look for new products in the field of elementary mathematics as the prices become more competitive. The use of telecommunications was mentioned earlier in conjunction with computers. This is another area that will become more popular as more schools have the equipment available. Teachers should read journals such as the *Arithmetic Teacher* and the many journals in the area of technology to keep abreast of other new developments in technology.

KEY IDEAS

Advances in technology give you the means to elevate your mathematics teaching beyond what can be taught using only paper and pencil. Calculators are recommended for school mathematics programs to help develop number sense, patterns, operations, graphics, problem solving, and mental computation and estimation. Computers may be used as a tool for teaching programming, data manipulation, drill and practice, simulations, problem solving, tutorials, and spatial visualization. There are other devices that can serve the teacher, including overhead projectors, lettering machines, videotapes, cameras, slides, interactive video, and telecommunications.

REFERENCES

BOBIS, JANETTE F. "Using a Calculator to Develop Number Sense." *Arithmetic Teacher*. Vol. 38, No. 5 (January 1991), pp. 42–45.

CARSON, JOAN C., AND BOSTICK, RUBY N. *Math Instruction Using Media and Modality Strengths*. Springfield, Ill.: Charles C. Thomas, 1988.

EDWARDS, NANCY TANNER, AND BITTER, GARY G. "Changing Variables Using Spreadsheet Templates." *Arithmetic Teacher*. Vol. 37, No.2 (October 1989), pp. 40–44.

HEMBREE, RAY, AND DESSART, DONALD J. "Effects of Hand-Held Calculators in Precollege Mathematics Education: A Meta-Analysis." *Journal of Research in Mathematics Education*. Vol. 17, No. 1 (March 1986), pp. 83–99.

HIATT, ARTHUR A. "Activities for Calculators." *Arithmetic Teacher*. Vol. 34, No. 6 (February 1987), pp. 38–43.

NATIONAL COUNCIL OF TEACHERS OF MATHEMATICS. *Curriculum and Evaluation Standards for School Mathematics*. Reston, Va.: NCTM, 1989.

PAGNI, DAVID L. "Teaching Mathematics Using Calculators." *Arithmetic Teacher*. Vol. 38, No. 5 (January 1991), pp. 58–60.

RYOTI, DON E. "Processing Data." *Arithmetic Teacher*. Vol. 34, No. 8 (April 1987), pp. 42–44.

SPIKER, JOAN, AND KURTZ, RAY. "Teaching Primary-Grade Mathematics Skills with Calculators." *Arithmetic Teacher*. Vol. 34, No. 6 (February 1987), pp. 24–27.

STARKEY, MARY ANN. "Calculating First Graders." *Arithmetic Teacher*. Vol. 37, No. 2 (October 1989), pp. 6–7.

USNICK, VIRGINIA E., AND LAMPHERE, PATRICIA M. "Calculators and Division." *Arithmetic Teacher*. Vol. 38, No. 4 (December 1990), pp. 40–43.

WATSON, CHARLES D., AND TROWELL, JUDY. "Let Your Finger's Do the Counting." *Arithmetic Teacher*. Vol. 36, No. 2 (October 1988), pp. 50–53.

WHEATLEY, GRAYSON H., AND CLEMENTS, DOUGLAS H. "Calculators and Constructivism." *Arithmetic Teacher*. Vol. 38, No. 2 (October 1990), pp. 22–23.

WIEBE, JAMES H. "Order of Operations." *Arithmetic Teacher*. Vol. 37, No. 3 (November 1989), pp. 36–38.

———. "Teacher-Made Overhead Manipulatives." *Arithmetic Teacher*. Vol. 37, No. 7 (March 1990), pp. 44–46.

YOUNG, SHARON L. "Ideas." *Arithmetic Teacher*. Vol. 38, No. 3 (November 1990), pp. 23–32.

ZOLLMAN, ALAN. "Low Tech, 198, and the Geometry of the Calculator Keys." *Arithmetic Teacher*. Vol. 37, No. 5 (January 1990), pp. 30–33.

3

Children and Mathematics

Elementary and middle school children are natural learners. Their potential and energy for learning are considerable. They are exposed to enormous amounts of information, much of it outside the school.

Mathematics touches many subjects that are popular among children. For example, children manipulate plastic toys of varying sizes to transform earth or space vehicles into heroic robots—an exercise in spatial visualization. These toys also demonstrate the notion of transformation, which is fundamental to the concept of an operation. Other aspects of children's lives affected by mathematics include correspondences between family members and meal portions, routes taken to and from school, time and television programming, pricing and the ability to purchase items at the store, and baking cookies using measured ingredients. These are but a few examples. The ways mathematics touches the experiences of children stretch the imagination.

Space, number, shape, puzzles, time, distance, and computers provide a rich milieu in which elementary and middle school children grow. A teacher enables children to understand mathematics by providing an instructional environment. In Chapter 3 we explore three topics that influence how children learn mathematics:

1. The children's world.
2. Psychological considerations in teaching mathematics.
3. Teaching children mathematics.

The Children's World

As a teacher, you must know and understand the world from a child's perspective. In this section, we describe the ways in which children experience the world mathematically. Their learning of mathematics, now and in the future, is built upon this foundation.

Children Have Many Number Experiences

Number experiences are a part of children's lives from the moment they begin to communicate. Communication and physical movement include intensity of sound; varying duration of activity; exploration of space; embrace and separation; sequence of occurrences; and similarities and differences among humans, objects, places, and emotions. As youngsters record these relationships, they learn to quantify their world.

In the future, the lives of children will increasingly be affected by the computer. Today, children can skillfully draw and design using computer power. Their lines are straight and their angles are precise. Color, texture, and shading can be applied at the touch of a key. Children can prepare graphic animation with the assistance of clever software packages. Outlining and writing with word-processors assist the thinking processes. Computers tease and baffle the imagination. Tomorrow's teachers will use these magnificent tools to an even greater extent.

As children begin their schooling, they do so with enthusiasm, energy, and a willingness to participate. They usually enjoy success in their early mathematical work because they have been exploring relationships for years. Mathematics is, after all, the study of relationships, or of how things are connected. Among the teacher's challenges will be increasing the likelihood of success by helping children see meaning and sense in their mathematics.

Children Are Active in Their World

Teachers and parents rarely have to instruct children on how to be active or how to play. School is one of the first places where children are asked to be passive and quiet. Some orderliness and conformity is necessary for significant learning to take place. But, at the same time, there must be opportunities for spontaneous response and divergent thinking.

Children should be physically involved in mathematics. Materials such as stacking blocks, pattern blocks, colored cubes, attribute blocks, puzzles, Cuisenaire rods, geoboards, sand, clay, water, various containers, computers and appropriate software, and calculators should be available in the school. These and other learning aids are introduced in later chapters of this book. Children use these materials for counting, developing patterns, creating, observing, constructing, discussing, and comparing. From the manipulations and observations come the abstractions of quantitative ideas and the communication of these ideas in pictorial and, later, symbolic form.

Children Observe Relationships in Their World

Children and adults make connections among pieces that seem separate at first glance. Language is often closely tied with the expression of these relationships. Relationships may be simple. For example, because, again and again, an infant hears a certain sound and then receives attention, he begins to learn the relationship between his name and himself. Later it becomes apparent that other individuals have names. Distinguishing between a person's name and the name of the position that person holds in the family is more difficult. That is, Linda and Julie are names, whereas mommy and sister, although used as names, state family relationships. Eventually, children discover their relationships to grandparents, uncles, aunts, and cousins.

Fifth grade students can readily differentiate between a square and a rectangle but may be confused as to how they are related. That is, it may not be apparent to them that all squares are rectangles but not all rectangles are squares. Some may firmly believe that squares and rectangles share no common characteristics. Experience and discussion, then, help them to understand some less obvious relationships.

As children quantify their world, they become aware of arithmetical, spatial, logical, and collective relationships through their active participation with their natural environment. Manipulative materials in the school classroom provide an effective basis for mathematical learning. Some relationships are obvious. Others are not, and it becomes necessary for teachers to provide experiences to link the more subtle relationships. Eventually mathematics makes sense, because the learners understand how most of what they have been learning is connected. Learning how things relate is sharply distinguished from learning by memorizing many disconnected facts. When things are related, mathematics is presented as possessing structure; when memory is stressed, structure is generally ignored.

Relationships can be expressed in visual form. In moving from purely concrete work using physical objects to more abstract work, visual representations can effectively be employed. For example, Figure 3–1 is a visual representation of the relationships in one family drawn by a five year old. This representation links the concrete world to the abstract idea of family relations. Teachers should be aware of relationships and are encouraged to develop experiences and representations to further the learning of how things relate.

Children Learn Mathematics in Concert with Other Subjects

By the time children enter school, they are proficient at learning interrelated skills and concepts. Walking, talking, toilet habits, rote counting, and language are learned without being isolated from other life experiences. Packaging bits and pieces of knowledge or separating skills from their applications defines artificial environments alien to the unified world of all humans. Integrated experiences in school are important in providing a more natural and balanced setting for learning.

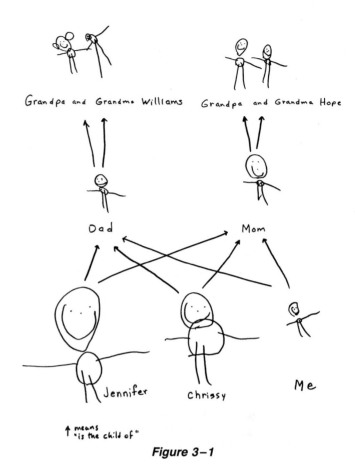

Figure 3–1

The skillful teacher is afforded the opportunity to integrate mathematics with other subjects and the world in general. The interrelatedness of mathematics with subjects such as art, language, literature, social studies, and science illustrates the close ties of all bodies of knowledge. Just as mathematics is embedded in other subjects, other subjects are embedded in mathematics. The educative process should bring many connected experiences to children in a lifelike environment. It is crucial that children see the connections.

There are few thinking skills unique to mathematics. Rather, most thinking skills transcend specific knowledge or a specific discipline. When taught in a setting of integrated learning, thinking skills provide the opportunity for children to learn how to learn. Observing, describing, conjecturing, questioning, judging, valuing, and communicating are skills of life. They are also the foundation upon which problem solving is built.

Children's Feelings Affect Their Ability to Learn

Children's feelings about themselves and their ability to succeed, as well as their feelings about others, home, school subjects, and life affect their behavior. We can ill afford to treat children as if they did not matter. Not all children will attack mathematics or anything else with the same energy and enthusiasm. Students should be exposed to the historical and cultural aspects of mathematics as well as its structure. Some students are interested in theoretical mathematics. Others are motivated by a historical approach. Still others are excited by the relationships between mathematics and art, mathematics and music, or mathematics and language. Applications of mathematics interest many children.

Much has been written about **math anxiety,** described as reluctance to engage in, and fear of, mathematics-related activities. Individuals who exhibit such anxiety do not enjoy doing arithmetic, particularly in public. They agonize over mental arithmetic, apologize for their lack of skill, and avoid activities associated with mathematics. In short, they are dysfunctional in mathematics.

Most individuals with math anxiety are beyond elementary school age. Research shows that mathematics is liked and enjoyed by a majority of elementary students. We are unconvinced that large numbers of elementary school children suffer math anxiety.

However, we are convinced that many children have learned not to enjoy mathematics. These children have experienced considerable failure in their attempts to learn concepts and skills. They have been asked to learn certain mathematical ideas that they were not ready to learn; they have been moved through a curriculum, "learning" mathematics for which they did not have the prerequisites and struggling with new concepts that did not make any sense. They may have been pressured to memorize hundreds of unrelated basic addition and multiplication facts and subjected to timed tests in front of their peers. They believe that success in mathematics is knowing a certain "magical process" that results in correct answers. As a result, some children begin to dislike mathematics and do not want to do mathematics. Failure and humiliation are powerful forces that cause children to be reluctant to engage in mathematics.

Research provides some insights into what teachers can do to develop and maintain positive attitudes toward mathematics. According to Suydam, these measures include:

1. Showing that you like mathematics.
2. Making mathematics enjoyable so that children develop positive perceptions of mathematics and of themselves in relation to mathematics.
3. Showing that mathematics is useful in both careers and everyday life.
4. Adapting instruction to students' interests.
5. Establishing short-term goals that students have a reasonable chance of attaining.
6. Providing experiences designed to help children be successful in mathematics.
7. Showing that mathematics is understandable by using meaningful methods of teaching (1984, p. 12).

Psychological Considerations in Teaching Mathematics

In the broadest sense, learning mathematics serves as both a means and an end. Learning mathematics is a means of developing logical and quantitative thinking abilities. The key word is *thinking.* Thinking children are liberated from the dull routine that sometimes characterizes school. Learning mathematics is an end when children have developed basic computational skills and can apply mathematics to their world, that is, when mathematics becomes functional in the lives of children. At least a part of a child's environment can be explained by simple mathematical principles.

At every level, learning mathematics should be a natural outgrowth of the children's lives. Learning should be interesting for the children, should challenge their imagination, and should beget creative solutions in their art, dance, music, movement, and conversation. Learning mathematics should be devoid of boredom, meaninglessness, and coercion.

Logical and Psychological Approaches to Mathematics

Approaches to teaching mathematics have generally followed the logical structure of mathematics presented by most children's textbooks. Thus, counting is followed by adding at the pictorial and symbolic levels. Subtracting, multiplying, and dividing follow. Later or concurrently, children learn the properties of these operations. To augment learning computation, courses of study include patterns and relationships, spatial sense, probability, statistics, and simple measurement. Understanding is developed to the extent that children see meaning in what they are doing. Some children readily understand. Many others do not or cannot understand. Presenting mathematics as an organized, logical structure does not ensure children's understanding.

To complement the logical structure of mathematics, teachers should weigh the psychological aspects of learning mathematics. Considering how children learn mathematics makes it possible for teachers to develop activities that blend what is known about children and mathematics. Thus, developing an initial understanding of number involves classifying, relating, and ordering, along with discussion. Objects and groups of objects are used to illustrate and enhance the learning of number, operation, and addition. Psychological considerations that help children learn mathematics are the focus of the next several pages.

Sources of Information About How Children Learn Mathematics

The study of how children learn mathematics is not new, but until relatively recently, little had been written that was directly applicable to classroom instruction. Useful works include those by Bruner (1977), Burger and Shaughnessy (1986), Copeland (1984), Ginsburg (1983), Kamii (1985, 1989), Piaget (1973), and Skemp (1987). These and other sources have been included in the bibliography at the end of Chapter 3. Much of what appears in the following chapters rests on the foundation provided by these authors.

The work of Piaget has greatly influenced the way teachers view cognitive development. Among the most well-known ideas associated with Piaget are those dealing with the evolution of thought through a series of four stages. Children move through the stages of cognitive development as a result of the interaction between internal forces (maturation) and external forces (environment). The first stage, **sensorimotor,** generally occurs in the first two years of life. Here, the child begins to imitate sounds and actions, and recognizes that objects still exist when they are out of sight. The second stage, **preoperational,** generally lasts from age two to age seven. Here, the child gains an initial use of language and the ability to think in symbolic terms. The third stage, **concrete operations,** generally lasts from age seven to age eleven. Here, concrete objects provide the medium for learning. Children discover that objects can be changed or moved and still retain many of their characteristics, and these changes can be reversed. The fourth stage, **formal operations,** may begin at age eleven, although many adults never operate fully at the formal level. Here, students can think logically about abstract problems. The age ranges are only approximations, but all individuals progress through these stages in the order they are presented.

An important idea associated with Piaget's work and noted in the stage of concrete operations is **conservation.** Conservation of number means that the number of objects in a set does not change if the objects in the set are placed in different positions. Conservation of quantity means that the amount of liquid poured from a tall narrow container into a short wide container remains constant. Conservation of length means that an object retains the same length if the object is moved. Until children are about seven years old, they do not conserve number. Again, seven is just a benchmark; children vary in their ability to conserve number. Teachers do not, nor should they, teach conservation, for being able to conserve is a result of children having mentally constructed logico-mathematical knowledge (relationships).

Two additional and complementary processes that Piaget described are **assimilation** and **accommodation.** Assimilation is the process by which an individual takes in information. The information may come from any source, such as playing with blocks, watching a bird fly, or listening to an explanation. Assimilated information may cause individuals to adjust or modify their understanding of an idea or event, in such cases the process of accommodation takes place. In the process of learning, assimilation and accommodation are constantly taking place. You should recognize the importance of these processes in learning mathematics. Further mention is made of assimilation and accommodation during the discussion of how children form mathematical concepts.

Piaget's work has influenced the authors, and you will find references to his work throughout the book. Perhaps Piaget's most important idea is that children can and should be involved in inventing mathematics. It is through experiences that children discover relationships and solve problems.

Investigations on how children think mathematically have shown that young children, in the preoperational stage, use mental counting procedures to solve arithmetic problems. This has led to revised theories about number understanding (Resnick, 1983). Certain types of simple mathematical thinking can occur before the Piagetian stage of concrete operations. According to Carpenter (1986, p. 114), "Contrary to popular notions, young children are relatively successful at analyzing and solving simple word problems. Before receiving formal instruction in addition and subtraction, most young children invent informal modeling and counting strategies to solve basic addition and subtraction problems."

Mathematics education was particularly affected by Brownell, who set forth his **meaning theory** of arithmetic instruction in the 1935 yearbook of the National Council of Teachers of Mathematics. According to Brownell (1935, pp. 19, 31), ". . . this theory makes meaning, the fact that children shall see sense in what they learn, the central issue in arithmetic instruction." He went on to call for an "instructional reorganization" so that arithmetic would be ". . . less a challenge to the pupil's memory and more a challenge to his intelligence." Brownell supported his theories with his research throughout his professional career. One such study, *Meaningful vs. Mechanical Learning: A Study in Grade III Subtraction,* suggested that retention, transfer, and understanding are enhanced by teachers using a "meaning method" as opposed to a "mechanical method" of instruction (Brownell and Moser, 1949). In the ensuing years, general agreement has been reached among psychologists and educators that teaching with meaning or understanding tends to be richer and longer lasting than other teaching. It has also been suggested that when learning is seen as a function of personal meaning, teaching centers on the children and their interpretations of what is being taught. Thus, teaching the meaning of mathematics provides an extra incentive for the teacher to know how children learn mathematics.

How Children Form Mathematical Concepts

A **concept** is an idea or mental image. Words and symbols are used to describe or label concepts. For example, *potato* is a collection of sounds that brings to mind an image representing some generalized form of a garden vegetable. Exactly what image appears depends on the experiences, heritage, geographical location, and language of the listener. The symbol 5 represents a mental image of all groups containing ●●●●●, or five, things. Again, the precise image that appears depends on the background and experiences of the listener.

Concepts are learned. Virtually all children from the time of birth can learn concepts. Concept formation begins immediately. The language and symbols that name concepts lag behind concept formation but eventually emerge. As children grow and mature, language and symbols are introduced to name mental images already formed and are used later to teach new concepts. To learn a concept, children require a number of common experiences relating to the concept. Initially, a parent introduces potatoes to a child by spooning a white, strained substance into the child's mouth, perhaps exhorting the child to, "Eat your potatoes." As this procedure continues for several months, the child begins to associate the word *potato* with the mushy substance. Obviously, the concept of potato is very limited at this time. Soon, mashed potatoes from the parent's plate may be introduced to the child with the same plea, "Eat your potatoes." Over time, potatoes prepared in many ways are given to the child and, finally, after two or three years, the child is informed that the vegetable the parent is washing, peeling, and cutting is a potato and can be prepared in numerous ways. The concept of potato begins to emerge as an accurate, generalized mental image.

Two aspects of this example have clear implications for teaching children. First, the concept of potato did not become *known* until potatoes had been seen, felt, smelled, and tasted in many ways, that is, until the child experienced potatoes in numerous guises. When the child was able to discern the common property among the various ways in which potatoes were prepared, namely, each dish originated from a certain recognized vegetable, the concept of potato was formed. Second, the word *potato* had to occur in concert with or had to follow the experiencing of the vegetable. The sounds that make up the word *potato* were not helpful before the experience. Hearing only the word or a definition, the child would not have learned what a potato was.

The two processes just described provide the foundation for the learning of mathematics. Children who find the common property of several seemingly disconnected examples are **abstracting.** The abstraction that is made is a **concept.** Children learn their mathematics by abstracting concepts from concrete experiences. The language is developed during or after concept formation, never before. Objects and events that are a part of children's lives and are easily observed are less abstract than objects and events that are not easily observed. Thus, dogs, automobiles, houses, toys, and mothers are less abstract than are color, height, number, time, and multiplication.

Skemp has stated two principles of learning mathematics that relate directly to the notion of concepts:

1. Concepts of a higher order than those which people already have cannot be communicated to them by a definition, but only by arranging for them to encounter a suitable collection of examples.
2. Since in mathematics these examples are almost invariably other concepts, it must first be ensured that these are already formed in the mind of the learner. (1987, p. 18)

Children learn the meaning of number by experiencing number in many varied situations—through a suitable collection of examples. The same holds true for addition, subtraction, multiplication, division, fractions, geometry, measurement, and so forth. The activities presented throughout the following pages are typical examples of activities for learning mathematical concepts.

Also, building mathematical concepts requires constantly building foundations on which to base further mathematical learning. Attempting to develop mathematical concepts on a foundation of previously memorized, vague notions results in frustration for both children and teacher. Skemp again notes that ". . . before we try to communicate a new concept, we have to find out what are its contributory concepts; and for each of these, we have to find out *its* contributory concepts, and so on, until we reach either primary concepts [derived from sensory and motor experiences] or experience which we can assume." (pp. 19–20)

Dealing with Concepts Once Formed

The concepts and experiences acquired by a person make up the knowledge that person possesses. As new experiences occur, they are fitted into a person's existing mental structure. This is the Piagetian process of assimilation. Depending on the familiarity of the experiences and the learning style of the learner, the experiences are received or rejected because of a person's mental structure or **schema.** The schema is a part of the mind used to build up the understanding of a topic. Thus, to increase or alter what is already known, the schema takes in new ideas and fits them with what is already known. For example, if children are still learning about potatoes and have experienced only strained, mashed, and boiled potatoes, the schema of *potato* may be limited. When french-fried potatoes are introduced, it may be difficult for children to immediately recognize the new food as potatoes. Although they are told that what they see, smell, feel, and taste are potatoes, children initially may not be convinced that they are experiencing potatoes. The schema of potato must be changed to accept potatoes in this new form. Once this change or accommodation has taken place, the schema has adjusted to accept french-fried potatoes as potatoes. French-fried potatoes are understood. Understanding a concept means an appropriate schema has accommodated that concept.

The idea of schema and how it functions provides a powerful tool for teaching mathematics. That a mental framework can be identified and developed means that mathematical relationships, patterns, and ideas can be understood rather than merely memorized; in the long run, children will have the ability to build up mathematical knowledge. When rules are memorized, children reach a point in their mathematical learning at which they are unable to remember the rules and are unable to continue learning. Understanding has long since vanished. As mathematical knowledge is introduced, its understanding is predicated on children's having already developed appropriate early schemas. The implications are clear. Teachers should provide mathematical experiences in a form that will ensure that the mathematics is understood. Such a foundation provides a basis for all later mathematical understanding.

Children's Thinking

The teacher's goal should be to provide experiences so children can progress from a concrete, intuitive level of learning dependent on the teacher toward a more symbolic level of learning independent of the teacher. This is a life's work for both learner and teacher. **Intuitive thinking,** the process of knowing without conscious reasoning, is not a stage through which an individual passes on the way to abstract thinking. Even when an individual has reached the Piagetian stage of formal operations, intuitive thinking may be necessary to gain a preliminary understanding of a new concept. For a child yet unready for formal thinking, intuitive thinking is the primary source of learning.

Learning by intuitive thinking means learning by experimenting with concrete materials, through experiencing ideas in various concrete ways, and by visualizing ideas. Because of intuitive thinking, a concept makes sense to a child before full understanding has taken place. For example, children who are constructing patterns using rods of two colors and who stumble on the pattern shown in Figure 3-2 are not learning much about the commutative property of addition, but they are certainly gaining an intuitive grasp of what the commutative property means. Later, when the commutative property of addition is presented, these children should be able to understand the property. The bulk of children's early learning takes place at the intuitive level. Teachers should encourage such thinking.

Reflective thinking comes later. Reflective thinking means being able to reason with ideas without needing concrete materials. The processes of reflective thinking include reflecting, inventing, imagining, playing with ideas, problem solving, problem creating, theorizing, and generalizing. Reflective thinking allows individuals to know how something is accomplished rather than merely being able to perform a task. Not only is it important for seven-year-old children to add two numbers such as 18 + 17, but it is also important that they be able to effectively explain how they accomplished the task.

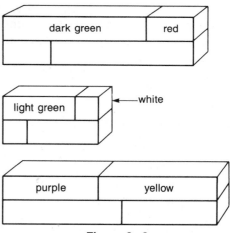

Figure 3-2

Reflective thinking also allows individuals to alter or correct schemas. Individuals may consider their perception of an idea or process and compare it with a new interpretation of the same idea or process and thus alter the original schema. Piaget has noted that children develop the ability for some verbal thought with sufficient concrete representations by the age of seven or eight. From seven or eight to eleven, children can manipulate concrete ideas in their heads, but complete facility with reflective thinking comes later, as children approach adolescence. Many adults never perfect their ability to think reflectively.

Children's Communicating of Mathematical Concepts

Children think quantitatively long before they engage in their first school activities. They have explored their personal space, and they begin to think about the **proximity** of objects, that is, they become aware that some objects are near to them while others are farther away. They notice that fingers are close to a hand or an arm, eyes are near the nose, and grandparents live far away (even if they live down the street).

Order is another spatial relationship about which children think. A child may have noticed the order of the cars on the toy train in the playpen or may be aware of the sequence of significant events. When the child cries, a parent appears, then holds and comforts the child.

Children classify objects as belonging together or not belonging together, for example, close family members versus neighbors and friends. They begin to judge objects as being few or many, big or small, tall or short, fast or slow. Obviously, they are not studying mathematics per se. The children are, however, thinking about their world quantitatively.

As children experience quantitative events and develop language to express these ideas, they are able to communicate with other people. They are developing the ability to classify objects and events more precisely. Although the language that emerges may not sound mathematical, it does represent the foundation on which the more exact language of mathematics is built. When children discriminate by *volume,* they may use the following words:

much	lots	some	empty
more	all	full	huge
less	little		

When discriminating by *size,* children may use the following words or phrases:

big	little	tall	bigger than
short	biggest	wide	smaller than
thin	fat	long	fatter than

When indicating *time,* children may use the following words or phrases:

before	now	spring	when the
after	later	winter	bell rings
yesterday	tomorrow	last	when it gets
		summer	warmer

When discussing the *location* of objects, children may use the following words or phrases:

here	there	inside	on top of
up	down	outside	in the box
over	under	above	below

When describing *how many,* children may use counting strategies that they have developed through a variety of contexts. Recent investigations have provided fresh insights into how children develop the ability to count and how they use counting to solve simple problems. For example, Fuson and Hall reported that children acquire a variety of number word meanings by their use in sequential, counting, cardinal, measure, ordinal, and nonnumerical contexts (1983, pp. 49–107). Counting is an important part of quantitative learning.

Children develop language in concert with their experiences. The experiences are crucial for the language to make sense. In the initial stages of mathematics learning, the quantitative experiences must be closely connected to the language that describes those experiences. Elementary school children need physical experiences as models before they are exposed to the language from which the concepts are abstracted. A serious mistake occurs when addition is taught before the child can bridge the gap between the intuitive notion of addition (usually involving counting and manipulating objects) and the symbolic representation of addition.

The home provides early, extensive language development, but the teacher is responsible for providing school experiences and the concomitant language that allow for a natural wedding of experience and language. A textbook or workbook alone cannot perform this function. Children must be physically active. Teachers enhance language development by frequent discussions with individuals, small groups, or the entire class about a particular activity or

discovery. As a natural extension of mathematical and language growth, the teacher or children may write experience stories to describe quantitative experiences.

Mathematics is an area of knowledge in which language often causes distortions and misunderstandings. The confusion most often occurs when unfamiliar language is presented before the experiences described by the language. Once introduced, the language may conjure up mental images that are distorted. Children may decide not to attempt to develop images and cease to listen, or they may simply memorize the verbal description and repeat it verbatim as a response to a stimulus. Language that is introduced by drawings, pictures, or diagrams is generally more helpful for children; however, the illustrations sometimes make little sense.

By way of contrast, the language may be introduced after children have experienced the ideas that the language describes. Thus, children using state-operator-state machines may discover that they can develop an understanding of the term *operation.* For example, consider a machine with an input, an operator, and an output. The machine accepts as inputs various attribute blocks. The rule for operating is to change the color of the incoming block (Figure 3–3). A red block placed in the input may be blue when it reaches the output, that is, after the operation has been carried out.

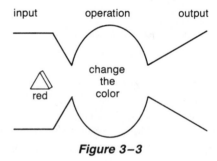

Figure 3–3

A join machine, in which a group of objects is placed in the input, might cause children to join three objects to the input group as in Figure 3–4. In such a case, the output would be the input group joined with the group described in the operation.

Many other machines can be devised to allow children to manipulate and discover the notion of operating or transforming and to help them understand the term *operation* when it is later applied to this transforming. Weaver pro-

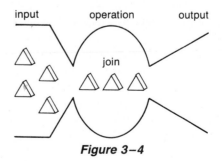

Figure 3–4

vided support for using change-of-state situations to introduce symbolic sentences (1982, pp. 60–66).

Language and symbols for language stand for abstractions; thus, it is by language and symbols that we communicate what has been abstracted. In order to be meaningfully communicated, the abstraction must have taken place. Unfortunately, students are often introduced to language and symbols before they are ready to make the appropriate abstractions. To abstract a concept, students must perform a variety of concrete operations. From these activities, which have a common structure, youngsters draw mental pictures, or images. The images of the elements common to the structures are the abstractions of concepts to be learned. Once concepts are understood, they can be verbalized and symbolized meaningfully.

Teaching Children Mathematics

The main purpose of this textbook is to provide some suggestions about how to successfully teach mathematics. You have already learned some ways to be an effective elementary or middle school mathematics teacher. You have strengthened your mathematics background by taking courses. Your attitudes about mathematics will grow more positive as you gain more experience with mathematics and children. A psychological foundation for teaching mathematics should include teachers' attitudes about children, teaching style, a constructivist view of teaching young children, and enhancing and guiding learning.

Teachers' Attitudes About Children

Much of this book focuses on children as learners of mathematics, but it is important also to look at just the children themselves. The teacher's beliefs about children will affect the children's performance in many academic cases, including mathematics.

The beliefs adults have about children, how children should be treated, and how children should be taught vary dramatically. Some believe that children must be left alone, to grow and develop with little interference from adults, and others believe that children must be closely watched and directed. Surely the optimum treatment of children includes some combination of laissez-faire and strict direction. The overall degree of teaching success depends more on what teachers believe about children than on how they organize to teach them, because teachers interact with children in ways that reflect their beliefs about children.

One example has been described in the research literature. When teachers believe children are low achievers, the children tend to receive marks indicating low achievement. When teachers believe comparable children are high achievers, the children tend to receive marks indicating high achievement. Teachers are often unaware that they treat children according to their personal beliefs about the children. They do so through both verbal and nonverbal interaction. As beliefs about children become more positive, a greater amount of the children's potential can be realized (Rosenthal and Jacobson, 1968). Teachers need to be keenly aware of the children with whom they work. A well-articulated set of beliefs provides the basis on which to develop a sound style of teaching.

Teaching Style

Teaching means directing, channeling, providing, suggesting, expecting, and encouraging children. It also means managing a classroom of diverse individuals, some who are happy to be there and cooperative, and a few who are reluctant to be there and combative. Some people consider teaching an art; others consider teaching a science. It must surely be some of both. Skillful teaching is difficult, and tiring, but it is rewarding, too. Although much direction is given to teachers in carefully designed guides to textbooks and workbooks, mathematical teaching is improved immeasurably if teachers:

1. Enjoy teaching and appreciate mathematics.
2. Continue to learn mathematics in formal or informal settings.
3. Extend their teaching beyond the basal textbook.
4. Seek out and communicate with colleagues who are excited about and current in elementary mathematics.
5. Pay close attention to how children learn mathematics.

There are also numerous learning environments, and within them are many teaching-learning strategies. The learning environment consists of physical settings in which teaching and learning take place. Within the context of schooling, these physical settings are generally at or near the school building. Thus, the learning environment may be the physical organization of a classroom, the confines of the playground, a nature trail, an urban neighborhood, or an auditorium.

Within the learning environment, a teacher's behavior toward a particular group of children is determined by that teacher's beliefs, experiences, education, and feelings. In developing teaching-learning strategies, the teacher should be aware that many alternatives exist. Some of those alternative strategies are described in Chapter 15. Each strategy contains the way students are to participate in the learning venture, the particular teaching behavior that participation requires, and the way students are organized for instruction during that time.

Developing a particular strategy is the function of an individual teacher. Although it is recognized that there is no one best way to teach, if you teach in only one way (that is, using but one strategy continuously) you are not as likely to succeed over a period of time as you are if you use several approaches. Teachers should be aware of more than one way of teaching and learning. Variety makes teaching and learning more interesting and enjoyable. This is true of teaching mathematics as well as teaching reading, science, or social studies.

A Constructivist View of Teaching Young Children

From the literature on teaching mathematics to young children comes an approach that is influenced by the principle of **constructivism.** Constructivism is a belief that children best gain knowledge by inventing it, that they construct knowledge for themselves. According to Piaget (1973, p. 20), ". . . to understand is to discover, or reconstruct by rediscovery, and such conditions must be complied with if in the future individuals are to be formed who are capable of production and creativity and not simply repetition." Based heavily on the work of Piaget, constructivism guides teachers to interact with children through questioning and discussion, skillfully responding to the children's ideas, and allowing children to discover relationships and predict future events. Children engage in manipulating materials, playing games, and interacting with one another.

A proponent of teaching based on the principle of constructivism, Kamii (1989, p. x), notes that "Encouraging children to construct knowledge from within is the diametric opposite of trying to impose isolated skills from the outside." The approach that Kamii (1989, p. 184) advocates contrasts with that of more traditional educators, who ". . . assume that the job of the teacher is to put knowledge into children's heads. They also assume that the proof of this transmission of knowledge is a high score on standardized tests. Both of these assumptions . . . are erroneous and outdated."

Kamii's work has been based, in part, on her work with primary teachers of grades 1–3. Besides stressing the importance of children creating their own mathematical relationships, Kamii emphasizes that word problems are an important part of teaching. Children are able to construct mathematical ideas from problems arising from their personal lives; first grade children can solve verbal problems without formal instruction. This leads Kamii to note that

> If children add numerical quantities repeatedly, actively, in the context of everyday classroom occurrences, games, and problems that they understand, they *will* remember the results of these mental actions and *will* become able to read and write conventional mathematical signs. The focus of the teacher's concern should be on children's *thinking* rather than on their ability to *write* correct answers. (1985, p. 94)

Kamii suggests that the overriding aim of education is to develop moral and intellectual autonomy among learners. As a result, children should be encouraged to develop their own opinions and to judge when another opinion is better. Kamii notes that:

> This is not to say that children do not learn from workbooks and transmission. They do, and they usually acquire the truth faster by being told than by constructing it them-

selves. But we must think of learning in a larger context than the memorization of sums and the ability to produce high test scores. In other words, we need to see autonomy as the ultimate aim of education. . . (1985, p. 36)

We believe that you should consider constructivist teaching methods as a teacher of elementary and middle school mathematics. Recent discussions and guidelines presented by the National Council of Teachers of Mathematics and the National Research Council are having a significant impact on school mathematics (see Chapter 1). They strongly support a shift away from a teaching model based on the transmission of knowledge and toward a model based on student-centered experiences. Thus, the opportunity to employ alternative teaching approaches, including constructivist approaches, is at hand.

The following were taken from a list of constructivist teaching practices developed by teachers in an inservice education program and research project in the Shoreham-Wading River (New York) School District:

1. Encourage and accept student autonomy, initiation, and leadership.
2. Whenever possible, use raw data and primary sources, along with manipulative, interactive, and physical materials.
3. Allow student thinking to drive lessons. Shift instructional strategies or alter content based on student responses.
4. Ask students for their theories about concepts before sharing your understanding of those concepts.
5. Encourage students to engage in dialogue, both with the teacher and with one another.
6. Seek elaboration of students' initial responses.
7. Encourage student inquiry by asking thoughtful, open-ended questions and encouraging students to ask questions of others.
8. Allow wait-time after posing questions.
9. Provide time for students to discover relationships and create metaphors.
10. When designing curriculum, organize information around conceptual clusters of problems, questions, discrepant situations. (Brooks, 1990, p. 70)

You are encouraged to further explore independently how children learn mathematics. Only when teachers begin to understand the aspects and stages of mathematical learning will children receive the kind of instruction most appropriate to their individual learning styles. In addition, exploration will help you to understand why this textbook emphasizes manipulation of concrete objects, multiple embodiments of mathematics ideas, active participation of learners, use of alternative teaching strategies, use of mathematical relationships, and building mathematical ideas according to the developmental characteristics of children.

Enhancing and Guiding Learning

To recommend a single approach to teaching elementary or middle school mathematics is inappropriate. Because of the various ways in which individuals learn and the different personalities of learners and teachers, teachers must vary their strategies and adapt them to their own specific needs and those of their students. The process of fitting instruction to children's differing needs and styles is discussed at some length in Chapter 15, where a number of options are presented. It is appropriate at this juncture to consider a framework in which successful teaching can take place. Recent documents, mentioned in Chapter 1, have made strong statements about mathematics teaching.

Duckworth (1987, p. 64) remarks, "In most classrooms, it is the quick right answer that is appreciated. Knowledge of the answer ahead of time is, on the whole, more valued than ways of figuring it out." Duckworth continues, suggesting that

> If a child spends time exploring all the possibilities of a given notion, it may mean that she holds onto it longer, and moves onto the next stage less quickly; but by the time she does move on, she will have a far better foundation—the idea will serve her far better, will stand up in the face of surprises. (1987, p. 71)

We, too, want to encourage you to help your students figure out mathematics. Based on the recommendations of the National Research Council (1989, 1990) and the National Council of Teachers of Mathematics (1991), as well as our own reading, study, and teaching experience, we have developed the following guidelines for forward-looking teachers. There are ten principles to which we adhere in designing mathematics instruction. Each is discussed briefly in the list that follows:

1. *Provide developmental instruction.* Although most of the following principles help define developmental instruction, it is appropriate to keep your attention focused on this principle. Developmental instruction suggests that teachers should attend to the cognitive growth of their students. Learning how children think and the levels of their thinking is crucial to developmental instruction. In planning your teaching, keep the children foremost in your mind. You must continually assess the understanding and progress of the students.

2. *Engage the children in active learning.* Using manipulative materials has been a cornerstone of successful mathematics instruction for a number of years. One part of constructing knowledge is exploring materials freely or in problem situations. Active learning may also mean engaging in cooperative learning projects or spirited exchanges of mathematical ideas. You will find many examples of activities in this textbook that are intended to sug-

gest ideas for your teaching. You are challenged to develop activities of your own.

3. *Lead discussion and question children about their thinking.* Children should be encouraged to explain their thinking, offer opinions, and exchange thoughts with other students. The teacher plays an important role in this process. The types of questions you ask can lead you to discover what and how a child understands. Duckworth suggests various kinds of questions:

> What do you mean? How did you do that? Why do you say that? How does that fit with what she just said? I don't really get that; could you explain it another way? Could you give an example? How did you figure that? In every case, those questions are primarily a way for the interlocutor to try to understand what the other is understanding. Yet in every case, also, they engage the other's thoughts and take them a step further. (1987, p. 97)

The teacher seeks to discover what the students understand by carefully questioning the children and sharing opinions with them. Students are able to share their thinking in writing as well as orally. Writing should be encouraged throughout the mathematics program.

4. *Employ calculators and computers.* Calculators provide children with a means to explore mathematics and enjoy the challenge of problem solving. Children at all levels should have calculators available as tools in the learning process as well as during examinations. Computers also serve as powerful tools in the learning process and provide children with opportunities to be problem solvers. Logo programming serves as a creative outlet for children in learning aspects of geometry.

5. *Utilize student-centered instruction.* The ideas, opinions, and interests of the children are focal points around which instruction is centered. Word problems based on the children's environment help begin the guided learning process. State and local content goals for mathematics along with textbooks are necessary for a sound mathematics program. Coupled with the interests and curiosity of children, the content of mathematics becomes fertile ground for the intellectual growth of children.

6. *Develop children's mathematical power.* Children gain power in mathematics when they understand the concepts and procedures that they have constructed. Having power means that children can apply mathematics to tasks and problems because they understand what they are doing. Having power in mathematics also means that children are able to reconstruct concepts and procedures when they have been forgotten. Mathematical power provides children with the confidence to attack problems and to persist when challenges arise.

7. *Encourage higher-level thinking.* Children should be challenged with problems, puzzles, and patterns throughout their learning of mathematics. The place to start is with examples from the students' own experiences. Stories that incorporate problems can be intriguing and can stimulate considerable divergent thinking. Warm-up activities at the beginning of a mathematics lesson can lead to creative thinking. Encouraging children to pose problems will help the children think differently about problem solving.

8. *Provide opportunities for children to construct and communicate mathematics.* When children construct mathematics through their experiences and subsequent interactions with teachers and peers they develop schemas that serve them well as they continue to learn mathematics. Their understanding of mathematics is powerful in the sense mentioned above. You have the opportunity to be a constructivist teacher for at least part of the time you teach mathematics. Be prepared for an explosion of ideas as children invent mathematics. Encourage them to share their findings both orally and in writing.

9. *Teach diagnostically.* As children learn mathematics, teachers need to constantly monitor their progress. Assessing children's work and assisting them in overcoming their misconceptions are major teaching responsibilities. The primary techniques for gathering information are observing children and discussing their thinking processes with them. You should be aware of the personal logic of students and ways to interpret their thinking. By discovering how children think you will be a more effective teacher.

10. *Introduce new techniques.* Throughout your teaching career you will be invited to participate in workshops and conferences where exciting ideas for teaching mathematics will be introduced. The presenters will be successful teachers and experts. Take advantage of these opportunities and employ those techniques that you believe will enrich your classroom. Read journals such as the *Arithmetic Teacher* to discover the trends and new directions in mathematics education. If you wish, accept the role of change agent.

In addition to these principles, there are other considerations to be taken into account when teaching. For example, the way in which students are organized for instruction is important. Johnson and Johnson (1987), in *Learning Together and Alone: Cooperative, Competitive, and Individualistic Learning*, develop a strong argument for providing cooperative learning experiences for children. Cooperating and sharing a common goal foster personal growth and identification. When cooperation is encouraged, competitive learning and individualistic learning are decreased. In many

of the activities presented in this textbook, cooperative learning is suggested. A discussion of cooperative learning is included in Chapter 15, along with a presentation of other useful suggestions for teaching.

KEY IDEAS

A fundamental reason for learning mathematics is so we can intelligently function in society. The environments in which children are born and grow provide many experiences that help establish a foundation for learning mathematics. The years before school are important for mathematical learning. Children are active learners. They observe and establish relationships among people and objects. These relationships become an important basis for learning mathematics, which is a study of patterns and relationships. Children learn mathematics in concert with other areas of knowledge. Children's attitudes about mathematics and their perceptions about themselves play a role in how they learn mathematics.

Psychological considerations in teaching mathematics include knowing about how children develop concepts in general and mathematical concepts in particular, what happens once a concept is learned, how to foster the thinking process, and the role of language. In teaching mathematics to children, teachers' attitudes about children affect the learning environment. There are many successful teaching styles.

Constructivism holds that students should have the opportunity to invent mathematics. Ten principles were suggested in designing mathematics instruction.

As you read the remaining chapters of this textbook, we hope you will gain a sense of excitement and opportunity. The excitement of active, meaningful mathematics will be reflected in the attitudes and enthusiasm of the children you teach. The opportunity to enliven the classroom environment will provide both you and your children many enjoyable hours of mathematical exploration.

REFERENCES

ALLARDICE, BARBARA S., AND GINSBURG, HERBERT P. "Children's Psychological Difficulties in Mathematics." In Ginsburg, Herbert P., ed. *The Development of Mathematical Thinking.* New York: Academic Press, 1983.

BAROODY, ARTHUR J., AND GINSBURG, HERBERT P. "The Relationship Between Initial Meaningful and Mechanical Knowledge of Arithmetic." In Hiebert, James, ed. *Conceptual and Procedural Knowledge: The Case of Mathematics.* Hillsdale, N.J: Lawrence Erlbaum Associates, 1986.

BROOKS, JACQUELINE GRENNON. "Teachers and Students: Constructivists Forging New Connections." *Educational Leadership.* Vol. 47, No. 5. (February 1990), pp. 68–71.

BROWNELL, WILLIAM A. "Psychological Considerations in the Learning and the Teaching of Arithmetic." *The Teaching of Arithmetic.* National Council of Teachers of Mathematics, The Tenth Yearbook. New York: Bureau of Publications, Teachers College, Columbia University, 1935.

BROWNELL, WILLIAM A., AND MOSER, H. E. *Meaningful vs. Mechanical Learning: A Study in Grade III Subtraction.* Duke University Research Studies in Education, No. 8. Durham, N.C.: Duke University Press, 1949.

BRUNER, JEROME S. *The Process of Education.* Cambridge, Ma.: Harvard University Press, 1977.

BURGER, WILLIAM F., AND SHAUGHNESSY, J. MICHAEL. "Characterizing the Van Hiele Levels of Development in Geometry." *Journal of Research in Mathematics Education.* Vol. 17, No. 1 (January 1986), pp. 31–48.

CARPENTER, THOMAS P. "Conceptual Knowledge as a Foundation for Procedural Knowledge." In Hiebert, James, ed. *Conceptual and Procedural Knowledge: The Case of Mathematics.* Hillsdale, N.J.: Lawrence Erlbaum Associates, 1986.

COPELAND, RICHARD W. *How Children Learn Mathematics.* New York: Macmillan Co., 1984.

DIENES, ZOLTAN P. "An Example of the Passage from the Concrete to the Manipulation of Formal Systems." In Freudenthal, H., ed. *Educational Studies in Mathematics* (Vol. 3). Dordrect, Holland: R. Reidel Publishing Co., 1971, pp. 337–352.

DUCKWORTH, ELEANOR: *"The Having of Wonderful Ideas" and Other Essays in Teaching and Learning.* New York: Teachers College Press, 1987.

FUSON, KAREN, AND HALL, JAMES W. "The Acquisition of Early Number Word Meanings: A Conceptual Analysis and Review." In Ginsburg, Herbert P., ed. *The Development of Mathematical Thinking.* New York: Academic Press, 1983.

GINSBURG, HERBERT P., ed. *The Development of Mathematical Thinking.* New York: Academic Press, 1983.

GINSBURG, HERBERT, AND OPPER, SYLVIA. *Piaget's Theory of Intellectual Development.* Englewood Cliffs, N.J.: Prentice-Hall, 1969.

JOHNSON, DAVID W., AND JOHNSON, ROGER T. *Learning Together and Alone: Cooperative, Competitive, and Individualistic Learning.* Englewood Cliffs, N.J.: Prentice-Hall, Inc. 1987.

KAMII, CONSTANCE KAZOKO. *Young Children Reinvent Arithmetic.* New York: Teachers College Press, 1985.

_____ . *Young Children Continue to Reinvent Arithmetic.* New York: Teachers College Press, 1989.

_____ . "Constructivism and Beginning Arithmetic (K–2)." In Cooney, Thomas J., and Hirsch, Christian. *Teaching and Learning in the 1990s.* Reston, Va.: National Council of Teachers of Mathematics, 1990.

NATIONAL COUNCIL OF TEACHERS OF MATHEMATICS. *Curriculum and Evaluation Standards for School Mathematics.* Reston, Va.: NCTM, 1989.

_____ . *Professional Standards for Teaching Mathematics.* Reston, Va.: NCTM, 1991.

NATIONAL RESEARCH COUNCIL. *Everybody Counts: A Report to the Nation on the Future of Mathematics Education.* Washington, D.C.: National Academy Press, 1989.

_____ . *Reshaping School Mathematics: A Philosophy and Framework for Curriculum.* Washington, D.C.: National Academy Press, 1990.

PIAGET, JEAN. *The Child's Concept of Number.* New York: W.W. Norton & Co., 1965.

_____ . *To Understand Is to Invent.* New York: Viking Press, 1973.

RESNICK, LAUREN B. "A Developmental Theory of Number Understanding." In Ginsburg, Herbert P., ed. *The Development of Mathematical Thinking.* New York: Academic Press, 1983.

ROMBERG, THOMAS A., AND CARPENTER, THOMAS P. "Research on Teaching and Learning Mathematics: Two Disciplines of Scientific Inquiry." In Wittrock, Merlin C., ed. *Handbook of Research on Teaching.* New York: Macmillan Co., 1986.

SKEMP, RICHARD R. *The Psychology of Learning Mathematics.* Hillsdale, N.J.: Lawrence Erlbaum Associates, 1987.

STEFFE, LESLIE P. "Adaptive Mathematics Teaching." In Cooney, Thomas J., and Hirsch, Christian. *Teaching and Learning in the 1990s.* Reston, Va.: National Council of Teachers of Mathematics, 1990.

SUYDAM, MARILYN N. "Attitudes Towards Mathematics." *Arithmetic Teacher.* Vol. 32., No. 3. (November 1984), p. 12.

TOBIAS, SHEILA. *Overcoming Math Anxiety.* New York: W.W. Norton & Co., 1978.

WEAVER, J. FRED. "Interpretations of Number Operations and Symbolic Representations of Addition and Subtraction." In Carpenter, Thomas P., Moser, James M., and Romberg, Thomas A., eds. *Addition and Subtraction: A Cognitive Perspective.* Hillsdale, N.J.: Lawrence Erlbaum Associates, 1982.

YAKEL, ERNA, ET AL. "The Importance of Social Interaction in Children's Construction of Mathematical Knowledge." In Cooney, Thomas J., and Hirsch, Christian. *Teaching and Learning in the 1990s.* Reston, Va.: National Council of Teachers of Mathematics, 1990.

4

Solving and Creating Problems, Estimating, and Mental Calculating

NCTM *Standards*

GRADES K–4

Standard 1: Mathematics as Problem Solving

In grades K–4, the study of mathematics should emphasize problem solving so that students can—

- use problem-solving approaches to investigate and understand mathematical content;
- formulate problems from everyday and mathematical situations;
- develop and apply strategies to solve a wide variety of problems;
- verify and interpret results with respect to the original problem;
- acquire confidence in using mathematics meaningfully.

Standard 2: Mathematics as Communication

In grades K–4, the study of mathematics should include numerous opportunities for communication so that students can—

- relate physical materials, pictures, and diagrams to mathematical ideas;
- reflect on and clarify their thinking about mathematical ideas and situations;
- relate their everyday language to mathematical language and symbols;
- realize that representing, discussing, reading, writing, and listening to mathematics are a vital part of learning and using mathematics.

GRADES 5–8

Standard 1: Mathematics as Problem Solving

In grades 5–8, the mathematics curriculum should include numerous and varied experiences with problem solving as a method of inquiry and application so that students can—

- use problem-solving approaches to investigate and understand mathematical content;
- formulate problems from situations within and outside mathematics;
- develop and apply a variety of strategies to solve problems, with emphasis on multistep and nonroutine problems;
- verify and interpret results with respect to the original problem situation;
- generalize solutions and strategies to new problem situations;
- acquire confidence in using mathematics meaningfully.

Standard 2: Mathematics as Communication

In grades 5–8, the study of mathematics should include opportunities to communicate so that students can—

- model situations using oral, written, concrete, pictorial, graphical, and algebraic methods;
- reflect on and clarify their own thinking about mathematical ideas and situations;
- develop common understandings of mathematical ideas, including the role of definitions;
- use the skills of reading, listening, and viewing to interpret and evaluate mathematical ideas;
- discuss mathematical ideas and make conjectures and convincing arguments;
- appreciate the value of mathematical notation and its role in the development of mathematical ideas.

(NCTM, 1989, pp. 23, 26, 75, 78. Reprinted by permission.)

Every day, children and adults alike are faced with problems. A seven year old may be faced with the problem of saving enough money to buy a coveted baseball trading card, or a thirteen year old may have the problem of budgeting her time wisely to be sure there is enough time for homework, piano lessons, soccer games, and talking with her friends. She may need to determine the most important activity if she finds there is simply not enough time in a day to do everything. Frequently, when we solve our daily mathematical problems, we do not use paper and pencil or a calculator. Either we make an estimate, which will give us an approximate answer, or we calculate mentally to determine the exact answer. In this chapter, we will look at the importance of problem solving and the uses of estimation and mental calculation. These topics will be expanded further in the following chapters.

Problem Solving

Problem solving has been recognized as one of the most important mathematical processes by both the National Council of Supervisors of Mathematics (NCSM) and the National Council of Teachers of Mathematics (NCTM). In 1976, the NCSM listed problem solving as number one on a list of the ten basic mathematics skills (NCSM, 1976). In 1988, when this list was updated, problem solving again was listed as the principal reason for studying mathematics. In 1980, the NCTM published *An Agenda for Action*, which consisted of a set of recommendations for elementary and secondary mathematics instruction in the 1980s. The first of eight recommendations was that "problem solving be the focus of school mathematics in the 1980s" (NCTM, 1980). When the NCTM published the *Curriculum and Evaluation Standards for School Mathematics* in 1989, problem solving was listed as one of the four strands that should receive increased attention at all grade levels from kindergarten through twelfth grade (NCTM, 1989). Frequently, students need to define or create problems on which to work. Outside of school, problems are not generally spelled out for us in advance. We need to determine the problem before we can find a solution.

This emphasis on problem solving and problem creating raises some very important questions: What is a problem? What are problem solving and problem creating? What are some characteristics of good problem solvers and good problem creators? What influences a person trying to solve or create problems? What are some heuristics and strategies one should use when solving or creating problems? What types of grouping should be used for problem solving and problem creating? How should the processes and results of problem solving and problem creating be communicated? Answers to these questions will be explored in the remainder of this chapter and throughout this book, for it is important to make problem solving and problem creating integral parts of all the topics in a math-

ematics program, not separate topics to be investigated only at special times during the school year.

What Is a Problem?

A problem may be thought of as a perplexing question or situation. It should be a question or situation that does not suggest an immediate solution or even an immediate method of solution. For our purposes, a problem should involve some aspect of mathematics, but that does not mean that it must involve numbers. Some excellent mathematics problems involve spatial or logical reasoning but do not involve numbers. A good problem is one that interests the problem solver and one that the problem solver makes an attempt to work. Notice that this definition does not include many of the so-called word problems or story problems that are seen in many textbooks because the student generally has an immediate method for solving the problem and is frequently not interested in the outcome. This definition refers to those nonroutine problems for which the students do not have an immediate method of solution and those problems that arise commonly in everyday life for which there may be several methods of solution.

Problem Creating and Problem Solving

We can think of our goals in mathematics as forming a continuum or hierarchy as shown in Figure 4–1. At the bottom of the hierarchy is the elimination of mathematical illiteracy. Many people in the United States today seem almost proud of the fact that they do not understand or use mathematics. It is not uncommon to hear an adult say, "Oh, I always hated mathematics. I never was good at it." This attitude rubs off on students, who assume it must be all right not to be able to do mathematics. This is a very dangerous belief, however, as our world becomes ever more dependent on a mathematically and technologically literate society.

Just above illiteracy is the ability to do some computation with whole and rational numbers; students at this level are called *doers*. These students have memorized rules for addition, subtraction, multiplication, and division and generally do fairly well on tests of computation from the math text, as long as they simply have to repeat a process over and over again. Generally, however, they do not understand why they are performing an operation in a certain way.

Above the level of the doers is the student who can compute well with all types of rational numbers and who understands the structure of the number system and the concepts of the operations. However, we certainly cannot be satisfied with students who can merely compute, when any $5 calculator would be faster and more accurate than most of the best human calculators.

Beyond the ability to compute is the ability to apply mathematical concepts to solve everyday problems. For

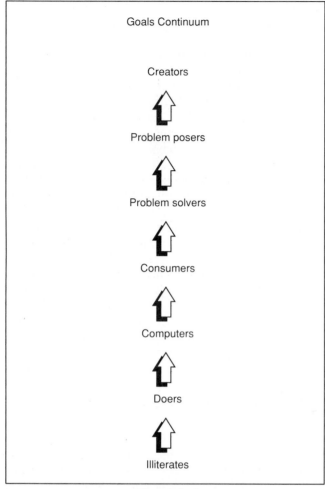

Goals Continuum

Creators

Problem posers

Problem solvers

Consumers

Computers

Doers

Illiterates

Figure 4–1

our society to function, we must be able to use mathematics every day, both at work and at home. Students need to learn how to use mathematics in stores, in restaurants, in painting a house, in balancing a checkbook, and in a multitude of other situations. They must use mathematics to be intelligent consumers in today's society.

It is at the next level that students solve the problems mentioned above. These students are able to apply their knowledge of mathematics in new situations when the answer is not obvious and they have no pre-set rule to fall back on. They frequently use a method that they have not tried before or apply a method they have used to solve a completely different type of problem.

Beyond the ability to solve problems that someone else has suggested is the ability to create, define, or pose problems. This requires an ability to see important aspects of a situation and ask questions about it. Most of the mathematics known today has been discovered in the last 50 years, and new solutions would never have been found if someone had not suggested new problems upon which to work.

At the top of the continuum is the creation of new mathematics. This requires first the creation of new questions upon which to work and then the discovery or invention of the mathematics to answer the questions. Even young children can discover or create mathematics that is new to them, and they should be encouraged to do so. They will understand and remember the mathematics they have constructed for themselves much better than any of the mathematics we try to teach them.

Characteristics of Good Problem Solvers and Good Problem Creators

In *Problem Solving—A Basic Mathematics Goal: A Resource for Problem Solving*, the following are listed as traits that good problem solvers usually possess:

- good estimation and analysis skills
- ability to perceive likenesses and differences
- reflective and creative thinking
- ability to visualize relationships
- strong understanding of concepts and terms
- ability to disregard irrelevant data
- capability to switch methods easily, but not impulsively
- ability to generalize on the basis of few examples
- ability to interpret quantitative data
- strong self-esteem
- low test anxiety (1980, p. 13)

The ability to see relationships seems to be one of the main characteristics that separate expert problem solvers from novices. Students who are not good at solving problems tend to try to memorize rules and facts as unrelated bits of information. Good problem solvers look for the underlying structure and try to relate any new problem to information they already possess.

In creating new problems, good problem solvers use information they understand and problems they have already solved as jumping off points for new questions. They view mathematics as a topic to be explored, with rich new ideas waiting to be discovered or invented, rather than as a series of rules that they must memorize.

Good problem solvers may not be the fastest computers. They take time to think about the problem before they begin to write. They do not give up easily if a problem is difficult. They view the problem as a challenge and enjoy working on it. These are skills that all children can learn. If we give students time to think about what they are doing and let them experience the joy of solving a difficult problem, all students can become better problem solvers.

Influences on a Person Trying to Create or Solve Problems

The ability to solve and create problems and to construct new mathematics is closely related to the student's attitude toward mathematics, the student's beliefs about the nature of mathematics and about his or her ability to do

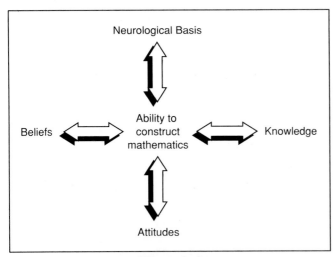

Figure 4-2

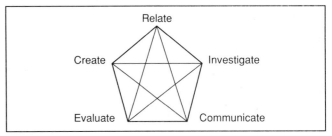

Figure 4-3

mathematics, the student's knowledge of mathematics, and the neurological makeup of the student's brain. All of these components may change as the student gains more experience in the construction of mathematics. The arrows on the diagram in Figure 4-2 go both ways, indicating that as the student constructs mathematical knowledge, attitudes and beliefs about mathematics and the neurology of the brain change. In addition, attitudes, beliefs, and neurology affect the ways in which the student creates mathematics.

Students with strengths in areas affecting the learning of mathematics may have different brain patterns than students with weaknesses in these areas. For example, students who process information well visually have different brain processing patterns than those who do not. As students with weak visual skills practice visual tasks, their brain patterns change to more closely resemble those of strong visual processors. Attitudes toward mathematics may also be reflected in the physical makeup of the brain. Students who are afraid of mathematics may emit chemicals in the brain that inhibit the higher cognitive functions. On the other hand, a student who enjoys solving mathematical problems may emit chemicals in the brain that enhance learning.

Beliefs about mathematics that affect learning include the belief that mathematics is a series of unrelated facts to be memorized as opposed to the belief that mathematics is the study of patterns and relationships. Students with one belief will approach the learning and construction of mathematics much differently than students with the other. Students' beliefs about themselves also greatly affect their learning. A student who believes that he or she cannot do math will generally work to show that this is true, while a student who is confident of success will generally be much more successful. Again, the arrow goes both ways, showing that students who succeed at constructing mathematical concepts change both their beliefs about the nature of mathematics and their beliefs about themselves.

Attitudes toward mathematics affect the construction of mathematical knowledge in much the same way as beliefs. Students who hate or fear mathematics do not do as well as students who love it. Again, the arrow goes both ways, showing that students who do construct mathematics learn to enjoy it.

Background knowledge, of course, also affects a student's ability to construct new knowledge. Students need to develop fundamental concepts before they can build on that foundation to construct new information. Again, as students construct new knowledge, they naturally add to their store of knowledge. This knowledge is generally much longer lasting than anything they have tried to memorize.

Heuristics or Strategies Used in Problem Solving and Creating

A heuristic is a general method of solving a problem. A heuristic that has proven to be useful for both the creation and the solution of problems is the model in Figure 4-3. Notice that this is not a linear model. Students can move from any point on the star to any other point.

Students may begin at the RELATE step. It is here that a student uses all available information that relates to the mathematical area on which he or she is working. For example, a student studying prime numbers may study the sieve of Eratosthenes, greatest common divisors or factors, least common multiples or denominators, composite numbers, even and odd numbers, and other number theory topics. After a student has investigated several related areas, he or she may create a new question on which to work. This question is then investigated, although the student may look at other relationships or create other questions to study during this process. After a thorough investigation, the solution or solutions are evaluated and promising solutions are reported to any interested individuals. These individuals may include classmates, younger students, professional mathematicians or mathematics educators (through journals or conferences), interested parties in industry, and teachers. Results from any investigation should then be recycled to stimulate the creation of other questions to study.

Among the best-known general approaches to problem solving is that of Polya (1957). He outlined four steps in the problem-solving process: (1) **understand the problem,** (2) **devise a plan,** (3) **carry out the plan,** and

(4) **look back.** The first step seems obvious, yet children are often frustrated because they do not understand what the problem asks. The second step suggests that reflection and planning will be rewarded later on. The third step requires children to apply one or more problem-solving skills. The fourth step calls for reviewing the process to make sure the problem is solved and there are no loose ends.

The skills that serve children as they tackle mathematical problems have been enumerated in many publications, some of which are listed at the end of the chapter. One exemplary program, Problem Solving in Mathematics (Lane County Mathematics Project, 1984), suggests these five skills:

1. **Guess and check.** Individuals using this skill make an educated guess and check the guess against the conditions of the problem. The result allows the problem solver to make a new, more refined guess. The process continues until a solution is reached. Here is an example.

   Problem: The Ridefun Toy Store sells only wagons and bicycles. On a particular day, it sold 12 items, with a total of 32 wheels. How many wagons and how many bicycles were sold that day?

   - *Understanding the problem.* Perhaps we can visualize the wagons and bicycles. The wagons each have 4 wheels; the bicycles, 2. A total of 12 items are sold on this day.
   - *Devising a plan.* We can draw pictures of wagons and bicycles and count the number of wheels. This will take a while. Perhaps we can guess a number of wagons, then find the number of bicycles by subtracting. Then, we can multiply the number of wagons by 4 and the number of bicycles by 2 and add together the products to find the total number of wheels.
   - *Carrying out the plan (guess and check).* We observe that 8 wagons alone have 32 wheels, so fewer than 8 wagons are purchased. Our first guess is 6 wagons and 6 bicycles. We find that those 12 items produce 24 + 12, or 36, wheels. Because the total is too many wheels, we refine our guess to 5 wagons and 7 bicycles. Now we have 20 + 14, or 34, wheels. Finally, we guess 4 wagons and 8 bicycles. Thus, we have 16 + 16, or 32, wheels, meeting the requirements of the problem.
   - *Looking back.* We make sure our 4 wagons and 8 bicycles make a total of 12 items. And because we know there are 32 wheels, we are satisfied with the results.

2. **Look for a pattern.** In some problems, we try to find patterns. These patterns may be visual, numerical, or sometimes both. Once a pattern is recognized, the problem will most likely be quickly solved.

Problem: Numbers describe the triangular dot patterns shown below. What are the next three numbers that follow 15?

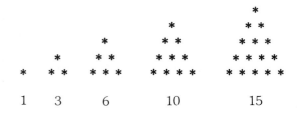

As we reflect on this problem using Polya's guidelines, we look for a pattern. We notice that each dot pattern has one more dot in its bottom row than the previous dot pattern: 1 has one dot in the bottom row; 3, two dots in the bottom row; 6, three dots in the bottom row; 10, four dots in the bottom row; and 15, five dots in the bottom row. We also notice that as we move up each triangle, each dot row has one less dot than the previous row. Using this pattern, the next triangle would have 6 dots in the bottom row for 6 + 5 + 4 + 3 + 2 + 1, or 21, total dots. Next would be a triangle with 7 dots in the bottom row for 7 + 6 + 5 + 4 + 3 + 2 + 1, or 28, total dots. The final triangle would have 8 dots in the bottom row for 8 + 7 + 6 + 5 + 4 + 3 + 2 + 1, or 36, total dots.

3. **Make a systematic list.** This skill is used when it is necessary to describe all possibilities for an event. A list is developed systematically to decrease the chance of omitting an item.

   Problem: Using only quarters, dimes, nickels, and pennies, how many different ways can you pay for an item that costs 25 cents?

   Using Polya's framework for problem solving, we decide to list all the ways to produce 25 cents with the coins given. Here is the list that we developed:

Quarters	Dimes	Nickels	Pennies	Total
1	0	0	0	$.25
0	2	1	0	.25
0	2	0	5	.25
0	1	3	0	.25
0	1	2	5	.25
0	1	1	10	.25
0	1	0	15	.25
0	0	5	0	.25
0	0	4	5	.25
0	0	3	10	.25
0	0	2	15	.25
0	0	1	20	.25
0	0	0	25	.25

Notice that as we moved from left to right in the list we started with the largest number of each coin we could use to produce 25 cents. (You may have thought of another, equally effective way to make this list systematic.) As a result, we found 13 different ways to pay for an item that costs 25 cents.

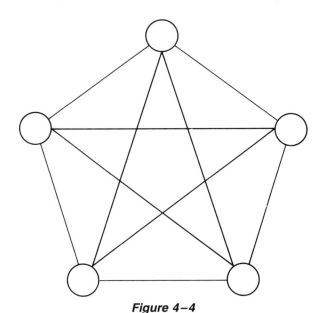

Figure 4-4

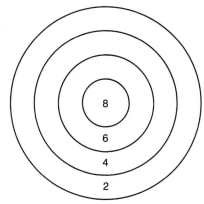

Figure 4-5

4. **Make and use a drawing or model.** Some problems can be solved easily with a drawing or model. The drawing includes the conditions of the problem and allows the solver to see the solution.

   Problem: Adams School has 11 members on its volleyball team, 6 on the court and 5 substitutes. Whenever the team scores a point, the 5 substitutes jump up and give each other a *high five* to celebrate, that is, each substitute jumps up and slaps the upheld hand of each of the other substitutes. How many high fives are given for each point scored?

   After understanding the problem, we decide that drawing a representation of the substitutes will be helpful. Figure 4-4 shows circles to represent the five substitutes. The line drawn between each pair of substitutes represents the hand slap between those two. When we count all the lines on the drawing, we find that there are ten high fives for each point scored.

5. **Eliminate possibilities.** Using this skill allows the problem solver to reduce the number of possible responses that a problem may suggest. When possibilities are eliminated, the solution becomes more manageable.

   Problem: Jill threw 4 darts at a dartboard target like the one shown in Figure 4-5. Each dart hit the target and none landed on a line.
   Which of the scores below could Jill have earned?

14	23	26	8	16
34	19	32	6	30

   As we thought about how this problem might be solved, it occurred to us to find the largest and smallest scores possible with 4 darts. If they all landed in the region worth 8 points we would have 32 points, the largest possible score; if they all landed in the region worth 2 points, we would have 8 points, the smallest

possible score given that all the darts hit the target. We can now eliminate any score below 8 or above 32.

Next, we notice that all regions have even-numbered scores. Even scores mean that any combination of 4 scores will produce an even total. We can now eliminate the odd numbers. After eliminating the odd numbers, those below 8, and those above 32, we have 14, 26, 8, 16, 30, and 32. With a little checking, we find that Jill could have earned any of these scores.

The skills just discussed are but a sampling of the problem-solving skills you are likely to find as you further investigate problem solving. The skills fit well into Polya's general approach.

There are several other lists of problem-solving strategies, some of which are variations of these and some of which are a bit different. For example, you might see "make a table or chart," which would be similar to "make a systematic list," or you might see "use a simpler problem" or "use logical reasoning," which may be similar to "eliminate possibilities." Other useful problem-solving strategies include working backwards, acting out the problem, and changing your point of view.

Working backwards is useful when you know the desired outcome of a problem, but are having difficulty getting started working toward it. You may be able to work to the beginning from the end more easily than working to the end from the beginning.

Problem: Hank is selling boxes of raspberries from his garden. The first person he meets buys half of his boxes. The next person buys 4 boxes of berries. When he meets a third person, he sells half of the remaining boxes. The next customer buys 7 boxes of berries. Hank now has only one box of berries left, and he gives this to his father for a fresh raspberry tart. How many boxes of berries did Hank have to begin with?

After understanding the problem, we may try the strategy of guess and check. It is very difficult to guess the original number of boxes, however, and this strategy may go on for quite some time before the answer is guessed. However, if we start at the end and work back to the beginning, the problem is fairly simple. We know that at the end, Hank had one box left. If customer number 4 bought 7 boxes, he must have had 8 boxes left after the third customer. If the third customer bought half of his boxes, he must have had 16 boxes left after the second customer. The second customer bought 4 boxes, so he must have had 20 boxes left after the first customer. This was half of the original number. Therefore, Hank began with 40 boxes. **Acting out the problem** is a fun strategy as well as a useful one. Children frequently need to work through the problem with their peers both to better understand the problem and to solve it.

Problem: Jack said to Andreas, "You are my father." Andreas said to Judy, "You are my mother." Judy said to Tyson, "You are my son." Tyson said to Jennifer, "You are my daughter." What is the relationship between Jack and Jennifer?

Students would probably have difficulty figuring out the relationship between Jack and Jennifer by just reading the problem. Some students may be able to figure out that the two are cousins by drawing a picture of the relationships, but others may need to actually see children taking the roles of the people in the problem and discussing their relationships.

Changing your point of view is a useful strategy to use with some tricky problems. You might think you understand a problem when in fact you are reading into it conditions that do not exist. The following problem is a well-known example.

Problem: Connect all nine dots in the following diagram using only four line segments without lifting your pencil.

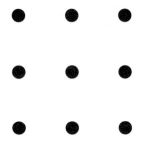

You might think that you are not allowed to break out of the restrictions of the 3 × 3 array. You have to change your point of view to realize that if you go beyond the array, you can connect the dots as follows.

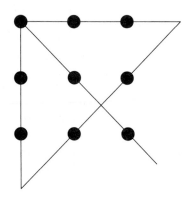

As useful as they may be, there is more to teaching problem solving than teaching about heuristics and strategies. Schroeder and Lester (1989, p. 32) distinguish between three different approaches to teaching problem solving: 1) teaching about problem solving, 2) teaching for problem solving, and 3) teaching via problem solving. In teaching about problem solving, teachers help students learn strategies and heuristics such as those discussed here and students are encouraged to be aware of these processes as they solve problems. In teaching for problem solving, teachers help students learn to solve problems so that they may use their problem-solving abilities outside of school. In teaching via problem solving, teachers help students realize that problem solving is not only the goal of learning mathematics but also the means to learn mathematics. In this way, students begin a mathematical topic with a problem to solve and learn mathematics through the solution of the problem. They construct their own knowledge of mathematics as they work through the problem. Throughout this book, we will present problem solving as a means to learn mathematics and as a goal and a topic to be learned in its own right.

Estimating

Another of the topics that will be presented throughout this book is that of estimation. Estimation may involve amount, length, weight, area, volume, or size. Computational estimation may involve using addition, subtraction, multiplication, or division of whole or rational numbers when an approximate answer is needed. Computational estimation involves many of the same skills as mental calculation, but is used when an exact answer is not needed.

Coburn (1989, p. 44) distinguishes between six categories of computation. Each category combines one of three methods of computation—mental, written, or calculator—and one of two reasons for computation—to calculate an exact sum, difference, product, or quotient, or to obtain an estimate. You should discuss with the children which of the methods of computation is appropriate for the type of problem on which you are working.

Estimation can assist children as they calculate using both paper and pencil and calculators. When estimation is applied, overall accuracy should improve. For example, a fifth grader who is multiplying 38 × 27 should be able to determine that the answer will be close to but less than 40 × 30, or 1,200. If that student multiplies and gets 10,206 or 102.6, it should be apparent that an error was made. Estimating will not eliminate all errors, but it will help in many cases.

Just as there are strategies in problem solving, there are also strategies in estimating. Reys (1986) reports five such strategies from her research. They include the front-end strategy, clustering, rounding, compatible numbers, and special numbers. The **front-end strategy** has two steps: (1) perform the operation using only the most significant digits and (2) adjust or refine the estimate by performing the operation on the remaining digits. For example, to estimate 193 + 428 + 253, children should think that 1(hundred) + 4(hundred) + 2(hundred) = 7(hundred), using the most significant digits first. Then, to adjust the estimate, they should think that 93 is about 100 and 28 + 53 is about 75. The final estimate is 700 + 100 + 75, or 875.

The **clustering strategy** is used to estimate the sum of several numbers that cluster around a particular value. The strategy involves estimating the average of the numbers and then multiplying the average by how many numbers there are. For example, to estimate 23 + 28 + 22 + 25 + 29 + 27, children can estimate the average of the numbers as 25 and then multiply by 6. The result, 150, is a good estimate of the actual sum.

The **rounding strategy** may be used for any operation. The process involves rounding the numbers being used and then performing the operation. It is important to round carefully in order to provide the best estimate. For example, to estimate 43 × 57, children can round and then multiply 40 × 60 to give an estimate of 240. The 43 is rounded down and the 57 is rounded up. If both numbers are rounded down, the result is an underestimate. If both numbers are rounded up, the result is an overestimate. In such cases, mental adjustments up or down should be made.

The **compatible numbers strategy** refers to looking for numbers that seem to fit together. For example, to estimate 2 + 8 + 5 + 2 + 3 + 9 + 4, children can search for pairs of numbers whose sums are close to 5 and 10. Thus, 2 + 8 = 10, 5 + 2 + 3 = 10, 9 is close to 10, and 4 is near 5, so the estimated sum is 10 + 10 + 10 + 5, or 35.

The **special numbers strategy** refers to seeking values that are easy to use in mental computation. This strategy is best used with fractions, decimals, and percentages. For example, to estimate 0.9 + 5.8, children can think that 1 + 6 = 7. Children should look for values near $\frac{1}{2}$, 1, 10, and 100 as part of the special numbers strategy.

To become competent estimators, children must be carefully taught the estimation strategies and must be given time to practice the strategies. Periodic checks should be made to confirm that students are remembering estimation strategies.

Weekly Estimations

We have suggested ways to incorporate estimation with each topic presented in this text. Another successful procedure is to have a special estimation activity once a week at the beginning of a math lesson. For example, using the same glass jar, one with a capacity of 250–450 milliliters, place a different type of material in the jar each week.

Begin with larger objects, such as colored cubes, shell macaroni, marshmallows, individually wrapped candies, unshelled peanuts, and bottle caps. On estimation day, distribute slips of paper to the children and have each child write down his or her name and estimation of how many objects are in the jar. Make up a small certificate to award to the student or students with the closest estimate.

As time passes, decrease the size of the objects placed in the jar. For example, the following items work well: pennies, beans, buttons, various sizes of Cuisenaire rods, breakfast cereal, and paper clips. Toward the end of the year, try rice and split peas. Every now and then, vary the type of estimation. For example, for three or four weeks, use different sizes of jars but use the same objects.

One week, put a golf ball in the jar with instructions to estimate the number of dimples on the ball (336). Another week, tape the torn-off edges of several pages of computer paper along the chalkboard and have the children estimate the number of holes in the strip. Place a small box from some household product on a table and ask how many centimeter cubes it would take to fill the box (the volume of the box).

Keep looking for objects to challenge the estimation skills of your students. It will not be too difficult to provide something for each week of the year.

Mental Calculating

The skill of mental calculation not only helps students become better at computation, it also enhances the development of concepts in a number of different areas. Students who are good at mental computation have good number concepts, whether the numbers are whole numbers, integers, or other rational numbers. They understand the properties of numbers and operations, can switch from one form of a number to another, and generally can be quite flexible in their thinking. These are all skills we wish to develop in good problem solvers. Teaching mental calculation supports the curriculum for teaching via problem solving. A student who is a good mental calculator, when faced with the problem of finding 75% of 280, would not try to mentally multiply 75 × 280. This student would probably recognize that 75% is $\frac{3}{4}$ and would then find $\frac{1}{4}$ of 280. After finding this answer, 70, the student could either multiply 70 × 3 to find $\frac{3}{4}$ of 280, or subtract 70 from 280 since $\frac{3}{4}$ is $\frac{1}{4}$ less than one. Perhaps the student will work the problem both ways to double check her answer. Students working in different groups might solve the problem using different methods and then compare answers to determine that there are a number of correct methods of solution and that some might be more efficient than others. Students who have had experience with mental calculation find that this method is frequently faster and sometimes more accurate than using pencil and paper or even a calculator. It also may help build spatial visualization abilities since mental calculation skills frequently build upon skills with manipulative materials. Students who usually construct concepts using manipulatives may learn to use mental images of manipulatives to solve problems.

Reys lists several skills that should be included in a mental calculation curriculum:

- Development of, and proficiency with, efficient mental algorithms
- Attention to place value (especially how certain mathematical operations affect the place value of an answer)
- Inclusion of, and a healthy respect for, the creation and use of alternative algorithms
- Emphasis on thinking about an efficient solution strategy before attacking the problem
- Attention to various types of numbers, including not only whole numbers but fractions, decimals, and percentages as well
- Attention to oral mental computation as well as to visual presentation of problems (1985, p. 46)

These skills for mental calculation will be presented in each chapter in the "Estimating and Mental Calculating" section.

Warm-Ups for a Math Lesson

We believe that nearly every math lesson should be preceded by some sort of brief activity to get the mind prepared for doing the mathematics of the day. When possible, this mental warm-up should match the content to be presented, thus anticipating what is to come. But this is not always practical, and other warm-ups work as well.

Warm-ups should last from 5 to 10 minutes. They should be fast-paced, involve all the students, and require thinking. Warm-ups may be problem-solving activities, mental arithmetic exercises, or short games.

We present three examples of warm-ups below, the first intended for first grade students, the second intended for fourth grade students, and the third intended for eighth grade students.

Illustrative Warm-Up (grade 1).   Begin by making sure the students are quiet and ready. Then turn to the chalkboard and make three loops connected by lines, as in Figure 4–6a. Then write the numeral 5 in the top loop, make five tallies in the second loop, and draw five triangles in the third loop, as in Figure 4–6b.

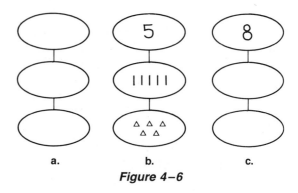

Figure 4–6

Do this carefully so the children can observe what you are doing. Then move to a new space on the board and begin again, this time starting with the numeral 8, as in Figure 4–6c. Instead of completing the loops, turn and offer the chalk to a student you think may be able to fill in the second loop (tallies).

The student may ask, "Do you want me to go up there?" pointing to the board.

Nod. It will become clear that talking is unnecessary. The student goes to the board, makes eight tallies in the second loop, and gives the chalk back to you.

Then point to the third loop and offer the chalk to another student. This student knows what to do and draws eight triangles in the third loop. After that, the next two examples are completed rapidly. Time is up. Thank the class for being quiet during this game of silent math.

Illustrative Warm-Up (grade 4).   Before the activity starts, place five cards containing numerals and letters like those in Figure 4–7 above the chalkboard. Begin the lesson by saying, "As we start today, I am going to give you a signal, and I would like you to give me a letter of the alphabet." Clap three times then tap the top of a nearby table two times. Wait. In a moment, two hands shoot up. Call on Julie, who says, "The letter *h*."

Respond, "Sorry, not this time." Then call on Lori.

Lori suggests, "The letter is *m*." Walk over, shake her hand, and nod.

By this time, other hands are going up. Next, clap four times and tap the table four times. Most of the hands go up. The first student called on says the letter is *t*.

Give several other examples. All but two students easily identify the letters. Ask Margo to explain how the letters are identified.

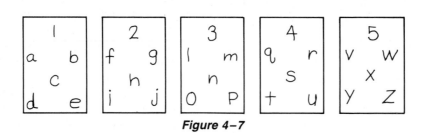

Figure 4–7

"Well," Margo replies, "the number of claps tells which card the letter is on and the number of taps tells which position on the card is being used. Five claps mean card 5 and three taps means the third position, or the letter *x*."

Then ask the students what value they think the letter *i* will have.

One student points out that if the letter is identified by two claps and four taps, the value could be 2×4, or 8.

That is the value you want to use today, so agree. Then ask what the value of the word *cat* will be. Shortly, the answer 20 is given. Not all students catch on, but an explanation is saved for another day. Time is up.

Illustrative Warm-Up (grade 8).   The teacher begins by using the overhead projector to introduce the letter combinations shown in Figure 4–8. The information is derived from basic musical chord "inversions." To provide practice for the students, the teacher asks each student to write three three-letter words as "roots" and to write the first and second inversions of the words.

INVERSIONS		
Root	First Inversion	Second Inversion
cge	gec	ecg

Figure 4–8

Several students are asked to share their inversions. Dan says that one of his root words is "top," the first inversion is "opt," and the second inversion is "pto." Sarah notes that Dan's first inversion is the word "opt" and suggests that there might be other root words with words in their inversions. The teacher asks if there are other students who found words in their inversions. Six students raise their hands and share their words. Then the teacher asks if anyone found a root word that produced words in both the first and second inversions. No one has found such a word and the students are challenged to find one. Within a minute, Darlene finds the word "ate" and its inversions, "tea" and "eat."

Once it is apparent to the teacher that the students understand how inversions work, she asks the students to see if they can find a "root" number whose first inversion is an even number and whose second inversion is divisible by five. The students are then asked to find a number whose first inversion is a palindrome and whose second inversion is an odd number. Then the students are asked to find the root number whose first inversion is greater than 500 and whose second inversion is less than 50. Finally, the students are asked to pose a number inversion problem for the rest of the class.

These three episodes represent just the tip of the iceberg. These activities can be reused and modified or extended. They can all be made more or less difficult. It will take a while to collect enough suitable warm-ups for a given grade level, but the effort is well worth it. The opportunity for mental activity should be taken advantage of. The bibliography at the end of the chapter includes several sources of warm-up activities.

Grouping Students for Problem Solving, Estimating, and Mental Calculating

If you use problem solving as the means of instruction, you will find that students can learn to solve problems in many different types of learning groups. Solving a problem with one or two other students is often very effective. It gives the student a chance to discuss the problem and learn from others' mistakes and successes. Cooperative learning needs to be carefully planned, however. You cannot simply let the students work with their friends and hope for the best. If students have never worked in cooperative learning groups before they will probably need some instruction on how best to work together. They need to learn that everyone should be actively involved in problem solving and that they cannot leave all the work to one person. Students must learn to be accepting and supportive of each other's ideas; telling one of the group members that he is stupid is counterproductive. It may help to assign roles in the group, such as an organizer, a checker, a relator, and a confidence builder, as suggested by Johnson and Johnson (1989, p. 242). Make sure that the children understand that they are each individually accountable for understanding and explaining the problem. Students then become peer coaches as they make sure that everyone in the group understands the problem. Cooperative learning is described in detail in Chapter 15.

Some problems may involve estimation and mental calculation. A group may be asked to make the best possible estimate of the number of jelly beans in a jar or to find as many ways as possible to add 289 and 345. When a question is open ended, it allows for much more mathematical reasoning and creativity on the part of the students. Students may also be given a situation such as the jar full of jelly beans and asked to find both their own questions and the answers.

Roper (1990) has developed a series of books for exploring cooperative problem solving using a variety of manipulative materials such as unifix cubes, pattern blocks, tangrams, and attribute materials. Each problem is written on a set of four cards, which are dealt to four students. Each student keeps the card he or she was dealt and does not show it to anyone else. Students do, however, read their clues to each other. All four cards are essential to solving the problem. The students must work cooperatively to piece all the clues together and solve the problem.

Not all problem solving goes on in small groups. There are times when the teacher will want to present a problem to the entire class at once. Be careful when using problems with a number of children at once. These problems

1. These are stories written by children.
   Tell how you could find the answer to each story.

The plane cost 54 dollars and the car only cost 23 dollars How much more did the plane cost?

I bought a pair of Daniel glosses and they cost $23 and I bought some Batman cards they cost $16. Then I had 0 money. How much did I spend altogether?

I had 39¢ AND I GOT SOME money from mom AND I had 52¢ altogether. How Much did mom give me?

I had 60 cents and I bought a Present and I had 25 cents left What did my Present cost
ELISABETH

I had $39 and I cleaned the Car and the house and I worked hard in the garden and I got $13.

One day a boy had 85¢ and went shopping. He bought a flower pot that cost 60¢.

2. Use money or tens and ones blocks to act out each story.
   Tell what you did.

3. Write an addition or a subtraction story using money.
   Draw a picture for your story.

Figure 4–9
Calvin Irons. Moving into Math. *San Francisco, Ca.: Mimosa, 1991, p. 63. Reprinted by permission.*

Figure 4—9 illustrates a page from a second grade Core Book. It is designed to get children started writing their own word problems by looking at word problems written by other children, solving those, and then writing their own. It is a good illustration of students creating their own questions. Notice that even though all the questions involve money, they do not all involve the same operation. Even the problems involving subtraction include a missing addend problem, a comparison problem, a missing subtrahend problem, and a take-away problem. Students would have to reason through each problem individually to determine how to solve it. They are asked to demonstrate their solution to each problem using money or base ten blocks and not just to use paper and pencil to come up with an answer. It is likely that each subtraction problem would be solved differently on a concrete level.

This series integrates a great deal of language arts into the mathematics program. In addition to this Core Book, the series includes sixteen Read Together books that illustrate mathematics concepts with children's stories, and an Activity Book that includes a wide variety of concrete activities to go along with the pages in the Core Book. The Read Together books are designed to build a base for problem solving where children relate their own language and experiences to the language of mathematics. In the Activity Books, the students are engaged in hands-on problem solving; they analyze patterns and create mathematical situations as well as practice skills.

The program also includes a kit of manipulative materials and staff-development videotapes that go along with the Teacher's Resource Book and other materials.

should have lots of correct answers, so that when one child finds an answer, the rest of the children can continue to work on the problem to discover other correct answers or methods.

When children work on problems individually, be sure to leave time for a discussion of the solutions. Children may need time to construct information for themselves without the distractions of others, but after a solution has been found, discussion with others opens up many other possibilities.

Communicating in Problem Solving

As students work together to solve problems, they are naturally communicating. This should be encouraged whenever possible. It changes your classroom from a setting with one teacher and 25 students to a setting with 26 teachers and 26 learners. Do not limit yourself to oral communication, however.

Students should be encouraged to draw pictures or build models to demonstrate the solutions to problems whenever possible. Students of all ages benefit when they see how someone else has solved a problem. You might even challenge students occasionally to demonstrate their solutions without using words.

Problem solving is also an area in which students can use writing. Not only can they write out a description of how they went about solving a problem, they can write out word problems for other students to solve. These problems are very motivating for the other students. It is much more fun to solve problems written by your friends than to solve the ones in the book.

Be sure not to restrict problem solving to mathematics. The communication of mathematical learning enhances the language arts curriculum, strengthens reading skills, and is a necessary skill in science and social studies. Ask the students to be on the lookout for opportunities to express their mathematical problem-solving abilities in all of these areas. For example, students may keep a daily journal in which a discussion of the solutions of any new problems might be a prominent part.

Evaluating Problem Solving, Estimating, and Mental Calculating

The evaluation of a student's ability to solve problems, estimate, and calculate mentally will require many methods other than the traditional paper and pencil test. Although paper and pencil tests seem to be getting better at

testing skills other than computation and rote memorization, there are better ways to determine if a student is competent at problem solving, estimating, or calculating mentally.

Two types of **holistic scoring** suggested by Charles, Lester, and O'Daffer (1987) provide examples of alternative assessment techniques for teachers. Holistic scoring refers to evaluating the solution to a problem by considering all the written evidence produced by the student in solving the problem. The focus of holistic scoring is the process the student used to reach a solution. This approach is different from the typical approach: first check the answer, then, if the answer is incorrect, try to figure out what the student did wrong. Holistic scoring helps the teacher assess the student's overall performance.

The first example is of **analytic scoring.** With this approach, the teacher assigns a certain number of points to each aspect of the problem-solving process. The aspects of the problem-solving process that are scored are those you have specified. You establish the criteria, and this scale is used to assess the student's work. For example, Charles, Lester, and O'Daffer (p. 30) provide the analytic scoring scale shown in Figure 4–10.

Let's apply this scoring scale to a problem presented earlier in this chapter.

Problem: Adams School has 11 members on its volleyball team, 6 on the court and 5 substitutes. Whenever the team scores a point, the 5 substitutes jump up and give each other a *high five* to celebrate, that is, each substitute jumps up and slaps the upheld hand of each of the other substitutes. How many high fives are given for each point scored?

One Possible Solution: Make a drawing. Each connecting line segment represents a high five from one substitute (circle) to another. A total of 10 high fives are represented in the drawing.

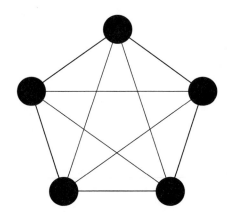

The following is a solution to the problem by a fourth-grade and a fifth-grade student who worked together:

Analytic Scoring Scale		
Understanding the Problem	0:	Complete misunderstanding of the problem
	1:	Part of the problem misunderstood or misinterpreted
	2:	Complete understanding of the problem
Planning a Solution	0:	No attempt, or totally inappropriate plan
	1:	Partially correct plan based on part of the problem being interpreted correctly
	2:	Plan could have led to a correct solution if implemented properly
Getting an Answer	0:	No answer, or wrong answer based on an inappropriate plan
	1:	Copying error; computational error; partial answer for a problem with multiple answers
	2:	Correct answer and correct label for the answer

Figure 4–10

Randall Charles, Frank Lester, and Phares O'Daffer. How to Evaluate Progress in Problem Solving. *Reston, Va.: National Council of Teachers of Mathematics, 1987, p. 30. Reprinted by permission.*

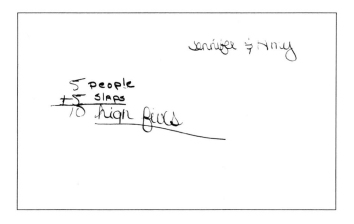

Jennifer and Amy decided that there were 5 people and that they each got 5 slaps. Then they added the number of people and the number of slaps to reach the answer 10. Using the analytic scoring scale in Figure 4–10, we find that part of the problem was misunderstood or misinterpreted, so 1 point is awarded for *Understanding the Problem*. Next, there was a partially correct plan based on part of the problem being interpreted correctly, resulting in a score of 1 for *Planning a Solution*. Finally, the answer is correct and is labeled correctly, so the score for *Getting an Answer* is 2. The overall score for Jennifer and Amy for this solution is 1 + 1 + 2, or 4. Additional information about the solution could be gathered by interviewing the girls and having them further explain their thinking.

The second type of holistic scoring is referred to as **focused holistic scoring.** This type of scoring results in a single score for the entire problem. The scoring is based on a set of specific criteria sometimes called a **scoring rubric.** It is called a focused procedure because a single score is determined using specific criteria to represent the processes involved in thinking through the solution to a problem. Figure 4–11 shows a focused holistic scoring scale presented by Charles, Lester, and O'Daffer (p. 35).

The California State Department of Education (1989) has developed a procedure for constructing a scoring rubric for an open-ended question in mathematics that is specific to the problem. Thus, each problem being assessed has its own scoring rubric. A generalized rubric has been developed from which the specific rubrics are derived.

Holistic scoring techniques provide alternative assessment tools for teachers who wish to collect information that is helpful both in designing problem-solving instruction for students and in helping students become aware of their own thinking processes.

If students are keeping journals, one method of evaluating their work would be to ask the students to record not only the methods of solution, but also how well they are doing at understanding concepts or mastering particular skills. Even young students are often quite capable of self-evaluation and can fairly accurately judge their own strengths and weaknesses. You could collect three or four journals each day and keep your own log of the students' progress. Be sure you get to everyone's journal at least once every two weeks.

Students working in groups can often do a peer evaluation. They know which students are strong on particular skills, and they can help each other and you pinpoint weaknesses, as well as work to help each other strengthen those areas.

There is no substitute for direct observation in evaluating some problem-solving, estimation, and mental calculation skills. If you want to know if a student can mentally add two two-digit numbers or multiply a percent by a whole number, you need to give the student a problem and listen for a response. If you want to know if students are working well in cooperative learning groups, you need to

Focused Holistic Scoring Point Scale

0 Points

These papers have one of the following characteristics:

- They are blank.
- The data in the problem may be simply recopied, but nothing is done with the data or there is work but no apparent understanding of the problem.
- There is an incorrect answer and no other work is shown.

1 Point

These papers have one of the following characteristics:

- There is a start toward finding the solution beyond just copying data that reflects some understanding, but the approach used would not have led to a correct solution.
- An inappropriate strategy is started but not carried out, and there is no evidence that the student turned to another strategy. It appears that the student tried one approach that did not work and then gave up.
- The student tried to reach a subgoal but never did.

2 Points

These papers have one of the following characteristics:

- The student used an inappropriate strategy and got an incorrect answer, but the work showed some understanding of the problem.
- An appropriate strategy was used, but—
  a) it was not carried out far enough to reach a solution (e.g., there were only 2 entries in an organized list);
  b) it was implemented incorrectly and thus led to no answer or an incorrect answer.
- The student successfully reached a subgoal, but went no further.
- The correct answer is shown, but—
  a) the work is not understandable;
  b) no work is shown.

3 Points

These papers have one of the following characteristics:

- The student has implemented a solution strategy that could have led to the correct solution, but he or she misunderstood part of the problem or ignored a condition in the problem.
- Appropriate solution strategies were properly applied, but—
  a) the student answered the problem incorrectly for no apparent reason;
  b) the correct numerical part of the answer was given and the answer was not labeled or was labeled incorrectly;
  c) no answer is given.
- The correct answer is given, and there is some evidence that appropriate solution strategies were selected. However, the implementation of the strategies is not completely clear.

4 Points

These papers have one of the following characteristics:

- The student made an error in carrying out an appropriate solution strategy. However, this error does not reflect misunderstanding of either the problem or how to implement the strategy, but rather it seems to be a copying or computational error.
- Appropriate strategies were selected and implemented. The correct answer was given in terms of the data in the problem.

Figure 4–11

Randall Charles, Frank Lester, and Phares O'Daffer. How to Evaluate Progress in Problem Solving. *Reston, Va.: National Council of Teachers of Mathematics, 1987, p. 35. Reprinted by permission.*

sit in on the group and observe. Checklists and a teacher's journal help you remember what you have seen.

You might also want to enlist the aid of parents in this evaluation. You could send home a note explaining that you are interested in evaluating the student's ability to solve a problem involving making change or estimating volumes, and ask parents to observe the students as they go to the store or cook and to question them on their understanding.

The skills you are evaluating are as important as the evaluation methods you use. In problem solving, you should be concerned with more than whether or not students get the correct answer. The processes they use and the strategies they attempt are also important. You should observe if students can work on problems independently or if they always seem to need help, if they give up before really trying or if they are persistent, if they will try more than one method of solution or if they quit if the first method does not work, and whether or not they check their answers.

In evaluating estimation and mental calculation, look beyond the answer to determine if the students have a good understanding of the numeration system, if they can use a variety of techniques, if they can apply the principles and properties of operations, and whether or not they double check to determine whether answers are reasonable.

Something for Everyone

Problem solving is an excellent topic for all types of learners. Auditory learners do well in cooperative learning groups, where they can discuss their ideas and listen to those of others. Tactile-kinesthetic learners do well using concrete materials and models to solve problems, and visual learners may draw sketches to illustrate their solutions. It helps all types of students to realize that there are lots of ways to solve problems, and that they can use their areas of strength to attack a problem.

Auditory learners frequently enjoy mental calculation, but visual and tactile-kinesthetic learners may find it difficult. Be sure that problems are often presented visually as well as orally. Let the students use concrete materials when they are first developing the mental algorithms, for they may later manipulate these images in their heads.

Cooperative learning groups give students of all ability levels a chance to participate with others of varying abilities. Students who are good at paper and pencil computation, who may have been the shining stars in the classrooms of yesterday, will learn to appreciate students with other abilities in areas such as spatial reasoning or mental estimation and calculation. All students can learn to be accepting of differences and they will find there is much to be learned from those with different learning styles.

KEY IDEAS

In this chapter, a problem is defined as a perplexing question or situation that does not suggest an immediate method of solution. A problem should also interest the learner. Goals in problem solving include not only helping the students to become proficient problem solvers, but also helping them to pose problems and to create mathematics to solve those problems.

Characteristics of good problem solvers were listed and discussed, as were factors that influence a person who is constructing new mathematical concepts. Strategies for solving problems include guess and check, look for a pattern, make a systematic list, make and use a drawing or model, eliminate possibilities, work backwards, act it out, and take a different point of view. Polya has outlined a four-step problem-solving heuristic: 1) understand the problem, 2) devise a plan, 3) carry out the plan, and 4) look back. Students may also use the following five-point heuristic: relate, create, investigate, evaluate, and communicate.

Estimation and mental calculation involve similar skills, but estimation is used when students want to find an approximate answer and mental calculation is used when an exact answer is required. Students should discuss when it is appropriate to use mental skills, and when using paper and pencil or a calculator would be more efficient. Estimation strategies include the front-end strategy, the clustering strategy, the rounding strategy, the compatible numbers strategy, and the special numbers strategy.

Students should work in a number of different types of groups as they solve problems, estimate, and calculate mentally. Students should be encouraged to communicate their learning in these areas orally, in writing, and through pictures and models.

Each of these areas poses a number of issues for the teacher to consider in evaluation.

Each of these areas will be explored more fully throughout the remainder of this book. As you read each chapter, ask yourself how you will encourage the students to use their problem-solving and mental skills in each area.

REFERENCES

BERNSTEIN, BOB. *Monday Morning Magic.* Carthage, Ill.: Good Apple, 1982.

———. *Friday Afternoon Fun.* Carthage, Ill.: Good Apple, 1984.

BROWN, STEPHEN I., AND WALTER, MARION I. *The Art of Problem Posing.* Hillsdale, N.J.: Lawrence Erlbaum Associates, 1983.

CALIFORNIA STATE DEPARTMENT OF EDUCATION. *A Question of Thinking: A First Look at Students' Performance on Open-Ended Questions in Mathematics.* Sacramento, Ca.: Bureau of Publications, California State Department of Education, 1989.

CHARLES, RANDALL; LESTER, FRANK; AND O'DAFFER, PHARES. *How to Evaluate Progress in Problem Solving.* Reston, Va.: National Council of Teachers of Mathematics, Inc., 1987.

COBURN, TERENCE. "The Role of Computation in the Changing Mathematics Curriculum." In Trafton, Paul R., and Shulte, Albert P., eds. *New Directions for Elementary School Mathematics.* Reston, Va.: National Council of Teachers of Mathematics, 1989.

GREENES, CAROLE; GREGORY, JOHN; AND SEYMOUR, DALE. *Successful Problem Solving Techniques.* Palo Alto, Ca.: Creative Publications, 1977.

HOPE, J. A.; LEUTZINGER, L.; REYS, B. J.; AND REYS, R. E. *Mental Math in the Primary Grades.* Palo Alto, Ca.: Dale Seymour Publications, 1988.

HOPE, J. A.; REYS, B. J.; AND REYS, R. E. *Mental Math in the Middle Grades.* Palo Alto, Ca.: Dale Seymour Publications, 1987.

———. *Mental Math in Junior High.* Palo Alto, Ca.: Dale Seymour Publications, 1988.

JOHNSON, DAVID W., AND JOHNSON, ROGER T. "Cooperative Learning in Mathematics Education." In Trafton, Paul R., and Shulte, Albert P., eds. *New Directions for Elementary School Mathematics.* Reston, Va.: National Council of Teachers in Mathematics, 1989.

KRULIK, STEPHEN, AND RUDNICK, JESSE A. *Problem Solving in Math, Books G and H.* New York: Scholastic Book Services, 1982.

———. *Problem Solving: A Handbook for Elementary School Teachers.* Boston: Allyn & Bacon, 1988.

KULM, GERALD (ED.). *Assessing Higher Order Thinking in Mathematics.* Washington, D.C.: American Association for the Advancement of Science, 1990.

LANE COUNTY MATHEMATICS PROJECT. *Problem Solving in Mathematics.* Palo Alto, Ca.: Dale Seymour Publications, 1984.

LENCHNER, G. *Creative Problem Solving in School Mathematics.* Boston: Houghton Mifflin, 1983.

McFADDEN, SCOTT. *Math Warm-Ups for Jr. High.* Palo Alto, Ca.: Dale Seymour Publications, 1983.

MOSES, BARBARA; BJORK, ELIZABETH; AND GOLDENBERG, E. PAUL. "Beyond Problem Solving: Problem Posing." In Cooney, Thomas J., and Hirsch, Christian R., eds. *Teaching and Learning Mathematics in the 1990s.* Reston, Va.: National Council of Teachers of Mathematics, 1990.

NATIONAL COUNCIL OF TEACHERS OF MATHEMATICS. *Curriculum and Evaluation Standards for School Mathematics.* Reston, Va.: NCTM, 1989.

OHIO DEPARTMENT OF EDUCATION. *Problem Solving . . . a Basic Mathematics Goal: A Resource for Problem Solving.* Columbus, Oh., 1980.

_____ . *Problem Solving . . . a Basic Mathematics Goal: Becoming a Better Problem Solver.* Columbus, Oh., 1980.

POLYA, GEORGE. *How to Solve It.* Garden City, N.Y.: Doubleday & Co., 1957.

REYS, BARBARA J. "Mental Computation." *Arithmetic Teacher.* Vol. 32, No. 6 (February 1985), pp. 43–46.

_____ . "Teaching Computational Estimation: Concepts and Strategies." In Schoen, Harold L., and Qweng, Marilyn J. *Estimation and Mental Computation.* Reston, Va.: National Council of Teachers of Mathematics, 1986.

REYS, BARBARA J., AND REYS, ROBERT E. "Estimation— Direction from the *Standards.*" *Arithmetic Teacher.* Vol. 37, No. 7 (March 1990), pp. 22–25.

ROPER, A. *Cooperative Problem Solving.* Sunnyvale, Ca.: Creative Publications, 1990.

SCHROEDER, THOMAS L., AND LESTER, FRANK K., JR. "Developing Understanding in Mathematics via Problem Solving." In Trafton, Paul R., and Shulte, Albert P., eds. *New Directions for Elementary School Mathematics.* Reston, Va.: National Council of Teachers of Mathematics, 1989.

SHEFFIELD, LINDA JENSEN. *Problem Solving in Math, Books C, D, E, F.* New York: Scholastic Book Services, 1982.

SOWDER, JUDITH T. "Mental Computation and Number Sense." *Arithmetic Teacher.* Vol. 37, No. 7 (March 1990), pp. 18–20.

5 Thinking Mathematically

Children attending elementary school today need an emphasis in their mathematics classrooms different from that of one hundred, fifty, or even twenty years ago. The rapid advances in computer technology, the proliferation of inexpensive calculators, the explosion of the amount of data to be dealt with every day, and the ever-increasing rate of change require that children develop new skills. It is no longer sufficient that children develop proficiency in computation and in applying that computation to their day-to-day problems. By the time these children reach adulthood in the twenty-first century, they will be faced with problems that no teacher can foresee. It is crucial, therefore, that these children be taught how to think. Children of different ages think on different levels, but all children are capable of rational thought. Mathematics is an ideal subject in which to develop the thought processes, beginning at a very early age.

The following quotes from two recent reports emphasize the need to change our usual style of mathematics teaching drastically to improve the level of reasoning ability of the average American mathematics student. First, Dossey and others (1988, p. 7) found some cause for concern: "Findings from the 1986 mathematics assessment parallel those from recent assessments in reading, writing, and literacy, in which the Nation's Report Card has documented a critical shortage of effective reasoning skills among our young people. Although more students appear to have mastered basic mathematical skills and concepts in recent years, few achieve the higher range of mathematics proficiency. Our nation must address this deficit if it is to thrive in the technological era facing us."

The second quotation comes from the National Research Council's publication *Everybody Counts: A Report to the Nation on the Future of Mathematics Education* (1989, pp. 1–2): "More than ever before, Americans need to think for a living; more than ever before, they need to think mathematically. . . . Wake up, America! Your children are at risk. Three of every four Americans stop studying mathematics before completing career or job prerequisites. Most students leave school without sufficient preparation in mathematics to cope either with on-the-job demands for problem solving or with college requirements for mathematical literacy."

To help students strengthen their ability to think mathematically, four major thinking processes are emphasized in this chapter:

1. **observing and inferring** (encouraging children to describe objects both orally and pictorially)
2. **comparing** (asking children to note likenesses and differences)
3. **classifying** (asking children to sort objects on the basis of one or more attributes)
4. **sequencing** (asking children to order elements in a set on the basis of one or more given characteristics)

These processes lead to problem solving and do not require the use of numbers. Gibb and Castaneda (1975) and Mueller (1985) have noted similar processes as necessary for developing a solid understanding of number concepts. These processes should begin before formal work with numbers is introduced and should continue to develop throughout the elementary school years.

Prenumber concepts and the problem solving that utilizes those concepts occur frequently in children's everyday lives. Ask children to suggest ways that they use these ideas outside of class. Send home notes briefly explaining to parents some of the concepts you are teaching and giving some suggestions for the parents to continue the teaching of those concepts at home. This communication with parents may help prevent complaints from parents wondering when you are going to start mathematics, since the children are not bringing home worksheets full of numbers and the children report that all they do is play with materials during math time.

Some of the ways in which children naturally use the thinking processes at home include observing their surroundings and making predictions about the future, comparing and sorting everyday objects, and following and creating patterns. Enlist the support of the parents in building these skills.

You may encourage the parents to play a guessing game in which they ask the children to observe the room and then close their eyes. While the children's eyes are closed, the parents move one object. When the children open their eyes, they try to guess which object was moved.

Parents can play other simple observation games such as "I'm thinking of something. . . ." The parent chooses an object in plain sight and picks one word to describe it, such as its color or shape. The child then tries to guess the object. When the object is guessed, the child chooses a new object.

Several household events can encourage comparing and classifying. Parents should encourage children to set up a classification scheme for putting away their toys or for categorizing favorite books. Children can practice sorting as they put away the silverware and compare teaspoons and tablespoons, or as they fold the clothes from the laundry and separate their own clothes from those of their siblings. Children follow simple patterns as they set the table and realize the fork goes on the left of the plate and the spoon and knife go on the right. They note a sequence in the events of the day as they find that in the morning they wake up, go to the bathroom, get dressed, eat breakfast, brush their teeth, and get ready for school. Teachers and parents alike should discuss these events. Teachers should point out the thinking processes being learned and the relationship of these processes to problem solving and later to learning number concepts.

In this chapter, we explore some of the ways in which to develop thinking skills. The emphasis is on assisting children to become creative mathematicians. Children should be encouraged not only to solve problems but also to create them. Teachers should allow children to use their own natural thought processes as often as possible and should aid children in expanding their thinking. This chapter focuses on skills and processes that do not require number concepts. It begins with activities appropriate for very young children but also includes activities challenging for older children and adolescents. We suggest ways to encourage children to look at the relationships between objects and sets and to begin work with formal logic. In Chapter 6, these properties are related to number concepts. In Chapter 7, number processes are expanded to include the processes of joining and separating as addition and subtraction are introduced.

Many of the activities work with sets of objects. These activities may involve the use of commercial materials such as attribute blocks and People Pieces, computer games such as Gertrude's Puzzles and Moptown, or materials collected around the home or constructed by the teacher and the students. Seeing, manipulating, and constructing materials are necessary for children of all ages to develop solid prenumber concepts.

Developing Concepts

As mentioned earlier, the concepts necessary for a solid foundation upon which to build later number concepts include observing and inferring, comparing, classifying, and sequencing. These concepts develop as children recognize relationships between objects and sets and later as they develop number concepts. These concepts are not exclusive to mathematics, but rather are necessary in all subject areas.

Observing and Inferring

To make observations and inferences, children should be encouraged to use all five senses. As children gather information about the world, they should describe what is being observed and inferred. Language is a powerful tool for gathering and disseminating information, so children should be encouraged to talk to each other and the teacher while they are engaged in these activities. The activities presented here are generally sequenced from simplest to most complex. Grade levels are listed as a general guide. Choose activities that best meet the needs of your students and adapt them as you see fit.

ACTIVITIES

Primary (K–1)

Objective: *to develop the ability to observe and describe using the five senses.*

1. Choose several objects that are safe to touch, smell, and taste, such as sugar, fruit, crackers, and cookies. Put one object in a clean bag and ask the children to feel the object without looking inside. Have the children describe what they feel. Then let them sniff the object without peeking and describe what they smell. Encourage the children to take a small bite and to describe the taste. Shake the bag and ask the children to describe what they hear. Finally, let the children look at the object and describe what they see. Allow the children to bring in objects to place in the bag.

Objective: *to develop inference skills based on the sense of hearing.*

2. Have the children close their eyes and listen to familiar sounds, such as a door closing, someone writing on the board with chalk, and a chair that squeaks. Let the children describe what they hear and make a guess as to what it was. Have them make noises for the other children to guess.

Objective: *to develop inference skills based on the sense of touch.*

3. Let the children feel several geometric figures placed in a feely box and guess what they are. See if they can fit a figure into a frame of the same size and shape without looking. Children may be able to match the shape to the frame without knowing the name of the shape.

Objective: *to develop inference skills based on the senses of sight and touch.*

4. Outline several familiar objects. Have the children guess which object matches each outline. Give the children the objects to fit onto the outlines to see if their guesses were correct. Let the children make outlines of their own to exchange with each other.

Objective: *to distinguish between observations and inferences.*

5. Repeat the activity for developing the ability to observe, but let the children guess what the object is at each step. Discuss the difference between observing using the senses and guessing based on observations.

For all the activities, be sure to discuss with the children the strategies they used to make their guesses. They can learn from each other better ways to make inferences.

ACTIVITIES

Primary-Intermediate (2–5)

Objective: *to develop the ability to observe a three-dimensional figure and match it to a two-dimensional model and vice-versa.*

1. Using a set of geoblocks, a set of solid wooden blocks of varying shapes and sizes available commercially, or another set of solid figures in which the blocks are different shapes, draw the outline of each face of a block on a sheet of paper. Let children choose from a pile of blocks the block which has all those faces. Have them pick up the block and check each face against the outline to see if it matches. (See Figure 5–1.)

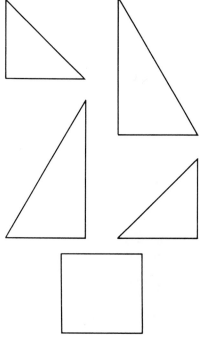

Figure 5–1

Objective: *to develop the ability to make a sketch from an observation.*

2. Using the geoblocks, have the children sketch each face of a block while looking at it. Have the children also sketch each face of a block while feeling the block but not looking at it. Let the children exchange sketches with one another to see if they can select the block that was sketched.

Objective: *to describe and recognize three-dimensional geometric figures.*

3. Use the geoblocks again. This time, have one child describe a block while another child listens. Let the second child select the block that was described. Change roles so the selector has a chance to be the describer.

Experiences such as these in observing, inferring, and describing give children an opportunity to develop and refine many mathematical concepts. Children may use vague or emotional words rather than specific, descriptive words. Children who describe something as *good,* may

realize after some of these activities that it is more effective to use words such as *soft, warm,* and *fuzzy.* Let the children discuss which words give better descriptions.

Comparing

Once children learn to observe and describe objects, they should begin to compare two or more objects. Often, children begin to compare objects even before they know the names of the objects. They may say they want more or fewer even if they do not know the name of what they have. They may be able to tell you what is the same or different about two objects whether or not they know the names of the objects. They may also know that comparisons do not always remain constant. Children themselves may be small when compared to adults but large when compared to a favorite doll. A child may describe a set as having more when it really has fewer but larger objects. Teachers should assist children in developing difficult comparison concepts. Being able to compare individual objects, and later sets of objects, will help children when they are deciding whether 3 is more or less than 5.

ACTIVITIES

Primary (K–2)

Objective: *to compare two or more objects using all the senses.*

1. Discuss the terms *alike* and *different* with the children. Then collect a group of objects from the children. Select two objects at random and ask the children to list all the ways that the objects are alike or different. List their responses on the chalkboard or on a large sheet of paper. Encourage the children to use all their senses.

2. Play line-up with the children. One child is the leader. The next child in line must name one way in which he or she is like the leader and one way in which he or she is different. Each subsequent child then names one likeness and one difference between himself or herself and the child directly before him or her. After playing the game, ask the children to tell you what *alike* and *different* mean. Encourage them to suggest other objects to use to play the same game.

Classifying

After children learn to compare objects, they should begin to categorize, or classify, them. **Classifying** is the process of grouping or sorting objects into classes or categories according to some systematic scheme or principle. The children must use specific properties of the objects to be classified and must make comparisons between objects in order to decide on proper categories. Students should be encouraged to construct their own categories and not just match objects to labels or categories that the teacher has decided upon.

The groups into which objects are sorted are called sets. A **well-defined set** is a collection of objects defined

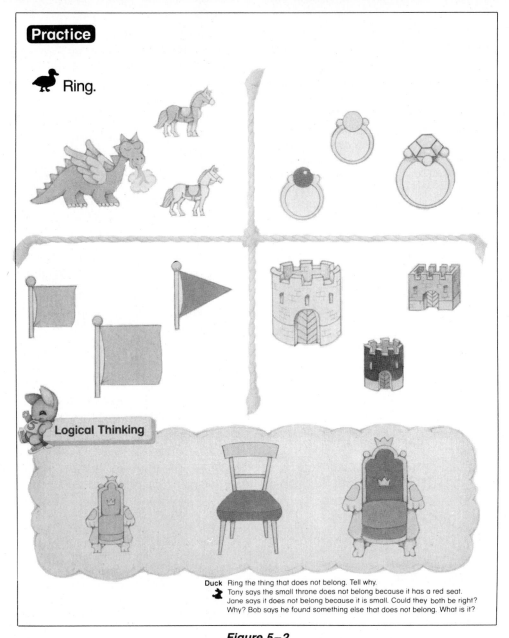

Practice

🦆 Ring.

Logical Thinking

Duck Ring the thing that does not belong. Tell why.
🦆 Tony says the small throne does not belong because it has a red seat. Jane says it does not belong because it is small. Could they both be right? Why? Bob says he found something else that does not belong. What is it?

Figure 5–2
Ginsburg, Herbert P. et al. Mathematics: Exploring Your World. *Morristown, N.J.: Silver Burdett & Ginn, p. 42. Reprinted by permission.*

THE MATH BOOK

The page shown in Figure 5–2, from a kindergarten textbook, is part of a unit on sorting and classifying. The teacher's manual suggests that before beginning this practice page, students experience a number of activities involving concrete manipulatives such as attribute blocks, connecting cubes, buttons, and punchout figures of toys that are included in the book. Students are encouraged to group objects in a number of different ways and to discuss their reasons for the groupings. Notice that the problems on this page have several possible answers. Students discuss why there can be more than one right answer.

The teacher's manual suggests connecting this lesson to a science lesson about leaves. Students may collect leaves of different types, sizes, and colors, and then decide on different ways in which to group them. Other subject area connections include ideas for writing, poetry, reading, and physical education. The manual also contains a number of suggestions for working with different types of students such as learning disabled, gifted, or at-risk students, or students with limited English proficiency. For example, the manual suggests that gifted and talented children might sort leaves into Venn diagrams by color. If students have one loop of the Venn diagram for red leaves and one loop for green leaves, the students can be challenged to decide where to put the leaves that are both red and green.

This unit ends with a page for the students to take home to work on with their families. It includes a letter to let the family know which skills the children have learned and a game to play with cutouts of jewels that can be sorted by size, shape, and color. The adults are then encouraged to help the children find objects around the house that can be sorted according to several criteria. Note that in this series, as in many others, the important activities are not those that are found in the book, but those that precede and follow the work done with the textbook.

so that given any object, it is possible to determine without question whether or not that object is in the collection. Classification systems also serve to describe an object that is not given. By observing the position in a classification scheme the object would occupy, children can give the properties of the object.

As in other types of mathematical problems, children must find the missing part. This requires that they analyze the structure of the problem much as they will later analyze the structure of a numerical equation. Carpenter (1985) has noted that children naturally attend to the structure of a problem and that expert problem solvers put more emphasis on structure than less-capable problem solvers. The activities in this section are designed to help children focus on essential structures or relationships.

Young children should begin by classifying objects that differ in only one way, such as shape or color. Use structured materials such as attribute blocks or People Pieces or use collections of materials found around the home or classroom such as buttons, shells, or baseball cards. **Attribute blocks** usually are made from either wood or plastic and vary in attributes such as color, size, shape, and thickness. Each set of blocks contains only one block with each possible combination of attributes. For example, a set of blocks with two sizes (large and small), three colors (red, blue, and green), and three shapes (circle, square, and triangle) would have the following eighteen pieces:

large red circle
large red square
large red triangle
large blue circle
large blue square
large blue triangle
large green circle
large green square
large green triangle
small red circle
small red square
small red triangle
small blue circle
small blue square
small blue triangle
small green circle
small green square
small green triangle

Notice the pattern in the listing of pieces. Other sets may have more or fewer shapes, colors, or sizes and may add other variables, such as thickness. You may make a set of attribute shapes of your own by copying the set in Appendix B onto colored construction paper or posterboard or by using the attribute shapes die to punch out pieces of railroad board with a machine such as the Ellison letter machine. These attribute shapes are used for several of the activities in this chapter.

People Pieces are a set of attribute materials consisting of sixteen wooden or plastic tiles with a different person stamped onto each one. The people are of two heights (tall and short), two weights (stout and thin), two colors (red and blue), and two sexes (male and female). List the sixteen possible combinations for yourself. Other structured sets you might use include the Animal Tiles from Creative Publications and Zogs from the Addison-Wesley Explorations 2 book.

The following activities do not require the purchase of any commercial materials.

ACTIVITIES

Primary (K–2)

Objective: *to sort materials according to one property.*

1. Give the children a set of buttons that are alike except that some are black and some are white. Ask them to sort the buttons into two piles. Encourage them to create their own classification schemes. How many different ways can they sort the buttons?
2. Give the children a large magnet and a group of materials, some of which are made of iron and some of which are not. Ask the children to use the magnet to classify the materials according to whether or not they are attracted by the magnet.
3. Give the children a basin of water and a variety of materials that will not be harmed when they are placed in the water. Ask the children to sort the materials. Put everything that will float in one pile and everything that does not float in another.

Let children collect their own sets of materials and set up their own classification schemes. Children may trade materials with each other to see if they all define the sets in the same ways. Encourage them to classify in many different ways and to discuss their methods of classifying with you and each other.

After the children are proficient at classifying objects into two categories, the activities should be made more difficult. You may increase the number of materials to be classified, increase the number of categories into which the materials are grouped, or increase the abstraction of the categories, such as classifying pictures of people as happy or sad rather than as male or female. All of these activities should involve categories that are mutually exclusive.

As children become more mature in their reasoning abilities, they can begin to categorize materials into overlapping categories. For instance, children may group a set of toy vehicles by placing all the trucks in one group and all the red vehicles in another. For some children, it will be difficult to decide what to do with the red trucks. The teacher should provide opportunities for the children to construct the concept that they can overlap the circles as shown in Figure 5–3 and then place the red trucks in the intersection of the two circles. The teacher may need to ask a few leading questions to create disequilibrium in the children. For

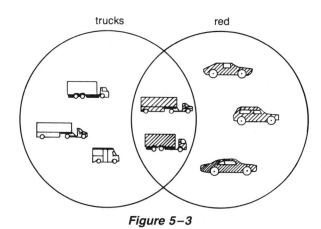

Figure 5–3

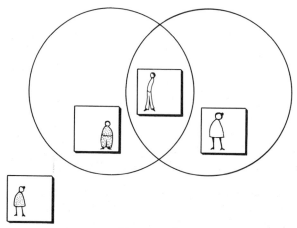

Figure 5–4

example, if the children are content placing all the red trucks in the circle for the trucks, the teacher might ask why they are not in the red circle. The following activities encourage children to use slightly more complicated classification schemes and to create problems as well as solve them.

ACTIVITIES

Primary (1–3)

Objective: *to classify using multiple parallel categories.*

1. Ask the children to cut out pictures of animals from magazines or newspapers or to draw different animals. Then have them classify the animals in different ways. They might classify them by type of animal, type of food they eat, or according to their natural homes, such as grassland, forest, and desert.

2. Give the children a set of buttons or toys with a number of properties or attributes. Ask the children to sort the materials into three categories. See if they can sort the materials into four or more categories. Ask them to tell you the property or properties of each category and how they decided upon the properties.

After the children have set up classification systems, they should be able to abstract properties that a group of objects have in common. **Abstracting** may be thought of as the reverse of classification. Children should be able to look at a set of objects that have been classified and discover the similarity among the elements of the set. They should be able to summarize this similarity in a single statement or rule. Some examples of this type of activity follow.

ACTIVITIES

Primary (1–3)

Objective: *to abstract a property of a set.*

1. Select a small group of children who are alike in some way, such as all are wearing tennis shoes, have blonde hair, or are wearing glasses. Do not tell the class how the children are

alike. A child who thinks he or she knows the attribute chosen may name a child who has not yet been chosen but who also has the given attribute. When all children with the given attribute have been named, the children may guess the attribute. Select one of the children to name a new group of children with a new secret attribute.

2. From a set of the attribute shapes from Appendix B, secretly choose an attribute such as small objects. Pick three or four shapes that have the attribute. Let the children guess which of the other shapes belong in the set. After all the shapes have been chosen, ask the children to tell you the attribute they all have in common. Let the children take turns choosing their own sets. Allow the children to make the game more complex by choosing the union or intersection of two or more attributes such as large or red, or small and triangular.

Objective: *to identify the attributes of two sets and their intersection.*

3. Secretly choose two intersecting attributes for the People Pieces, such as male and tall. Set up two overlapping circles, and place one piece of each type in the correct section as shown in Figure 5–4.

   Let the children take turns picking up a piece and guessing in which section the piece belongs. If the child guesses the correct section, leave the piece in that section. If the guess is incorrect, another child may guess. Don't forget that some pieces will not fit into either section. For example, in Figure 5–4 the label for the left circle might be male and the label for the right circle might be tall; short females would not go inside either circle. Continue until all the pieces are placed correctly. Ask the children what the proper labels are for each section. Have the children describe the pieces in the intersection. Discuss with the children how they knew where to place each piece and why some pieces are outside the sets.

   If the activity is too difficult, let the children play with nonintersecting circles. If the children are good with two intersecting circles, try the activity with three circles. After the children understand the concept of abstracting attributes of intersecting sets, let them develop their own problems with other attribute materials.

Activities should include a variety of materials. The **multiple embodiment principle,** described by Dienes

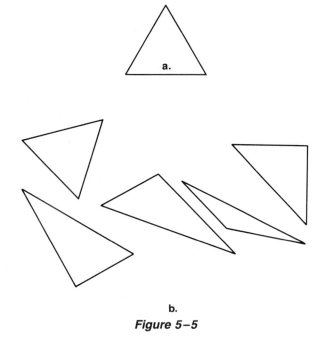

Figure 5–5

and Golding (1966), illustrates a given concept in many different forms. The principle helps children to abstract the essence of the concept so their perception is not based on only one or two specific concrete examples. In looking at only one or two examples, children may be confused by irrelevant features, especially if they occur frequently, and may reject or accept an item as an example of the concept on the basis of some irrelevant feature (Wilson, 1986; Tennyson, 1973). For example, many children believe that a figure is a triangle only if it is an equilateral triangle in the position shown in Figure 5–5a because equilateral triangles in this position are presented frequently in textbooks. The features of congruent sides and a base parallel to the bottom of the page are irrelevant, but they are so common that children believe them to be relevant. The triangles shown in Figure 5–5b may be rejected as triangles if children have not experienced the triangle in many forms.

In studying the van Hiele levels of development in geometry, which are discussed in Chapter 12, Burger and Shaughnessy (1986) found that when they asked young children to draw a triangle and then draw another one that was different in some way, the drawings often featured irrelevant attributes and ignored relevant attributes. For example, one child drew an equilateral triangle with one base parallel to the bottom of the page and then rotated it to make "different" triangles, which were pointing right, left, or down. Another "triangle" had crooked lines for one side. Congruent triangles were thought to be different and nontriangles were included as triangles.

As children move from classifying to sequencing objects and looking for patterns, the ideas of multiple embodiments, irrelevant features, and negative examples should

be continued. Teachers should not only introduce a number of materials but should also encourage children to find examples of their own. To ascertain that the children are abstracting the correct concepts and not focusing on irrelevant details, teachers should discuss with the children what they are learning.

Sequencing

Children live with sequences and patterns. They may notice patterns in nature such as the symmetry of a leaf or patterns such as the tessellations of tiles on the bathroom floor. To aid young children in recognizing sequences or patterns, form a pattern and ask the children to copy it. Use commercial or teacher-made materials similar to those used for other attribute activities or materials collected by you and the children. Several activities for creating and copying designs are given in Chapter 12.

After children have seen and copied designs, they should work with designs for which they can predict the next object. The following activities focus on completing or extending a definite pattern. Sequencing forms the foundation for children to see the pattern in the counting numbers and to complete counting sequences such as skip counting or counting backwards.

ACTIVITIES

Primary (K–2)

Objective: *to recognize and complete a simple pattern.*

1. Use a set of materials that differ in only one way, such as a set of beads that are alike except for color. String the beads in a definite pattern such as red, blue, blue, red, blue, blue, red, and so on. Let the children tell you what the pattern is and continue it.

2. Use a set of materials such as wooden or plastic blocks of different shapes and colors to set up a pattern where the materials differ in shape and color. Set up a pattern such as the one shown in Figure 5–6, and ask the children to complete it. Encourage the children to construct other patterns of their own for other children to complete or fill in.

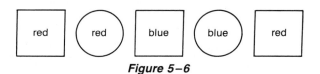

Figure 5–6

Objective: *to recognize and complete the pattern in a two-dimensional array.*

3. Use the attribute materials to set up a two-dimensional array such as the one shown in Figure 5–7. Ask the children to find the missing block.

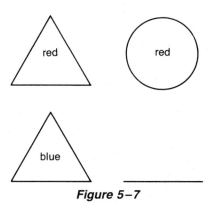

Figure 5–7

After children are proficient at 2 × 2 arrays, extend the arrays to include more blocks. Ask the children to set up arrays for each other. In the beginning, arrays should involve only two attributes, such as color and shape or shape and size. Arrays using three or more attributes are discussed later in this chapter. Ask the children to discuss the strategies they used to place the shapes in the array.

Relationships

As children compare, classify, and sequence objects, they look at relationships among objects. **Relationships** are rules or agreements used to associate one object or concept with another. Mathematics is a collection of relationships among objects or concepts. Children should have a wide variety of experiences exploring relationships among objects. As children develop more abstractions, the same relationships can be explored on a more abstract level, including using numbers. Two of the most important types of relationships are **order** and **equivalence.**

To understand order and equivalence, the children must first understand three properties of relationships: reflexive, symmetric, and transitive. Children use properties of an object in making observations about it, such as that the object is a red circle made of wood and it rolls. To investigate the properties of relationships, the rule for the relationship and the set of objects upon which the relationship is defined must be clearly stated. For instance, to compare the relative lengths of objects, the relationship "is longer than" should be stated and the set of objects, such as a set of Cuisenaire rods, should be defined.

The three properties of relationships are defined as follows:

1. **Reflexive property.** If A is an element of the set, then A is related to itself.
2. **Symmetric property.** If A and B are both elements of the set and A is related to B, then B is related to A.
3. **Transitive property.** If A, B, and C are all elements of the set and A is related to B and B is related to C, then A is related to C.

Children need not know the names of the properties, but they should have many experiences with relationships and their properties. The following activities let children discuss a variety of relationships and their properties.

ACTIVITIES

Primary (1–3)

Objective: *to explore the properties of relationships.*

1. Choose children to act the parts of a grandmother (Mrs. Jensen), a mother (Mrs. Smith), a father (Mr. Smith), a brother (Danny), and a sister (Maureen). Discuss relationships among the family members and ask questions. For example, Mrs. Smith is the mother of Maureen. Is Maureen the mother of Mrs. Smith? Maureen is the sister of Danny. Is Danny the sister of Maureen? If we learn that Mrs. Jensen is Mrs. Smith's mother, then we know that Mrs. Jensen is the mother of Mrs. Smith and that Mrs. Smith is the mother of Maureen. Is Mrs. Jensen the mother of Maureen? Is Maureen the mother of herself?

   Many more family relationships can be discussed, including such ideas as age and height. If Mrs. Jensen is older than Mrs. Smith and Mrs. Smith is older than Maureen, then is Mrs. Jensen older than Maureen?

   Even though young children may use words for family relationships, they may have difficulty exploring the properties of these relationships; they may need to substitute the names of their own family members. Some children may be able to understand the relationships by drawing pictures of their families or by using dolls to represent each person. Others will need to act out the relationships, and still others will not yet be ready to understand the properties, especially those children in the Piagetian preoperational stage.

2. For this activity, a set of Cuisenaire rods may be used. **Cuisenaire rods** are wooden or plastic rods in ten different colors; they progress in length from one to ten centimeters. Each rod has a cross section of one square centimeter. All the rods one centimeter in length are white; the two-centimeter rods are red, and so on up to the orange ten-centimeter rods.

   Get out a set of Cuisenaire rods and discuss the relationship "is the same color as." Is every rod the same color as itself? If rod A is the same color as rod B, then is rod B the same color as rod A? If rod A is the same color as rod B and rod B is the same color as rod C, then is rod A the same color as rod C?

   Now look at the relationship "is longer than." Is rod A longer than itself? If rod A is longer than rod B, then is rod B longer than rod A? If rod A is longer than rod B and rod B is longer than rod C, then is rod A longer than rod C? Because the answer to the first two questions is no, the relationship "is longer than" is not reflexive or symmetric. The answer to the last question was yes; therefore, "is longer than" is transitive.

On many occasions during the school day, you can discuss relationships between objects or between children themselves. Encourage the children to ask questions about relationships. Sue is behind John in line. Is John behind Sue? The homework paper is under the book and the book is under the desk. Is the homework paper under the desk?

The reflexive, symmetric, and transitive properties are each independent of the others. It is possible for any of the three properties to be either true or false for any given relationship. Determine for yourself if each property is true or false for these relationships:

"lives next door to"   "knows the name of"
"has eaten dinner with"   "is congruent to"

Think of other relationships and determine which, if any, of the properties are true for each one.

Relationships that are reflexive, symmetric, and transitive are called **equivalence relationships**. These include such relationships as "is the same age as," "is the same color as," or "is the same shape as." When classifying objects into sets, equivalence relationships are sometimes used.

Equivalence relationships are part of a classification system where every object belongs in a set and sets do not overlap. The attribute shapes may be sorted into sets using the relationship "is the same shape as." Triangles are in one set, squares in another, and circles in another. Each piece belongs to one and only one set.

An equivalence relationship can be used in exploring properties of numbers. Sets may be sorted using the relationship "has the same number of objects as." All the sets containing two objects are placed together to aid children in abstracting the concept of the number two.

The following activities encourage children to form equivalence classes.

ACTIVITIES

Primary (K–1)

Objective: *to form equivalence classes of structured concrete materials.*

1. Using a set of materials such as the attribute shapes, ask the children to form sets using the rule "has the same shape as." Let the children decide on other rules to use. Discuss whether each of the properties is true. Ask the children why the shapes satisfy the rules.

Objective: *to abstract the rule used to form equivalence classes.*

2. Secretly choose a rule by which to form equivalence classes of students. It may be "has the same color eyes as." Choose one child to represent each group. Let the children guess your rule for setting up the classes and suggest other groupings of their own. After the children have abstracted the rules, ask them to share their strategies for solving the problem.

An **ordering relationship** is one in which the reflexive and symmetric properties are not true, but the transitive

property is true. Consider the relationship "is older than" for the Smith family described earlier. No person is older than himself or herself. If Mrs. Smith is older than Maureen, then Maureen is not older than Mrs. Smith. If Mrs. Jensen is older than Mrs. Smith and Mrs. Smith is older than Maureen, then Mrs. Jensen is older than Maureen. The relationship is transitive only. All members of the family can be ordered from the oldest to the youngest using this relationship.

Ordering relationships can also be used with numbers. For example, when the natural numbers are placed in order using the relationship "is less than," you obtain the counting sequence 1, 2, 3, ...

The following activities for ordering do not require the use of numbers.

ACTIVITIES

Primary (K–2)

Objective: *to order objects using various characteristics of the objects.*

1. Give the children a set of baby food jars, each containing a different amount of colored liquid. Let the children order the jars from the one with the least liquid to the one with the most. Include both an empty jar and a full one. Some of the jars may be easy to order, but in some cases students may need to compare the jar to a group of two or more jars to determine the exact placement. The activity may be varied by giving the children a number of identical plastic cups and asking them to pour colored water into the cups so that each cup contains more water than the last one.

2. To play Something Bigger, one child begins by naming a small object. The next child must name an object bigger than the first one. Play continues as long as the children can name something bigger. So you can check whether the object named is really bigger, you may wish to restrict the play to objects visible in the classroom.

3. Select five or six students and ask them to order themselves according to an attribute of their own choosing. Let the other children in class try to decide what criterion was used for the ordering. The children may select something obvious, such as height, or something less conspicuous, such as the number of pieces of jewelry being worn.

4. Cut out a cartoon from the Sunday paper and mount each frame separately on a piece of tagboard. Let the children try to put the cartoon back in its proper order. Choose a cartoon with no reading involved.

   For a variation, let children draw pictures to tell a story. Ask them to trade pictures with a friend and try to put each other's stories in order. Does everyone agree on the proper order?

After children have experienced activities dealing with relationships between individual objects, they should explore relationships between sets of objects. Equivalence and ordering relations on sets form the basis for understanding number concepts. Children should explore sets to determine if the objects in one set can be placed in one-to-one correspondence with the objects in another set. If there is exactly one element in the second set for each element in the first set and no elements are left over in either set, the two sets are said to be equivalent.

Developing and Practicing Skills

After children have had sufficient experience with the processes of observing, comparing, classifying, and sequencing and have had experience with relationships and their properties, they are ready for more formal activities involving these concepts. Activities in this section require children to use these early concepts to further develop thinking skills. Children are given the opportunity to practice earlier skills and to both solve and create problems. Even though most of the problems do not require the use of numbers, they do require some fairly sophisticated reasoning abilities. In most instances, children are encouraged to use concrete materials to explore teacher-posed problems and to use these materials or to develop new ones to create their own problems. Computers can be used to simulate some of the same types of problems that children may solve with concrete, manipulative materials.

Carroll Diagrams

After children can classify materials easily according to one or two attributes, they may begin to use **Carroll diagrams,** or charts, to classify materials according to two or more attributes. Carroll diagrams are named after the mathematician/author Lewis Carroll. Used to classify materials according to more than one attribute, Carroll diagrams are a good introduction to later work with data tables and multiplication. An example of a Carroll diagram is given in Figure 5–8, which may be used with the attribute shapes.

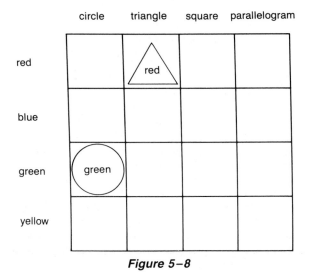

Figure 5–8

ACTIVITIES

Primary (K–3)

Objective: *to abstract properties using a Carroll diagram.*

1. Place about half of the shapes in a Carroll diagram for which the labels are not given, as shown in Figure 5–9. Let the children try to detect your pattern and place the rest of the pieces in the correct places. After all the pieces have been correctly positioned, ask the children to tell you what the labels should be. Let them discuss the strategies they used to solve the problem.

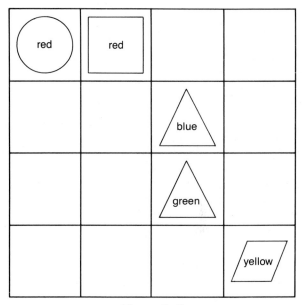

Figure 5–9

2. Divide small groups of children into two teams each. Give each group a set of four People Pieces, such as the tall males. Ask one team to draw a 2 × 2 Carroll diagram for the People Pieces and to label it but not to show the diagram to the other team. (The diagram must be labeled according to the differences in the pieces, in this case the color and weight.) One possible diagram for these pieces is shown in Figure 5–10.

   The group with the labeled diagram should then place one piece in the correct place on an unlabeled diagram. The other group must decide where to place the other three pieces and then correctly name the labels for the diagram. Keep track of the number of incorrect guesses as the team places the pieces

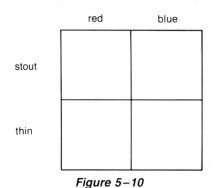

Figure 5–10

in the diagram. After one team has correctly placed the pieces and determined the labels, it is their turn to make a secret Carroll diagram with another set of four People Pieces.

After students become proficient at solving Carroll diagrams involving two sets of attributes, they should begin to work with diagrams involving three or more sets of properties. Figure 5–11 shows a Carroll diagram for the shape, color, and size of attribute blocks; Figure 5–12 shows a Carroll diagram for the height, weight, sex, and color of People Pieces.

Let the children place the pieces in Carroll diagrams that are already labeled, such as those in Figures 5–11 and 5–12. Explain to the children that they must use all of the labels for each piece.

Let the children make up their own Carroll diagrams and have other students place the pieces in the diagrams. Discuss the construction of the diagrams and the fact that you

must use parallel labels, such as male and female, in the same type of position. For instance, male could not label a column if female labeled a row, because you would not be able to find a piece that was both male and female to place in the intersection of that row and column.

The following activities are designed for children who are proficient at filling in and creating Carroll diagrams with three or more sets of labels.

ACTIVITIES

Intermediate (4–6)

Objective: *to set up Carroll diagrams involving four sets of characteristics and to abstract properties of sets from their placement in a Carroll diagram.*

1. Divide the class into groups of four or five and give each group a set of People Pieces. Ask each group to sketch a Carroll diagram using the People Pieces and showing the color, height, weight, and sex of the characters. Each group should then place the pieces on the table in the order shown on the Carroll diagram.

   After the teacher has checked all the Carroll diagrams to determine if the pieces are in the correct places, each team should turn all but three of the pieces over while leaving them in the same positions. The teams should then trade places with each other and guess what the hidden pieces look like in each other's diagrams. Students should take turns pointing at a hidden piece and giving the color, weight, height, and sex. When a student guesses correctly, the piece should be turned face up and left in that position. If the guess is incorrect, the piece should be left face down. Once all of the pieces have been correctly identified, the team that has been guessing should draw a Carroll diagram of the set and compare it to the Carroll diagram drawn by the team that created the problem.

2. The same teams used for the activity just described may be used again for this one. Give each team a set of attribute blocks, and ask the teams to devise a Carroll diagram using the shape, size, and color of the blocks, such as in Figure 5–13. Each team should draw the Carroll diagram and place the pieces on the table in the correct places.

   After all the pieces have been positioned and checked by the teacher, the students should exchange the positions of three of the pieces. Teams may then trade places and attempt to discover which pieces are in the incorrect positions. Each team should draw the Carroll diagram for the set at which they are looking and tell where the pieces should go. The team should then check with the original team to see if the diagram is correct.

3. Students may work individually, in pairs, or in small groups to create their own sets of attribute materials. Students should first decide on a theme for the materials to be made. Themes may arise from special days or seasons, such as valentines, snowmen, pumpkins, or spring flowers. After a theme is chosen, the children should decide on three or four characteristics to vary on their attribute materials. Remind the children that all other attributes must stay the same. For example, a group of children may decide to make hearts for Valentine's Day that differ in size, color, and arrows. They

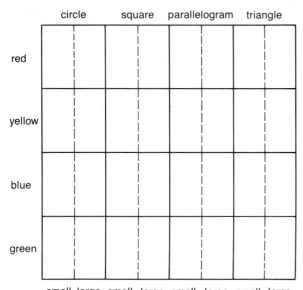

Figure 5–11

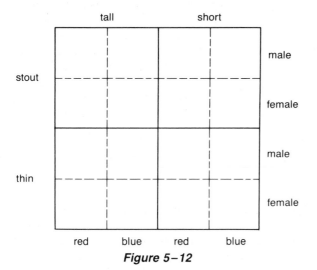

Figure 5–12

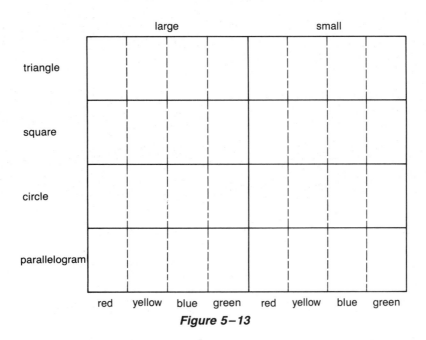

Figure 5–13

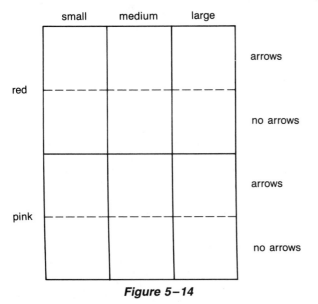

Figure 5–14

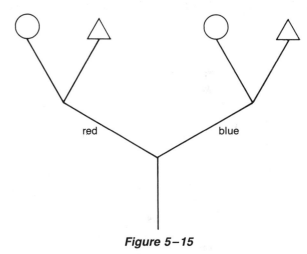

Figure 5–15

might have three sizes (small, medium, and large), two colors (pink and red), and with or without arrows. This set of materials would have 3 × 2 × 2, or 12, pieces in order to have one piece of each type. After the children have decided upon the pertinent attributes for the set, they should draw a Carroll diagram to illustrate all the pieces. The diagram in Figure 5–14 is one possibility.

After the diagram is drawn, the children may make the pieces necessary for the set. The pieces may be drawn on index cards or be cut from posterboard. After the set is made, the children may use this set for activities described in this chapter or may develop new activities of their own. Children may exchange sets with each other and draw Carroll diagrams for the other sets.

Tree Diagrams

After the children are comfortable with activities involving Carroll diagrams, they may use other types of diagrams

for classifying sets of attribute materials. A **tree diagram** is another useful way of classifying materials. On tree diagrams, the branches at each level indicate the characteristics of a different attribute. For example, to show attribute shapes with two colors and two shapes, you could use the diagram in Figure 5–15. Tree diagrams can be used for sequencing as well as classifying, and the activities described below involve both concepts. Variations of tree diagrams will be used later for multiplication, prime factorization, and activities involving probability.

ACTIVITIES

Primary (K–3)

Objective: to classify objects using a tree diagram.

1. Draw a tree diagram such as the one in Figure 5–15 and give the children a set of attribute shapes to place on the tree. The children should begin by placing one block at the foot of the

tree. Then they should move it to the first branching of tree limbs. The children should decide which branch to take according to the attributes of the piece. They should continue moving the piece upwards, deciding on the proper branch at each intersection. After the first piece is in place, continue by letting other children decide the proper position for the next block. Continue until all the blocks are in their proper locations or until all the children have had an opportunity to place a block.

When children are first learning to use tree diagrams, you should begin with a simple diagram, such as the one in Figure 5–15, and gradually increase the complexity of the task. Ask the children to compare the tree diagram to the Carroll diagram. Use the same set of attribute materials on both diagrams. Encourage the children to create other diagrams for the same or different sets of materials. Discuss the strategies they use to place the materials on the tree.

2. Draw a tree diagram on the floor and make label cards for the branches. The positions of the cards may be changed at the discretion of the children. Use labels such as those shown in Figure 5–16. Ask each child to start at the bottom of the tree and to walk until he or she comes to an intersection. At each intersection, let the child decide which way to go. The child should continue until the end of the last branch. Ask the children how they decided which way to go at the intersections.

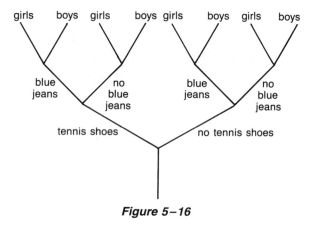

Figure 5–16

3. Use the tree diagram you have drawn on the floor or one drawn on a large piece of posterboard on a table. Collect small trucks and cars to use with this activity (the children may

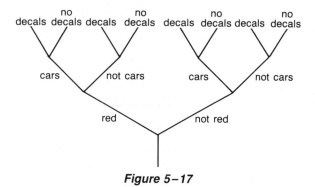

Figure 5–17

bring in vehicles for this activity). Using attributes of the vehicles you have, label the intersections on the tree. Try using attributes and their negations, such as red and not red, cars and not cars, and with decals and without decals. Let the children take turns driving the vehicles up the tree, deciding the proper turn to take at each intersection. One such tree diagram is shown in Figure 5–17.

After the children become familiar with the activity, let them decide on other labels and sets of materials to classify using tree diagrams. They may develop their own sets of materials or suggest other activities to use with familiar sets such as People Pieces and attribute blocks.

Older children who have had experience classifying materials on a tree diagram may wish to carry the concepts even further. The following activities are designed to further develop concepts using tree diagrams. If older students have not worked with tree diagrams before, they should first try some of the activities described above.

ACTIVITIES

Intermediate (4–6)

Objective: *to use a tree diagram to classify objects.*

1. Show the children an unlabeled tree diagram on which the pieces have already been placed on the correct limbs, such as in Figure 5–18. Ask the children to tell you what the labels must be.

    After the children have decided on the proper labels, have them label the tree and remove the pieces. Then have them start the pieces at the bottom of the tree and follow them to their correct branches. Discuss with the children the differences between deciding on the labels after the pieces are in place and placing the pieces when the labels are in place.

2. Have the children use a tree diagram to create a set of attribute materials in the same way that they used a Carroll diagram. Ask the children to decide on a theme for a new set of attribute materials and to name three or four categories of properties for these materials. For example, children may decide to create a set of flowers, using roses or daisies with three, four, or five leaves and yellow or white petals. This will

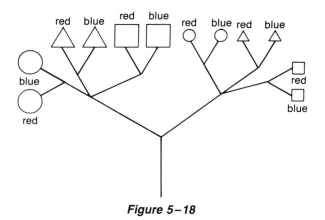

Figure 5–18

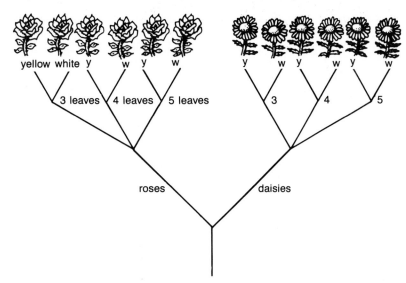

Figure 5-19

give them a set of $2 \times 3 \times 2$, or 12, pieces. Ask the children to sketch a tree diagram to show all the possible pieces, such as the one in Figure 5–19.

After the tree diagram is drawn, have the children sketch the proper flower at the end of each branch. These sketches may then be transferred to index cards or posterboard so that each piece can be individually manipulated. After the sets are made, children may use them for activities in this section or may make up new activities of their own.

3. Tree diagrams may be used for sequencing if a value is assigned to each branch. The branch on the far left is assigned the highest value and the values decrease as you go to the right. Elements are first sequenced according to the branches on the bottom, and the importance of the branches decreases as you move up. This may be illustrated by alphabetizing a set of nonsense words. Put each of the following words on a separate index card:

cat	cab	cot	cob	cut	cub
lat	lab	lot	lob	lut	lub
mat	mab	mot	mob	mut	mub

Sketch a tree diagram in which the bottom branches are labeled with the first letters of the words in alphabetical order from left to right, the middle branches are labeled with the second letters in alphabetical order, and the top branches are labeled with the last letters, again in alphabetical order, as shown in Figure 5–20.

Ask the children to take the words and hang them from the proper branches. Then ask the children to tell you what they observe about the order of the words. This activity is helpful for children who have difficulty alphabetizing words when the first few letters of two or more words are the same.

4. When studying binomial nomenclature in science, use a tree diagram to separate kingdoms, phyla, classes, orders, families, genuses, and species. Let the students locate the proper position on the tree for a variety of plants and animals. This is a good activity for a bulletin board, with students drawing or finding pictures of objects to place on the diagram.

5. Children may use tree diagrams to sequence any set of materials from the greatest to the least. Children may decide to set up an

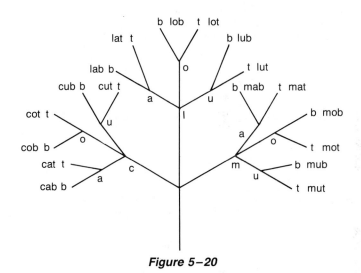

Figure 5-20

imaginary kingdom with the People Pieces. They may decide that all females are more powerful than males, that the stout ones are more powerful than the thin ones, that the short are more powerful than the tall, and finally that the red are more powerful than the blue. The children should set up a tree diagram such as the one in Figure 5–21 and line up all the People Pieces from the most powerful to the least powerful.

This activity may be done in groups of four or five. Each group may line up the pieces and then show the line-up to another group. The other group should not be shown the original criteria for the line-up or the tree diagram. This other group must then decide on the criteria selected and draw the tree diagram. After the diagram is drawn, ask the original group to determine the accuracy of the drawing. Discuss with the students whether individuals are ever ranked in real life and what types of criteria are used.

As you can see from these activities, skills learned in math class often carry over into other subject areas. Carryover should be encouraged whenever possible. Cer-

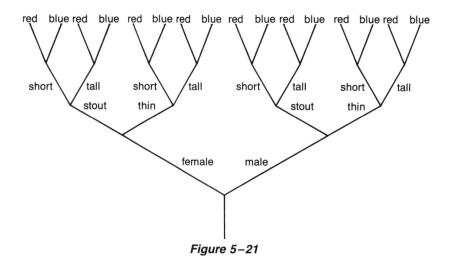

Figure 5-21

tainly in real life, people do not do math for the first 50 minutes of the day, then spelling, and then social studies. These skills must be used together. Encourage the children to find applications for new skills regardless of the subject area.

Venn Diagrams

Another diagram used to classify materials with more than one variable attribute is a **Venn diagram.** Unlike tree or Carroll diagrams, Venn diagrams are not used to give a unique position to each element of a set, but they are useful for classifying by both attributes and the negations of attributes. Young children should begin by using only one or two attributes, while older children may classify using three or even more intersecting categories. Simple intersecting Venn diagrams were introduced earlier in this chapter. A few more examples of Venn diagrams are introduced here for additional practice.

ACTIVITIES

Primary (K-3)

Objective: to classify using Venn diagrams and parallel categories.

1. When you first introduce children to Venn diagrams, use circles with parallel categories, such as colors. Give the children a set of attribute shapes and several large (150-centimeter circumference) loops of yarn or heavy cord. Make three separate circles with the cord and ask the children to put all the red blocks in one circle, all the blue ones in another, and all the yellow ones in another. Do any pieces belong in none of the circles? Ask the children to create other ways to classify the blocks. Discuss their methods of deciding where to place the pieces.

Objective: to form Venn diagrams using two intersecting sets.

2. Ask the children to form two circles of yarn and to put all the yellow pieces in one loop and all the triangles in another. Let the children discuss what to do with the yellow triangles. Lead them to discover that they can overlap the two loops of yarn and put the yellow triangles inside the section where the loops overlap. This is called the **intersection** of the two sets. Note the intersection of the sets in Figure 5-22.

Let the children suggest other labels for the two circles. Discuss how they know whether or not there will be pieces in the intersection. Will there be any pieces in the intersection if the loops are labeled triangles and squares? If the loops are labeled small and triangles, where would a small green square go?

3. Using attribute materials such as People Pieces, ask one group of students to secretly draw a Venn diagram with two intersecting loops and to add labels to them such as red and male. Form two intersecting loops with the yarn and put one correct piece in each section. Ask children who did not see the Venn diagram to guess where the other pieces should go. After all the pieces have been placed in the correct sections, ask the children to identify the labels on the secret Venn diagram.

Let a new group of children decide on another diagram. Discuss with the children such things as whether or not all of the pieces go inside the loops. If any pieces do not belong in the loops, do they help you decide on the proper labels for the loops?

Objective: to properly use the terms and *and* or *when referring to the intersection or union of two sets.*

4. Set up two intersecting loops as in the first activity in this group and place the attribute blocks in the proper sections. Using the loops labeled yellow and triangles, ask the children where to find the pieces that are yellow *and* triangles. Note that these pieces are all found in the intersection. Some children may be confused by the word *intersection* and by ending up with a set smaller than either the set of yellow pieces or the set of triangles.

Later, for addition, the children may read "3 + 4" as "3 and 4." These are not the same concepts, even though they

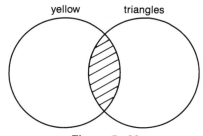

Figure 5-22

use the same familiar word, *and*. Addition is actually based on the concept of union (or), not the concept of intersection (and). **Addition** may be defined as the number of items in the union of two disjoint sets.

Let the children make up several intersecting sets and describe the pieces in the intersection using the word *and*. Be sure to use mutually exclusive attributes such as color and shape or size and thickness. After the children are comfortable using the word *and,* ask them to find the pieces that are yellow or triangles. Note that these are all of the pieces in the union of the two sets. The mathematical use of the word *or* includes those pieces that are yellow *and* triangles.

Give the children several chances to explore the uses of these words with the attribute shapes and in everyday life. Use statements such as "Today we will take attendance *and* say the Pledge of Allegiance before math class" and "Tomorrow we will go to gym *or* music class." Children may wish to discuss whether the promise in your second statement will be broken if you go to gym *and* music tomorrow.

Later, the use of the terms *and* and *or* with Venn diagrams can lead to a more formal study of logic. A **conjunction** (denoted by p $\wedge$ q) consists of any two statements joined by *and*. A **disjunction** (denoted by p $\vee$ q) consists of any two statements joined by *or*. A **negation** (denoted by $\sim$ p) is the statement "it is not true that p." These very basic concepts of formal logic have their beginning in the study of sets.

Objective: to use symbols for union and intersection.

5. After the children are comfortable with the terms *and* and *or* for union and intersection, introduce the symbols for these operations. The symbol $\cup$ is the symbol for **union.** The union of sets A and B is the set consisting of all the elements in A *or* B, including those in both A and B. The symbol $\cap$ is the symbol for **intersection.** The intersection of sets A and B is the set consisting of all elements common to both A *and* B.

Using the same materials used in the previous two activities or attribute materials that the children have created for themselves, let the children begin to work problems such as "Shade the sections in your Venn diagram for yellow objects $\cup$ triangular objects" or "Point to the section for yellow objects $\cap$ triangular objects." Remember that the concepts of union and intersection are more important than the symbols. Don't overemphasize the symbols with young children.

Objective: to form Venn diagrams using negations of attributes.

6. After children can form Venn diagrams with two intersecting loops, try forming Venn diagrams using the negations of the attributes. Put all the pieces that are not male in one section and the pieces that are not red in the other. Ask the children to describe the intersection. Is it the same thing to say "the pieces are not male and not red" as it is to say "the pieces are not male and red"? Where do you find the pieces that are not male or red? Try this activity for yourself. You will find that it is not easy to properly use the familiar terms *and, or,* and *not* in combinations.

After children are familiar with Venn diagrams and the terms and symbols for union and intersection, use Venn diagrams with more than two loops. These diagrams will be difficult for children in the Piagetian preoperational stage, so they are better used with older children. If older

children have not had previous experience with Venn diagrams, let them first experience the activities described earlier.

ACTIVITIES

Intermediate (4–6)

Objective: *to classify attribute materials according to three intersecting characteristics.*

1. Set up a three-loop Venn diagram for the attribute shapes, such as the one shown in Figure 5–23. Ask the children to place the pieces in the proper locations. Are all the pieces inside the loops? Does each section have more than one piece?

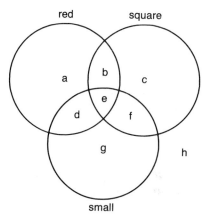

Figure 5–23

Completely describe the characteristics of all the pieces in each section. Note that you can tell the color, shape, and size of the pieces in each section if you use the negations of some of the attributes. For example, the pieces in section (a) in Figure 5–23 are red, not squares, and not small. Ask the children to describe the pieces in each section in a similar fashion.

Encourage the children to create other Venn diagrams using attribute materials they have created. Describe the pieces in each section of the Venn diagrams.

2. Ask a group of children to secretly draw a three-loop Venn diagram. Arrange the yarn into three intersecting loops and ask the children with the secret diagram to place four or five pieces in the correct sections. The children who have not seen the diagram should attempt to place each of the other pieces in a loop by asking the children with the diagram if they have chosen the correct section. After all the pieces are correctly positioned, the children who placed the pieces should tell the proper labels for the three loops.

In the beginning, use only positive attributes for all three loops and choose attributes so that all sections contain elements of the set. For children who understand the concepts well and really want a challenge, use negations and nonintersecting sets.

Objective: *to understand the meaning of the complement of a set and to use the symbol for complement.*

3. A **complement of a set** consists of all the elements in the universal set that are not elements of the set under consideration. For example, the complement of the triangles in the

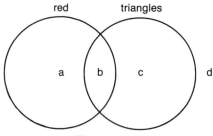

Figure 5–24

set of attribute blocks are all the attribute blocks that are not triangles. The children should be familiar with this concept through the use of negations. Two different symbols may be used to denote the complement. The complement of set A may be shown as A′ or as Ā. This notation is usually read as "the complement of A."

The following activity uses the A′ notation. Set up a Venn diagram for the attribute blocks such as the one in Figure 5–24, and label each section as shown.

Ask the children to list the letters designating the sections for various sets and their complements. For example, give the sections for each of the following:

red pieces	red pieces ∩ triangles
(red pieces)′	(red pieces ∩ triangles)′
red pieces ∪ triangles	red pieces′ ∪ triangles′
(red pieces ∪ triangles)′	red pieces′ ∩ triangles′

Which of the above are the same sections? State the characteristics of each of the above sections in words rather than symbols, such as "not red and triangles" is the same as "not red or not triangles." This may be stated formally as "the union of the complements is equal to the complement of the intersections" and "the intersection of the complements is equal to the complement of the union." These are known as **de Morgan's laws.** These concepts are fairly complicated and students will need a great deal of experience working with concrete materials before being able to state the properties abstractly.

The activities using either People Pieces or attribute shapes may also be performed using any set of structured attribute materials, such as those made for the seasons or holidays. Encourage children to use their own sets of materials whenever possible. The children will probably enjoy reading about a set of materials developed by Frances Lettieri called **Zorkies,** which are creatures from an alien planet that differ in color and the number of eyes, arms, and legs. You may read about them in the September 1978 issue of the *Arithmetic Teacher.*

Children may construct their own sets of Zorkies and try the suggested activities. As with concepts discussed earlier, the use of multiple embodiments helps children develop a solid foundation in the thinking skills and logic necessary to perform these activities.

Computer Programs

Another embodiment of these concepts exists in some computer programs. Four programs that reinforce several of

the concepts in this section, including Venn and Carroll diagrams, are Gertrude's Secrets, Gertrude's Puzzles, Moptown Parade, and Moptown Hotel, which are available from The Learning Company (the address is in Appendix A).

Gertrude's Secrets is designed for children ages 4–9 and uses elements similar to the attribute blocks. The elements are four different shapes and four different colors. Students place these shapes in one- and two-difference trains, one- and two-loop Venn diagrams, and 3×3 and 4×4 Carroll diagrams.

Gertrude's Puzzles uses the same elements with slightly more difficult problems. Children are asked to solve problems involving two- and three-loop Venn diagrams and problems similar to the network game described later in this chapter.

The moptown programs involve Moppets, which are similar to the People Pieces. The Moppets are tall or short, fat or thin, red or blue, and Bibbits or Gribbits. **Moptown Parade** is designed for students ages 6–10 and involves finding likenesses and differences; opposites; sequences; one-, two-, and three-difference trains; and rules for joining a club. **Moptown Hotel** is designed for students ages 9 and up and involves Carroll diagrams, two-difference trains, negations, guess my attributes, and a hotel puzzle.

Estimating

Even though the activities described in this chapter often do not require the use of numbers, children should still be asked to estimate and predict. Estimation and prediction skills prove to be invaluable for a wide variety of problem-solving activities, including those involving numbers. The activities in this section describe some ways in which the teacher may encourage the children to estimate and predict.

ACTIVITIES

Primary (K–3)

Objective: *to predict the total set of attribute materials when shown a few.*

1. Choose a set of attribute materials the children have not seen before. Put the pieces in a paper bag and take them out one at a time. Ask the children to observe the properties of each piece. After the children have seen three or four pieces, ask them to describe a piece they believe is still in the bag. Remind the children that in this set of attribute pieces, no two pieces are identical but there is one piece for each possible combination of crucial attributes. For example, you may have a set of shapes that are red, blue, and green; large, medium, and small; and circles, triangles, and squares. After the children have seen a large blue square, a medium red triangle, and a small green circle, they may predict that there is still a small green square in the bag. Continue the activity until the children have predicted all the pieces in the set. Ask them how they knew what pieces were still in the bag.

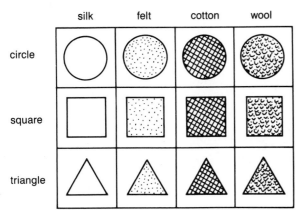

	silk	felt	cotton	wool
circle				
square				
triangle				

Figure 5–25

Objective: *to predict what the distinguishing characteristics will be in a set of attribute materials.*

2. Again, choose a set of attribute materials the children have not seen. Choose attributes that can be distinguished by feel, such as shape, texture, and size. You may wish to make your set from scraps of material. Cut three different shapes using four different types of material, such as silk, felt, cotton, and wool. Paste the materials onto a sheet of cardboard in a Carroll diagram such as the one shown in Figure 5–25.

    Have the children feel three or four of the pieces without looking and predict what pieces will be in the other positions. After they have predicted one other piece, let them feel that piece if they were correct and then predict another piece. The children may look at the Carroll diagram only after all the pieces have been predicted. Let them discuss their strategies and determine whether or not they could have used any better ones.

Intermediate (4–6)

Objective: *to predict the total number of objects in a set.*

1. Even though the emphasis in this chapter is on activities that do not require the use of numbers, work with sets and Carroll diagrams is a good introduction to the Cartesian product concept of multiplication. Older children who have worked with attribute materials should not only be able to describe the pieces that should be in an attribute set, but should also be able to predict the total number of pieces in a set after they are aware of all the distinguishing attributes. Again, choose a set of attribute materials the children have not seen.

    Put the pieces in a paper bag and take them out to show the children one at a time. Ask the children to observe the attributes and to guess how many pieces are still in the bag. Encourage the children to draw Carroll or tree diagrams to aid in their predictions.

    After the children have tried predicting a few times, they may be able to give you a formula for determining the total number of pieces in a set of attribute materials. To do so, they must find the number of attributes in each pertinent category and multiply those numbers together. This is the **Cartesian product** concept of multiplication. To find the Cartesian product of two sets, match each element of the first set with each element of the second set. For instance, if a set has four shapes and three sizes, the total number of pieces in the set would be 4 × 3, or 12, pieces.

Objective: *to estimate and use the Cartesian product to find the total number of combinations.*

2. Children may wish to find the total number of different combinations for several familiar circumstances. Tell the children stories about events in everyday life that involve Cartesian products, and ask the children to estimate the number of possible combinations. Let the children figure the exact number using their formula, a tree, or a Carroll diagram after they have estimated. Some ideas for stories follow:

    - Amy got 3 new pairs of pants and 4 new blouses for Christmas. She can wear each of her pants with each blouse. She plans to wear one pair of the pants to school each day with one of the blouses. How many days can Amy go to school wearing a new combination each day? How many different combinations would Amy have if she also got 2 new sweaters, and each sweater goes with each outfit?
    - José got a job working in an ice cream parlor. His favorite task is making ice cream sundaes. He likes to make up new combinations. The store has 12 kinds of ice cream, 4 kinds of toppings, and 3 kinds of nuts. How many different sundaes can José make if he puts one dip of ice cream, one topping, and one kind of nut on each sundae?
    - Suzanna is in charge of making up names for a new kind of doll. She has decided on 20 good first names, 15 middle names, and 25 last names. How many different names can she make up if each doll gets a first, a middle, and a last name?

Be sure the children estimate before they actually figure the number of combinations. Many of the children will be surprised at the large number of possibilities. Let the children make up their own stories involving combinations for each other.

Problem Creating and Solving

As you have probably noticed, this chapter contains suggestions for both problem solving and problem creating. Throughout the book, you will see emphasis on the development of children's ability to think. In this section are even more ideas for helping children create their own mathematical problems as well as for encouraging them to solve problems created for them.

ACTIVITIES

Primary (K–3)

Objective: *to sequence attribute materials according to likenesses and differences.*

1. Give the students a set of People Pieces and tell them that the pieces are going to have a parade. The people have strict rules for their parades. They must march in single file. Anyone may lead the parade, but the next person in line must have one attribute the same as the first person and three attributes different. Each person in line must have one attribute the same and three attributes different when com-

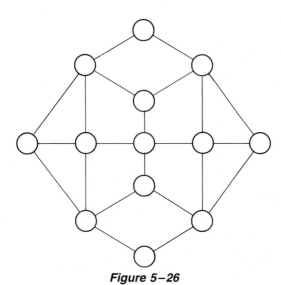

Figure 5-26

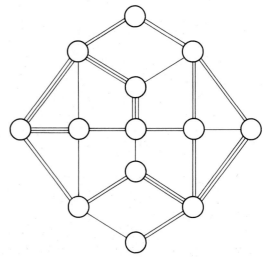

Figure 5-27

pared with the person directly in front. The only attributes they may use are sex, height, weight, and color.

- *Understanding the problem.* Each person in the parade matches the person that follows in only one of the four attributes. If only one attribute is the same, the other three will be different. For example, the tall, skinny, blue male could follow the tall, fat, red female because both are tall.
- *Devising a plan.* Try to use all the tall people first, then change to short people (guess and check).
- *Carrying out the plan.* Look for a tall, fat, red female. There are no more, so change the attributes that will be the same for each new piece. Experiment by putting a short, fat, red male next. That works! Continue to add pieces by changing the attributes that are the same.
- *Looking back.* Check to make sure that all of the People Pieces have been used, and that each piece is the same in one way and different in three.

To make the activity even more difficult, have the paraders march in a circle so that the first and last people in line also match according to the rules. After the children have tried the parade with the People Pieces, ask them to line up sets they have made using their own rules. They may challenge other students to discover the rules they used and to find one or two pieces out of order.

Objective: to place attribute materials on a network according to a given number of differences among various pieces.

2. Give the children a set of attribute blocks and a network similar to the one in Figure 5–26. The children should play this game in two teams. The play starts with one attribute block placed in the center of the diagram. The first team then places an attribute block adjacent to the first one and scores one point for each difference from the first block in color, shape, or size. Teams alternate placing blocks on the diagram and add the number of different attributes on each turn. If a block has lines connecting it to two or more blocks already in place, that team's score is the total number of differences from all the adjoining blocks. Play continues until all the spaces on the board are filled. The team with the most differences wins.

Students may vary the game by using their own sets of attribute materials. After students have played the game a few

times, discuss with them strategies for winning, which should include offensive as well as defensive moves.

3. Network solitaire may be played using a network board similar to the one shown in Figure 5–27. In this diagram, the number of lines between two positions indicates the number of differences there must be between the two connecting attribute blocks. Students may work on this activity alone or in small groups.

The students begin by placing an attribute block anywhere on the diagram and then trying to place other blocks, moving away from the first one according to the number of differences. The activity is complete when all the spaces on the diagram have been correctly filled with blocks.

Students may check each other or the teacher may check them. As with many problem-solving activities, there are several correct answers. Students may make up networks for each other to solve. Some networks may be impossible, and students should discuss why certain combinations do not work.

Children who have worked with the activities just described and wish for a greater challenge may try some of the following activities. Children who have not tried the previous activities should work through the easier activities before attempting these more difficult ones.

ACTIVITIES

Intermediate (4–6)

Objective: to place attribute materials in an array according to the number of likenesses and differences between adjoining pieces.

1. This activity is similar to the parade described above. In this case, the People Pieces all wish to move into a 4 × 4 apartment house such as the one in Figure 5–28. The People Pieces are very particular about their neighbors. Each piece can differ in only one attribute from anyone living to the right, to the left, above, or below; its other three attributes must be the same as those of its neighbor. As the children work on placing the People Pieces in the array, they may realize that simply using trial and error can become quite frustrating.

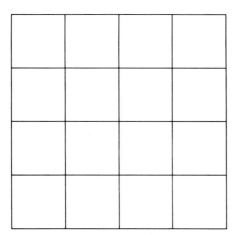

Figure 5-28

Remind them that there are other effective problem-solving strategies, such as looking for a pattern.

This activity is also effective with the attribute blocks or the Zorkies. The students may vary the rules so the pieces have two or three differences. Be aware that not all rules are possible, especially with the People Pieces, which only have two choices for each category of differences.

***Objective:** to choose the correct attribute block with the use of deductive reasoning and questioning.*

2. Using a set of attribute blocks, the leader secretly chooses one of the blocks and the rest of the class tries to guess which block it is by questioning the leader. The class may ask only questions of the form, "Does it have any of the attributes of ___ ?" (describing the size, color, and shape of one block).

Suppose the block is a small red triangle. If someone asks whether the block chosen is small, red, or a triangle, the leader will answer, "Yes, it has at least one of those characteristics." The leader does not tell the class if they guess the secret block. If the block has none of the characteristics guessed, the leader will answer, "No, it does not have any of those characteristics." Play continues until the class is sure which one is the secret block. One person in the class must announce, "I know which block it is." If the rest of the class

agrees, then that person may describe the block. If the person is correct, he or she may become the next leader.

After the children have played one or two games, let them discuss the strategies they use. Would they prefer to get a no or a yes answer? How do they eliminate a size, shape, or color?

***Objective:** to use the properties of isomorphic sets to solve problems involving patterns.*

3. Two systems are **isomorphic** if they have the same structure and the same internal set of relationships. Isomorphic sets must be equivalent and the patterns in two isomorphic sets must be the same. The sets shown in Figure 5–29 are isomorphic. The pieces may be put in one-to-one correspondence and they relate to each other in the same way.

After the children have worked with simple isomorphic sets, they may work with more detailed sets. Construct a set of attribute materials that are isomorphic to the People Pieces. This means that the pieces must have four sets of distinguishing characteristics with two choices for each. Make a set of sailboats with purple or green sails, with small or large sails, with a sailor or without, with an anchor or without. After the sailboats are constructed, place the sailboats in a 4 × 4 Carroll diagram next to the People Pieces, which are also in a 4 × 4 Carroll diagram. The diagrams should be similar to the ones shown in Figure 5–30.

Do not put labels on the diagrams. Ask the children to decide on the labels for the Carroll diagram for the People Pieces. Tell the children that the sailboats are isomorphic to the People Pieces and you want them to discover the matching labels for the set of sailboats. For instance, the positions that have males in the diagram for the People Pieces all have purple sails in the diagram for the sailboats. Ask the children to finish the following chart, which asks them to match all the characteristics:

males	purple sails _____
females	_____
red	_____
blue	_____
tall	_____
short	_____
skinny	_____
fat	_____

	small	medium	large
cat			
dog			
rabbit			
monkey			

Animals

	red	blue	green
car	red	blue	green
truck	red	blue	green
bicycle	red	blue	green
tricycle	red	blue	green

Vehicles

Figure 5-29

Figure 5–30

After the children have matched the characteristics, ask them to develop their own sets of materials that are isomorphic to either the People Pieces or the attribute blocks. A group of students making up a new set should write down the corresponding characteristics and then turn over each piece of the newly created set so that the characteristics of the set are not visible but the set remains in the same Carroll diagram. A group of students who have not seen the new set should then turn the pieces over one at a time and attempt to predict the matching characteristics. Keep track of how many pieces have been turned over before all the characteristics have been identified and matched. The fewer pieces used, the better.

This section gives only a brief idea of the many problems that may be developed using mathematical concepts that often do not involve numbers. Children should be encouraged to develop problems of their own for each other to solve. The classroom should always have a special table or bulletin board that contains problems for the children and a place for the children to suggest new ideas.

Grouping Students for Mathematical Thinking

Many of the activities described in this chapter are best done in a small group or with a partner. While students are solving problems and learning to think mathematically, they need to express their reasoning to other students as well as to the teacher. Young children may work best with only one or two others, while older students might work in slightly larger groups.

Many of the games described that use attribute materials require two teams of students. Each team sets up problems and puzzles and then gives them to the other team to solve. It is helpful if each team has at least two students so they may discuss the strategies and tech-

niques that they are using to solve the puzzle. Teams larger than four students would probably be unwieldy and might not give everyone a chance to fully discuss their thinking processes. This discussion is very useful not only in helping students learn from each other but also in helping them clarify their own reasoning.

Many of the activities can be presented to the students as a single group initially with time then given for them to work in groups on the solution. The teacher should move from group to group as students are working to monitor student progress and to ask questions that cause the students to think about their solutions. After groups have arrived at one or more solutions, they can gather together as a class to compare the methods they used.

Communicating in Learning Mathematical Thinking Processes

Kindergartners and first graders may communicate much of their reasoning verbally. When the class gets together as one group to discuss processes used, the teacher might assist younger groups in recording their results on an experience chart or graph. Second through fourth graders should record not only the solutions to problems, but also the methods that they used. They can trace around attribute blocks and draw pictures of People Pieces to record their solutions to those problems. Middle school students may want to generalize their solutions to specific attribute problems to include general strategies for any similar problems.

If you have a video camera available, students may want to record some of their work visually. Good problems to videotape include those in which the students themselves are an integral part of the problem, such as making pat-

```
┌─────────────────────────────────────────────────────────────────┐
│                           Sample                                  │
│              INDIVIDUAL PORTFOLIO RECORD SHEET                    │
│  Attach this sheet to a sample of a problem you have created or solved either by yourself or in a group. │
│  Comment on your strengths and weaknesses and on any other student who may have been helpful in │
│  your problem-solving process.                                    │
│                                                                   │
│  STUDENT COMMENTS                                                 │
│                                                                   │
│  I FEEL I DID WELL WHEN I . . .                                   │
│                                                                   │
│                                                                   │
│                                                                   │
│  I HAD SOME PROBLEMS WHEN I . . .                                 │
│                                                                   │
│                                                                   │
│                                                                   │
│  I AM STILL UNSURE OF . . .                                       │
│                                                                   │
│                                                                   │
│                                                                   │
│  _____ HELPED ME BY . . .              │
│                                                                   │
│                                                                   │
│                                                                   │
│  TEACHER'S COMMENTS: (include clarity, appropriateness, complexity, accuracy, persistence, and │
│  variety of solutions attempted)                                  │
└─────────────────────────────────────────────────────────────────┘
```

Figure 5–31

terns and sets using the attributes of the children in the classroom or discussing relationships as they role play the parents and children in a family.

Encourage children to look for examples of observing and inferring, comparing, classifying, and sequencing at home or to discuss with adults in their lives how these skills are used in their work. Students can discuss their finding with others in the class. You might make a bulletin board to display a list of ways people use these skills at work, such as the librarian's use of the Dewey Decimal System, a store manager's inventory system, or a restaurant manager's method of inferring the amount of hamburger to order each week.

Evaluating Mathematical Thinking

There are many ways in which to evaluate learning. A paper and pencil test is one way to evaluate learning; however, it may not be the most appropriate method of evaluation for young children. Other methods should be used.

As children are solving the problems mentioned in this chapter, the teacher should make regular observations and keep track of them in a log or on a chart. The teacher should make notes of which skills children have mastered and which the children still need to work on. This may be done by using codes on a chart to show the level of mastery for each skill and by keeping anecdotal records for individual children. In addition to watching children work problems, the teacher should note the problems that the children have created. These may be evaluated for clarity, appropriateness, and complexity. Children may also aid in the evaluation by keeping personal logs noting which tasks they feel confident in doing and which tasks they are challenged by. Children may also mention the other children who have been helpful in explaining certain tasks and who they feel make up good problems. A portfolio containing samples of student work from the beginning, middle, and end of a unit is very useful for both the teacher and the student in noting progress. The form in Figure 5–31 may be included in the portfolio.

The teacher should test the children on certain Piagetian tasks to determine readiness for some of the problems mentioned in this chapter. The children's ability to understand the logic of classification is necessary for many of these problems. This includes testing the children's ability

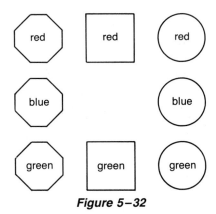

Figure 5–32

to use additive, hierarchical, and multiple classifications, testing their understanding of class inclusion, and testing their ability to use seriation.

Additive classification requires children to place objects in categories according to their likenesses and differences. Children should be given a set such as the attribute blocks and asked to group them so each group is alike in some way. If they can group the blocks according to one criterion, such as shape, they should be asked if there is another way to do it. Preoperational children may use trial and error to group the objects and may not be able to think of more than one or two ways to group them. Concrete operational children will not use trial and error and will be able to group the objects according to shape, size, and color. Preoperational children need many experiences sorting a variety of objects according to one or two attributes. Concrete operational children are ready to move on to hierarchical classification.

For **hierarchical classification,** it is necessary for children to consider a whole and its parts at the same time. Children must realize that a yellow daisy is both a daisy and a flower. If children are shown eight daisies and two roses and asked if there are more daisies or more flowers, preoperational children will respond that there are more daisies. The children are comparing the daisies to the roses, rather than comparing the daisies to all the flowers, of which the daisies are one part. Concrete operational children respond correctly.

In **multiple classification,** each item belongs to two or more classes at once. Multiple classification may involve solving a matrix problem such as the one in Figure 5–32. The children must notice both that the color is constant in a row and that the shape is constant in a column in order to find the missing piece.

Young preoperational children are often able to solve both two- and three-property matrices, but cannot justify their solutions. They appear to solve the problems based on the visual clues. Slightly older preoperational children have more difficulty solving the problems because they try to reason out the answer rather than simply placing the piece that "looks right." They find two-property items

easier than three-property ones. Concrete operational children can solve both two- and three-property problems and can give you the reasons for the solutions.

Children of all ages can be given matrix problems and encouraged to discuss their solutions. Teachers can judge from the discussions whether the children are thinking logically (they can give a reason for their solution) or graphically (they simply say that it looks right).

Class inclusion includes being able to distinguish between objects that belong to and objects that are similar to, being able to correctly use the terms *some* and *all,* and being able to distinguish the object that is different. These terms are used extensively in the problems in this chapter, and the teacher needs to determine if the children have a solid understanding of each term before proceeding with the activities.

Seriation is arranging objects in order and must be understood before children can proceed with the activities involving patterns and sequencing. Young preoperational children may be able to distinguish size differences in two objects but are unable to sequence a whole line of objects. Older preoperational children may be able to sequence objects in a line but cannot place objects in an array, which requires them to sequence both vertically and horizontally at the same time. Many concrete operational children can sequence two variables simultaneously in an array.

Teachers may see Copeland (1982) for more ideas on both testing and teaching children based on their Piagetian stages. A sample recordkeeping form for use with these Piagetian tasks is shown in Figure 5–33.

Something for Everyone

The activities described in this chapter are for children with different learning styles and abilities. Children should use materials that suit their strengths and should also explore other modes of learning in order to develop all their abilities.

For visual learners, seeing and drawing what they are learning is quite helpful. Visual learners enjoy the computer activities that show prenumber concepts such as classification and seriation. They should be encouraged to draw the answers to problems with attribute materials and to create new sets of materials in which they draw the elements they have created. They should draw diagrams such as Carroll, tree, and Venn diagrams and add the elements that solve the problems. They may also pose new problems for other children by drawing them.

Tactile/kinesthetic children prefer to solve attribute problems by actually manipulating the materials. If they are working on the computer, they may wish to have a set of concrete materials available to manipulate that match those on the computer screen. They can recreate the problem from the computer with their own pieces. Even without the actual pieces, they enjoy the computer activ-

Sample Class Evaluation Form

Name	Additive Classification	Hierarchical Classification	Multiple Classification	Class Inclusion	Seriation	Comments

Key
N—not begun
B—begun
M—mastered

Figure 5–33

ities if they can move the pieces on the screen with a joystick or other device, although they prefer Carroll, Venn, and tree diagrams made from string or drawn on a large sheet of posterboard, so that they may place the attribute shapes or People Pieces directly on the diagram. They may wish to create new attribute sets from clay or collect household objects for their sets. They may pose problems for other students by creating a model.

Auditory learners wish to solve attribute problems by talking about the problems and their solutions. They prefer written or verbal directions to drawings or models. They prefer writing down "small yellow circle" to drawing a picture of one. They prefer computer programs that contain detailed written directions to pictures and diagrams. They feel more comfortable working with computer programs if they have read or discussed the manual first. They may wish to create problems by writing them or taping them on an audio tape rather than by drawing pictures or making a model. They may need to talk to themselves as they work on difficult problems.

Gifted children should be encouraged to explore topics in greater depth. One good topic is that of formal logic. Children who have a good understanding of the concepts of union and intersection may move from studying those ideas to the concepts of disjunctions and conjunctions. Computer programs such as Rocky's Boots, from The Learning Company, and What's My Logic, from Midwest Publications, and numerous activity books from companies such as Midwest Publications, Dale Seymour Publications, and Creative Publications, include interesting problems and activities designed to teach children to think logically

(the addresses are given in Appendix A). Many of the activities can be used by all children, but many others are challenging for even the most gifted students. Gifted children should enjoy the challenge of solving the problems and creating new ones for each other. Don't restrict them to the age levels suggested for the problems. Children should be encouraged to move at their own speed with these activities. Let the children's interests and abilities determine the problems that they work on.

This should also be true for any children having difficulty. There is no reason for a ten year old not to do activities suggested for the primary grades if the child is more comfortable on that level. The child may simply need more maturation or more experience before he or she is ready for more advanced activities. To determine the cause of the problem, take time to talk to any children having trouble. It could be that the children have not reached the proper level of Piagetian development or that they are not operating in their strongest mode. Children whose strengths lie in holistic or spatial processing may be frustrated trying to work problems in a logical step-by-step sequence. These children may need to see the total picture and then "mess around" with the materials. They may need to be allowed to solve a problem using a method or methods quite different from any the teacher has suggested.

Children who are stronger at linear, sequential processing may be frustrated by a very broad, open-ended problem. They may wish for a very specific problem with a definite sequence of steps to follow. All children should be encouraged to develop their strengths but should also work with children with different strengths in order to develop an appreciation and understanding of other styles. Good problem solvers can choose and use the style best suited to the problem.

KEY IDEAS

The thought processes that children should develop in their mathematics classes are very important. In Chapter 5, we have presented four major thought processes that should be developed beginning even before children learn number concepts and continuing throughout the elementary and middle grades. Children should observe, compare, classify, and sequence, beginning with a variety of concrete materials such as attribute blocks, People Pieces, or teacher- and student-made collections of objects. Computer programs for developing the concepts can also be quite useful. Later, children should abstract concepts from the work that set a solid foundation for the development of number concepts.

Ideas learned concretely in the primary grades can be formalized and practiced using models such as Carroll, tree, and Venn diagrams. Properties of non-numerical relationships are useful in later work with numbers. Throughout the activities, children should be encouraged to propose and solve problems. The emphasis should be on levels of thinking that go beyond memorization. The children's own development of thought should be the basis for instruction.

Skills should be informally identified, with a great deal of attention paid to the children's developmental levels. Individual learning styles should also be taken into account when planning lessons.

REFERENCES

BARATTA-LORTON, MARY. *Mathematics Their Way.* Palo Alto, Ca.: Addison-Wesley Publishing Co., 1976.

BOLSTER, CAREY L., et al. *Invitation to Mathematics. Level K.* Glenview, Ill.: Scott, Foresman & Co., 1985, p. 87.

BURGER, WILLIAM F., AND SHAUGHNESSY, J. MICHAEL. "Characterizing the van Hiele Levels of Development in Geometry." *Journal for Research in Mathematics Education.* Vol. 17, No. 1 (January 1986), pp. 31–48.

CARPENTER, THOMAS P. "Research on the Role of Structure in Thinking." *Arithmetic Teacher.* Vol. 32, No. 6 (February 1985), pp. 58–60.

COOMBS, BETTY, AND HARCOURT, LALIE. *Explorations 2.* Don Mills, Ontario: Addison-Wesley Publishing Co., 1986.

COPELAND, RICHARD W. *Mathematics and the Elementary Teacher.* 4th ed. New York: Macmillan Co., 1982.

CRUIKSHANK, DOUGLAS E.; FITZGERALD, DAVID L.; AND JENSEN, LINDA R. *Young Children Learning Mathematics.* Boston: Allyn & Bacon, 1980.

DIENES, ZOLTAN P., AND GOLDING, E.W. *Learning Logic. Logical Games.* New York: Herder & Herder, 1966.

DOSSEY, JOHN A; MULLIS, INA V.S.; LINDQUIST, MARY M.; AND CHAMBERS, DONALD L. *The Mathematics Report Card: Are We Measuring Up? Trends and Achievement Based on the 1986 National Assessment.* Princeton, N.J.: Educational Testing Service, 1988.

DOWNIE, DIANE; SLESNICK, TWILA; AND STENMARK, JEAN KERR. *Math for Girls and Other Problem Solvers.* Berkeley, Ca.: Math/Science Network, Lawrence Hall of Science, University of California, 1981.

FENNELL, FRANCIS, ET AL. *Mathematics Unlimited.* New York: Holt, Rinehart & Winston, 1987.

GIBB, GLENADINE, AND CASTANEDA, ALBERTA. "Experiences for Young Children." *Mathematics Learning in Early Childhood.* National Council of Teachers of Mathematics, 37th Yearbook. Reston, Va.: NCTM, 1975.

LETTIERI, FRANCES M. "Meet the Zorkies: A New Attribute Material." *Arithmetic Teacher.* Vol. 26, No. 1 (September 1978), pp. 36–39.

"MANIPULATIVES (FOCUS ISSUE)." *Arithmetic Teacher.* Vol. 33, No. 6 (February 1986).

MAROLDA, MARIA. *Attribute Games and Activities.* Palo Alto, Ca.: Creative Publications, 1976.

"MATHEMATICAL THINKING (FOCUS ISSUE)." *Arithmetic Teacher.* Vol. 32, No. 6 (February 1985).

MUELLER, DELBERT W. "Building a Scope and Sequence for Early Childhood Mathematics." *Arithmetic Teacher.* Vol. 33, No. 2 (October 1985), pp. 8–11.

NATIONAL COUNCIL OF TEACHERS OF MATHEMATICS. *Curriculum and Evaluation Standards for School Mathematics.* Reston, Va.: NCTM, 1989.

NATIONAL RESEARCH COUNCIL. *Everybody Counts: A Report to the Nation on the Future of Mathematics Education.* Washington, D.C.: National Academy Press, 1989.

NUFFIELD FOUNDATION. *Beginnings.* New York: John Wiley & Sons, 1967.

PAYNE, JOSEPH N., ED. *Mathematics Learning in Early Childhood.* National Council of Teachers of Mathematics, 37th Yearbook. Reston, Va.: NCTM, 1975.

TENNYSON, ROBERT D. "Effect of Negative Instances in Concept Acquisition Using a Verbal-Learning Task." *Journal of Educational Psychology.* Vol. 64, No. 2 (April 1973), pp. 247–260.

TRIVETT, JOHN V. *Exploring Cubes, Squares and Rods.* New Rochelle, N.Y.: Cuisenaire Co. of America, 1975.

WILSON, PATRICIA S. "Feature Frequency and the Use of Negative Instances in a Geometric Task." *Journal for Research in Mathematics Education.* Vol. 17, No. 2 (March 1986), pp. 130–139.

6 Attaching Meaning to Numbers

Number concepts begin early in a child's life and extend far beyond being able to count or to recognize numerals. A good foundation in number concepts is crucial in the primary years because it is the basis for much of the work in mathematics throughout the school years and indeed throughout one's life. According to Piaget, children do not acquire number concepts when these concepts are taught to them. They cannot learn arithmetic by internalizing rules or algorithms. They must acquire knowledge by constructing it themselves, not by internalizing it from the environment. He distinguishes between physical knowledge, which is present in external objects and includes facts such as weight and color, and logico-mathematical knowledge, which consists of relationships that must be constructed by the individual and includes the concept of number. The concept of number takes many years to develop, and children who have number concepts up to ten may not necessarily have number concepts up to 50 or 100 (Kamii, 1990).

It is almost impossible to live in today's world without encountering numbers. Being able to use and understand numbers is a basic skill that no child or adult can ignore. Using numbers intelligently is as important, if not more important, than being able to read critically and intelligently. According to the National Research Council:

Without the ability to understand basic mathematical ideas, one cannot fully comprehend modern writing such as that which appears in the daily newspapers. Numeracy requires more than just familiarity with numbers. To cope confidently with the demands of today's society, one must be able to grasp the implications of mathematical concepts—for example, chance, logic, and graphs—that permeate daily news and routine decisions. Literacy is a moving target, increasing in level with the rising technological demands of society. . . . Mathematical literacy is essential as a foundation for democracy in a technological age. (1989, pp. 7–8)

There are three number concepts with which a child should be familiar: cardinal numbers, ordinal numbers, and the nominal use of numbers. The **cardinal number** of a set tells how many objects are in the set. The **ordinal number** of an object refers to its order in a set, such as first, second, and last; and the **nominal use of a number** is simply the use of a number to name something, such as putting a numeral on a football jersey.

Children encounter these number concepts when very young but often attach little meaning to the numbers they hear or recite. Try this experiment: Close your eyes and think of a tree. Now think of autumn. Think of good-looking. Now think of seventeen. Have you done it? When you thought of a tree did you see a picture in your mind? Did you see pictures for autumn and good-looking? Did you see a picture for seventeen or did you see the numerals *17*? Why is it that we do not see the letters *g-o-o-d-l-o-o-k-i-n-g* even though good-looking is a fairly abstract concept, and yet we see the numerals for number concepts? Perhaps as children we never truly developed a good foundation in these concepts.

Children should be able to use numbers in their cardinal, ordinal, and nominal senses as well as understand the use of numbers for such measurement ideas as money, time, temperature, length, area, and volume. This chapter contains ideas for using numbers in their cardinal, ordinal, and nominal senses. The measurement chapter includes ideas for those uses of number.

Both parents and teachers should encourage children to use numbers informally whenever possible. You may ask the children to keep a scrapbook of the ways in which numbers are used. You can help the children separate the uses into cardinal, ordinal, nominal, and measurement, with a bulletin board to show the uses the children find.

Encourage children to find the number of objects in a set (the cardinal number). Ask the children to take attendance or the lunch count. Children can help inventory books or count the Cuisenaire rods to make sure none are missing. Parents can ask the children to count the plates, knives, spoons, and forks while setting the table, or to count out prizes for everyone at a party.

Children can use ordinal numbers to find the location of something. Julio can note that he sits in the first row in the fourth seat. This skill can be extended to finding the location of the car in a large parking lot or the seats at the circus. Nominal numbers may be noted on the jerseys of the players on the basketball team or used to identify anonymous drawings posted in a display.

Older children can look out for very large numbers. The newspaper reports such things as the national debt or the distance to a newly found star. The *Guinness Book of World Records* reports many interesting facts that the children can challenge each other to discover. They can interview store owners to find how numbers are used to predict future sales.

Children can think of many other interesting projects themselves. Discussion of the uses of numbers should arise naturally throughout the children's day.

As children work with numbers, give them the opportunity to recognize numbers by sight as well as to count. When there are three or four children in a group, the children should be able to tell you the number without stopping to count each child individually. Give the children plenty of opportunities to count larger amounts. Counting can be used for real problems such as those just noted or for contrived problems such as counting the number of times a child can jump rope without missing or the number of stop signs between school and home. Let the children think of other things they would like to count.

Children learning to count often have misconceptions. Here are some examples that typify a child's misunderstanding:

1. If a child sees a group of buttons in a pile and then sees the same buttons spread out, he or she may think that there were fewer buttons when they were piled up.

2. If a child is counting six buttons on the table, the child may point at each of the buttons, but may count to ten before reaching the last button.
3. A child may miss certain items or recount them, especially when counting items placed in a circle or spread out randomly.
4. If you ask a child to give you three apples, he or she may give you only the third apple counted.

Parents and teachers should be aware of these difficulties as they help children attach meaning to numbers. Many good commercial materials exist, but parents and teachers should also take advantage of materials in the child's environment such as dried beans or peas for counting and coffee stirrers for regrouping. Useful commercial materials include Cuisenaire rods, multibase blocks, chips for trading, abacuses, dot cards, wooden cubes, counters, counting sticks, some computer programs, and calculators. The specific material is not as important as using some physical or visual material to represent the concept. We must realize that seventeen is not just the numeral we visualize or just the word we say after sixteen.

Developing Number Concepts

The concepts that underlie early number ideas include conservation, one-to-one correspondence, classification, comparison, patterns, and sequences. Later number concepts include place value and matching sets to numerals and number words. To understand the difficulties a young child has with number concepts, you must first understand the way a child views the world of numbers.

Very young children do not use or understand number ideas the way they do terms for objects in their immediate environment, such as dog, mommy, and cup. Around the age of two, they begin to understand the difference between being allowed to have two cookies and being allowed to have only one cookie. At this time, numbers begin to have some meaning for the child.

It is not uncommon for a two or three year old to be able to count to three or four or even to ten and beyond and yet not know the meaning of six. It is up to the teacher or parent to help the child understand the cardinal usage of numbers. The concepts of numbers beyond two or three were not developed until relatively recently in history, and the concept of zero came even later. We should not expect children to learn these ideas automatically. Some children enter kindergarten with an unclear notion of what numbers really are beyond being words to recite in order.

The following activities are designed to help young children understand early number concepts. They begin on the concrete level with the manipulation of actual objects and the oral discussion of number words. Later, numerals and other written symbols are associated with objects and pictures.

ACTIVITIES

Primary (K–3)

Objective: *to develop early number concepts through observations and the use of one-to-one correspondence.*

1. Go on a scavenger hunt for the number two. Have the children hold up their hands and discuss the fact that they each have two hands. If they pick up an object in each hand, how many objects will they be holding? Let the children see how many sets of two they can find in the classroom. Repeat the activity for other amounts.

2. Play the alike and different game introduced with sets in Chapter 5, only this time make the differences and likenesses be the number of objects on each card. You may have a set of cards such as those pictured in Figure 6–1.

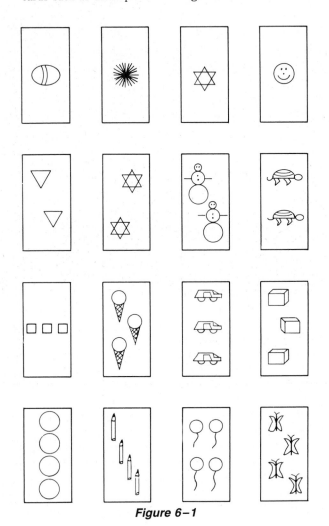

Figure 6–1

If you are working on the concept of the number 2, have the child place all the cards with two objects in the alike pile and those that have more or less than two objects in the different pile. Repeat the game for other amounts.

3. Use the cards from the alike and different game to play several variations. Show the child a set of cards in which all of the cards but one show the same number of objects. Ask the child which card does not belong. If the child responds with a card

you do not expect, ask the child the reason for his or her response; the answer may be a legitimate correct response.

Show the child a set of cards that show the same number of objects. Ask the child to give you another card that shows this number.

Ask the child to sort the cards so that cards in each pile show the same number.

Objective: *to reinforce early number concepts on a semi-concrete level.*

4. Let the children draw pictures of all the body parts they can think of that come in twos. Encourage responses that we may not think of immediately, such as elbows and thumbs, as well as the more conventional arms and legs.

Notice that in these activities the numerals are not introduced. The numbers are discussed orally with the children, and counting is not used. The children learn to associate a number with a set by looking at the set, but they can check the answers that they have gotten by sight by counting the objects in the sets. After children are comfortable with these activities, add numerals to the activities.

Conservation

For kindergarteners, the number concepts up to five should be stressed. Piagetian research has shown that most five year olds do not conserve numbers beyond five. To see if a child can conserve, try the following experiment: Show the child two groups of seven beans each. First, line the groups up so there is an obvious one-to-one correspondence as in Figure 6–2. Ask the child if the two groups have the same number of objects. If the child says yes, spread the objects in one group apart. Ask the child if the two groups still have the same number of objects or if one group now has more. If the child believes that one group has more objects, then this child is not conserving number.

If the child is a nonconserver, try the same activity with smaller numbers of objects. Some children can conserve when there are only three or four objects but are overwhelmed by the visual configuration when there are more. A teacher should judge the readiness of each child individually, however, and not rely on a child's birth date. Some

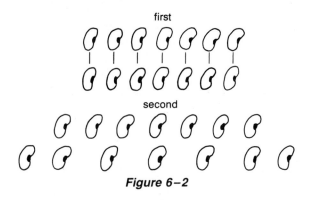

Figure 6–2

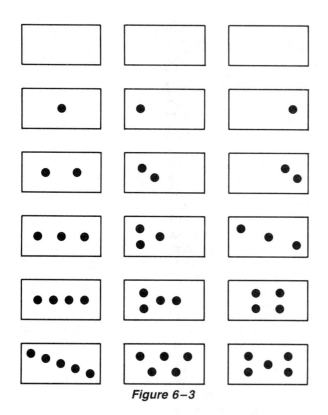

Figure 6–3

five year olds are able to conserve large numbers, and some seven or eight year olds are not able to conserve at all.

Teachers should incorporate the use of dot cards without numerals in their classes with young children. A set of cards with up to five dots in several different configurations, such as the ones in Figure 6–3, can be constructed. Later, cards with numerals may be added to the set so that children may match numerals to dots.

ACTIVITIES

Primary (K–3)

Objective: *to develop early number concepts using dot cards.*

1. Let the children play a matching game with the cards. Have them sort the cards into piles so that all the cards in one pile have the same number of dots.
2. Play the alike and different game, only this time use the dot cards.
3. Play a game of go fish or old maid in which the children try to make books of three cards with the same number of dots.
4. To encourage the problem-finding abilities of children, let them make up their own games with the cards. Be sure that each child understands the rules before play starts. Let the children change the rules as they go along if everyone playing agrees. Similar games may be played with sets of objects.

Objective: *to develop sight number concepts.*

5. Play a magic number game in which small objects are hidden under a box. When the objects are uncovered, the children must tell you how many objects there are. Uncover the objects for only a short while so the children do not have time

to count them. Children should learn to recognize up to four or five objects on sight.
6. To expand sight number concepts to the semi-concrete level, play the magic number game with the dot cards. Use the cards as flash cards, and have the children tell you the number of dots on a card without counting. You may wish to try this with more dots, although most adults can recognize only about five things without having to count.

Do not put children into too many testing situations. The activities should be fun for the children and should not always be tests. Go through the cards frequently, reciting the numbers yourself without making the children repeat them. Just let the children see the dots and listen to you say the words.

7. Children's dice and board games often give them the opportunity to use sight numbers. Any game in which a child rolls one or two dice and then moves the number of spaces shown encourages him or her to recognize the number of dots on the dice and then to match the number to moves that are similar to moves on the number line. These games may be played at home and during indoor recess as well as during the mathematics class.

Combinations

As numbers larger than 5 are introduced, the children may begin to see them as combinations of groups of smaller objects. For example, seven may be seen as three and four or as two and five. This grouping should be encouraged because it will greatly facilitate the later learning of addition and subtraction facts. If a child has learned to recognize eight as five and three, there will be no need to memorize $5 + 3 = 8$ or $8 - 5 = 3$. The child will simply need to learn the symbols $+$, $-$, and $=$; the facts will already be known. The following activities are designed to help children learn larger numbers by partitioning and combining sets.

ACTIVITIES

Primary (K–3)

Objective: *to use partitioning of sets to develop concepts of larger numbers.*

1. Make a shake box out of a small box with a partition in the middle. Place six to ten beans in the box (see Figure 6–4).

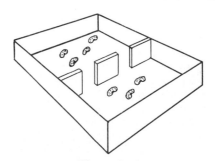

Figure 6–4

Put the lid on the box and shake it. Then remove the lid. Record the number of beans on each side of the partition. For instance, with seven beans, you may have six and one, four and three, five and two, etc.

2. Give each child six to ten buttons. Let the children separate the buttons into groups as many ways as possible. For instance, six may be five and one, two and four, three and three, etc. Encourage the children to recognize these amounts without counting, but they may use counting to check the answers, if necessary.

Objective: to increase visual number concepts.

3. Using a set of dominoes with up to five dots on each side, let each child sort the dominoes into piles with the same number of total dots. There may be piles such as those in Figure 6–5. The magic number game described in #5 of the previous activity set may also be played with dominoes.

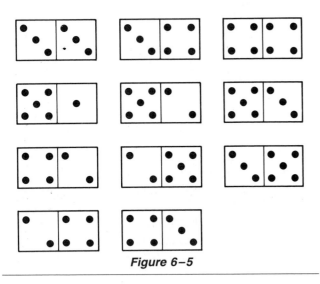

Figure 6–5

Zero

Zero is often a difficult concept for children. It is easier to recognize that there are no elephants in the room than it is to realize that there are zero elephants. Having zero seems to be a more difficult concept than not having any. The following activities are designed to help children learn the concept of zero.

ACTIVITIES

Primary (K–3)

Objective: to introduce the concept of zero concretely.

1. Let the children name things they have zero of in the classroom. The answers may range from pink elephants to a rug on the floor to sit on.

2. Collect several small empty boxes and put pennies in every box but one. Leave that box empty. Tape the boxes shut. Ask the children to guess which box has zero pennies without peeking inside. If you used cotton balls instead of pennies, could they still guess?

3. When you take attendance, talk about the number of children absent. How many are absent if you have perfect attendance? What will happen if zero children enter the room? What if zero children leave?

4. During physical education time or recess, talk about the number of times you jump rope or bounce a ball. Can you jump zero times? Bounce the ball zero times?

Objective: to introduce the concept of zero on a semi-concrete level.

5. Add blank cards to either the set of dot cards or the set of picture cards used before. Have the child sort them as before. Tell the child the number of dots on the blank card is zero.

One-to-one Correspondence and Counting

As the number of children attending preschool and watching programs such as "Sesame Street" increases, the number of children entering kindergarten with some ability to count also increases. These children may be counting either **rotely,** just reciting words memorized in order, or **rationally,** with understanding. Some children with the ability to rote count can complete a counting sequence only if they use the "sing-song" pattern that they used in learning to count. Other children chant the counting sequence, becoming progressively faster at repeating the number names. Initially, children develop the ability to count groups of objects, at first pointing and reciting the number names, sometimes skipping numbers or, at other times, skipping objects. Rational counting occurs when a child can count a set of objects and realizes that the last number name spoken in the counting sequence tells how many objects there are all together—for example, "One, two, three, four, five, six, seven. There are seven buttons." And, when asked again how many objects there are, the child can confidently repeat just the last number name in the sequence, "Seven, there are seven buttons."

The following activities give children a variety of counting experiences based upon putting two sets of objects or objects and pictures in one-to-one correspondence. In this way, the children connect the counting numbers to objects and do not just recite them rotely. They learn that each word goes with only one object and that the final word spoken when counting aloud gives the number of objects in the entire set.

ACTIVITIES

Primary (K–1)

Objective: to use one-to-one correspondence and counting to determine the number of objects in a set.

1. Draw and cut out five garages and collect five toy cars or trucks. Tell the children that one vehicle may park in each garage. Randomly set out from one to five garages on the table. Ask the children to match a vehicle to each garage and

to count as they make the matches. After the vehicles have been matched to the garages and counted, ask the children to tell you how many vehicles there are altogether. How many garages are there?

The activity may be varied to use boats and docks or airplanes and hangars. Let the children suggest other things to match. As the children gain the ability to match and count, increase the number of objects to be counted.

2. Using dot cards used for earlier activities and some bingo chips, tell the children to cover each dot on the card with a chip and to count as they make the matches. Ask them how many dots are on each card. Is that the same as the number of chips on the card? Increase the number of dots on the cards when the children are ready.

3. Collect five small cans for flower pots, such as juice cans (be sure there are no sharp edges). Make fifteen flowers using pipe cleaners for the stems and construction paper for the petals. Put one to five dots on the outside of each can. Ask the children to match the number of flowers to the number of dots on the can. Let the children count to tell you how many dots and flowers there are. Later, you can increase the numbers of flowers and dots.

In all of these activities, it is fine if children can tell you the numbers without counting; they should not be required to count. After the children are proficient at matching two sets of objects or dots and objects, numerals may be substituted.

Measuring Length

Thus far, the discussion has focused on number as related to sets of discrete or individual objects. Number may also refer to the length of an object, a measurement concept. Number lines, rulers, and Cuisenaire rods depend upon this measurement idea. We also use number in a measurement sense when we say "He is four years old" or "She is three blocks from home." Measurement concepts are discussed in detail later but are mentioned here briefly as they relate to number concepts.

Difficulties arise as children move from associating numbers with sets of discrete objects to associating numbers with lengths. First, children may not see two as two units long rather than as two objects. Second, many children do not conserve length. They believe a rod changes in length when its position changes. They have difficulty understanding that two is twice as long as one. To introduce number concepts using measurement, the teacher should first ascertain whether the children can conserve length.

One good model for length is Cuisenaire rods. For young children, Cuisenaire rods may be used to discuss such ideas as longer than, shorter than, and the same length. After discussing these concepts, children who can conserve length may begin to associate the rods with numbers. The children may use the white rod, which is the shortest one, to represent one. Using this white rod to measure, the children may then determine the length of each of the other rods. They will find that if the white rod is one unit long, then the red is two, the light green is

three, the purple is four, the yellow is five, the dark green is six, the black is seven, the brown is eight, the blue is nine, and the orange is ten. Once children have determined these lengths, they should try some of the following activities to reinforce the number concepts.

ACTIVITIES

Primary (K–3)

Objective: *to introduce ordering concepts involving length.*

1. Make a staircase using one rod of each color. Construct the staircase so that the rod on top is a white rod, the rod on the bottom is orange, and the rods between are arranged in order of length (see Figure 6–6). What will happen if you put a white rod next to each rod in your staircase?

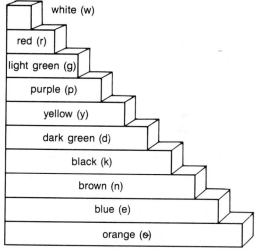

Figure 6–6

Objective: *to reinforce the concept of equivalent lengths as an introduction to addition.*

2. Pick out the yellow rod. Make trains of rods that are the same length as the yellow rod, such as those shown in Figure 6–7.

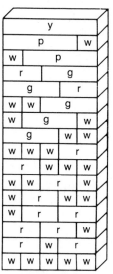

Figure 6–7

Have the children record every combination whose length equals the length of the yellow rod, beginning with the colors and later using the numbers to represent the rods.

3. Have one child pick a rod other than the white one and challenge another child to find two rods that will make a train the same length as the first rod. Children may determine the number values of their trains after they become proficient using the colors.

4. Have one child make a train of two rods and challenge another child to find one rod that is the same length as the train. What happens with a train such as brown and black? Let the children play the game for several days before asking them to associate numbers with the rods.

The rods may also be used in an introduction to the number line. Construct a number line so that the numbers are one centimeter apart. Mark zero, and number the line up to 10 or so, as shown in Figure 6–8.

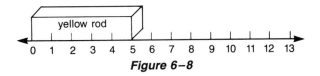

Figure 6–8

Ditto number lines for the children, and obtain a transparency of a number line and translucent Cuisenaire rods made for the overhead projector to use yourself. Because children often do not associate the numbers on the number line with lengths, using Cuisenaire rods helps them see that two is the distance from zero to two and not just the point half way between one and three. For this reason, it is important to have zero on the number line.

Points to the left of zero need not be marked at this time, but there should be an arrow pointing to the left from zero as well as an arrow pointing to the right from the last number marked. If the children ask the meaning of the arrows or ask if there are other numbers to the left of zero, answer that the numbers go on forever in both directions.

Decide if the children are ready for a detailed discussion of infinity or negative numbers at this time. If they are, show them such uses of negative numbers as low temperatures on the thermometer, or let them use a calculator to see what happens if they try to subtract 4 − 6, but generally do not introduce a detailed study of negative integers or infinity in the primary grades.

The following activities are designed to introduce the child to the use of the number line.

ACTIVITIES

Primary (K–3)

Objective: *to introduce the concept of measurement on the number line.*

1. Let the children choose a Cuisenaire rod at random. Place the rod on the number line with the left end of the rod on zero.

Where is the right end? How does this number compare to the value of the rod? Try this with several different rods.

2. Choose two rods to line up on the number line in a train, as shown in Figure 6–9. Place the left end of the first rod on zero. Can you predict where the right end of the second rod will be?

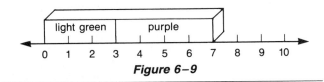

Figure 6–9

After working with the Cuisenaire rods on the number line, children should understand the concept of length on the number line and should be able to move on to number line activities that do not require the use of the rods. The following are examples of other number line activities.

ACTIVITIES

Primary (K–3)

Objective: *to develop number line concepts through body movements.*

1. Make a large number line on the floor with masking tape. Make sure the starting point is marked zero to show that no steps have been taken at this point. On the number line, mark intervals that are about the size of a child's step (about 25 centimeters). Make sure that each interval is the same size.

   Have the children take turns starting at zero and walking a given number of steps. Have them confirm that the point where they finish corresponds to the number of steps they took. In this way, the number of discrete steps is associated with the distance walked on the number line.

2. Using the one-centimeter number lines, let the children use their fingers to count a certain number of spaces, beginning at zero. What number is on the point where the children land? Emphasize counting spaces rather than points.

Objective: *to practice skills on the number line.*

3. Use Freddie the frog with the number line. Let Freddie start at zero and hop a certain number of spaces. Where does he land?

Grouping

Grouping is an important concept in the Hindu-Arabic system of numeration. Many older systems of numeration, such as the Roman and the Egyptian, did not use grouping, and thus they were awkward for writing and manipulating symbols for large numbers. Because place value is so important in our system of numeration and because grouping is essential to place value, young children should begin grouping even while they are learning the numbers from 1 to 9. They should have experience grouping by twos, threes, fours—all the way to tens—for two good reasons. First, the notion of an exchange point is a key to understanding place value. Second, regrouping using smaller

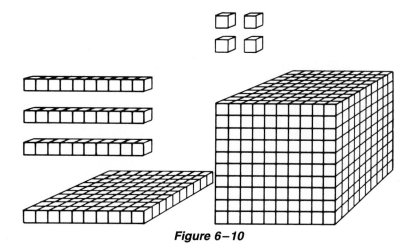

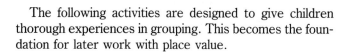

Figure 6-10

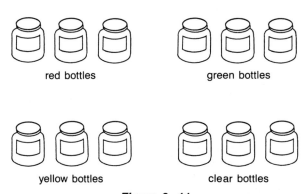

blue bottles

red bottles   green bottles

yellow bottles   clear bottles

Figure 6-11

numbers provides more practice than waiting until ten objects have been collected to make an exchange.

There are two basic types of grouping materials: proportional and nonproportional. Proportional materials are constructed so that if the grouping is by tens, the material that shows 10 is ten times as large as the material that shows 1 and the material for 100 is ten times as large as the material for 10, and so forth. Proportional materials include multibase blocks (see Figure 6–10), tongue depressors, coffee stirrers or straws, counting cups and beans, and Cuisenaire cubes, squares, and rods.

Nonproportional materials, such as money, do not show consistent size changes. A dime is not ten times as large as a penny and a dollar is not ten times the size of a dime. Nonproportional aids include chip trading, the bottle game (see Figure 6–11), money, and the abacus. Allow children to freely explore with the proportional grouping materials before you begin formal instruction. Let the children discover the size differences and the trades possible as they build houses or make designs with the blocks.

The following activities are designed to give children thorough experiences in grouping. This becomes the foundation for later work with place value.

ACTIVITIES

Primary (K–3)

Objective: *to introduce trading games using proportional materials.*

1. **Multibase blocks** are proportional grouping materials that consist of various sizes of wooden or plastic blocks representing the powers of particular grouping points. Multibase blocks are commercially available in sets with grouping points of two, three, four, five, six, and ten. A set of blocks with a grouping point of three is shown in Figure 6–12. Regardless of the grouping point, the smallest pieces are called units; the remaining pieces are called, in order, longs, flats, and cubes. You may make your own set of multibase blocks using self-adhesive paper (such as contact paper) printed with a grid approximately 1 cm square and railroad board. Cut out units, longs, and flats to match the grouping point desired.

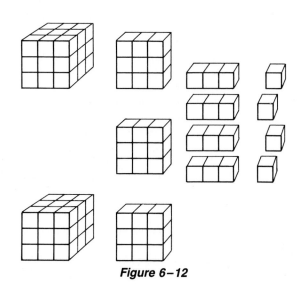

Figure 6-12

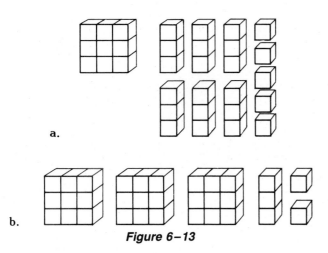

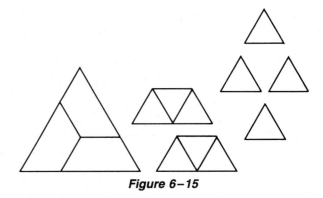

Figure 6-15

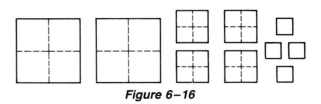

Figure 6-13

they are writing down? This activity will give them a good basis for learning addition or subtraction with regrouping.

3. Teachers may construct proportional materials from railroad board or other firm construction material. Figure 6-15 illustrates one such set, with a grouping point of three, and Figure 6-16 illustrates a set with a grouping point of four.

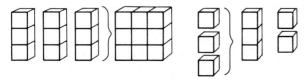

Figure 6-16

After the children have had an opportunity to play with the materials and to discover the grouping point, give them a handful of units and ask them to make all the trades possible. For example, the blocks in Figure 6-13a can be traded for the blocks in Figure 6-13b because the blocks represent the same amount of material. If children have difficulty making the exchanges, instruct them to make groups of three units until all units are used and then to exchange each group of three for a long, as shown in Figure 6-14.

Figure 6-14

2. After the children have played some grouping games, let them play the build a house game. For this game, children form two families. Each family is trying to build a house before the other. To begin the game, the children agree which city they will live in, Three for One, Four for One, or Five for One, and choose multibase blocks to match the city.

To play, each team rolls one die to determine the paycheck for the week. The dots showing on the die tell the child the number of single units earned that week. Children should trade in single units for a long as soon as possible, trade longs for a flat, and trade flats for a cube. A cube is the house, and a team wins as soon as it completes a house.

Children should also play the game in reverse. In this game, they begin with a large cube and try to spend it. The dots showing on the die tell the child the number of single units to spend on that round. In this game, the first team to spend all its money loses.

The second game is usually much more difficult for the children than the build a house game, especially in the beginning. The first move is the most difficult. Having one large cube and needing to spend two units is similar to the problem 1000 − 2. Children who realize that they must make several trades before they can spend two units have very good place value concepts. Discuss with the children the trades that they must make, both as they play the game and after they have finished.

After the children have played the games for a while, ask them to record their plays. How does the game relate to what

Use dot cards with these triangular materials. Place the dot cards face down in a pile in the center of the table. Let the children take turns drawing a card and picking up that many small triangles. Make trades whenever possible. The first team or child to get two large triangles wins.

Repeat the game with squares. Activities suggested for the multibase blocks may also be used with these materials.

4. Another material that the teacher may use is the counting cup and beans. For a grouping point of five, a cup with five beans represents a first grouping, five cups of five in a larger container represent a second grouping, and so forth, as shown in Figure 6-17.

Figure 6-17

The counting cups and beans may be used to play a store game. A single bean represents a penny, a cup with five beans is a nickel, and five cups of five beans represent a quarter. Put price tags on small objects, such as 37 cents for a hair ribbon and 7 cents for a pencil.

Ask the children to count out individual beans to represent a price and then to make any trades possible. Is the pencil more than a nickel? Is the ribbon less than a quarter? Encourage the children to set prices and make up questions of their own. This game may also be played using pennies, nickels, and quarters in place of the beans.

Again, each of the activities described for any of the grouping materials may be used with the counting cups. So far, each grouping activity suggested has focused on only one or two exchanges. Teachers should suggest activities requiring several exchanges and using a variety of grouping points. Discuss with the children why they make the trades that they do.

Objective: *to introduce grouping concepts using sets of discrete objects and charts and tallies to keep track of experimentations.*

5. Begin with eight buttons. Have the child place them in groups of three (***) and record the results on a chart, as shown in Figure 6–18.

groups	ones
2	2

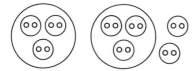

Figure 6–18

Repeat the activity with grouping points of two (**), four (****), five (*****), and six (******). Chart the results. Discuss with the children what happens as the grouping point changes.

Repeat the activity but use more buttons, say twenty-five. Put all the groups of three (***) together and label these groups. Next, group together three groups of three and label these super-groups. Record the results in a tally box with three sections, such as the one shown in Figure 6–19.

Repeat the activity with other grouping points and starting amounts. Record the results. There may be times when you will need super-super groups, or more, especially for small grouping points such as two.

Objective: *to introduce grouping concepts using nonproportional materials.*

6. Collect about 120 empty small infant-formula bottles with screw-on caps (hospitals will often provide you with bottles). Fill the bottles with water and use food coloring to tint the water different colors. You should have about 40 clear bottles, 30 yellow bottles, 20 green bottles, 20 red bottles, and 10 blue bottles.

Choose a magic number for trading. Let the children take turns rolling a die and collecting the clear bottles. When the magic number is reached, exchange clear bottles for yellow bottles. For example, if the magic number is 3, three clear

super–groups	groups	ones
2	2	1

Figure 6–19

bottles may be traded for one yellow, three yellows for one green, three greens for one red, and three reds for one blue. Set the goal as either a red or a blue bottle. Discuss the children's trades as they play and ask questions such as, "How many more will you need to reach the goal?" or "Who is winning and by how much?"

The bottles may be used for the same games as the proportional materials.

Colored chips are convenient nonproportional teaching aids. They may be purchased commercially or constructed by the teacher. The values for the chips may be established in the same manner as the values used in the bottle game. For example, four yellow chips are equivalent to one blue, four blue are equivalent to one green, and four green are equivalent to one red. The activities suggested for the other trading materials are also appropriate for the chips.

After the students have used several different grouping points with the materials, expand the activities to include base 10. Be sure to encourage the children to record their activities as they work and to discuss the strategies they use.

Developing and Practicing Number Skills

Once children can orally associate numbers with sets, then introduce written numerals. The fact that written numerals are often introduced too early may be one reason we picture numerals instead of sets when we hear a number; therefore, we should be sure children can orally associate the numbers with sets before they are asked to learn written numerals.

Matching Numerals to Sets

Children should be familiar with seeing written numerals before being asked to write them. Numerals are not something children can discover on their own. If we want children to learn Hindu-Arabic numerals, we must teach them.

The following activities are designed to help the child match the numerals from 1 to 5 with sets. These activities may be expanded later to include other numerals as well as the written words for the numbers.

ACTIVITIES

Primary (K–1)

Objective: *to match written numerals to sets.*

1. Play a treasure hunt game. Give each child a numeral and have the child look around the room for sets of objects that match the numeral. For instance, the child with the numeral 3 may find three cars, three pencils, and three chairs in the learning center.
2. Make several sets of cards for different card games similar to old maid or go fish. The cards should give children a chance

to practice matching numerals, sets, and dots, such as those in Figure 6–20.

3. Make a set of puzzles such as those in Figure 6–21, matching the numerals from 1 to 5 or 9 with dots or pictures. Puzzles of any type help students check themselves.
4. To make another type of puzzle, cut a picture from a magazine, mount it on posterboard or triwall, and cut it into puzzle pieces with scissors or a jigsaw. Draw around the pieces in their proper positions in a box lid. On each section of the box lid, write a numeral. On the back of the matching puzzle piece, put that number of dots. The children must match the dots to the numerals to make the puzzle pieces fit. The activity is self-checking because if the children do the puzzle properly, they will see the picture when they are finished.
5. A bingo game can help children learn colors as well as numerals. A typical bingo card can look like the one in Figure 6–22a. The caller holds up a color-coded dot card such as the one in Figure 6–22b. Each child with that number then places a matching number of colored chips on the numeral on the bingo card. When a child calls bingo, the caller confirms that the numerals are covered with the proper number of chips.
6. Dominoes with numerals on one side and dots on the other are useful for matching numerals and dots. Children can play dominoes, matching the numerals on the dominoes to corresponding dots.

Writing Numerals

After children can match numerals to sets, they can begin to learn to write numerals. Children at this age often do

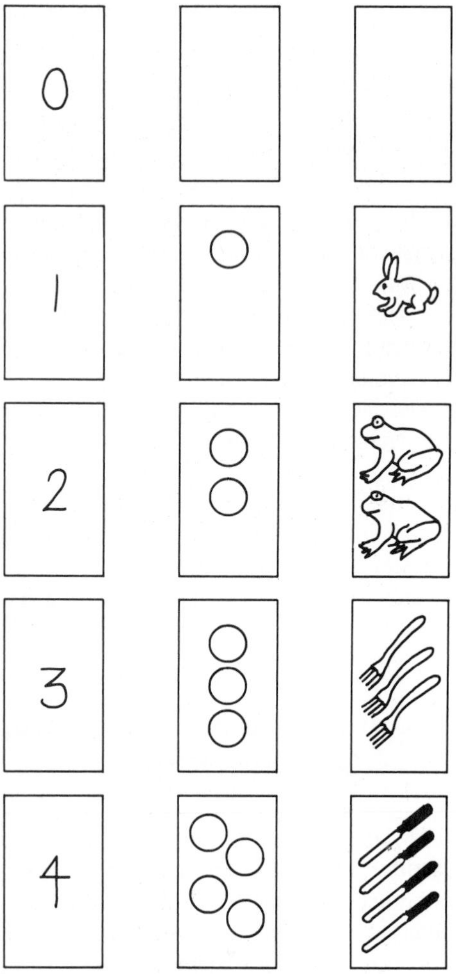

Figure 6–20

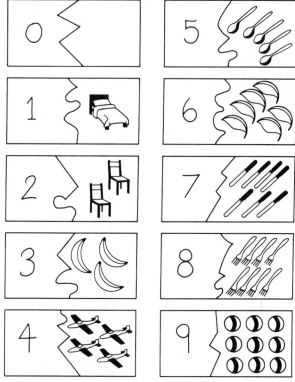

Figure 6–21

blue	red	orange	yellow	green
2 ○○	5	3	1	4 ○○○○
6	4	2 ○○	3	9
7	8	free	4	3 ○○○
5	2	7	9	1 ○
1 ○	9	6	2	6

a.

green

b.

Figure 6–22

not have good fine motor control and may need to use very large materials in the beginning. The following are a few suggestions for writing numerals with materials other than paper and pencil.

ACTIVITIES

Primary (K–3)

Objective: *to use tactile-kinesthetic abilities to write numerals.*

1. Cover the bottom of a box lid with sand or salt. Let the children practice writing numerals in the sand. After you check the work, have the children simply shake the box lid to erase their efforts and begin again.
2. Cut large numerals out of sandpaper and let the children run their fingers gently over the numerals. Have them do this blindfolded and try to guess what the numeral is. Then have them draw what they have just felt.

Objective: *to reinforce writing numerals.*

3. Draw numerals on large squares of posterboard. Use arrows to show the direction to move to write the numerals. Laminate the numerals and let the children copy them with a grease pencil or washable crayon.

Rational Counting

After children have had several ordering experiences such as those described in Chapter 5, are able to associate the numbers from one to five with sets of objects, and are able

to write numerals, they should be ready to order numbers. In this way, counting is developed in a rational, not rote, manner. It is important for children to realize that three comes after two when they are counting because a set of three has one more object than a set of two, not just because they have heard the numerals recited in that order several times. The activities in Chapter 5 with sets should be repeated with numbers. In summary, these include:

- *Observations and inferences.* By observing a variety of sets of objects, pictures, and dot cards, children learn to associate numbers with sets, first orally, then using the written numerals, and finally using the written number words.
- *Classifications.* The children group together all sets containing the same number of objects. This is a good time for the teacher to determine if children can conserve numbers. If a child realizes that the number of objects in a set stays the same regardless of the configuration, then the child can conserve.
- *Comparisons.* The children use one-to-one correspondence to determine if two sets have the same number or if one set has more or less than another.
- *Sequencing.* The children order sets of objects from those containing the fewest to those containing the most. The increase may be by one object each time but it does not have to be.

The following activities are designed to help children to rationally order numbers. When numbers increase by one each time, the child may begin rational counting. Rational counting should not always begin with one. Children should practice counting beginning with any number. They should count both forwards and backwards. The symbols >, <, and = may be introduced. Ordinal number concepts should also be developed as children understand the concepts of sequencing.

ACTIVITIES

Primary (K–3)

Objective: *to practice rational counting.*

1. Children should practice counting several times during the day, whenever the opportunity arises. For example, if you ask the children to set up the chairs for the reading group, ask them to count the number of children in the group and to count out the same number of chairs.

   They can also practice counting on from a given number. If there are already five chairs in the circle and they need eight chairs, ask them to count beginning with five. As they get more chairs, they can count on—6, 7, 8. If there are too many chairs in the circle, they can count backwards as they remove chairs. If there are ten chairs and they need only seven, they can count back from ten as they take the chairs away—9, 8, 7. Counting on and counting back will help later when they learn addition and subtraction.

2. Children do not always need to count physical objects. They may count the number of days until a special holiday, the number of seconds in the countdown on a microwave oven, or the number of claps of thunder they hear. Again, encourage them to count forwards as well as backwards, beginning with different numbers. Later they should learn to count by whole numbers other than one as well as by fractions and decimals.

Objective: to introduce the concepts greater than and less than.

3. Make a set of dot cards with numerals on one side and dots on the other. Have the children put the cards in order from the fewest to the most dots. Turn the cards to the numeral side to check. Reverse the process by ordering the numerals and checking on the dot side using one-to-one correspondence. Reinforce the concepts of one less and one greater.

4. An adaptation of the card game war may be played with a deck of cards having from zero to nine dots on a card. When constructing the cards, make four or five cards for each number of dots. The game is played by two or more children. The cards are shuffled, and all are dealt. Each player turns over one card and places it face up on the table. The player with the card showing the most dots wins all the cards on this round.

If there is more than one card showing the greatest number of dots, the two players with those cards have a war. Each of these players then places a card face down on his or her first card and another card face up on top of the second card. The new face-up cards are compared. The player whose card has the most dots wins all the cards in the war as well as any other players' cards on this round. Play continues until one player has all the cards or a time limit expires. If the children question which card has the most dots, they should establish a one-to-one correspondence between the dots to decide.

The second phase in this game is to use a regular deck of cards or the numerals without the dots. Play continues as before.

5. After the children have had experience with the words *greater than* and *less than,* the symbols may be introduced. A popular

idea for introducing the symbols is the use of a hungry fish (see Figure 6–23). The fish always wants to eat as much as possible, so its mouth is always open toward the greater amount. The children may use blocks to represent the fish's food. Set up two groups of blocks with between one and nine blocks in each group and let the children place the corresponding numeral under each set. The children then decide which set is greater, using one-to-one correspondence if necessary, and place the fish between the numerals with its mouth open toward the larger numeral. If the activity is done on paper, the children may trace inside the fish's mouth to keep a permanent record of the larger number. Discuss the terms *greater than* and *less than* as the symbols are introduced.

The symbol for equals (=) may be introduced at the same time. The fish cannot decide which group to eat because both are the same size. It keeps its mouth neither wide open nor closed (see Figure 6–24).

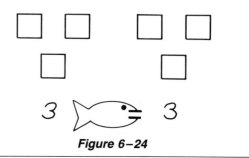

Figure 6–24

After children are comfortable putting numbers in order to count the cardinal number of objects in a set, they may begin to use numbers in an ordinal sense. Ordinal numbers should be used throughout the day just as cardinal numbers should. As children line up for any activities, discuss who is first in line, second, last, etc. Talk about the first thing on the schedule each morning. Let the children suggest other times they use ordinal numbers.

Several computer and calculator activities are designed to help a child with early number concepts. The following examples include programs and activities for both number recognition and counting.

ACTIVITIES

Primary (K–3)

Objective: to use the computer to reinforce matching numerals to pictures of sets of objects.

1. Sticky Bear Numbers, available from Weekly Reader Family Software, is a program for preschool and kindergarten children. It consists of sets of objects that appear in response to a child's pressing either the space bar or a numeral on the keyboard. The child may make a new set of objects appear by pressing a new numeral or may increase or decrease the set by one by pressing the space bar. For example, by pressing the space bar once, the child makes a train appear. Each additional time the child presses the space bar, another train appears until nine trains are showing on the screen. After nine trains have appeared, the trains disappear one by one as

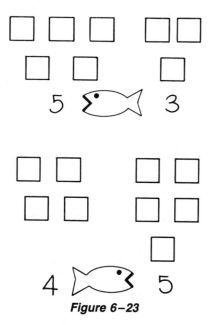

Figure 6–23

the space bar is pressed until all trains have disappeared. The numeral 0 then appears on the screen. A numeral accompanies each set shown on the screen. When the space bar is pressed again, new sets of objects appear. Children as young as eighteen months can enjoy the graphics, and older children can begin to associate numerals with sets.

Objective: to develop simple programming concepts and to reinforce counting skills.

2. Older children may write programs for younger children, instructing the computer to count by ones, by tens, backwards, etc. Fairly simple programs in BASIC can accomplish this. For example, Program A counts forward by twos beginning at 2 and ending at 40 and Program B counts backward from 50 to 0 by fives.

```
                    Program A
        10 NEW
        20 FOR I = 2 TO 40 STEP 2
        30 PRINT I
        40 FOR J = 1 TO 500: NEXT J
        50 NEXT I
        60 END
                    Program B
        10 NEW
        20 FOR I = 50 TO 0 STEP −5
        30 PRINT I
        40 FOR J = 1 TO 500: NEXT J
        50 NEXT I
        60 END
```

Even very young children can use a calculator to develop counting skills. As children first learn to count they may use a calculator to help them keep track of their counting. For example, a child counting a pile of chips can push +1 on the calculator each time he or she moves a chip. Having finished counting the pile of chips, the child can push = on the calculator to see if the calculator will show the numeral just said.

Children can explore with a calculator to see what happens when they push +1 = = = =. Later, children may start with any number and then try +5 or +2. What happens when you start with a larger number and push − 2 = = = ? Children may encounter negative numbers doing this. (Calculators work differently, and you may need to explore with yours before you try this with the children.)

Number Names

After children have learned to match numerals to sets and have learned to count rationally to at least five, introduce written words for the numerals. Expand many of the same activities used to introduce the numerals to include the number names. In card games, add cards with the number names on them for the children to include in the matches. Mark the number line with the number names as well as the numerals.

Children's literature is a good source for teaching numerals and number names. Several good children's books

introduce numerals and number names. Encourage children to find the number names in stories and to check the pictures to see if the pictures show the correct number of objects.

Children also enjoy finger plays such as "Ten Little Indians" and "Five Birds on a Fence," in which they recite a poem and show numbers with their fingers. Children's music also is a good source of ideas for reinforcing counting. Music can be a great memory aid for many children. Librarians and music teachers can help you find excellent children's books, stories, finger plays, and songs.

Place Value

Be certain that children have had many experiences with grouping activities before you formally introduce the numeral 10. It is difficult for young children to understand that 10 represents ● ● ● ● ● ● ● ● ● ● objects and not ● object. Children may logically assume that if you put a one and a zero together, you should have one plus zero objects, not ten. Early experience with grouping smaller amounts should help children understand why ten is written as 10. The 1 here means one group and the 0 means zero units.

Extensive work on this concept when it is first introduced can help children avoid problems later. Many teachers in the intermediate grades can tell you that place value remains a major difficulty as children learn operations that require regrouping. Even high school students often repeat lessons on addition and subtraction in general math classes because they never fully understood the meaning of place value.

As place value in base 10 is introduced, the previous activities that involved grouping materials in other bases should be expanded to grouping by tens. Children should have extensive experience using both proportional and nonproportional materials to group by tens, hundreds, thousands, etc. Encourage the children to record the results of their work using such things as place value charts. The following are some additional activities to be used in base 10.

ACTIVITIES

Primary (1–3)

Objective: to reinforce concretely place value concepts in base 10.

1. With a large box of coffee stirrers or tongue depressors, have a contest to see who can pick up the largest handful of sticks. Have the children group the sticks by tens and secure each group of sticks with a rubber band. If a child has picked up more than one hundred sticks, bundle each ten groups of ten to form groups of one hundred. Let the children record results on a chart such as the following:

Bundles of 100	Bundles of 10	Singles

DEVELOPING A CONCEPT
Building Tens and Hundreds

Sue went to the post office to buy stamps.
Stamps are sold in:

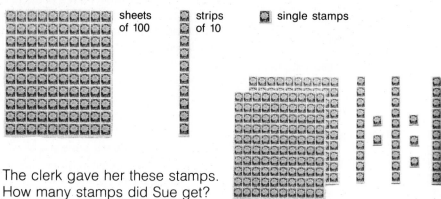

sheets of 100   strips of 10   single stamps

The clerk gave her these stamps.
How many stamps did Sue get?

WORKING TOGETHER

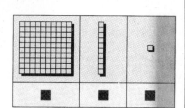

Hundreds	Tens	Ones
■	■	■

1. Use place-value models to show how many stamps Sue got. Arrange them on a chart like this.

2. Record on a place-value chart how many of each model you used.

3. Complete.   ■ hundreds ■ tens ■ ones

4. How many stamps did Sue buy? Write the number.

5. Read the number aloud. Then write its word name.

6. *What if* Sue bought 2 sheets and 7 strips of stamps? Build the model to show this. How many stamps would she get? Write the number.

7. *What if* Sue bought 207 stamps? Build the model to show this. Read the number aloud. Then write the number in words.

Figure 6–25
Reproduced with permission of Macmillan/McGraw-Hill School Publishing Company from Mathematics in Action, *Grade 3, page 12 by Alan R. Hoffer et al. Copyright 1991.*

THE MATH BOOK

The numeration activity shown in Figure 6–25 is from the beginning of a third grade textbook. Notice that it introduces the concept using a familiar example from the children's environment. Students use base ten blocks to model the stamps and work the problems in small groups. The teacher's manual suggests cooperative learning groups of three children for this particular activity. The three children can take turns saying, modeling, and writing the numbers.

The teacher's manual recommends several other activities to practice place value concepts. These include playing card games in which numerals are written on cards and the students must model the amount shown using base ten blocks and correctly write the name of the amount. Other group activities suggested include a similar activity in which students roll number cubes and build a model of the amount shown.

This page is followed by activities for regrouping, including an activity that asks the students to pretend that they are textbook authors writing a lesson that explains regrouping using dimes and pennies. After they write the outline for the lesson, they are asked to teach the lesson to a group of classmates. In this way, students gain needed experience using communication skills and strengthen their own concepts of place value.

The teacher's manual also includes ideas for recognizing and remediating common errors, working with students with limited English proficiency, using mental mathematics, and integrating mathematics into social studies lessons. Other resources included with the series are worksheets for reteaching, practice, and enrichment, ideas for a problem of the day, and suggestions for an activity center. The discerning teacher must decide which of the many resources to use each day in addition to his or her own ideas for ways to best help the children construct the concepts.

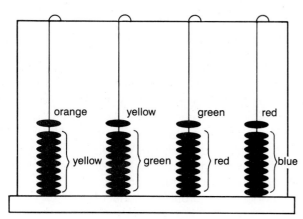

Figure 6-26

2. Purchase or construct several simple base 10 abacuses. Among the most useful are those with ten beads on each bar, with the first nine beads of one color and the tenth bead the color of the first nine beads on the next bar (see Figure 6–26).

   When the children fill up one bar with ten beads, the color of the tenth bead should remind them to trade in all ten beads for one bead on the next bar. Thus, 10 ones are traded for 1 ten, 10 tens for 1 hundred, etc. Let the children use the abacus to record results from work with other place value materials.

Objective: to extend sight numbers to amounts between 10 and 20.

3. Use chips placed on a grid such as the one shown in Figure 6–27. You should have a transparent grid for the overhead projector, and each student should have a grid of his or her own. Ask the students to explore by placing varying amounts of chips on the grid and determining the total. When children are comfortable finding the total, place a secret amount of chips on the grid and flash it briefly on the overhead projector. Discuss with the students their methods of determining the total.

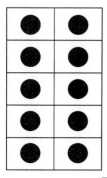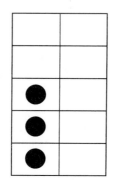

Figure 6-27

After the children have had much experience with concrete materials and with recording base 10 numerals, concepts on the abstract level may be reinforced with materials that use only numerals. These materials may include hundreds charts, place value charts, bingo games, flash cards, and playing cards. The following activities are designed to be used after the children have a good concrete understanding of grouping in base 10.

ACTIVITIES

Primary–Intermediate (2–4)

Objective: *to explore patterns on a hundreds chart.*

1. Make a large hundreds chart out of pegboard, with cup hooks from which to hang numerals (see Figure 6–28). Hang all one hundred numerals in their proper places. Let the children discuss the patterns they notice.

Figure 6-28

Now turn the cards over so that they are still on the proper hooks but the children cannot see the numerals. Ask a child to guess where the 3 should be and then to check by turning the card over. If the child is correct, leave the numeral showing. If the child is incorrect, turn the card over and let another child guess. Continue by guessing other numbers such as 23 or 30. Let each child get a turn to guess.

Discuss the patterns as you go along. This activity can also be used with dittoed hundreds charts and small squares of paper to cover the numerals. (A hundreds chart is included in Appendix B for you to copy.) Let the children remove the square of paper when they guess where the number is. Discuss the strategies used to find the numbers.

Objective: *to provide for practice in forming numerals as they are spoken and to allow for rapid checking of children's responses.*

2. Let each child make a place value chart that includes the hundreds or thousands, depending on the level of practice you need. This chart may be made from heavy manilla paper by folding up the bottom part of the paper and stapling pockets to hold numeral cards. Label the pockets ones, tens, hundreds, etc. Let each child make three cards for each numeral from 0 through 9.

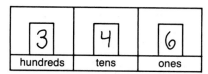

Figure 6-29

Read a numeral you wish the children to form, such as 346 (three hundred forty-six), and let the children form the numeral by placing cards in the correct pockets and then holding the place value chart up for you to see (see Figure 6-29). Quickly check to see whether or not each child has the correct response. This way, each child can get immediate feedback on the response and will not practice incorrect techniques. You may also use magic slates or individual blackboards; the children write their responses, hold them up for you to check, and then erase them to prepare for the next problem.

Objective: *to provide for practice in recognizing numerals and place values.*

3. Use 3 × 5 cards to make a deck of playing cards with values such as:

600	six hundred	6 hundreds
30	thirty	3 tens
4	four	4 ones

These cards may be used in a variety of card games that require the children to make matches, such as go fish, rummy, and old maid. After children become proficient with these cards, add others such as 3 tens, 14 ones, and 44; 2 hundreds, 15 tens, 6 ones, and 356. These cards may also be used for games such as bingo. The bingo cards should look like regular bingo cards, only with numerals to match the calling cards. The rules are the same as those for regular bingo.

Calculators may also be used to reinforce place value concepts. Let children predict what will show on the display if they press 3 × 100 or 7 × 1000. Be careful if you try 3 × 100 + 5 × 10 + 7 = . On a scientific calculator, you will get 357, but on many other calculators you will get 3057. Children may wish to discuss why this happens and try several examples to test their hypotheses. This is a good time to discuss the order of operations and the use of parentheses.

Some calculators for children, such as Speak and Math, read a numeral aloud to the child and ask the child to press the correct keys to show the numeral on the display. The calculator then either tells the child that the answer is correct or asks the child to try again. If the second response is also incorrect, the calculator shows the child the correct answer. The games have three levels, beginning with numbers in the hundreds on the first level and going to numbers in the millions with decimals on the third level. The games encourage listening skills as well as reinforce reading and writing of numerals.

Children may also play games such as go fish on the calculator. Three or more children may play this game at a time. Play begins with each child putting a secret six-digit number into his or her calculator. All six digits should be different to begin the game. To play, one child asks another for a number, say 6. If the child asked has a 6 in any place, then it is given to the child who asked for it. If the 6 is in the tens place, the child being asked will say, "Take sixty." That child will then subtract 60 from his or her number and the other child will add 60 to his or her number.

After the game has progressed, it is possible that the child being asked may have two 6s showing in the number. If this happens, the child has to give up only one digit and should choose to give up the 6 in the place with the least value. For example, a child with 236 762 should say, "Take 60" and subtract only 60, not 6060. Play continues until one player's score goes above six digits, or 1 000 000, or until one player's score goes to zero. The player with the highest score is declared the winner.

Estimating

Often, it is more important for a child to be able to estimate a reasonable response than it is to have an exact answer. Politicians and newspapers may estimate the number of people in a crowd at a rally, or a store manager may estimate the number of pounds of hamburger to have on hand for the big sale, but in both of these cases an exact answer may not be necessary or even possible. As we continue to utilize calculators and computers, the estimation skills become even more crucial. Calculators may give an exact answer, but children need to be able to estimate to determine whether that response makes sense. Estimating, therefore, should be a major part of all strands of mathematics.

The following are a few examples of estimating activities related to number concepts.

ACTIVITIES

Primary (1-3)

Objective: *to practice estimating large numbers of concrete objects.*

1. Fill a large jar with small objects such as jelly beans. Have a contest to let the children guess how many jelly beans are in the jar. The winner of the week's contest may win a bag of jelly beans. Use another container and another small object the following week. Let children use smaller containers and small numbers of jelly beans to assist them in making intelligent guesses. Their guessing and their number concepts should improve as the weeks go on.

2. Make a set of large flash cards (about 25 centimeters × 25 centimeters) and put various numbers of bright dots on each one. Doman (1980) uses up to one hundred dots on a card for children under the age of four, but most elementary school children (and most adults) have difficulty actually recognizing more than about five dots. Using larger numbers of dots and flashing the cards briefly force the children to use estimation and grouping skills to guess the number of dots on a card.

With only a few odd minutes a day of practice, children can become quite accurate in their guesses, and although exact answers are not a goal of estimation activities, a few children may be able to tell you the exact number of dots each time. With this activity, all children should improve visualization skills and develop a better concept of number.

3. Tactile-kinesthetic children often are not given the opportunity to use some of their best abilities. This activity gives tactile-kinesthetic children a chance to shine and helps other children develop tactile skills.

     Children should work in pairs. Let the children put a pile of Cuisenaire rods on the table. One child should close his or her eyes and the other child should select a Cuisenaire rod and place it in the hand of the first child. The first child should put that hand behind his or her back and open his or her eyes. The child with the rod should try to guess the color or length of the rod. Looking at the rods on the table, the child tries to match the rod felt with the rods seen.

     Let the partners change roles and try the activity again. Children are often better at this activity than adults after only a few tries.

Older children can also develop estimation skills with numbers. For these children, the numbers should include thousands and even millions.

ACTIVITIES

Intermediate (4–6)

Objective: *to develop concrete ideas of large numbers.*

1. Send the children on a scavenger hunt for a million of something. They may decide to look for grains of sand in the sandbox or blades of grass in the yard. This activity will cause them to search for methods other than counting. They may measure 5 milliliters (ml) of sand and count the grains in that amount and then estimate how many milliliters it takes to make one million grains, or they may count the blades of grass in 4 square centimeters (cm^2) and estimate the number of square centimeters it takes to make a million blades of grass.

2. Children may try to collect one million of something, such as bottle caps or twist ties. They must devise a method of keeping track of their collection. They may put ten caps or ties in a small bag and ten small bags in a larger bag. Ten bags of one hundred may be placed in an even larger bag, and so on. After collecting for awhile, the children may try to predict how long it will take to get one million and to estimate how much space it will take up.

3. Children can do research to discover the current national debt. They then may determine how high a stack of $1 bills the debt would make or how long a row of $100 bills the debt would make if the bills were laid end to end. Children will probably wish to use a calculator to aid in computations but will discover that the numbers may be larger than what the calculator can display. Then they will need to discuss what to do.

     The children's book *If You Made a Million* by Schwartz (1989) contains some excellent ideas for teaching the concepts of large numbers, money, banking, and interest. It could be used as an introduction to these topics. The book *Innumeracy* by Paulos (1988) also has several suggestions for estimation

topics that would help students (and adults) better understand today's world.

4. Let the children predict how many seconds they have been alive and then use a calculator to check their guesses.

5. Check with a local manufacturer or a restaurant. Ask how many gadgets they produced or hamburgers they sold in the last week. How many would this be per hour? per month? per year? If the restaurant is part of a chain, can you estimate how many hamburgers were sold by the entire chain in one year? Use your calculator to assist you.

6. Ask the children to predict how long it would take for all the children in the school to read a total of one million pages in trade books. Ask them to devise a method for keeping track of the number of pages read and for involving all the students.

Work with estimation will probably help children realize a need for rounding numbers. Often, exact answers are not required or even sensible. One cannot tell exactly how many grains of sand there are in a bucket or exactly how many stars are in the sky, but approximate answers involving rounded numbers may be useful.

Number lines may be used for rounding numbers less than 100 to the nearest 10. Children can locate a number such as 47 and determine whether it is closer to the 50 or to the 40. For a number such as 45, teach whether it is more sensible to round up or round down. Base 10 blocks are also useful for rounding numbers. Children may show a number such as 287 with the blocks and then determine whether the amount is closer to 280 or 290 if they are rounding to the nearest 10 or closer to 200 or 300 if they are rounding to the nearest 100. Notice that rules for rounding numbers often do not apply to real life. If 63 students and 3 teachers are going on a field trip in buses with a maximum capacity of 30, finding you need 2 buses when rounded to the nearest whole number will leave 6 people stranded. Use several real-life situations and discuss when it is reasonable to round up and when it makes more sense to round down.

Problem Creating and Solving

As children develop concepts and skills, they should use their problem-solving abilities to construct knowledge for themselves. Many of the activities presented earlier were presented in a problem format. Following are activities to encourage children to seek and define problems as well as to solve them.

ACTIVITIES

Elementary (K–6)

Objective: *to reinforce the concepts of greater than and less than and to encourage the development of strategies.*

1. The game guess my number can be played in several ways. One variation of the game found on many computer programs and on calculators such as Speak and Math can also be played

by two children without a calculator or computer. It involves one player who writes down a secret number from 1 to 100. The second player tries to guess the number with the fewest possible guesses. After each guess, the first player tells the second whether the guess was too high or too low.

- *Understanding the problem.* I need to know that I am trying to guess a number my partner has chosen. This number may be as small as 1 or as large as 100. I will be told if my guesses are too large or too small.
- *Devising a plan.* The number could be even or odd, prime or not prime, but guessing a particular number will not give this information. What I will do, then, is to guess the number 50 and eliminate half of the numbers (eliminate possibilities).
- *Carrying out the plan.* I guess the number 50 and find that I am too high. Therefore, I know the number is from 1 to 49. I have eliminated fifty-one numbers. Next, I will guess 25 and see if I can eliminate more numbers. After six guesses I discover that my partner's number was 17.
- *Looking back.* I was able to find the number quickly. I did not waste any guesses. I can use this strategy again. I can generalize this process to find that it should never take more than seven guesses to discover a whole number from 1 to 100.

Children can develop several strategies as guess my number progresses. They learn that wrong guesses can be valuable. The knowledge that guessing and wrong answers can be quite useful in mathematics is important to a child's willingness to tackle new problems.

One major skill in problem solving is finding a pattern. Working with numbers provides a good opportunity to practice finding patterns. Provide patterns such as 2, 4, 6, 8, −, − or 97, 94, 91, 88, −, −, −. After children understand the process, let them make up patterns for each other and place them in a learning center or copy them for everyone to work.

After children have worked with patterns of their own, they may be ready for such patterns as Pascal's triangle, Fibonacci numbers, and sequences involving finite differences. For ideas on presenting these topics, see the *Arithmetic Teacher* or the references for this chapter. Encourage children to do further research on their own both with concrete materials and in books.

Study the following:

Pascal's Triangle

```
        1
      1   1
    1   2   1
  1   3   3   1
1   4   6   4   1
1  5  10  10  5  1
1  6  15  20  15  6  1
1  7  21  35  35  21  7  1
```

Make a list of the patterns the children find in the table. The patterns may include the following:

- Each number is the sum of the two numbers diagonally above it.
- The sum of the numbers in each row is a power of two. The first row is just 1, 2^0; the sum of $1+1$ in the second row is 2, 2^1; the sum of $1+2+1$ in the third row is 4, 2^2; etc.
- Diagonals that intersect in the middle of the triangle are identical, and each of these diagonals has an interesting pattern. The diagonal on the outside is all ones. In the next diagonal, the numbers have a difference of one. The difference between numbers in a diagonal is equal to the closest number in the next diagonal to the outside.
- Each row is symmetrical.

There are many other patterns as well.

Many interesting activities that relate to the triangle can be found or created. The following are just a few. Children may create more of their own.

- Flip a penny five times. How many ways can you get five heads? Only one way, right? How many ways can you get one head and four tails? Five ways, right? List them. How many ways can you get two heads and three tails? Three heads and two tails? Four heads and one tail? Five tails? Do you see a pattern? Is it related to Pascal's triangle? Test your hypothesis to see if this pattern holds for flipping coins other numbers of times.
- How many ways can you select three children for the safety patrol from five volunteers? How is this related to Pascal's triangle?
- Study the map in Figure 6–30. Danny lives at Main and 1st Street. He has a crush on Maureen, who lives at Grand and 6th Street. Maureen says Danny may visit each afternoon as long as he takes a different path each time. He may travel only north and east and must stay on the labeled streets. For how many afternoons may Danny visit Maureen?

Try to discover how Pascal's triangle is related to each of these activities. Encourage children to make up other patterns by asking themselves such questions as "What would happen if . . . ?" or "What if this were not true?" or "How else may I look at this?"

Study the following pattern:

Fibonacci Numbers
1,1,2,3,5,8,13,21,34,55, ...

How was this pattern formed?
As Fibonacci studied natural objects such as spirals on a pine cone and petals on a flower, he discovered that the number of parts was often a number in this sequence. Let the children go on a scavenger hunt to find natural occurrences of Fibonacci numbers.

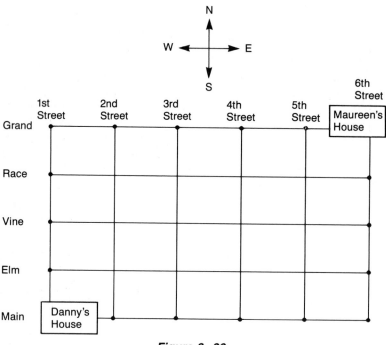

Figure 6–30

The following activity uses Fibonacci numbers. Try to discover how this pattern may help you solve the problem.

- Antonio works at a soda fountain. He mixes milk shakes after school. He makes either regular shakes or extra large shakes. An extra large shake is twice the size of a regular shake. How many different ways can Antonio fill shake orders to use the equivalent of ten regular shakes? For example, Antonio might first make two regular shakes and then four extra large shakes or he could first make four extra large shakes and then two regular shakes; these two combinations are considered different. List all the possibilities. Do you see a relationship to Fibonacci numbers? Try your hypothesis out on different shake orders.

Objective: to reinforce place value concepts and encourage logical thinking.

2. The game Pica-Fermi is an old one, but more recent versions of it may be seen in Mastermind and in computer and calculator games. This game is similar to guess my number in that it has two players and one writes down a secret number that the other tries to guess with the least possible number of guesses. This time, the first player may write down a two-, three- or four-digit number (the number of digits should be agreed upon ahead of time), which the second player tries to guess by naming a number with the number of digits. The first player then tells the guesser the number of correct digits in the wrong place and the number of correct digits in the correct place by saying *pica* for the digits correct and in the wrong place and *fermi* for the digits correct and in the correct place. A sample game follows.

The secret number is 482.

The player guesses	Pica	Fermi
123	1	0
456	0	1
789	0	1
147	1	0
519	0	0
736	0	0
482	0	3

The score is 7 for this round, the number of guesses used to find 482. Try the game yourself with a partner. Discuss the strategies you used.

Objective: to reinforce concepts of consecutive numbers and to encourage problem-solving strategies.

3. Number shuffle is played by placing the digits 1–8 in a diagram such as Figure 6–31. No two consecutive digits may be touching horizontally, vertically, or diagonally.

Encourage students to ask themselves questions such as "Are any positions basically the same because of symmetry?" and "Are any numbers special in terms of consecutive digits?" The last question should prompt the students to place the 1 and the 8 in the two center squares, which is a key to solving this problem.

Objective: to reinforce place value, give practice with a calculator, and encourage logical thinking.

4. Challenge the students to make their phone number show on the calculator using only the 1, 0, +, and =. No phone number should require pressing the + more than eight times. Many students begin by pressing 1 + 1 + 1 + 1 + 1 . . . , but should soon realize this will be futile. Others will try 1 000 000 + 1 000 000 + 100 000 . . . , but should quickly realize that while this may result in the phone number, the + was pressed more than eight times. Some hints may

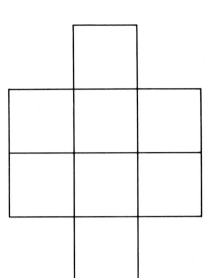

Figure 6–31

be needed before the child realizes that he or she may use 1 111 111 + 1 011 111 + ...

Objective: to develop a combination of spatial and numerical problem-solving abilities.

5. Estelle Dickens and Jeffrey Sellon (1981) have developed a set of activities called Cuisenaire Roddles, which include games and puzzles for use by one or more children. The children place the Cuisenaire rods on game boards to solve brain teasers or play strategy games that involve such skills as finding number patterns and completing magic triangles. Children enjoy solving these puzzles and creating puzzles of their own for each other.

Objective: to encourage historical research, to introduce non-place value systems of numeration, and to aid in the development of new systems of numeration.

6. Introduce systems of numeration that do not use a base 10 system. Roman numerals may be evaluated in terms of using both an additive and a subtractive system and compared to the Egyptian system, which was strictly additive. The Chinese system, which uses both multiplication and addition, may then be introduced. These systems may be compared to the Babylonian and Mayan systems, which were both place value systems although neither used base 10.

   Children may research how our Hindu-Arabic numerals were developed. Encourage children to make up systems of their own and share them with other students. Children may look at the use of binary numbers in computers and perhaps create a tertiary number system, which might be more efficient.

Grouping Students for Learning the Meaning of Numbers

Many of the activities in this chapter can be done by two children working as partners, while others may be done individually or in larger groups. Because meaning of numbers must be constructed by the children themselves,

they must be given time to work with materials and to think about questions that the teacher or other students have asked. It is important to give the students time to argue with each other about their work so that the group time is as important as the individual time to work on problems and to discuss meanings with the teacher.

The games described in this chapter may be led by the children as well as the teacher. They give the children the opportunity to discuss meanings and understanding without trying to match an answer in a book. Be sure to give children plenty of time to work together on activities to develop grouping concepts. These concepts are crucial to the understanding of place value, and they will take a great deal of time to develop.

Communicating Understanding of the Meaning of Numbers

As students work with a partner and in groups, the discussion that goes on is essential to the development of the concept of number and the idea of place value. Kamii (1990, p. 27) states, "A characteristic of logico-mathematical knowledge is that there is absolutely nothing arbitrary in it. Two plus two makes four in all cultures. It follows that in the logico-mathematical realm, children *will* arrive at the truth if they debate long enough. They need to talk about their number experiences in everyday life, during games, and as they work with the development of place value concepts. The social interaction that goes on is an important part of the construction of the number concepts."

It is important that the teacher focus on the children's understanding and not try to impose an arbitrary notion of number on the children. The only way the teacher will know what the children understand about numbers is to ask them. The teacher must take the time to listen to the children and to ask them questions that cause them to reach disequilibrium and construct their own concepts of number. In this way, the oral communication becomes an integral and necessary part of every lesson.

Older children can communicate their reasoning in written form. They can keep logs about what they have learned and write out their methods of solving different types of problems. They can write down the new concepts they have learned and what areas may still be confusing.

Evaluating Number Learning

As in any evaluation, begin evaluating number learning by looking at goals and objectives. Begin by looking at the goals and objectives from the NCTM *Standards*, those set by your state or school district, and those of the textbook series being used. Look at standardized tests that the students are required to take. These are often incomplete, however. Many of the conceptual and attitudinal goals and objectives of the teacher and learner are not tested by current standardized tests. Many of these tests are changing, however, to include more conceptual and performance-based items.

Take advantage of the tests provided in the children's textbook for evaluating achievement of the objectives of that series. Texts often have pre- and post-tests for the students in the teacher's manual or related materials and practice tests for the students in their own books.

Your observations of students should provide much additional information. Look for what the child does well and for what the child may be having difficulty with. It probably is more important to look at the child's processes than it is to look at the child's products. A teacher can learn much more about the child by watching him or her work than by looking at the answers to a test. To aid in keeping track of children's individual work, the teacher should jot down anecdotal records each day. The teacher will probably not be able to observe each child each day but should make a point of seeing each child at least once a week. Ask the children to keep a portfolio of some of their best work, also. They may keep a written log, noting what they have learned and what is still not understood.

When studying number concepts, look for the following as the children manipulate concrete materials:

1. *Conservation.* With young children, be sure to determine whether the child can conserve both number and length.
2. *Counting.* Again with young children, look for common counting errors. Can the child count rationally? That is, does the child say one number for each object counted or are numbers simply recited in order with no regard for the material being counted? Does the child realize that the sequence of the counting numbers does not change? Does the child understand that objects may be counted in any order? Does the child realize that the number named last when counting refers to the total number of objects in the set and not just to the last object? Can the child start counting with any number or only with 1? Can the child skip count by 2s, 5s, 10s, etc.? Can the child count past the "hard numbers," 29 to 30, 99 to 100, 999 to 1000, etc.? Can the child count backwards from any number?
3. *Number concepts.* Does the child associate a number with a set of objects or is the number only a word to recite? Does a child recognize amounts up to at least five by sight?
4. *Grouping concepts.* Does the child understand trading units for groups and groups for super-groups, and so forth? Can the child record the results of trading?
5. *Place value concepts.* Does the child understand the meaning of numerals written in standard Hindu-Arabic notation? Can the child model numerals with place value materials such as multibase blocks and chips for trading? Check this understanding by asking the child to count out 25 chips or other small objects. Circle the 5 in 25 and ask the child to show you that many chips. Then circle the 2 in 25 and ask the child to show you how many chips the 2 represents. If the child shows you 2 chips rather than 20, you will know that place value is not clearly understood.
6. *Estimation and approximation concepts.* Can the child use numbers to make reasonable estimates or approximate results?
7. *Problem creating and problem solving.* Can the child use numbers and strategies to create and solve problems?
8. *Uses in everyday life.* Is the child aware of numbers in the world around him or her and can the child use numbers rationally outside of school?

Something for Everyone

Concrete materials described in this chapter, such as Cuisenaire rods, multibase blocks, chips for trading, bean cups, the abacus, and bundling sticks, are excellent aids for tactile learners, while visual learners may wish to combine these materials with drawings. Auditory learners will wish to discuss what they are doing as they manipulate the objects. Music is an excellent aid for auditory learners, and they will enjoy counting songs and poems.

Students who learn well sequentially and in discrete units often like recipes and rules and tend not to estimate. They see the parts better than the whole. These students can count forward well and prefer discrete materials such as chips to spatial materials such as Cuisenaire rods and multibase blocks when learning number concepts.

Students who tend to be more spatial and holistic see the whole rather than the parts. They estimate well and see the answer without knowing how they got it. They prefer Cuisenaire rods and multibase blocks to chips or other discrete objects for learning number concepts. They often can count backwards better than forwards.

Gifted children should be given many opportunities for both problem solving and problem creating. Topics such as Pascal's triangle, Fibonacci numbers, and other numeration systems, such as the Egyptian, Roman, or Babylonian system, are a good starting point. Gifted students can find a wealth of information and interesting problems in these areas with some research in a good library, in the Ideas section of the *Arithmetic Teacher,* or in the student section of the newsletter of the National Council of Teachers of Mathematics (see Appendix A for the address). Encourage gifted students to make up problems of their own after solving some of these.

Children having difficulty with number concepts should not be rushed to work with numbers on an abstract level.

Whether the difficulty is with beginning number concepts, perhaps due to the inability to conserve number or to counting misunderstandings, or with later ideas such as place value, the children need to work with concrete materials. After the children feel comfortable with the materials and seem to have grasped the concepts, be sure to have the children write the numerals with the materials still in front of them. For many children, it is very difficult to make the transition from the concrete materials to the abstract numerals. Do not expect them to be able to remember what you did with the materials yesterday if you ask them to work solely with numerals today. The two must be used together.

KEY IDEAS

Work with numbers constitutes a major portion of all elementary and middle school math programs. In this chapter, suggestions for making the most of this work have been presented. It is important that number concepts be understood and not just rotely recited. This includes understanding the cardinal, ordinal, and nominal uses of numbers. Children should be able to recognize the number of objects in a small set by sight and should be able to use rational counting to solve problems involving numbers.

Concrete materials such as counters, Cuisenaire rods, multibase blocks, chips for trading, abacuses, wooden cubes, and counting sticks should be used extensively. Concrete work should be reinforced with semi-concrete and abstract work using number lines, charts, computers, calculators, and practice activities such as card games and board games devised by the teacher and/or the students themselves.

Because place value is such an important part of our numeration system and an understanding of it is crucial to later work with algorithms, children in kindergarten and first grade should work with grouping and trading such small amounts as those with a grouping point of three, four, or five. This practice should extend to base 10 by second grade, using both proportional and nonproportional grouping materials. Estimation skills should be practiced frequently. Children should be challenged to create problems involving interesting number patterns as well as to solve them.

Diagnosis of number concepts should be done individually with a focus on developmental levels and common counting errors in the early grades and misconceptions about place value as the children get older. Provisions should be made for children with different abilities and learning styles.

REFERENCES

BAROODY, ARTHUR J. "Basic Counting Principles Used by Mentally Retarded Children." *Journal for Research in Mathematics Education.* Vol. 17, No. 5 (November 1986), pp. 382–389.

DAVIDSON, PATRICIA I.; GALTON, GRACE K.; AND FAIR, ARLENE W. *Chip Trading Activities.* Arvada, Col.: Scott Scientific, 1972.

DICKENS, ESTELLE, AND SELLON, JEFFREY. *Cuisenaire Roddles.* New Rochelle, N.Y.: Cuisenaire Company of America, 1981.

DOMAN, GLENN. *Teach Your Baby Math.* Philadelphia: The Better Baby Press, 1980.

DOSSEY, JOHN A.; MULLIS, INA V. S.; LINDQUIST, MARY M.; AND CHAMBERS, DONALD L. *The Mathematics Report Card: Are We Measuring Up? Trends and Achievement Based on the 1986 National Assessment.* Princeton, N.J.: Educational Testing Service, 1988.

DRISCOLL, MARK J. "Counting Strategies." In *Research Within Reach: Elementary School Mathematics.* St. Louis: CEMREL, 1980.

FUSON, KAREN C., AND HALL, J. W. "The Acquisition of Early Number Word Meanings: A Conceptual Analysis and Review." In H.P. Ginsburg, ed. *The Development of Mathematical Thinking.* New York: Academic Press, 1983, pp. 49–107.

GINSBURG, HERBERT P. *Children's Arithmetic: The Learning Process.* New York: D. Van Nostrand, 1977.

_____. "Children's Surprising Knowledge of Arithmetic." *Arithmetic Teacher.* Vol. 28, No. 1 (September 1980), pp. 42–44.

HIEBERT, JAMES. "Children's Thinking." In Shumway, Richard J., ed. *Research in Mathematics Education.* Reston, Va.: National Council of Teachers of Mathematics, 1980.

HOLMBERG, VERDA; LAYCOCK, MARY; AND STERNBERG, BETTY. *Metric Multibase Mathematics.* Hayward, Ca.: Activity Resources Co., 1974.

KAMII, CONSTANCE. "Constructivism and Beginning Arithmetic (K–2)." In Cooney, Thomas J., and Hirsch, Christian. *Teaching and Learning in the 1990s.* Reston, Va.: National Council of Teachers of Mathematics, 1990.

LEUTZINGER, LARRY P.; RATHMELL, EDWARD C.; AND URBALSCH, TONYA D. "Developing Estimation Skills in the Primary Grades." *Estimation and Mental Computation.* National Council of Teachers of Mathematics, 1986 Yearbook. Reston, Va.: NCTM, 1986.

NATIONAL COUNCIL OF TEACHERS OF MATHEMATICS. *Curriculum and Evaluation Standards for School Mathematics.* Reston, Va.: NCTM, 1989.

NATIONAL RESEARCH COUNCIL. *Everybody Counts: A Report to the Nation on the Future of Mathematics Education.* Washington, D.C.: National Academy Press, 1989.

PAULOS, JOHN ALLEN. *Innumeracy: Mathematical Illiteracy and Its Consequences.* New York: Hill and Wang, 1988.

PIAGET, JEAN. *The Child's Concept of Number.* New York: W.W. Norton & Co., 1965.

REYS, ROBERT E., ET AL. *Keystrokes: Calculator Activities for Young Students: Counting and Place Value.* Palo Alto, Ca.: Creative Publications, 1980.

SCHWARTZ, DAVID M. *If You Made a Million.* New York: Lothrop, Lee, and Shepard Books, 1989.

SKEMP, RICHARD R. *The Psychology of Learning Mathematics.* Hillsdale, N.J.: Lawrence Erlbaum Assoc., 1987.

STOKES, WILLIAM T. *Notable Numbers.* Los Altos, Ca.: Covington Middle School, Los Altos School District, 1972.

TAVERNER, NIXIE. *Unifix Structural Material.* North Way, England: Philograph Publications Limited, 1977.

WAHL, JOHN, AND WAHL, STACEY. *I Can Count the Petals of a Flower.* Reston, Va.: National Council of Teachers of Mathematics, 1976.

WILCUTT, ROBERT; GREENES, CAROLE; AND SPIKELL, MARK. *Base Ten Activities.* Palo Alto, Ca.: Creative Publications, 1975.

7 Teaching Addition and Subtraction

NCTM *Standards*

GRADES K–4

Standard 7: Concepts of Whole Number Operations

In grades K–4, the mathematics curriculum should include concepts of addition, subtraction, multiplication, and division of whole numbers so that students can—

- develop meaning for the operations by modeling and discussing a rich variety of problem situations;
- relate the mathematical language and symbolism of operations to problem situations and informal language;
- recognize that a wide variety of problem structures can be represented by a single operation;
- develop operation sense.

Standard 8: Whole Number Computation

In grades K–4, the mathematics curriculum should develop whole number computation so that students can—

- model, explain, and develop reasonable proficiency with basic facts and algorithms;
- use a variety of mental computation and estimation techniques;
- use calculators in appropriate computational situations;
- select and use computation techniques appropriate to specific problems and determine whether the results are reasonable.

GRADES 5–8

Standard 7: Computation and Estimation

In grades 5–8, the mathematics curriculum should develop the concepts underlying computation and estimation in various contexts so that students can—

- compute with whole numbers, fractions, decimals, integers, and rational numbers;
- develop, analyze, and explain procedures for computation and techniques for estimation;
- develop, analyze, and explain methods for solving proportions;
- select and use an appropriate method for computing from among mental arithmetic, paper-and-pencil, calculator, and computer methods;
- use computation, estimation, and proportions to solve problems;
- use estimation to check the reasonableness of results.

(NCTM, 1989, pp. 41, 44, 94. Reprinted by permission.)

Addition and subtraction are widely used in our daily lives. Much of the justification for keeping a strong emphasis on computation in the school curriculum comes from its perceived usefulness. At the market, our purchases are added together. We add and subtract to determine if we have enough money and subtract to determine our change.

Children begin purchasing items early in their lives. Learning to go to the store is a common part of growing up. Role playing various parts in a classroom store is a valuable learning activity.

Computation, on the other hand, is seldom performed using pencil and paper any more. Cash registers scan universal price codes and use electronic voices to inform the customer what is being purchased, its cost, and the change due. Gasoline may be purchased by inserting a plastic card in a small reader at the station, activating a particular pump so the customer can fill the automobile tank; the amount of gasoline and its cost are registered in a computer file established for that customer, who will be billed at the end of the month. Even small businesses rely on calculators and small computers when determining purchase amounts.

It may be argued that such conveniences undermine the personal interaction of daily life. Still, the efficiency and accuracy of electronic computation is a persuasive argument for its continuing use.

Children growing up in a time of electronic computation will need to be skillful in estimation and mental arithmetic. They will need to know when answers are reasonable and when they are not. They will need to be able to double check purchase lists to assure accuracy and to know what operation is required to determine purchase amounts. They must learn how to comparison shop and how to manipulate numbers to their advantage in the marketplace. Students should be encouraged to make an important decision associated with calculation: whether to use paper and pencil, mental calculation, estimation, or a calculator or computer. This decision is made when a particular calculation is needed. Students skilled in using a variety of computational techniques have at their command the power and efficiency of mathematics. The instructional program should provide children many opportunities to make decisions regarding which computational technique to employ. This chapter is written to suggest ways in which you may help students think about numbers and create in students the habit of using their heads.

Children begin learning addition and subtraction long before they are introduced to the **basic addition facts,** the whole-number sums from $0 + 0$ to $9 + 9$. First, children develop understanding of everyday relationships. They gain some meaning of whole numbers before they attend school. Through their explorations, youngsters invent informal systems of mathematics. Children use counting procedures to find how many things are in a collection of objects and counting strategies to solve simple word problems.

Your job as a teacher is to determine how children think about numbers. Pose simple word problems and observe how the children solve them. For example, "Sarah has four tiles; David gives her two more. Now, how many does she have?" The responses can serve as a basis for beginning number work. Perhaps the child **counts all,** that is, counts one, two, three, four, using fingers to keep track, and then five, six, continuing to count fingers, and gives the answer six by reporting how many fingers were counted. Perhaps the child **counts on,** that is, says four, and then five, six, counting two more fingers. Perhaps the child uses tiles or beans to count out the answer. Perhaps all counting is done in the child's head with lip movement as the only sign of counting. Perhaps the child visualizes a set of four objects joined with a set of two objects and responds without counting. You are responsible for getting children ready in the areas in which a teacher has control. For children having difficulty, you can provide practice in counting and grouping. You can use activities to introduce the concepts of addition and subtraction.

This work with oral word problems is very important and should precede any written work with number sentences. Students should have extensive work in posing and answering each other's story situations orally and with materials before they ever see an abstract addition or subtraction number sentence. They should use many manipulatives; attribute blocks, multibase blocks, Cuisenaire rods, beans, and cubes are among the most useful. Other counting objects, like bottle caps, nuts and bolts, buttons, and washers, are ideal for illustrating operations. Calculators and computers help in learning basic arithmetic operations as well. Children beginning to understand the basic operations are active learners.

Developing Addition and Subtraction Concepts

The general concept of an operation serves as a foundation for understanding the concepts of addition and subtraction. The term **operation** refers to a process that involves a change or transformation. The process begins with an object in a particular state of affairs; then an operation occurs that causes a change in the object, resulting in a final state of affairs. Children have many intuitive experiences with operations or transformations. For example, a child is sleeping, a parent calls to awaken the child, and soon the child is awake. The initial state of affairs is the child asleep; the final state of affairs is the child awake. What happened in between, being awakened, was the operation or transformation.

This transformation does not appear to be linked with addition, but it is in an important way. To illustrate, a simple machine as a model for the addition process is shown in Figure 7–1. It is a two-dimensional model with

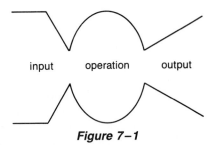

Figure 7–1

an input and an output. The operation takes place in the center of the machine. What enters at the input is an initial state of affairs, and the output resulting from the operation is the final state of affairs.

When this state-operator-state machine is used to teach addition, it may appear as in Figure 7–2. In Figure 7–2a, we have a machine similar to those introduced in Chapter 3. It is a single-input machine, sometimes called a function of one variable. It represents a unary operation. The join □ □ or +2 is the operator within the machine. The output is □ + 2. We cannot know the output until we know the input. Therefore, □ stands for a variable.

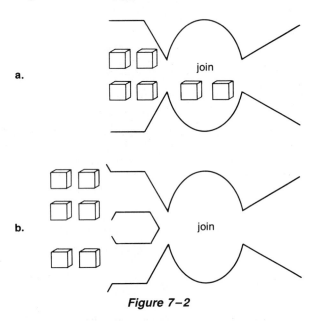

Figure 7–2

Figure 7–2b shows a double-input machine, sometimes called a function of two variables. It represents a binary operation. Mathematically, we would write □ + □. The two inputs are □ and □, and the operator is +. In our example, four cubes are in the first input and two cubes are in the second input. The operation is to join the cubes. For either machine, the resulting output is six cubes. Perform the operation of joining objects physically by moving the cubes through the machine, joining them, and moving them to the output. Our introductory work with operations uses the unary operation, but as we look at a model for addition, we employ the binary operation. The binary operation machine more closely resembles addition as it is presented in school.

Eventually, we move away from the concrete experiences with sets of objects toward the pictorial and abstract. Inputs can be numbers, and operators can be symbols. Children learn that addition renames a pair of numbers with a single equivalent number. For example, the pair (4, 2) is associated with 6 under the operation of addition. Learning how addition works is learning the concept of addition. Learning that 4 + 2 = 6 is learning a basic addition fact.

Each of the following sections, on operations, addition, and subtraction, presents activities for developing concepts. We use the machine model initially but also include other useful models.

Operations

To separate the concept of an operation from the concept of a numerical operation, activities to develop the concept of an operation should use materials that are not models for numbers. For these activities, we use the attribute blocks shown in Figure 7–3.

ACTIVITIES

Primary (K–3)

Objective: *to perform operations on objects.*

1. Begin with an introduction to the state-operator-state machine. On the floor, mark a large outline of a machine like that in Figure 7–1, using yarn or masking tape. Do not include the labels. Explain to the children that this special machine usually causes a change to occur. The machine has a starting place called an input and an ending place called an output. The change takes place in the middle.

    Invite a child to help demonstrate the machine. Have her stand in the input. Then ask her to step into the center and to raise one hand over her head. Tell her to step to the output region with her hand still over her head. While she is still in the output, question the class, "How did Marie start in the input? How did Marie end up? What change took place?"

    Next, put a sign in the center of the machine that says "raise your hand," as in Figure 7–4. Invite several more children to go through the machine and to do what the operator says when they get to the center. Encourage the children to discuss what happens as they pass through the center of the machine.

    To extend this activity, vary the operator. Include face the other way, hold your hands behind your back, start clapping your hands, and hug yourself. At some point, introduce do nothing to show that an operation can result in no difference from the input to the output. This is the first example of what will later be introduced as adding zero to a number.

2. Use the same type of machine that was used in the previous activity. This time, have a child hold an attribute block. The operator should be change color, change shape, or change size. Suppose Rich starts at the input with a large red triangle, and the operator is change color. This means to change only the color and no other attribute. In the center of the machine, let Rich decide how he can perform the operation. He may exchange the large red triangle for a large blue tri-

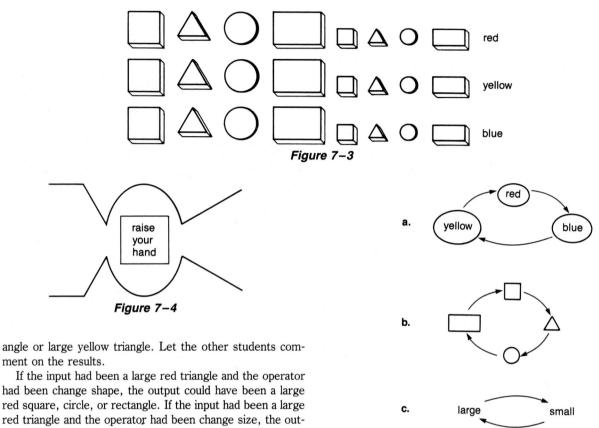

Figure 7–3

Figure 7–4

Figure 7–5

angle or large yellow triangle. Let the other students comment on the results.

If the input had been a large red triangle and the operator had been change shape, the output could have been a large red square, circle, or rectangle. If the input had been a large red triangle and the operator had been change size, the output could have been only a small red triangle.

3. The game Simon says, serves as an example of an operation. The leader may be a student or the teacher. The leader gives commands to the entire class, beginning some with "Simon says." Later the leader can give commands without saying "Simon says."

The children should follow only commands that are begun with "Simon says"; otherwise, they must sit out the remainder of the game. For example, with the children standing, hands to their sides, the leader commands, "Simon says, 'Touch your nose.' " The children should then reach up and touch their noses and remain in that position until another command is given. If the next command is "Put your hand down," no one should respond because the command did not begin with "Simon says." Discuss with the children how this game is like working with the state-operator-state machine. Let them decide when the operation is performed.

The children's position before a command corresponds to the initial state, or input. When a command beginning with "Simon says" is given and the children respond, an operation is performed. The new position of the children is the final state, or output. Giving a command without saying "Simon says" has the same effect as the operator doing nothing.

Objective: *to perform an operation on objects to produce a single, unique output.*

4. Now draw the machine on the chalkboard, place it on a flannel board, or sketch it on the inside of file folders. The operation for the state-operator-state machine appears in Figure 7–5a. Have the children explore the operation. For example, the operator in Figure 7–5a means that a red block will change to blue, a blue block will change to yellow, and a yellow block will change to red. Again, only the attribute mentioned will

change. Thus, if the input is a small yellow triangle, the output will be a small red triangle. No other block in the set shown in Figure 7–3 will satisfy the given operator.

Other operations for these attribute blocks are shown in Figures 7–5b and c. The operation in 7–5b would cause a small yellow triangle to change to a small yellow circle. The operation in 7–5c would cause a small yellow triangle to change to a large yellow triangle.

Extend this activity by putting two or three operators in the center of the machine as in Figure 7–6. When an object enters the machine, both operations must be performed. This set of operators would cause a small yellow circle to change to a large red circle. You can adjust the complexity of the operators to challenge the children. The common thread throughout these activities is that an operation is taking place, transforming an object from the initial state to the final state.

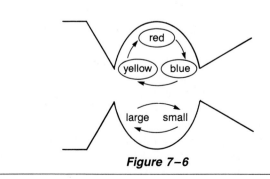

Figure 7–6

Addition

Sorting, classifying, and reversibility of thought are necessary for children to understand both addition and subtraction. Sorting and classifying were carefully discussed in Chapters 5 and 6. **Reversibility of thought** occurs when children are able to reverse their thinking process. An example from the Piaget task to check class inclusion is provided here: Suppose you have a set of ten colored cubes, seven green and three yellow. When asked if there are more cubes or more green cubes, children who have not achieved reversibility of thought respond that there are more green cubes. These children are comparing the green cubes with the yellow cubes. When their perception shifts from the whole set (cubes) to the subset (green cubes), they forget the whole set, the set of cubes, and cannot reverse their thought back to it.

The ability to recognize subsets of objects as included in a larger set occurs at about the age of seven. As a result, we sometimes try to teach the concepts of addition and subtraction to children before children can fully comprehend them. Take care to provide a variety of sorting and grouping experiences when the concept of addition is introduced.

A useful early addition experience is to have children respond to number stories. For example, give Missy three oranges and give Loren two. Start, "Loren has two oranges and Missy gives him three more. How many oranges does Loren have?" Prompt Missy to give Loren her three oranges. Some children will know the answer without counting oranges, while others will need to count. Let the children explain how they found the answer.

Along the same lines but subtly different is this story: "Jacob has three transformers and Margo has four transformers. How many do they have together?" Again, let students explain how they determined the answer.

The difference between the two number stories is that the first story suggests an operation and the second merely describes a state of affairs. While both stories are useful, the first one provides a somewhat stronger foundation for understanding joining, or combining, the basis of addition.

With encouragement, children can tell number stories of their own, using real or imaginary situations. Telling or writing these stories provides language experience. The final result may be a bulletin board or a class number-stories book.

Colored cubes, beans, bottle caps, multibase blocks, Cuisenaire rods, and other counting objects can serve as bases for developing the concept of addition. The operator initially presented is join or combine (an operator for sets of objects), to prepare children for addition (an operation on number). We use join or combine because we believe these words are easier for children to understand. State-operator-state machines may be used to present the concept of addition, but other models also should be used. Of particular importance is to have children discuss their thinking and the procedures they use to reach solutions. The following activities include various ways to present the addition concept.

ACTIVITIES

Primary (K–3)

Objective: *to join objects as a model for addition.*

1. Provide the children with individual state-operator-state machines drawn on the insides of file folders, on individual chalkboards, or on paper. Place a large model on the floor with yarn or masking tape. The operator for the machine is join. Let there be two inputs.

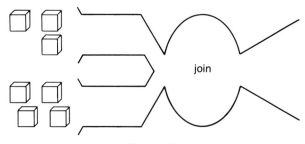

Figure 7–7

Figure 7–7 shows three cubes in one input and four cubes in the other input. Take the objects to the center of the machine and join them. Then move them to the output.

As in an earlier activity, have one child pick up several cubes—six, for example—and move to one of the inputs of the large machine. Then have another child pick up several cubes—four, for example—and move to the other input. Next, have both children move into the center of the machine and join their cubes. Finally, have the children move to the output with their larger collection of cubes.

Initially, perform this operation without counting the cubes. When the children have joined cubes successfully several times, have them describe how many cubes are in each input and how many cubes result in the output. For example, "Six cubes are in one input, four cubes are in the other input, and the output has ten cubes." Eventually, the children will write number sentences that describe their actions.

2. Once students have mastered the previous operation, extend the activity by introducing problem solving. Set up a machine like that in Figure 7–8 to challenge thinking. The machine in Figure 7–8 shows objects in one input and objects in the output. Have the children find the number of objects missing

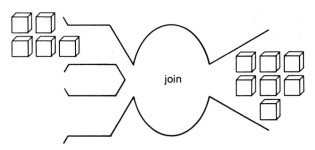

Figure 7–8

from the second input. In this case, a group of five cubes appears in one of the inputs. Ask the children if they know how many cubes should go in the other input so the number of cubes in the output will be correct. If the children find this question difficult, begin by asking, "What if we put one cube in the other input? Would we get the right number of cubes in the output?" One cube will not produce a large enough output. Next, try two cubes, the answer. This is a manipulative version of the abstract problem 5 + ? = 7. The children should be challenged to make up similar problem situations for members of their group or the whole class.

Vary this activity by using objects such as Unifix cubes, multicolored plastic cubes that snap together. Unifix cubes offer children a chance to combine cubes into long sticks or, even better, groups of ten.

Navy beans offer children a chance to work with large collections of objects. The beans can be conveniently grouped into tens as they are put into the input and then combined. Small portion cups serve well for each collection of ten.

Multibase blocks serve as an excellent model for addition. Children who have had earlier exposure to grouping with the multibase blocks can readily combine them and can exchange pieces in the output.

Objective: to use Cuisenaire rods to illustrate addition.

3. Two or more Cuisenaire rods placed end to end form a train. Invite the children to make a train using a light green rod and a yellow rod. Next, have them find another rod that is the same length as the light green and yellow rods combined; they will soon discover the brown rod. Have the children place the brown rod beside the light green-yellow train as in Figure 7–9. Explain that light green plus yellow equals brown.

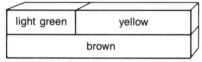

Figure 7–9

Ask the children to find another train that is the same length as the brown rod. Expect a variety of answers. Seven two-rod trains are possible if reversals, such as red plus dark green and dark green plus red, are considered different trains. There are also three-rod trains (for example, red plus red plus purple) as well as four-, five-, six-, and seven-rod trains, and there is even an eight-rod train. When the children have discovered the trains, have them say what the trains are.

At this point, numbers are not directly associated with the Cuisenaire rods. Number values will be assigned to the rods later.

Children soon find two-rod trains that are longer than the longest single rod, orange. Here, ask them to make a train using orange plus whatever rod is necessary to equal the length of the original two rods. For example, Figure 7–10

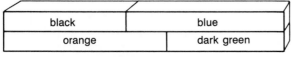

Figure 7–10

shows a two-rod train, black plus blue, which is equal to orange plus dark green. Other trains may be equal to more than two orange rods; for example, purple plus dark green plus yellow plus black equals orange plus orange plus red.

As children become more proficient in adding rods they produce more complex sums. To shortcut writing the names of all the rods, use the symbols used by the developers of Cuisenaire rods: w (white), r (red), g (light green), p (purple), y (yellow), d (dark green), k (black), n (brown), e (blue), ☞ (orange).

The first three activities present different embodiments for the concept of addition. Another embodiment begins the transition toward more abstract work. The transition to using symbols should be presented slowly and with concrete models representing the symbolic expressions.

ACTIVITIES

Primary (K–3)

Objective: to introduce the concept of addition using sets of objects.

1. Draw three loops on a large sheet of paper or place them directly on the floor with yarn or masking tape. Place several objects in the top two loops, as in Figure 7–11a. Indicate to the children that they are to combine or join the objects and put them in the third loop. Be sure you introduce the word *combine* or the word *join*.

Begin the transition from joining objects to adding numbers by asking the children how many objects are in the first loop. In Figure 7–11a, the first loop has three objects. The second loop has four objects. The loop with the objects combined

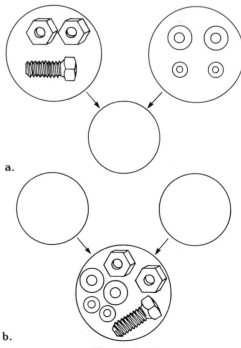

Figure 7–11

(Figure 7–11b) has seven objects. Explain that this diagram shows three plus four equals seven.

Let the children develop other examples and have the children tell what the various diagrams show. The children read the diagrams at this point rather than actually add the numbers.

Soon, you can use the symbols for addition. Introduce the symbols in an addition sentence, such as 3 + 4 = 7. Then introduce them in the vertical form:

$$\begin{array}{r} 3 \\ +4 \\ \hline 7 \end{array}$$

Help the children read these sentences and become familiar with them.

ACTIVITIES

Primary (K–3)

Objective: *to introduce the symbolism of addition.*

1. Review with the children the state-operator-state machine. Using Figure 7–12, ask the children, "How many cubes are there in the first input?" There are two. "How many are there in the second input?" There are three. "How many are there in the output?" There are five in the output. Summarize, "Then this machine shows us that 2 plus 3 equals 5. It is written 2 + 3 = 5."

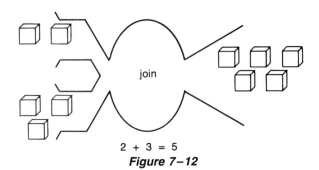

2 + 3 = 5
Figure 7–12

Let the children explore numerous examples of their own until they can easily determine the values illustrated in the machine and easily recognize the symbols of addition.

2. With the Cuisenaire rods, establish a value for the white rod. For example, if the white rod is one, what is the value of the red rod, the light green rod, the purple rod, and so on? Children may determine these rod values by finding how many white rods it takes to make the red rod (2), the light green rod (3), and the purple rod (4). Give children additional experience with the rods to allow them to be at ease in symbolizing the rod values.

Later, particularly in work with fractions, other rod values will be established. Be careful not to declare one the permanent value of white, two the permanent value of red, three the permanent value of green, and so on. If white equals one, the train in Figure 7–13 is 4 plus 2 and is equal to 6, or 4 + 2 = 6.

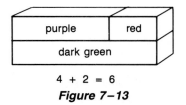

4 + 2 = 6
Figure 7–13

To this point, we have introduced the concept of addition using several embodiments and the symbols that describe addition. As we continue, we introduce an important property of addition, the commutative property, along with the new embodiment for addition.

ACTIVITIES

Primary (K–3)

Objective: *to use dominoes as a model for addition.*

1. Use sets of double-six dominoes for this activity. These may be the commercial sets or teacher constructed sets. A set of double-six dominoes consists of twenty-eight dominoes showing all combinations from blank-blank to six-six. To begin, explain that a domino such as the one in Figure 7–14a is read as "three plus five." If all of the dots are counted, we can complete the number sentence by supplying the sum, eight; thus, 3 + 5 = 8. If the same domino is picked up differently or rotated, it may be read as "five plus three equals eight," and will appear as in Figure 7–14b.

Soon, children will pick up a domino, read it as either "3 + 5 = 8" or "5 + 3 = 8," and will know that no matter which way it is read, the sum is still 8. This important characteristic of addition is the **commutative property of addition**. Knowing that order has no effect on the sum reduces the number of addition facts to be remembered.

When the students can read the dominoes, let them spread the dominoes face down, mix them up, and select them one at a time and explain to other group members or the whole class what they "say." Any domino portion that is blank is read as zero, meaning it has zero dots. The domino blank-four is read as "zero plus four" or "four plus zero," depending on how it is held. As the children become confident reading dominoes, change to a double-nine set.

To vary this activity, change the standard domino dot pattern to a more random pattern. This will require that you construct your own dominoes. Thus, the dot pattern for three may be three dots in a triangular configuration rather

a.

b.

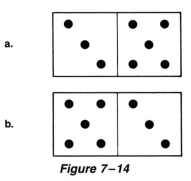

Figure 7–14

than the traditional diagonal line of three dots. Using different configurations helps children to visualize various dot patterns and to attach number values to those different patterns.

The concept of addition has been presented with a model that uses joining or combining sets of objects. While the children were still at the concrete level, number was attached to the operation to assist the children in seeing the relationship between joining objects and adding numbers. We continue by introducing the concept of subtraction in a similar manner.

Subtraction

Teaching the concept of subtraction is much like teaching the concept of addition. The materials, models, and approach are the same. The concept, however, is different. As teachers, we want children to be able to see that subtraction is the inverse of addition. To achieve this, we need to teach subtraction at the same time we teach addition. Subtraction can be fully understood only when children can classify and sort objects by their attributes and achieve reversibility of thought. Children must be able to see the relationship between a collection of objects and a subset of that collection in order to understand subtraction. When a subset is removed from a collection, the idea of subtraction is understood if children realize that returning the subset to the collection restores the collection to its original state. Thus, the student is doing more than merely following the teacher's direction to remove a subset and count the elements in the resulting collection. For the children who cannot yet classify and have not achieved reversibility of thought, you will need to provide additional material, time, and encouragement.

Number stories that introduce children to addition can also introduce children to subtraction. For example, "Cindy has six model horses. Trisha takes two of the horses to play with. Now, how many horses does Cindy have?" This story, an example of take away or remove, may be acted out to find the solution.

We begin another story: "Chris has five books. Ken has two books. How many more books does Chris have than Ken?" Here, the books can be compared and the difference can be determined.

Still another approach is this: "Jack has three cookies and Nicole has seven. How many more cookies does Jack need to have as many as Nicole?" This is an example of a missing addend approach to subtraction.

Include stories with too much or too little information. For example, "Robert has four pencils. Corrie has some pencils. How many more pencils does Corrie have than Robert?" Encourage children to develop their own number stories, and see if other children can solve them. You may place copies of number stories children have written in a learning center for other students to work on. The following activities offer additional ways to teach the concept of subtraction to students.

ACTIVITIES

Primary (K–3)

Objective: *to introduce the concept of subtraction using sets of objects.*

1. Draw a loop on a large sheet of paper or use a loop of yarn. Place nine objects in the loop (Figure 7–15a). Then, with yarn or string of a different color, form another loop inside of the first loop, surrounding some of the objects (Figure 7–15b). Ask the children to remove all objects that are inside the smaller loop (Figure 7–15c). Practice this activity a number of times and discuss the results with the children to see if they understand that they have removed some of the objects they had when they started, and that if they returned the objects, the collection of objects would be the same as the original collection.

   Next, ask the children to indicate how many objects there are at each step of the activity. For example, "How many cubes did we start with?" There were nine cubes. "How many did we take away?" We took away four cubes. "How many do we have left?" We have five cubes left. "How can we say what we just did in a number sentence?" Nine take away four equals five.

   Children can invent and work out many other examples of the subtraction operation as part of learning about subtraction. This "take away" model is the most direct way to show the meaning

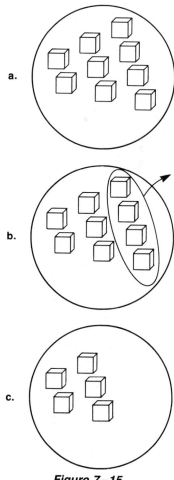

Figure 7–15

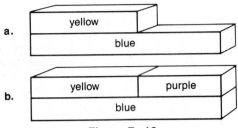

Figure 7-16

of subtraction. It should be used considerably in the beginning stages of learning subtraction.

Objective: to use Cuisenaire rods as a model for subtraction.

2. Have the children place a blue rod in front of them. Then have them put a yellow rod beside the blue rod, with one end even with an end of the blue rod, as in Figure 7-16a.

   Ask the children to find a rod that will form a train with the yellow rod and will make that train as long as the blue rod. You may say, "What rod will go here?" as you point to the space at the end of the yellow rod. Children will find that the purple rod will fill the space (Figure 7-16b). Explain that blue minus yellow equals purple.

   Ask, "What is blue minus green?" Have the children place the rods as described above and find the rod to fill the space.

   The Cuisenaire rods model used here emphasizes the **missing addend concept.** In the abstract version of the missing addends problem $9 + ? = 12$, children ask themselves, "What added to nine makes twelve?" With the rods, we are asking, "What added to yellow makes blue?" when solving the problem blue minus yellow equals what?

Objective: to use dominoes to illustrate the concept of subtraction.

3. When dominoes are used for the subtraction concept, the children are asked to find the difference between the number of dots on one side of the domino and the number of dots on the other side. Help children make a comparison. "How many more dots are there on one side than the other?" In the case of the two-six or six-two domino, shown in Figure 7-17, there are four more dots on the six side.

   Let the children describe how they are able to find a solution. For example, some children make the comparison by matching the dots on one side of the domino with those on the other side. The dots left over represent the difference. Other children may be more comfortable mentally removing the dots found on the lesser side (two) from the dots found on the

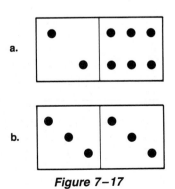

Figure 7-17

greater side (six). Of course, when each side has the same number of dots, as in Figure 7-17b, the difference is zero. How do the children explain this difference?

Children can describe this operation numerically after they learn to make the comparison and find the difference. The number sentence that describes the domino in Figure 7-17a is "six minus two equals four." Students should be able to pick up any domino and quickly give the number sentence that is shown.

Objective: to symbolize the subtraction concept.

4. Once the language describing subtraction has been introduced and used, the operation can be symbolized. The loops in Figure 7-15 may be used to illustrate the transition from the concrete to the abstract. Children are asked, "How many cubes are in the first loop?" They respond by counting them. There are nine cubes. "How many cubes are we removing or taking away?" We are removing four cubes. "How many cubes are there left?" There are five cubes. "We say this using numbers by writing $9 - 4 = 5$."

   The transition to the abstract symbols should take place as soon as the children are competent using the loops and using the language that describes the operation they are performing. Practicing subtraction and discussing and recording the results in learning groups will strengthen the children's abilities in subtraction.

5. Activity 2 above employs the Cuisenaire rods. The language the students are using is "blue minus yellow equals purple," thus the change to numbers is quick. If the white rod equals one, blue has the value of nine, yellow has the value of five, and purple has the value of four. Thus, the number sentence $9 - 5 = 4$ results directly from the sentence using colors.

   Again, be cautious by reminding the children, "In this case, the white rod is one." Later, the value of the white rod may change. Considerable experience is necessary for children to be skilled with the abstract numerical description of the rods.

While presenting the concept of an operation and the concepts of addition and subtraction, take the opportunity to include examples of important properties of these operations. We have already mentioned the commutative property of addition. Note that the join machine where one input contains no objects and the yarn loop activity where no objects are removed are models of the **identity element** for addition and subtraction; that is, any number plus zero or minus zero results in the number you started with. With this background, $6 + 0 = 6$, $0 + 6 = 6$, and $6 - 0 = 6$ will be easy to remember. Allow children to discover that $0 - 6$ (left-hand identity) does not hold for subtraction.

We have already demonstrated the commutative property of addition using dominoes; however, this property can be clearly shown with any of the preceding models. For example, with Cuisenaire rods, the train dark green plus purple can be reconstructed as the train purple plus dark green. Then both can be shown to be equal to orange, as in Figure 7-18. Thus, we have the numerical

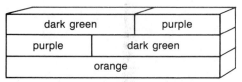

Figure 7-18

examples 6 + 4 = 10 and 4 + 6 = 10 when the white rod is equal to one.

Give children the opportunity to discover that in operations with whole numbers, the commutative property does not hold for subtraction. That is, 6 − 4 = 2, but 4 − 6 does not have a solution.

Another useful pattern is **adding or subtracting one.** Again, examples from any of the models already presented can illustrate this pattern. After a number of examples, children discover that adding one to a number results in the next number in the counting sequence. For example, 6 + 1 = 7 and 12 + 1 = 13. Likewise, children discover that subtracting one from a number results in the previous number in the counting sequence, 9 − 1 = 8 and 2 − 1 = 1.

Just as adding zero, using the commutative property of addition, and adding one can help in learning the basic addition facts, the **associative property of addition** can help children as they determine sums like 9 + 6. If children make mental groupings of ten, the answer is easier to determine. For example, 9 + 6 = 9 + (1 + 5) = (9 + 1) + 5 = 10 + 5 = 15. The associative property is shown in 9 + (1 + 5) = (9 + 1) + 5; how the numbers are grouped to add does not affect the sum.

A good way to show the associative property of addition is to have children place beans or chips on number strips like those in Figure 7-19. For a larger sum, more than two strips may be necessary.

Have the children put beans for the addends on the first and second number strips. Then have them fill up the first strip to complete the ten, using the last beans on the second strip. Here, we see 9 + 6 = 10 + 5 = 15.

This is a good procedure to use with problems with three or more addends and with larger problems that require regrouping. Use the associative property to make learning certain number combinations easier. Avoid presenting it as simply an abstract property of addition.

The time students spend discussing properties of addition and subtraction is very important to the learning

process. After the students understand the concepts of addition and subtraction and are comfortable posing story problems and solving each other's story problems using manipulative materials, they should begin to analyze the operations and discover properties that make it easier to remember the basic facts. The commutative property, the identity element, and the associative property are just a few of the patterns children should discover and discuss. They may notice many patterns of addition and subtraction facts other than the pattern of adding and subtracting one. They may notice that if they know the doubles, they can find the answer to the fact 6 + 7 by taking the double 6 + 6 and adding 1 or by taking the double 7 + 7 and subtracting 1. Encourage the children to keep a written record of the properties and patterns they have discovered. These may all be combined in a class book or bulletin board for the students to refer to.

Developing and Practicing Addition and Subtraction Skills

Children must learn the skills associated with adding and subtracting. These skills are called the basic facts and algorithms. The **basic addition facts,** of which there are one hundred, are those ranging from 0 + 0 to 9 + 9. The **basic subtraction facts,** again numbering one hundred, are those ranging from 18 − 9 to 0 − 0. An **algorithm** is any method used to solve a problem; mathematics consists of numerous algorithms for addition and subtraction. As children learn the basic facts and algorithms for addition and subtraction, they can make use of paper and pencil procedures, mental arithmetic, calculators, and computers.

Basic Addition and Subtraction Facts
For quick recall, children should be expected to visualize and/or memorize the basic addition and subtraction facts. Begin this memorization when children understand the concepts of addition and subtraction. Your efforts to develop the concepts will pay off as the facts are learned. Because the children have a manipulative model to which they can refer, they will be able to determine a basic fact temporarily forgotten. They can successfully use a variety of counters, including fingers. They can use Cuisenaire rods, Unifix cubes, number lines, state-operator-state machines, or calculators.

The addition and subtraction facts may be presented almost simultaneously. The join or combine model used with objects is somewhat easier to grasp than the take-away or remove model. We recommend you start first with join, later introducing take away. As the children understand the concepts, they will comfortably work with both addition and subtraction at the same time.

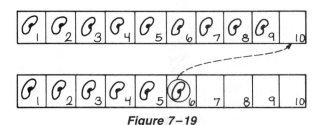

Figure 7-19

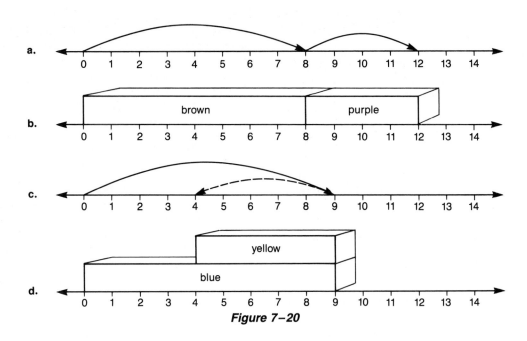

Figure 7–20

Many math textbooks show relationships between addition and subtraction facts. A family includes several facts, each of which relates the same three numbers through addition or subtraction. One such family follows:

$$8 + 5 = 13$$
$$5 + 8 = 13$$
$$13 - 8 = 5$$
$$13 - 5 = 8$$

The classroom should contain tools with which to perform simple calculations. The most basic tools for calculating are the many counters already mentioned. When children are first learning the basic addition facts, encourage them to use beans, cubes, bottle caps, nuts and washers, or chips to find the answer. Let them put some objects down, join others with them, and count the result or recognize the answer by sight. Let them use fingers as counters as well. As the work becomes more abstract and the children use paper and pencil or calculators, let them make marks on paper to help reinforce the basic addition facts.

The number line is useful for learning addition facts. For younger children, a walk-on number line provides large motor experiences. Its use in building the concept of number provides familiarity. For older children, a number line on the wall or chalkboard or attached to a desk top is useful. The procedure for use is the same. When children wish to add 8 + 4, they begin at 0 on the number line and take eight steps or move directly to 8 (Figure 7–20a). Next, they take four steps in the same direction along the number line. Where they stop, 12, is the sum of 8 + 4.

Another effective way to use the number line is to have it calibrated in centimeters. Cuisenaire rods may then be placed along the number line to help illustrate that the number line represents a continuous length and not just points where the numbers appear. For example, if the white rod has the value of one, 8 + 4 may be shown by placing a brown rod followed by a purple rod along the number line beginning at 0 as in Figure 7–20b. The result, 12, is clearly seen.

When students wish to subtract 9 − 5, they again begin at 0 on the walk-on number line and take nine steps or move directly to 9 (Figure 7–20c). Next, they take away by reversing their direction and moving five spaces back, to 4.

With the Cuisenaire rods, a blue rod (representing 9) is placed along the number line, beginning at 0. Then a yellow rod (representing 5) is placed along the blue rod beginning at the point marked 9. The difference, 4, can be seen by reading the number line, as in Figure 7–20d.

Children need careful instruction in working on a number line because they commonly forget to count spaces and instead count marks on the line. The walk-on and Cuisenaire rod number lines provide direct experience in counting spaces.

The calculator is another tool for learning addition facts. It should be used periodically to quickly produce facts that are forgotten or unlearned. Simple four-function (addition, subtraction, multiplication, division), light-activated calculators should be available for classroom use. Calculators speed up computations during games and activities intended to help children memorize basic addition facts. Set up races with basic facts in which some students have calculators and others do not. Students will quickly learn that it is faster to remember 8 + 7 than it is to push the buttons on the calculator.

The activities that follow are specifically designed for addition but may be used as effectively for subtraction by making simple changes in the materials.

ACTIVITIES

Primary (K–3)

Objective: *to help children memorize the basic addition facts.*

1. Commercially prepared printed flash cards have been used successfully for many years to help children reinforce basic addition facts. These cards have an addition problem on one side and the same problem with the answer on the other side. One card from a set is shown in Figure 7–21.

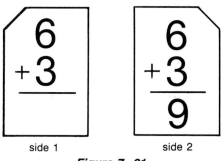

side 1   side 2

Figure 7–21

Show children the cards one at a time and ask them to recall the sum. Put aside the sums that the children can quickly recall but return to the stack any sums that they cannot recall. Vary this procedure to involve the whole class, several children and the teacher, several children and an aide, or children showing other children the cards. A major advantage of flash cards is that they are portable and convenient. One disadvantage to flash cards is that they turn many students off to learning basic facts. Students become frustrated if they are not fast enough or if they forget a fact. Be careful not to push a child to the point of frustration by overemphasizing flash cards.

Electronic flash cards are drill activities resembling printed flash cards. They are displayed on preprogrammed calculators or computers. The programs most attractive to children are those on computer software that not only present the problems but also display graphic designs and pictures. Children enjoy graphic presentations and learn basic addition facts as they interact with the computer. At the end of such electronic activities, the children are informed of the number of correct responses they have given. Then they are rewarded or admonished with a final graphic design.

Certain drill-and-practice programs keep a cumulative record of each child's progress. The chief advantage of using the computer is that it frees the teacher for other important tasks. On the other hand, only one or two children at a time can work at each computer.

Ideally, use both printed and electronic flash cards so that children can practice on problems in several ways.

2. Provide children with daily practice in mental arithmetic. Because the main focus of learning the basic facts is mental, this type of activity is particularly useful. Limit practice periods to five or ten minutes. Orally present addition facts to children while the children write on a piece of paper or a small individual chalkboard only the answer to the problem. The difficulty of the problems should be determined by the age and experience of the children. Initially, speedy responses are not necessary. Children need an opportunity to practice using their mental faculties.

Ask the children to number from 1 to 5 on their papers. Explain that you are going to give them a problem and they should think of what the answer is and then write the answer down; they should not use their pencils to find the answer. Here are examples of mental exercises: What is four plus seven? What is three plus three plus two? Answer yes or no; three plus eight is more than ten. Four plus five equals nine; what else equals nine? As the children become more proficient, increase the number of questions from five to ten, fifteen, or twenty. Expand the questions from just basic facts to other, related mathematical topics, such as place value.

3. **Pic-addition,** short for picture addition, appeals to children. Pic-addition requires construction of activity materials. The first step is to select a picture of an animal, cartoon character, or athlete that is popular with the children. A fuzzy kitten appeals to some children. A Saturday morning cartoon character appeals to others. An athlete in action attracts the attention of some.

Next, glue the picture to a piece of oaktag or posterboard. Spread the glue thinly over the entire back of the picture. There should be no border (see Figure 7–22a).

In the next step, line off square or rectangular regions with pencil on the back of the oaktag or posterboard holding the picture. Within these regions, draw circular and square frames. With marking pen, write an addition problem in each circular or square frame on the back of the picture. Make sure problems with the same answer, such as 1 + 2 and 2 + 1, are placed in different-shaped frames (Figure 7–22b).

Finally, line off regions on the inside of a box lid or on a piece of oaktag to make an answer board. The lines should form regions the same size and shape as those on the back of the picture. Within these regions, draw circular and square frames to match those on the back of the picture. In each circular or square frame, write the answer to the addition problem in the position opposite that of the problem (Figure 7–22c). (When the problem is correctly answered, the problem card is flipped over and the regions are moved to different positions.) Cut the picture apart using the lines on the back to form individual cards.

Have students select one of the individual cards. Instruct them to look at the problem and solve it, then place the card in the answer frame in the region corresponding to the answer, turning the card picture side up. As they complete the solutions, the picture emerges in its entirety.

4. Concentration games are another source of basic fact practice. A popular version is called peopletration because children are important participants. Provide twelve large (22 centimeters × 30 centimeters) cards constructed from oaktag or posterboard. Each card should have five problems or answers to problems written on one side and a large alphabet

a.

b.

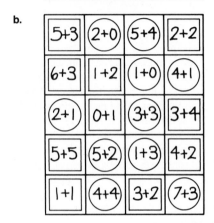

c.

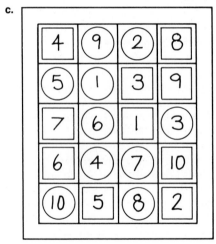

Figure 7-22

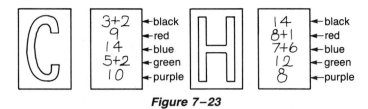

Figure 7-23

on the back of the card, in this case, "5 + 2." Whoever is holding Card H reads whatever is written in green on the back of that card, in this case, "12." Because 5 + 2 does not equal 12, there is no match. Play continues until there is a match. When a match occurs, the two cards are laid down or the children holding them sit down and the game continues. The game is over when all of the cards have been matched.

The conventional concentration game consists of two arrays of cards (10 centimeters × 12 centimeters). Each array contains sixteen cards. One side of each card is unmarked. The other side of one array has basic facts such as 2 + 9. The opposite side of the second array has answers such as 11. When an individual points to pairs of cards, the cards are turned over and compared. If the fact and the answer match, the child scores a point and gets another turn. Otherwise, the next player takes a turn. The game ends when all of the cards have been matched.

5. Bingo activities provide opportunities for children to review number facts. **Bugs bingo** is a popular activity for second and third grade students. Two to four players participate. Each group needs three regular dice or dice showing the numerals 1 to 6, as well as about twenty-five bingo markers (small squares of paper work well) and a bingo board. The board for this game is shaped like a ladybug, from which the name originates. Figure 7-24 shows one of the game boards and numeral patterns for three others.

Each player puts a marker on the FREE square and then rolls a die. The player with the highest number on the die begins. The first player rolls all three dice and adds the number of dots showing. The other players check that the sum is correct.

12	6	11	4	10
4	14	13	17	14
7	18	Free	15	8
13	9	17	3	16
15	12	6	10	5

8	10	9	17	7
16	9	11	4	13
3	15	Free	18	12
14	7	11	17	6
10	5	14	8	16

16	9	7	11	3
5	8	17	13	15
15	7	Free	10	6
10	14	5	16	12
13	17	18	9	14

Figure 7-24

letter *A* to *L* on the other side. To assure corresponding sets, the first set of problems and answers on each card should be written with black ink, the second set with red, then blue, green, and purple. Figure 7-23 illustrates two such cards, front and back.

Select twelve children to hold the cards. Give each child one of the cards identified with the large alphabet letters. Have the twelve children stand side by side across the front of the room, holding the cards so the letters face the rest of the class.

Designate which game color to use—green, for example. Instruct one of the other children to select two letters, hoping for a match. For example, a student might say "C" and "H." Whoever is holding Card C reads whatever is written in green

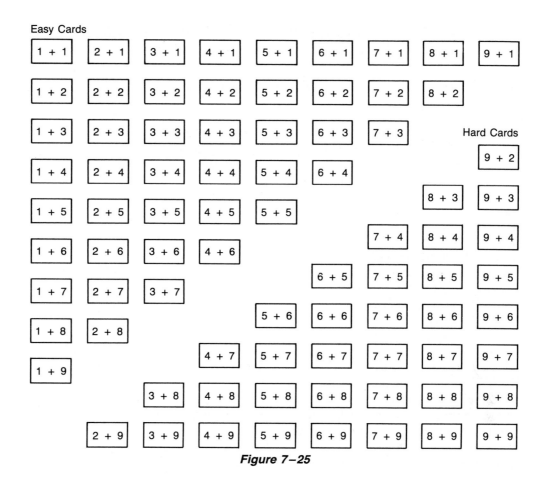

Figure 7–25

All players cover that numeral on their boards. Although the numeral may appear more than once on the board, only one numeral is covered at each turn. Once a marker is placed on a numeral, it cannot be moved to another region.

The dice are passed to the next player, and play continues until someone has five markers in a row, horizontally, vertically, or diagonally.

If additional boards are needed, they should be made with different numeral patterns. For each board, however, there is one FREE region and there are twenty-four regions with these numerals:

> one each of 4, 5, 6, 15, 16, 17
> two each of 7, 8, 9, 12, 13, 14
> three each of 10, 11

6. Top it uses a deck of cards representing the basic addition facts. Figure 7–25 shows a deck of easy cards and a deck of hard cards. Choose the deck that matches the children's ability levels, or combine the decks into a single deck representing nearly all of the basic addition facts.

The game is for two to four players. Begin play with the stack of cards face down. The first player turns over a card and places it face up, giving the sum. The second, third, and fourth players do the same. The player with the highest sum claims all of the cards played. If any players are tied with high cards, the cards remain in front of the players and those who are tied each draw another card. The player with the highest sum takes all of the cards. The winner of the game is the player with the most cards after all of the cards have been drawn.

This chapter presents a few of the many activities that can be employed to help children practice the basic addition facts. Each activity can be easily adapted to include subtraction practice. The goal is to have children able to accurately and quickly respond to all basic addition and subtraction facts. References to other activities and games are included in the chapter bibliography.

Addition and Subtraction Algorithms

Children progress more rapidly in learning the addition and subtraction algorithms when they already know the basic addition and subtraction facts. As well, modeling addition and subtraction using the base ten blocks or other manipulative materials prepares students for learning the algorithms and gives them the opportunity to discuss their work with one another and to write about their work. For example, give the students the base ten blocks and a problem such as: "Jack wants to buy a 39¢ top and a 59¢ airplane. How much will they cost altogether?" Let the children work in small groups to find a solution and have them write down their method using words and symbols. When all the groups are finished, ask each group to share their solution with the whole class. Discuss the pros and cons of each solution. This gives students the opportunity to explore and invent algorithms. Let the students decide which algorithm they prefer and when they no longer need to use the manipulatives. It is likely that the students will devise some algorithms similar to the standard addition and subtraction algorithms, but be ready for algorithms that are different. The process of inventing algorithms is an important part of learning how and why algorithms work.

In mathematics textbooks, the addition and subtraction algorithms are usually taught sequentially, beginning with the simplest problems, which require no regrouping (carrying or borrowing). The difficulty and complexity increase until multidigit problems with regrouping are presented. However, when students begin multidigit addition and subtraction using manipulative materials, there is no need to separate problems that require regrouping from those that do not. Do not assign page after page of laborious addition and subtraction problems. When children can demonstrate paper and pencil algorithms for addition and subtraction, it is time to let them use calculators to speed up computations. Using the calculator is the most efficient procedure for performing an arithmetic operation.

Paper and pencil algorithms for addition and subtraction are presented in turn. We begin with the standard and generally most efficient paper and pencil algorithm for each operation. Alternative algorithms are then presented. Alternative procedures often serve as teaching algorithms and help to bridge the gap between the concrete and abstract. Sometimes, alternative algorithms are the most efficient paper and pencil algorithms for children. It was mentioned above that students sometimes invent algorithms similar to the standard addition and subtraction algorithms. Likewise, they may invent algorithms similar to the alternative algorithms presented below. While in this chapter the standard algorithm is presented first followed by alternative algorithms, it is not suggested that children will invent algorithms in this order. You should not introduce standard algorithms until students have explored, invented, and discussed their own algorithms.

Addition. The standard addition algorithm is generally considered the most efficient paper and pencil procedure for adding. The algorithm has been applied to the six problems below, each of which has at least one two-digit addend. The problems are presented in order of difficulty. The first three algorithms involve no regrouping. The others have regrouping in one or more of the place value positions.

$$
\begin{array}{r@{\qquad}r@{\qquad}r}
10 & 34 & 22 \\
+\ 8 & +\ 10 & +\ 14 \\
\hline
18 & 44 & 36 \\
\end{array}
$$

$$
\begin{array}{r@{\qquad}r@{\qquad}r}
^{1}19 & ^{1}37 & ^{1}63 \\
+\ 2 & +\ 28 & +\ 59 \\
\hline
21 & 65 & 122 \\
\end{array}
$$

The first two problems involve adding a number to ten and adding ten to a number. Sums involving ten are important to know because they occur continually during computation. Being able to recognize sums involving ten and make groupings of ten while solving problems saves considerable time and energy during computation. The latter three algorithms display the regrouping numeral 1, indicating regrouping has taken place. In these cases, 10 ones have been grouped for 1 ten. While the algorithm would be simpler without the regrouping numeral, most who use this algorithm include it.

It is important that children understand what happens when regrouping occurs. Work with place value provides this understanding, and children can be taught the standard addition algorithm with little more than a set of multibase arithmetic blocks or a picture of them. Most math texts include illustrations of some sort of counting device and step-by-step procedures for teaching the algorithm that often encourage teachers to use manipulative materials as models for a textbook algorithm.

For example, Figure 7–26a shows both the numeral representation and the pictorial representation of 22 + 14. The joining of unit cubes and longs (Figure 7–26b) is accompanied by the teacher asking, "How many ones are there?" There are 2 ones plus 4 ones, or 6 ones. The children put the 6 beneath the 2 + 4 and make sure there are as many cubes on the place value board. Next, the teacher asks, "How many tens are there?" There are 2 plus 1, or 3. The children place the 3 beneath the 2 tens + 1 ten and make sure there are as many longs on the place value board. The transition from the concrete to the abstract should be made as often as necessary for the concepts to make sense to the children.

An algorithm that requires regrouping is only slightly more difficult than one that does not require regrouping when children have the proper foundation, that is, when they have learned about grouping various materials, including the multibase blocks, during the initial study of place value. When an algorithm requiring regrouping is accompanied by an illustration that reinforces earlier skills, children pick up the process quickly (see Figures 7–27a, b, and c).

The teacher asks, "How many ones are there?" There are 3 ones plus 7 ones, or 10 ones; that is, there is one group of tens and there are 0 ones. It is important here that the unit cubes be grouped and exchanged for one long, even if just pictorially. "How many units do we have after the exchange?" We have 0. The students write down the 0 beneath the 3 + 7. They record a 1 in the tens column to remind them that they have exchanged 10 ones and now have 1 ten. "How many tens are there?" There are 6 tens plus 5 tens plus 1 ten, or 12 tens; that is, there is one group of a hundred and there are 2 tens. The longs should be put together and exchanged for one flat and two longs. "How many tens do we have after the exchange?" We have 2 tens. The students write down the 2 beneath the 6 + 5. They also record the 1 hundred because they have no more place value positions in the problem. They check to make sure the multibase blocks show 120.

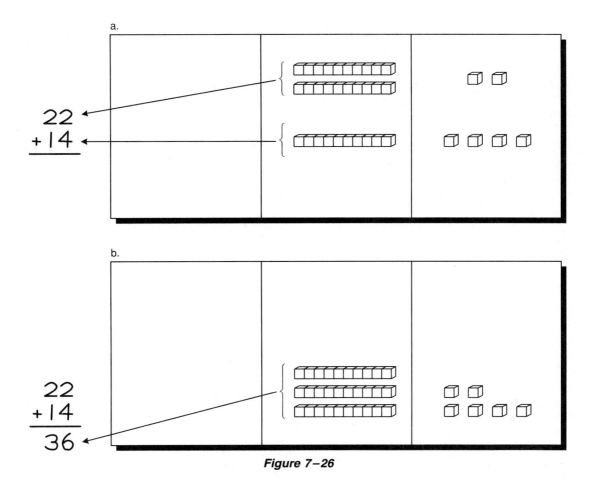

Figure 7–26

The addition algorithms for larger numbers are extensions of the above process. They take more time to perform but are not substantively different. Figures 7–28a, b, and c illustrate one such addition, showing both the standard algorithm and the pictorial model.

As soon as children can perform the algorithm without manipulative or pictorial models, they should be encouraged to do so. Math texts provide plenty of practice exercises. Children should practice for several days after they are able to perform the algorithm without materials. While there may be thirty problems on a page, ten to twenty problems will give children the necessary practice. If children continue to have difficulty reaching the correct answer, you should diagnose the difficulty. Assigning more problems to solve is unlikely to be the best strategy. We discuss diagnosing computational errors in the section on evaluation.

The following activities differ from those presented earlier. Their primary focus is not manipulative activities, but alternative ways to present addition algorithms.

ACTIVITIES

Primary (K–3)

Objective: *to add columns of three or more numbers.*

1. When children are competent with the basic addition facts, challenge them to add three digits. At first, the numbers

should be no larger than basic facts, such as 3 + 4 + 8 or 8 + 1 + 5. Present the problems in the format the children are used to seeing, most likely the vertical format. When the problems become more difficult, help students develop an algorithm that works for them. A column addition algorithm is shown in row a. shown below.

$$
\text{a.}\quad
\begin{array}{r} 6 \\ 7 \\ +4 \\ \hline \end{array}
\qquad
\begin{array}{r} 6 \\ 4 \\ +7 \\ \hline \end{array}
\qquad
\begin{array}{r} 10 \\ +7 \\ \hline 17 \end{array}
$$

$$
\text{b.}\quad
\begin{array}{r} 6 \\ 7 \\ +4 \\ \hline \end{array}
\qquad
\begin{array}{r} 6 \\ 7 \\ +4 \\ \hline \end{array}
\qquad
\begin{array}{r} {>}13 \\ +4 \\ \hline 17 \end{array}
$$

To approach this problem, find numbers that add to ten and reorder the problem to put those numbers together. The numbers adding to ten can then be combined and the final sum calculated. After some practice, most children can group the tens by reordering the problem mentally. They see groups of ten in the problem and group numbers mentally, shortcutting the algorithm.

A second approach is shown in row b. Beginning at the top or bottom of the column, add successive numbers, keeping

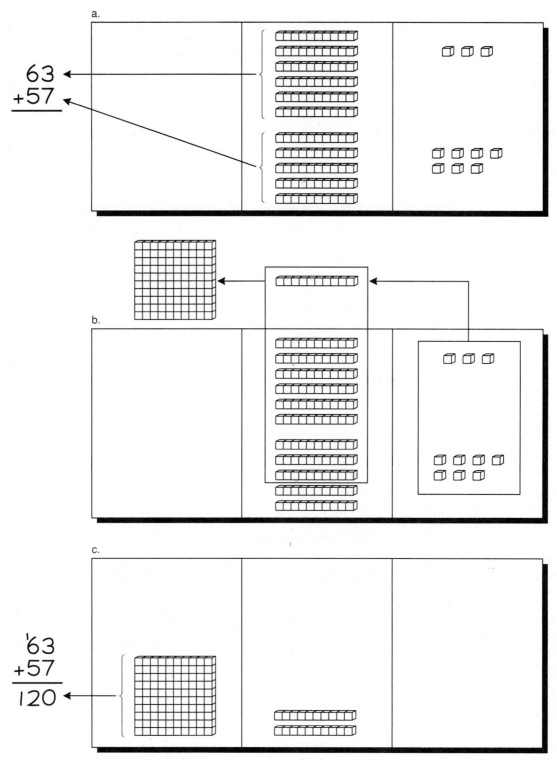

Figure 7-27

the intermediate sum in mind until reaching the last number. The last calculation provides the answer to the problem. Let children choose the algorithm they are most comfortable with when adding columns of numbers.

Objective: to use the expanded notation algorithm.

2. **Expanded notation** is a direct outgrowth of work with place value. Children recognize that two flats, four longs, and

six small cubes represent 200 + 40 + 6, or 2 hundreds + 4 tens + 6 ones. An alternative addition algorithm uses this knowledge. The problems below show two solutions using expanded notation. In problem a, no regrouping is necessary. The problem is rewritten from standard into expanded notation. The columns are added and the expanded notation is rewritten into standard notation.

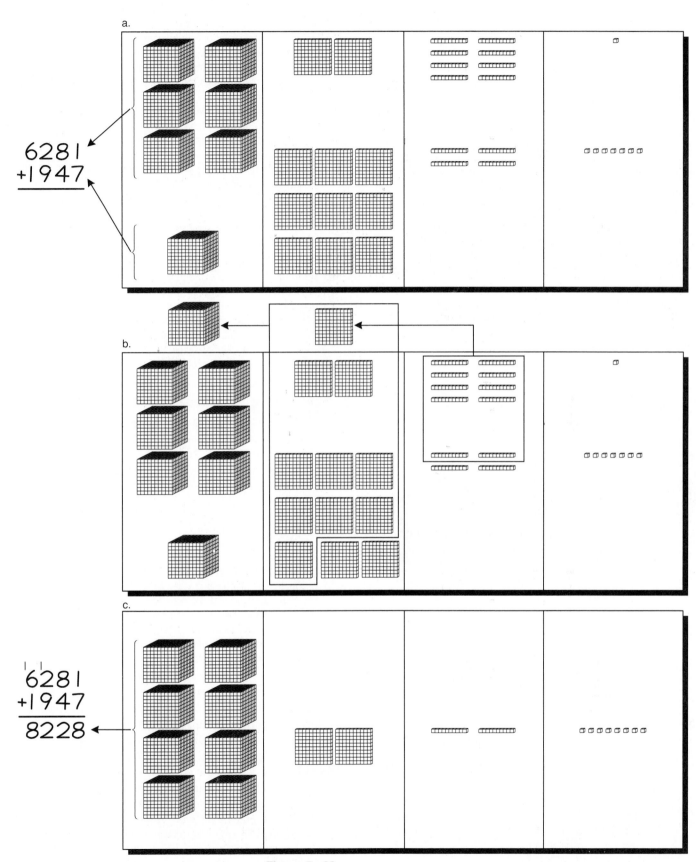

Figure 7–28

a. $\begin{array}{r} 63 \\ +24 \\ \hline \end{array}$ $\begin{array}{r} 60+3 \\ +20+4 \\ \hline 80+7=87 \end{array}$

b. $\begin{array}{r} 58 \\ +29 \\ \hline \end{array}$ $\begin{array}{r} 50+8 \\ +20+9 \\ \hline 70+17=80+7=87 \end{array}$

Problem b requires regrouping in the ones column. As in the earlier example, the problem is rewritten from standard into expanded notation. Each column is added. Finally, regrouping takes place as the answer is rewritten into standard notation. The algorithm used with problem b is advantageous because children never lose sight of the numbers they are regrouping.

Objective: to use the loop abacus as a tool in learning the addition algorithm.

3. The abacus is useful as a tool. As children learn about place value, the abacus is handy in teaching the regrouping process. Any time ten counters appear on a given loop, they must be exchanged for one counter on the adjacent loop. This is demonstrated as we solve a problem on the abacus. In the first frame of Figure 7–29, an abacus is ready to add

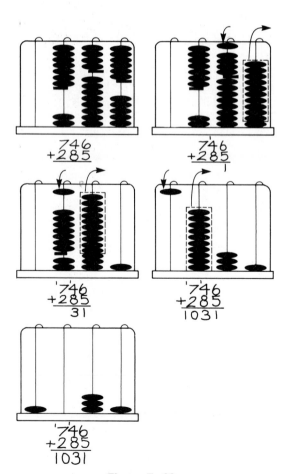

Figure 7–29

746 + 285. The counters below the holding clips represent 285; the counters above represent 746.

In the second frame, we remove the holding clip in the ones column and exchange ten counters on the ones loop for one counter on the tens loop. We have one counter left in the ones column.

In the third frame, we remove the holding clip in the tens column and exchange ten counters on the tens loop for one counter on the hundreds loop. We have three counters left in the tens column.

In the fourth frame, we remove the holding clip in the hundreds column and exchange ten counters on the hundreds loop for one counter on the thousands loop. We have no counters left in the hundreds column.

In the fifth frame, we see the abacus after all the exchanges have been made.

Objective: to use the partial sums algorithm for addition.

4. Another algorithm that helps children through the process of regrouping or carrying is the **partial sums algorithm.** It may be used whether or not regrouping is necessary. To illustrate this algorithm, note the problem below (48 + 39). The first step in the solution is to add the numbers in the ones column, 8 + 9. The answer, 17, is the first partial sum and is placed beneath the problem.

$\begin{array}{r} 48 \\ +39 \\ \hline \end{array}$ $\begin{array}{r} 48 \\ +39 \\ \hline 17 \end{array}$ $\begin{array}{r} 48 \\ +39 \\ \hline 17 \\ 70 \end{array}$ $\begin{array}{r} 48 \\ +39 \\ \hline 17 \\ 70 \\ \hline 87 \end{array}$

The next step is to add the numbers in the tens column, 40 + 30, or 4 tens + 3 tens. The result, 70, is the second partial sum and is placed beneath the 17.

Finally, the partial sums are added to arrive at the answer, 87. You may also add the tens first and then the ones using this algorithm. This is very useful for mental computation. Take care with this algorithm to assure that the correct place value positions in the partial sums are maintained.

Subtraction.   Again for subtraction, give the students a real-life situation such as spending 18¢ out of 25¢ and determining the amount of change you have left. Allow the children to work in groups to determine the solution and record their methods. Discuss the various methods used by the different groups before presenting the standard algorithm. Of course, the concept of subtraction is considerably different. The standard paper and pencil algorithm for subtraction relies on children's knowing the basic subtraction facts and being well versed in place value concepts. You recall that the basic subtraction facts are related to the basic addition facts. Thus, for the addition fact 4 + 7 = 11, we have the corresponding subtraction facts, 11 − 4 = 7 and 11 − 7 = 4. Techniques for helping children memorize these facts were discussed in the section on basic facts.

Six examples of the standard subtraction algorithm are presented below.

$$
\begin{array}{ccc}
18 & 44 & 36 \\
-10 & -34 & -22 \\
\hline
8 & 10 & 14
\end{array}
$$

$$
\begin{array}{ccc}
{}^{1}\!\!\not{2}1 & {}^{5}\!\!\not{6}5 & {}^{2}\!\!\not{3}{}^{11}\!\!\not{1}2 \\
-9 & -37 & -63 \\
\hline
12 & 28 & 259
\end{array}
$$

The first three examples involve no regrouping, or borrowing. The last three require that regrouping take place. The first two problems involve subtracting 10 and subtracting so that 10 is the difference. The third problem is solved by applying basic subtraction facts to the ones and tens columns in the problem.

The fourth example necessitates regrouping from the tens place to the ones place. This is shown by crossing out the 2 tens and replacing them with 1 ten. The ones place is increased from 1 to 11, reflecting the exchange of 1 ten for 10 ones. The subtraction is then carried out.

The fifth example is similar to the fourth because there is regrouping from the tens to the ones place. In this case, 6 tens and 5 ones are exchanged for 5 tens and 15 ones.

In the final example, there is regrouping from the tens to the ones place and from the hundreds to the tens place. Thus, 2 tens and 2 ones become 1 ten and 12 ones; then, 3 hundreds and 1 ten become 2 hundreds and 11 tens. The markings have been shown in the procedure as they are used by most who employ this standard algorithm.

When students develop a subtraction algorithm, it is important that they use manipulative materials. At this point, we demonstrate with multibase blocks. The first example is 65 − 37. Figures 7–30a, b, and c illustrate how to construct the problem using the blocks.

First, attempt to remove seven small cubes from five small cubes. Finding this impossible, perform an exchange (see Figure 7–30b). Trade one long for ten small cubes. Now, from the collection of fifteen small cubes, remove seven, leaving eight small cubes. Moving to the tens, remove three longs from five longs; this leaves two longs. Having finished the algorithm, you find the difference is two longs and eight small cubes, or 28 (see Figure 7–30c).

The first few times children encounter subtraction involving regrouping with the multibase blocks, they should use only the blocks. Once the procedure is mastered, use the written algorithm along with the blocks. Finally, as soon as the children are able, use only the written algorithm, bringing the blocks back if there is some difficulty in solving a particular problem. The multibase blocks, or any manipulatives, clearly show what happens at each step in an algorithm. Once each step is understood, children should practice the symbolic algorithm.

Children working with manipulatives or practicing paper and pencil algorithms occasionally discover clever shortcuts or original procedures. Children should be encouraged to create and demonstrate to the class their own algorithms.

As in the case of addition, the activities that follow include alternative algorithms and approaches for subtraction. If some children find an alternative that is superior for them, let them adopt it as their standard algorithm.

ACTIVITIES

Primary (K–3)

Objective: *to use the expanded notation algorithm for subtraction.*

1. When the **expanded notation algorithm** is used, children see what happens in the process of regrouping, or borrowing. The first example below, 43 − 22 (a), does not require regrouping. Both numbers are rewritten in expanded form. In the ones column, 2 is subtracted from 3, resulting in 1. In the tens column, 40 − 20 or 4 tens − 2 tens results in 20, or 2 tens. Then 20 + 1 is rewritten into standard form, 21.

a.
$$
\begin{array}{cc}
43 & 40+3 \\
-22 & -(20+2) \\
\hline
& 20+1=21
\end{array}
$$

b.
$$
\begin{array}{ccc}
43 & 40+3 & 30+13 \\
-27 & -(20+7) & -(20+7) \\
\hline
& & 10+6=16
\end{array}
$$

In the second example (b), regrouping is necessary. Children cannot subtract 7 from 3 after the problem is written in expanded form. They must rewrite 40 + 3 as 30 + 13. This process regroups 1 ten to 10 ones. Children can subtract 7 from 13, with the result of 6. They may also subtract 30 − 20 or 3 tens − 2 tens, with the result of 10. Then 10 + 6 is rewritten into standard form, 16.

You may find it necessary to help children work through several examples of the expanded notation algorithm with concrete objects such as multibase blocks, beansticks, and Cuisenaire rods before you move to symbols. Work first with problems that don't require regrouping, then advance to the more complicated regrouping problems.

Objective: *to use the abacus as a model to illustrate the standard subtraction algorithm.*

2. Children who have used the abacus when learning place value and addition have the exchanging skills necessary for subtraction. To solve 342 − 164, begin with 342 on the abacus as shown in the first frame of Figure 7–31. As in the second frame, exchange 1 ten for 10 ones, resulting in 12 ones. Then remove 4 ones, as shown in the third frame, resulting in 8 ones. Then exchange 1 hundred for 10 tens as in the fourth frame; there are 2 hundreds and 13 tens remaining. As in the fifth frame, remove 6 tens, leaving 7 tens. Because no further

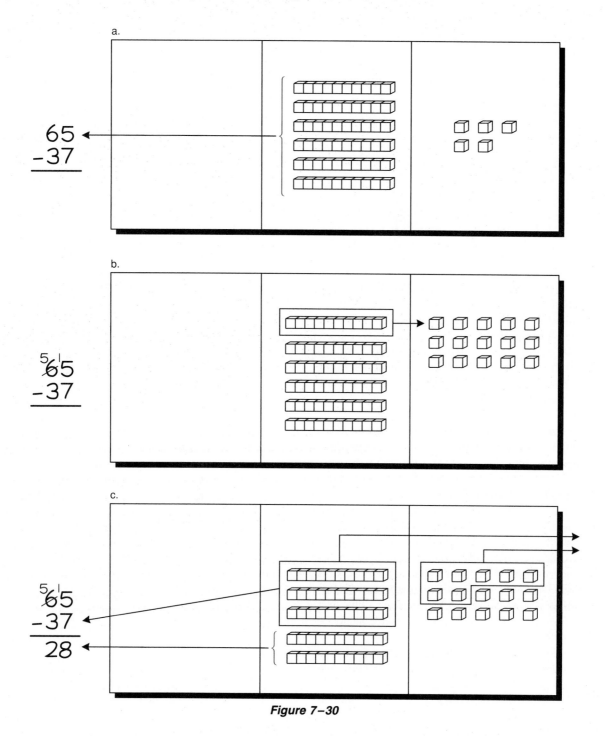

Figure 7–30

regrouping is necessary, remove 1 hundred, completing the problem. Frame six shows how the abacus looks after 164 has been subtracted from 342. The answer, 178, is easily read.

Objective: *to use the "How much more do I add" algorithm (HMMDIA) to subtract.*

3. The **"How much more do I add" (HMMDIA)** algorithm is a subtraction algorithm that uses an additive component. We demonstrate with the example below (168 − 49). The algorithm consists of adding to the subtrahend, 49, until you reach the minuend, 168. At each step of this adding process, the number added is recorded. Later, all of the written numbers are added together.

$$
\begin{array}{r} 168 \\ -49 \\ \hline \end{array}
\qquad
\begin{array}{r} 168 \\ -49 \\ \hline 1 \end{array}
\qquad
\begin{array}{r} 168 \\ -49 \\ \hline 1 \\ 50 \end{array}
\qquad
\begin{array}{r} 168 \\ -49 \\ \hline 1 \\ 50 \\ +68 \\ \hline 119 \end{array}
$$

Explain to the children using the example that they will be adding from 49 to 168. Begin with 49. Ask yourself, "How

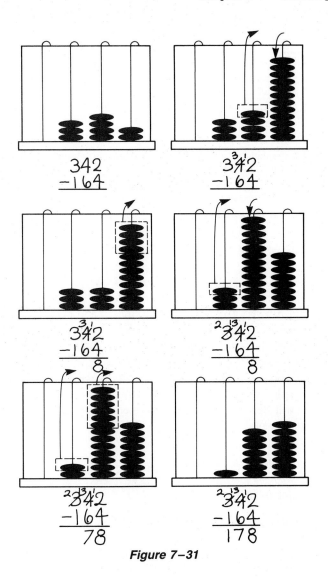

Figure 7–31

much more do I add to 49 to reach the next group of ten?" The next group of ten is 5 tens, or 50. The answer in this example is 1. Write down the 1 as shown and now think 50 because you have just added one to 49 and gotten 50.

Now ask, "How much more do I add to 50 to reach 100?" In this case the answer is 50. Write down the 50 as shown and think 100 because you have just added 50 to 50 and gotten 100.

Next, ask, "How much more do I add to 100 to reach 168?" The answer is 68. Write down the 68 as shown. You have reached the number you were adding to, 168, and can stop.

The final step is to add the three numbers that were recorded. The answer to 168 − 49 is 1 + 50 + 68, or 119.

You may find it necessary to review this algorithm several times before the process becomes clear. Once you have done this example, give other examples, particularly some that are simpler, such as 25 − 8. This unusual subtraction algorithm surprises and motivates children.

In this section, we have discussed developing two sets of skills. The first skills were the basic addition and sub-

traction facts. Children should have fairly immediate recall of these facts. The second skills were the algorithms for addition and subtraction. Current math texts present thorough instruction on how to develop the standard algorithms, and they contain an ample number of problems for practice. Be sure to let students invent their own algorithms before using those in the text. The next section provides a look at a typical mathematics textbook example of addition.

Estimating and Mental Calculating

Whether working with pencil and paper or working with calculators and computers, children should develop skill in estimating and mental calculating. To know if an answer seems reasonable, children should estimate the sum or difference either before or after working an algorithm and compare the estimate with the paper and pencil results.

Estimating skill should likewise be extended to work with calculators and computers. Trusting that a calculator will provide the correct answer is justified in most cases; however, we should be cautious about the numerals that appear in calculator displays. Incorrect answers are most commonly displayed when the operator instructs the calculator to perform the wrong operation, enters the wrong sequence of numbers and operations, or enters the wrong numbers. Because we cannot predict when those occasions will arise, it is important to have an estimate of the correct answer. Skill in mental calculation means that the child is able to answer number questions in his or her head without relying on paper and pencil or calculator. Facility with mental calculation helps the child perform computations more quickly and demonstrates the child's understanding of the basic arithmetic operations. The following activities are intended to strengthen children's skill in estimating and mental calculation.

ACTIVITIES

Primary (K–3)

Objective: *to practice estimating with addition and subtraction.*

1. Use a set of double-six or double-nine dominoes. Large-format dominoes work well if this activity is done with the whole class. To practice estimating with addition, hold a domino before the class for three to five seconds, then ask the children if they believe there are ten dots, more than ten dots, or fewer than ten dots on the domino. Let them signal to you that there are ten dots by putting a hand flat on the table or floor. A hand with thumb up means more than ten; a hand with thumb down means fewer than ten.

To practice estimating with subtraction, have the children find the difference between the number of dots on one end of a domino and the number of dots on the other end. Have them indicate whether the difference is equal to three dots, more

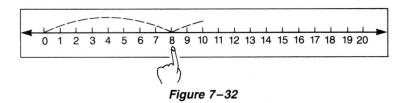

Figure 7-32

than three dots, or fewer than three dots. They can use the same hand signs as before.

Objective: *to use the number line to practice mental calculation with addition and subtraction.*

2. Use this activity with a wall number line, smaller desk number lines, or a walk-on number line. For addition, begin with an initial jump to the first addend and then explain the next jump, letting the children mentally calculate where the second jump will land. After the children answer, let them perform the second jump. Figure 7-32 shows a number line and the position of a finger after the children have been told to start at 0 and to jump to 8.

   Ask them, "If we make a jump of 6, where do you think we will land?" Let them mentally calculate. Then let them make the jump of 6 to find the answer.

   For subtraction, follow a similar procedure. Let the children make the beginning jump, for example, to 15. Then ask, "If we jump back (subtract) 4, where do you think we will land?" Let them mentally calculate, and then subtract 4 to find the answer. To vary this activity, have the children estimate if the answer will be equal to, more than, or less than a particular number, such as 10, 15, or 20.

Objective: *to use numbers that add to 10 for mental addition and subtraction.*

3. To begin, use multibase blocks, beans, or Unifix cubes to present combinations of objects that add to 10. For example, $0 + 10, 1 + 9, 2 + 8$, and $3 + 7$ each equals 10. Have the children estimate the sum of combinations of objects.

   Next, present combinations such as $9 + 7$ and challenge the children to make a grouping of 10 plus a second number to result in the same amount. For example, $9 + 7$ should be changed to $(9 + 1) + 6$, then to $10 + 6$. Beginning with manipulatives helps make the grouping to 10 more meaningful and slows the process so most children can successfully participate.

   For subtraction, round numbers to 10 for easier calculation. For example, write $12 - 5$ on the chalkboard. Explain that sometimes you forget a basic subtraction fact and need a way to figure the answer in your head. One way is to make the larger number a 10 or a 20, whichever is closer. Ask, "Is 12 closer to 10 or 20?" Twelve is closer to 10. Ask, "What do you have to do to 12 to make it 10?" Subtract 2 from 12. Explain, "If you subtract 2 from 10, you must also subtract 2 from 5 so the difference will be the same. What is $5 - 2$?" The answer is 3. Summarize, "We have changed $12 - 5$ to $10 - 3$. What is the answer?" The answer is 7; therefore, $12 - 5$ is 7. Review the process several times. The mental process of subtracting or adding to make the larger number (subtrahend) a multiple of 10 is much quicker than the oral explanation.

   The problem $12 - 5$ also may be solved by adding 5 to the minuend, 5, to reach 10. Then 5 must also be added to 12 to maintain the difference. The new problem becomes $17 - 10$.

By changing either the minuend or the subtrahend to 10, students may find problems easier to solve. These mental shortcuts are particularly useful when they are performed with larger subtractions in such problems as $36 - 18$; the problem may be restated as $38 - 20$.

Estimation and mental calculation activities for primary children should be informal and concrete. You can use more formal techniques with older children. The more formal procedures may require children to round numbers to the nearest 10, 100, or 1000 and to add and subtract numbers that have been rounded to 10, 100, or 1000.

As you move from topic to topic in the math text you can quickly prepare questions. Mental arithmetic problems should be presented to students three to five days a week, alternating with other warm-up activities. Over time, children will show considerable improvement in their ability to handle mental arithmetic.

ACTIVITIES

Intermediate (4-6)

Objective: *to use estimation to determine if certain purchases can be made.*

1. This activity consists of a series of questions that should be answered without calculating on paper. Tell students to number from 1 to 5 on a sheet of paper. As each question is presented, have them estimate the answer by rounding off; then have them record their answer. The questions below ask whether or not certain items can be purchased with a fixed amount of money. You can ask, "Answer yes or no. You have $50. Can you buy:

   1. a popcorn popper for $28.95 and a basketball for $24.00?
   2. an umbrella for $19.99 and a case of apples for $19.99?
   3. a ring for $37.50 and a chess set for $14.95?
   4. a radio for $21.89 and some tapes for $29.50?
   5. three books for $15.00 each?"

   Such questions take little time. Briefly discuss the solutions before going on to another topic. Ask students to explain to the class what they were thinking as they solved a particular problem. One student explained that as she worked on question 1 above, she thought the popcorn popper was about $30.00 and the basketball was about $25.00. The sum of the two items was $55.00. You can't buy that much with $50.00. Other students agreed that they had solved question 1 in the same way. Bill spoke up and said he thought the popper was about $28.00 and he knew the basketball was $24.00. He added the two amounts together and found the sum was $52.00. He agreed that you can't buy both with $50.

Objective: to use historical material to motivate estimation and mental arithmetic.

2. Textbooks from the past (usually available from municipal or college libraries) provide interesting and amusing mental exercises that students enjoy hearing and attempting to solve. One such textbook is Greenleaf's *Intellectual Arithmetic*, published in 1859. Its formal title is much more impressive: *A Mental Arithmetic, Upon the Inductive Plan; Being an Advanced Intellectual Course, Designed for Schools and Academies*. Examples of the exercises Greenleaf included in his book follow. The page on which each problem can be found is provided in parentheses.

- A farmer sold 6 bushels of wheat, 7 bushels of rye, and 8 bushels of corn; how many bushels did he sell? (p. 12)
- A lady expended for silk 4 dollars, for gloves 1 dollar, and for a bonnet 9 dollars; how many dollars did she expend in all? (p. 12)
- How many are 8 and 9? 8 and 19? 8 and 29? 8 and 39? 8 and 49? 8 and 59? 8 and 69? 8 and 79? 8 and 89? 8 and 99? (p. 15)
- George spent 19 cents for candy and 21 cents for fruit; how much more would he have to spend to make 50 cents? (p. 23)

Objective: to use the calculator to improve addition and subtraction estimation skills.

3. Select two teams of students. Each team may have as few as one member or as many as half of the class. Provide one calculator for each team.

As play begins, one member from Team A says a three-digit number. A player from Team B says another three-digit number. Both players silently write an estimate of the sum of the two numbers. Give a limit of 5 seconds to make estimates.

Then have both players use the calculator to determine the sum. The player whose estimate is closest to the actual sum scores a point for the team. In the case of a tie, both teams earn a point. The next player on each team should name, estimate, and calculate in the next round.

You may suggest other rules, depending on the age and ability of the students. For example, you may stipulate that only two-digit numbers be used or that the number must end in zero or have a zero in the tens place.

The rules for the subtraction activity are similar to those for the addition. One player each from Teams A and B names a three-digit number. Both players then write down their estimate of the difference between the two numbers. They use a calculator to determine the answer. Again, the player whose estimate is closest to the actual difference earns a point for the team. Children who engage in this activity for a while develop estimation strategies that benefit them in the game.

Many calculator activities that call for estimation not only strengthen estimation skills but also improve calculator skill. A sampling of calculator activity books is included in the chapter bibliography.

Problem Creating and Solving

Addition and subtraction are important skills in problem solving. Many problem situations require repeated additions and subtractions. Following are some activities that

a.

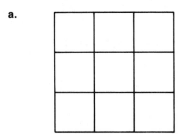

b.

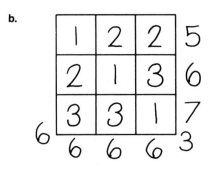

c.

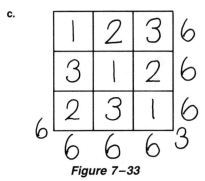

Figure 7–33

provide practice in problem solving and cause students to use addition and subtraction.

ACTIVITIES

Primary (K–3)

Objective: to use magic squares as problem-solving and computational practice.

1. Magic squares have been popular and engaging puzzles for many centuries. A magic square is a square array of numbers that produce the same sum when added along each row, column, and diagonal. For younger children, start with a 3 × 3 frame on the chalkboard or provide children with a worksheet with several such frames. Figure 7–33a shows how the frames look.

Ask the children to use the numbers 1, 1, 1, 2, 2, 2, 3, 3, 3 and to put one number in each region of the frame. It does not matter how the children decide to place the numbers. Initially, do not attempt to make the square "magic," as the sums of rows, columns, and diagonals are unlikely to be the same. Figure 7–33b shows how one child placed the numbers.

Next, ask the children to add each row and write each sum at the end of the row. Then have them add each column and

Name _____

Sums of 10

Work with a partner.

Share punchouts $\boxed{1}$ to $\boxed{9}$.

Share 10 two-color counters.

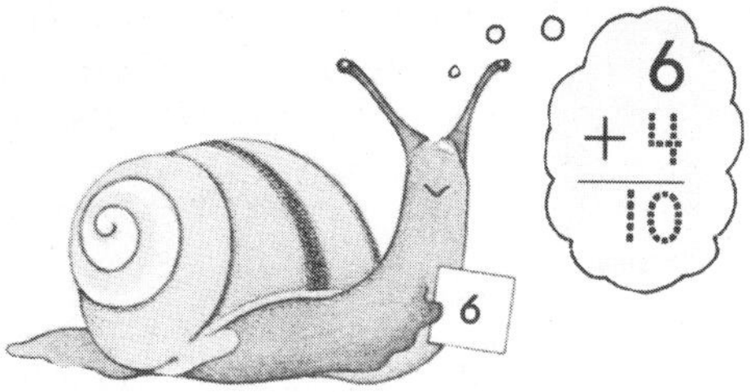

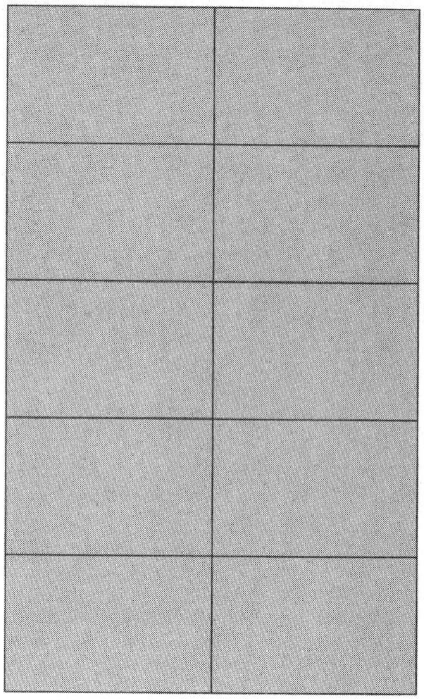

Take turns. Pick a card.
Use the number to write a
10 sum. Your partner uses
counters and ▦ to check it.

Write 10 sums here.

1.
$$\begin{array}{r} 6 \\ + \quad 4 \\ \hline 10 \end{array}$$
  +   +   +   +   +

2.
  +   +   +   +   +

Figure 7–34

Robert E. Eicholz et al. Addison-Wesley Mathematics Series (Grade 1). Menlo Park, Ca:
Addison-Wesley, 1991, p. 113. Reprinted by permission.

THE MATH BOOK

The first grade textbook page shown in Figure 7–34 illustrates a cooperative mental math activity for finding sums of 10. In this activity, children work in pairs and take turns finding two numbers that add to ten. One child picks a number card from one to nine and uses the number to write a sum of ten. The other child uses two-color counters and the 10-frame, which has been used previously, to check the addition fact. This is part of a unit on finding sums up to twelve. Throughout the unit, children are encouraged to use reasoning and not simply memorization to learn the addition facts.

This unit begins with a story about a snail that children find in a park. Eight of the ten children in the story touch the snail and the students are asked to determine how many children did not touch the snail. In addition to the math problem, the story includes scientific information about snails and a lesson in feeling good about your qualities, whatever they are. The story is available as a separate storybook and several lessons in the unit refer back to it.

Other concrete teaching ideas for this lesson include putting ten two-color beans in a cup, pouring them out on the table, and writing the addition sentence that corresponds to the colors shown, and using paper cutouts of ten circles and squares to make a picture. Lessons later in the chapter include analyzing addition facts to determine which strategy (doubles, 10-sum, or adding 1, 2, or 3) would best be used to find the sum, writing word problems for each other, finding all the different ways to write addition facts with sums of ten, using a calculator to find patterns in addition facts, determining the rule for a function machine, playing math bingo, and other mental math and cooperative learning ideas. In this series, as in many others, completing the workbook pages is only a small part of the program.

put each sum at the bottom of the column. Finally, have them add each diagonal and put the sums at the corners.

Talk to the children about some of their answers. Ask, "What is the largest sum you found in your square?" The largest would be 9. Ask, "What is the smallest sum you found in your square?" The smallest would be 3. Ask, "What number do you have the most of?" Six would be the sum found most often.

At the next stage, provide frames and ask the children to put a 1 in each row and to arrange the 1s so that no more than one 1 appears in each column. Ask the children to put a 2 in each row and to arrange the 2s so that no more than one 2 appears in each column. Then have the children put a 3 in each of the empty regions. Figure 7–33c shows one such arrangement.

Again, ask the children to add up each row, column, and diagonal. One of two possible results will occur. All but one diagonal will add to 6, or all rows, columns, and diagonals will add to 6. Discuss what numbers a diagonal needs to add to 6. See if the children can figure out what numbers to put in the diagonals. Complete the magic square so all sums are 6.

Extend magic squares by providing children with a new series of numbers to use in the square. For example, 2, 2, 2, 3, 3, 3, 4, 4, 4 can be used. This time the sum is 9. Challenge the children to put the numbers in the frame to make a magic square. Here is a possible solution using Polya's problem-solving steps.

- *Understanding the problem.* This is just like the problem we solved using 1s, 2s, and 3s, but now we are using three different numbers. The sum will be 9.
- *Devising a plan.* We will try the same plan that worked when we used smaller numbers. We will put a 2 in each row so that no more than one 2 appears in each column. Then we will do the same thing with 3 and 4. We need to make sure both diagonals add to 9 (guess and check).
- *Carrying out the plan.* When we fill in the numbers as we planned, we get a sum of 9 everywhere but in one diagonal, where we get 4 + 4 + 4, or 12. To get three of the same number that add to 9, the numbers need to be 3. So the diagonal needs to be 3 + 3 + 3. We will exchange the 4s for 3s and see if it works. It does!
- *Looking back.* We'll check again to make sure all the rows and columns and both diagonals add to 9. They do. For both the magic squares we have solved, the second of the three different numbers in the sequence fills one of the diagonals. It was 2 + 2 + 2 in the first magic square and 3 + 3 + 3 in the second magic square. We think it will be 4 + 4 + 4 in a magic square that uses 3, 3, 3, 4, 4, 4, 5, 5, 5 for its numbers. Let's try it.

Have the children name a number series, using three consecutive numbers three times each. Have the children solve the magic square.

Objective: to construct and solve problems involving addition.

2. Provide the children with a worksheet that contains several frames, each with nine numerals and an empty box at the top. Figure 7–35a illustrates one of these frames.

Ask the children to pick two numbers in the frame, add them together, and put the answer in the box at the top. Have them do the same thing for each frame on the worksheet. When the worksheet is done, the children will have made problem boxes for other children to solve.

Have the children exchange worksheets and see if they can find the pair of numbers that have the sum that equals the

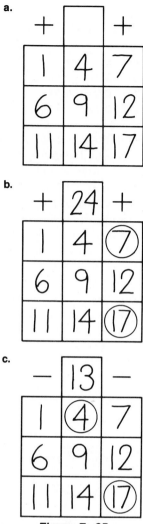

Figure 7–35

number on top. Instruct them to circle the two numbers. The children may check their solutions with a calculator. Figure 7–35b shows one solution. With the same set of numbers in a frame, many different problems can be made. It is also possible to have more than one solution for a single problem.

To extend this activity, have the children use subtraction instead of addition. Have them place the difference in the box at the top, with the operation sign on either side. Figure 7–35c shows an example. The circled numerals represent two numbers whose difference equals 13.

To further extend the activity, have the children determine all the numbers for the frame as well as the solution number and operation. You will find the problems become more difficult and challenging. Collect solutions and display them on the bulletin board.

Objective: to use a bingo activity involving problem solving, addition, and subtraction.

3. **Plus/minus bingo** may be played with two players or with the two halves of a class. Each player needs a plus/minus bingo card. The card shown in Figure 7–36a is for a game involving addition; that shown in Figure 7–36b, a game involving subtraction. Each addition card should be a different combination of the same numerals; subtraction cards should be similarly varied. Develop the bingo cards from the sums

a.

PLUS/MINUS BINGO

63	95	55	71	39	44
35	61	77	99	50	82
91	45	88	67	52	75
86	73	37	53	97	48
57	89	46	80	41	59
43	93	84	42	69	90

b.

PLUS/MINUS BINGO

21	48	5	57	31	12
3	10	58	23	50	37
43	63	14	54	7	25
20	35	9	65	29	52
61	39	56	16	59	13
45	11	27	67	18	41

c.

GROUP A	GROUP B
69 42 31 24 56 75	11 19 13 15 21 17

Figure 7–36

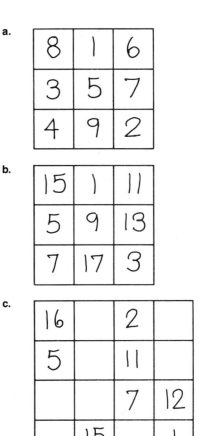

Figure 7–37

and differences of numbers found in Groups A and B (see Figure 7–36c). Select the numerals in Groups A and B to match the level of the children's arithmetic abilities.

In turn, players select one numeral from Group A and one from Group B. The players use calculators or paper and pencil to add the two numbers; then they put a marker (bean or bingo marker) on the numeral representing the sum. Four markers in a row, column, or diagonal represent a win.

To vary this activity, construct combination cards with some numerals taken from the addition card and some numerals taken from the subtraction card. Play proceeds as described above except players must announce whether they are adding or subtracting at the beginning of each turn. Because of the number of possible answers, let three markers in a row, column, or diagonal represent a win.

ACTIVITIES

Intermediate (4–6)

Objective: *to use magic squares as problem-solving and computational practice.*

1. Begin by challenging the students to complete a 3 × 3 magic square using the numbers 1, 2, 3, 4, 5, 6, 7, 8, 9. The sum for each row, column, and diagonal of that square is 15.

When that magic square has been solved, ask the students to try another 3 × 3 magic square using the numbers 1, 3, 5, 7, 9, 11, 13, 15, 17. The sum for that magic square is 27. Clever students will be able to use the solution pattern in the first magic square to guide them in solving the second magic square. Figures 7–37a and b show the completed magic squares.

There are solutions other than the ones shown, representing rotations and reflections of the square. Can you find another solution?

2. To extend work with magic squares, let the students select series of nine digits and try them in 3 × 3 magic squares. Have the students discover sequences that do not work. For example, any nine numbers in an arithmetic sequence can be successfully used in a magic square, but those in a geometric sequence seldom can be used.

See if students can determine what the sum of each row, column, and diagonal will be. Three times the middle number of a usable nine-number sequence is the sum for the magic square containing that sequence. Encourage students to make up magic square problems for other students.

Next, present challenge problems, such as a 5 × 5 magic square using the numbers from 1 to 25. The sum here is 65. Unless students discover a solution pattern, this is difficult to solve.

Try a 4 × 4 magic square, using the numbers 1 to 16. The sum is 34. A partially completed 4 × 4 magic square is shown in Figure 7–37c. Sometimes, providing partial solutions is an incentive for students who otherwise might not seek a solution.

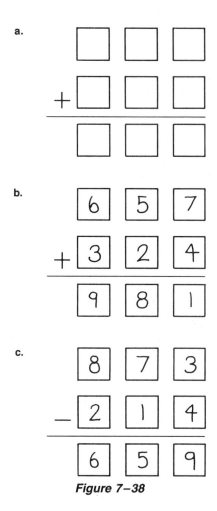

Figure 7–38

Objective: to use problem-solving skills in an addition and subtraction context.

3. The guidelines for this activity are simple. Present to the students a format for addition or subtraction with nine empty regions, as in Figure 7–38a. Explain that the object is to put each of the numbers from 1 to 9 in one of the regions so the correct sum results. Figure 7–38b shows a solution to the problem; Figure 7–38c is a solution to the corresponding subtraction problem.

   A number of solutions are possible for both the addition and the subtraction problem. To encourage students, begin a collection on a bulletin board or chalkboard of the different solutions. Have the students look for patterns to help them find solutions. For example, if the three digits in the sum add to 18, the other digits can be arranged in the addends to solve the problem. On the board, outline an area in which to place nonsolutions. Nonsolutions often provide insight to help solve the problem.

Grouping Students for Learning Addition and Subtraction

The ways that students are grouped for learning addition and subtraction are not different from the ways that students are grouped for learning multiplication and division.

To avoid repetition, we present the grouping discussion for all four basic operations in the Grouping Students section in Chapter 8.

Communicating in Learning Addition and Subtraction

For students to communicate in a variety of ways as they explore the concepts and practice the skills of addition and subtraction is an integral part of learning these topics. The discussion of communication for addition and subtraction does not differ from the same discussion for multiplication and division. To avoid repetition, we again present the discussion on communication in the Communicating section in Chapter 8.

Evaluating Addition and Subtraction Learning

Most of the addition and subtraction concepts and skills children learn have been presented by the time children finish the third grade. This is not to say that all children can add and subtract proficiently when they enter the fourth or even fifth grade.

What is a teacher to do in such cases? Individual or small group instruction is necessary. You will need to determine why there is an inability to add or subtract. Is it because the student does not know what addition or subtraction is (concept)? Does the student not know the basic addition or subtraction facts (skill)? Could it be that the student cannot perform the algorithm (skill)? Does the student lack interest because of past failure with addition and subtraction (affect)? Are assignments with too many problems causing discouragement (affect)?

There are many ways addition and subtraction learning is evaluated. Mathematics textbooks and supplemental materials accompanying the textbooks offer many options for evaluation. Chapter pretests and post-tests, midchapter checkups, and unit tests are common among math texts. Such tests provide information about how children add and subtract relative to the material contained in the chapter. They may provide page numbers in the text associated with various subsections of the test so that if children do not meet your standard for a particular section, you can quickly refer them to a particular page for additional work. Such analysis can serve as a basis for grouping or individual work with youngsters having difficulties.

While testing programs are well established for addition and subtraction, diagnostic teaching techniques are not so well established. Diagnostic procedures require observations from commercial and informal tests and, more importantly, from children's daily work to pinpoint the error patterns that lead to low computational success. Too often, children who make mistakes are told to redo a problem or to work more problems of the same type, but no one determines the source of the errors. Reworking incorrect problems may reinforce incorrect methods. Discover the cause of the problem to correct the process, not just the result.

Reasons that a child is making mistakes may include:

1. *Social, physical, or emotional problems.* Children may be hampered in cognitive skills by noncognitive problems such as a short attention span, hunger, fear of reprisal for getting the incorrect answer or not completing the work, or the desire to be "like everyone else." Ask for information about such children from the parents, previous teachers, or behavior specialists. Those who have worked with a particular youngster in the past may be able to offer advice about how to handle the youngster. Make your own observations. If you are the first to observe this behavior, report your concern to your principal or resource teacher. Generally, extra time, care, and patience are necessary to provide the environment in which mathematical growth can take place.

2. *Lack of prerequisite skills or appropriate state of development.* Children may not be ready to learn a particular concept because they have not mastered previous skills or because they have not reached the appropriate developmental stage. Children may not be ready to learn missing addends, the inverse of addition, because they have not yet reached Piaget's concrete operational stage of development and cannot comprehend the reversibility concept.

   By having children explain their thinking as they work, you will be able to pinpoint where their skills fail them. Reteach faulty prerequisite skills to students having difficulty. Give additional time and materials to children who are developmentally unready to learn.

3. *Weak knowledge of basic facts.* This is probably the most common diagnosis. For example, children may know how to perform an addition algorithm but be unable to recall basic addition facts. Children are then frustrated and unmotivated to even try. You can spot children with a weak command of basic facts by reviewing their paper and pencil work, administering fact tests, and listening to them as they explain how they perform operations.

   Work with children who need help. Provide activities that assist memorization of basic facts. Also, check to make sure children understand the concept of the operation for which they are practicing skills. The children may need to return to concrete materials to build necessary foundations.

4. *Incorrect or incomplete algorithm.* Teacher diagnosis is particularly important in this area. To help diagnose incorrect algorithms, have children show their work on all problems. If it is not apparent how the children reach an answer, ask them to explain. Their thinking is usually revealing. Children tend to make systematic errors. For example, in the subtraction algorithm, the most common error is subtracting the smaller digit from the larger digit regardless of whether it is in the minuend or subtrahend. VanLehn (1983) describes many of these faulty procedures, called **bugs.**

   Remember, an algorithm may be correct even if it is not the one you were taught. Many times, algorithms created by children who have manipulated materials are correct, whereas ones memorized by rote are not. Pinpoint the source of the error; have the children reconstruct the algorithm with manipulatives, and discuss their procedures.

5. *Wrong operation.* Children may use the wrong operation because they have misread the operation sign or because they have chosen the wrong operation in solving a word problem. The latter error is more serious. Continue the work with word problems. Discuss word problems with children having difficulty and encourage them to explain the operation to perform even if they do not work the problem. Practical situations and problems written by the children help children having difficulty understand the appropriate operations to use.

Once children's strengths and weaknesses have been diagnosed, group the children for at least part of their instructional time according to this diagnosis. At other times, let children with strong skills in a particular area help children with weaker skills; use cooperative learning groups; teach the class as a whole group or work with individuals. When grouping, keep the following points in mind:

- *Keep the groups flexible.* Do not group in October and expect to have the same groups in May. Groups should change as skills and concepts change.
- *Avoid labeling children.* Even if they are called eagles and seahawks, children know if the teacher thinks of them as slow and fast.
- *Avoid giving one group busy work while working with another group.* Let groups work independently with materials, games, the math text, or worksheets, but make sure the tasks are meaningful.
- *Have interesting tasks appropriate to the level of the group.* Each group may have different material, but all materials should be carefully thought out. What may be uninteresting for one group may be just what another group needs.

Your ability to diagnose the cause or causes of difficulty and to remediate the difficulty depends on your own familiarity with mathematics, the learning process, and the children. You will need patience. The children will need support and encouragement.

Children's activities must seem worth doing for them to gain the most. Fourth graders should be convinced that they are not doing just second- or third-grade work. To avoid boredom, try to choose aids or algorithms that are new to the child. Children who have previously failed may be visual or tactile/kinesthetic learners, so be sure to include experiences for visualizing and manipulating. Allow children to use calculators so that they may learn more advanced mathematics and not always be frustrated by their weaknesses with basic facts. Finally, be willing to set aside the regular textbook assignments that can pile up and, over time, overwhelm the slow, discouraged students.

Something for Everyone

Teaching operations with whole numbers requires attention to various learning modes. The styles of children learning addition and subtraction do not differ appreciably from the styles of children learning multiplication and division. To avoid repetition, we present a complete discussion of learning modes for all four basic operations in the Something for Everyone section at the end of Chapter 8.

KEY IDEAS

Addition and subtraction play prominent roles in our daily lives. Chapter 7 presents ways that the concepts of addition and subtraction may be taught. Children join, take away, and count objects in learning the processes of addition and subtraction. Objects such as cubes, rods, beans, and bottle caps serve the purpose well.

The skills of addition and subtraction require knowing the basic facts. Many activities help children learn the addition and subtraction facts. A variety of paper and pencil algorithms may be taught using traditional means.

Calculators and computers play increasingly important roles in addition and subtraction. Estimation and mental calculation need special emphasis as we shift from paper and pencil to electronic computation. It is important for students to decide which computational technique is most appropriate. Students should use oral and written communication skills to learn from one another as well as from the teacher. You should carefully monitor the progress of children as they learn addition and subtraction and adjust instruction according to your diagnosis.

REFERENCES

ASHLOCK, ROBERT B. *Error Patterns in Computation*. Columbus, Oh.: Merrill Publishing Co., 1990.

CARPENTER, THOMAS P.; MOSER, JAMES M.; AND ROMBERG, THOMAS A., EDS. *Addition and Subtraction: A Cognitive Perspective*. Hillsdale, N.J.: Lawrence Erlbaum Associates, 1982.

DRISCOLL, MARK J. "Estimation and Mental Arithmetic." In *Research Within Reach*. St. Louis: CEMREL, 1979.

_____ . "Mathematical Problem Solving: Not Just a Matter of Words." In *Research Within Reach*. St. Louis: CEMREL, 1979.

DAVIDSON, JESSICA. *Using the Cuisenaire Rod: A Photo Text Guide for Teachers*. New Rochelle, N.Y.: Cuisenaire Company of America, 1969.

_____ . *Idea Book for Cuisenaire Rods at the Primary Level*. New Rochelle, N.Y.: Cuisenaire Company of America, 1977.

GREENLEAF, BENJAMIN. *A Mental Arithmetic, Upon the Inductive Plan; Being an Advanced Intellectual Course, Designed for Schools and Academies*. Boston: Robert S. Davis & Co., 1859.

IMMERZEEL, GEORGE. *Ideas and Activities for Using Calculators in the Classroom*. Dansville, N.Y.: The Instructor Publications, 1976.

JACOBS, RUSSELL F. *Problem Solving with the Calculator*. Phoenix: Jacobs Publishing Co., 1977.

MOURSUND, DAVID. *Calculators in the Classroom: With Applications for Elementary and Middle School Teachers*. New York: John Wiley & Sons, 1981.

REISMAN, FREDRICKA K. *Teaching Mathematics*. Prospect Heights, Il.: Waveland Press, Inc., 1987.

REYS, ROBERT E., et al. *Keystrokes: Calculator Activities for Young Students*. Palo Alto, Ca.: Creative Publications, 1979–80.

SCHOEN, HAROLD L., AND ZWENG, MARILYN J. *Estimation and Mental Computation*. Reston, Va.: National Council of Teachers of Mathematics, 1986.

THIAGARAJAN, SIVASAILAN, AND STOLOVITCH, HAROLD D. *Games with the Pocket Calculator*. Menlo Park, Ca.: Dymax, 1976.

VANLEHN, KURT. "On the Representation of Procedures in Repair Theory." In Ginsburg, Herbert P., ed. *The Development of Mathematical Thinking*. New York: Academic Press, 1983.

8 Teaching Multiplication and Division

The NCTM *Standards* that refer to multiplication and division, concepts and skills, are the same as those that refer to addition and subtraction, which are listed at the beginning of Chapter 7. You are referred to that list.

Like addition and subtraction, multiplication and division are used extensively in our daily lives. For example, purchasing multiple items such as batteries, cans of motor oil, bars of soap, computer disks, and apples requires multiplication. Filling out tax forms requires some multiplication. Determining how many cookies and brownies are necessary for a group of children requires multiplication or division. Finding out how much you and a friend will each get when you split baby-sitting money requires division.

Computation, however, is seldom performed with pencil and paper except in school. Electronic tools, calculators, or computers do the work, or the computation is performed mentally. The efficiency and accuracy of electronic computation has been a convincing argument for its use. Children growing up in a time of electronic computation need to be skillful estimators and button pushers. They need to know what information a problem is presenting and what operations are required to solve that problem. Again, students should have the opportunity to decide whether to use paper and pencil, mental calculation, estimation, or the calculator or computer.

As with addition and subtraction, teaching multiplication and division begins before children actually start memorizing the basic facts, which are, for multiplication, the whole number products from 0×0 to 9×9. As children informally begin to manipulate objects and count, they learn about relationships and ideas associated with the concept of number. They are also establishing the foundations for learning the concepts of multiplication and division. The activities for learning the concepts of addition and subtraction mentioned in Chapter 7 are necessary for learning the concepts of multiplication and division as well. The skills of adding and subtracting are important prerequisites for performing the multiplication and division algorithms.

Teachers are responsible for determining if children are developing skills that provide a basis for multiplication and division. Word problems help to check children's thinking. For example, "Lori, Julie, and Greg decided to pick clover for a bouquet. Each of them picked two clovers. How many clovers were there all together?" Children with well-developed counting abilities will be able to count out the answer by **counting all,** using fingers or objects. Some children may **skip count,** that is, count by saying, "two, four, six." Other children may count in their heads. Opportunities to count and skip count should be provided as part of the mathematics program. For children having difficulty, provide practice in grouping objects and counting.

The manipulative materials that help children develop understanding and skill with multiplication and division include attribute materials, multibase blocks, Cuisenaire rods, beans, and cubes. Collections of objects such as bottle caps, shell-shaped macaroni, tiles, and buttons are useful in the early stages of learning about operations. Calculators and computers are helpful tools for learning multiplication and division. Children should be active as they learn about these operations.

Developing Multiplication and Division Concepts

The concept of an operation was carefully developed in Chapter 7. It serves as a supporting concept for all arithmetic operations, including multiplication and division. The idea of an operation as a change or transformation provides a consistent link from one specific operation to another. We introduced addition using real-life situations and a state-operator-state machine, and we introduced subtraction with a concrete take-away model. Thus, we introduce the concepts of multiplication and division using manipulative materials.

After a while, we move away from concrete experiences with sets of objects to the pictorial and abstract. Children learn that multiplication in renaming a pair of numbers by a single equivalent number. For example, the pair (3, 6) is associated with 18 under the operation of multiplication. Learning how multiplication works is learning the concept of multiplication. Learning that $3 \times 6 = 18$ is learning a basic multiplication fact, a skill that is dealt with later.

In the next two sections, activities to help develop the concepts of multiplication and division are presented.

Multiplication

Sorting, classifying by two or more attributes, and arriving at reversibility of thought are necessary in order for children to understand multiplication and division. Sorting and classifying were presented beginning in Chapter 5. Diagrams were used to assist children using attribute blocks in their sorting. Eventually, children were challenged to show ways of sorting using two attributes simultaneously, such as red and rectangular. **Reversibility of thought** occurs when children construct the knowledge that $2 \times 3 = 6$ also implies that $6 \div 2 = 3$. Achieving reversibility of thought means that children can understand

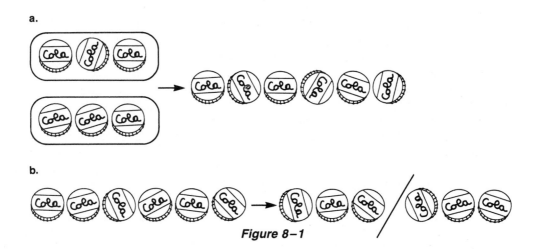

Figure 8–1

the processes of doing and undoing exemplified in the inverse operations of multiplication and division.

With bottle caps or cubes, let children construct two groups of three and indicate that there are a total of six as in Figure 8–1a. If at the same time, the children can begin with a group of six and separate it into two equivalent piles (Figure 8–1b), reversibility of thought is illustrated. Children need a variety of grouping experiences when multiplication and division are being introduced.

Have children continue to respond to number stories as they did while learning addition and subtraction. For example, each of four children is asked to select two poems about pets from those written by the class to put on a bulletin board. As they are about to put them up, the teacher says, "Genny, Ben, Theresa, and Angela have each picked two poems for the board. When they put them up, how many poems will be on the board?" Some children will know the answer without counting. Others will add two plus two plus two plus two. Still others will count the poems to find the answer. The counting or adding strategies will precede the response, "Four times two."

The operation presented initially to provide a basis for understanding multiplication joins multiple sets of equal size. Arithmetically, it involves adding several equal numbers to determine a product of two factors. For example, 3×4 may be thought of as three groups of four. The product, 12, can be determined by $4 + 4 + 4$. When first or second grade children participate in the following activity, they tend to count or use repeated addition to determine the answer. Repeated addition is a valuable model for multiplication. It is among the easiest models for young children to understand.

ACTIVITIES

Primary (K–3)

Objective: *to use repeated addition as a model for multiplication.*

1. Select a favorite flower such as a daffodil and display it. The flower may be real, artificial, or in a photograph. Ask the children how many petals are on a daffodil. If they are unable to tell or do not know, let them count the petals. Respond, "Yes, there are six petals on a daffodil. Here is another daffodil. How many petals are there on two daffodils?" The students may count the additional six petals, they may add 6 plus 6, or they may seem to just know that 2 sixes are 12. Ask, "How many petals would I have if there were three daffodils?" Continue this line of questioning for as long as it is fairly easy for the children to determine an answer. Encourage the children to discuss how they found their answers.

   As a variation of this activity, display other flowers with different numbers of petals. The National Council of Teachers of Mathematics has published an attractive book entitled *I Can Count the Petals of a Flower.* It provides many examples of flowers with varying numbers of petals. Garden catalogs usually picture many varieties of flowers. The children may have favorite flowers they would like to talk about. Children enjoy not only talking about flowers but also drawing and coloring them. A colorful display can result from a discussion that includes the foundations of multiplication.

2. You may extend the above activity by using materials that are found in small groups. Flashlight batteries are commonly found in groups of two, three, and four. Some pencils are packaged in pairs. Shoes and gloves come in pairs. Tennis balls are usually packaged in threes. Some soaps are found in groups of four. These examples and others that children will name provide opportunities to skip count and use equal addends.

Cartesian products, introduced in Chapter 5, may be models for multiplication. They employ rectangular arrays and intersecting lines. Cartesian products occur when we match, in ordered pairs, all members of one set with all members of another set. For example, if we have attribute blocks of three colors and four shapes, we can make an array with a row for each color and a column for each shape. The three rows and four columns form a rectangular array with twelve regions. Each region represents one of the twelve attribute blocks, designated by one color and one shape.

ACTIVITIES

Primary (K–3)

Objective: *to use rectangular arrays and intersecting lines as models for multiplication.*

1. Provide children with objects such as cubes, washers, or tiles. Ask the children to make a row with five tiles. Then have them make two additional rows, each having five tiles (Figure 8–2).

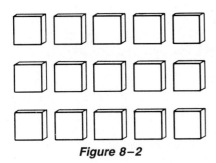

Figure 8–2

Explain to students that they have made three groups of five tiles; there are three tiles along one side and five tiles along the other. The children can describe their arrangement as three by five. Ask, "How many tiles are there altogether? Count them." There are fifteen tiles.

Next, ask the children to make a four by six group of tiles. They should have four rows with six tiles in each row. Some confusion initially may arise with children constructing six rows with four in each row. Careful explanation and patience will overcome this problem. Soon, the children will be able to construct groups that illustrate any product that is asked for.

2. For children to use intersections as models for multiplication, provide them with ten pieces of yarn, each about 25 centimeters long. Have the children lay three pieces of yarn on their desks side by side as in Figure 8–3a.

Next, have them lay four other pieces on top of the first three pieces, crossing the pieces as in Figure 8–3b. Then have the children point to each place where one piece of yarn crosses or intersects another piece of yarn. Let the children find how many intersections there are. Ask, "How many pieces of yarn were laid down first?" There were three. Ask, "How many were laid down on top of these pieces?" There were four. Ask, "How many intersections are there?" There

a.

b.

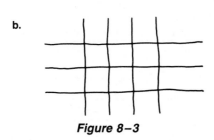

Figure 8–3

are twelve. Encourage the children to lay out different numbers of pieces of yarn and to record the number of pieces in each direction and the number of intersections. You may wish to have the children glue the yarn onto sheets of paper or to simply draw the pieces of yarn. In cooperative groups or in a whole-class setting, have the children discuss their findings and see if they can find the relationship between the number of horizontal pieces of yarn, the number of vertical pieces of yarn, and the number of intersections.

The array and intersection models may be used later as pictorial models for multiplication. The pictorial array in Figure 8–4a shows a three by six group, or the product 3 times 6. Figure 8–4b uses intersections to show 3 times 6. Both models work particularly well when children begin to develop skill with basic multiplication facts.

a.

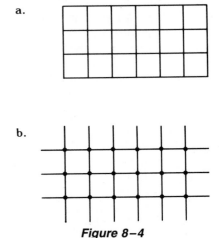

b.

Figure 8–4

Arrays and intersections are read the same symbolically. In Figure 8–4, the number of rows determines the first factor and the number in each row determines the second factor. The product is determined by counting the number of regions or the number of intersections. Both examples represent 3 times 6 equals 18, or $3 \times 6 = 18$.

Before long, children easily use the symbols for multiplication. Introduce the symbols in a multiplication sentence, $3 \times 6 = 18$, and in the vertical form

$$\begin{array}{r} 6 \\ \times\ 3 \\ \hline 18 \end{array}$$

Help the children read these sentences and become familiar with them.

Children who have had extensive work with the Cuisenaire rods can extend their work with the rods to show the meaning of multiplication. Capitalizing on the idea of repeated addition, students may begin by finding the value of three light green rods and then finding the single rod that will match this train. Thus, a blue rod is equivalent to three light green rods.

In the beginning stages of learning how to multiply with the Cuisenaire rods, do not assign number values to the rods. Only when the concept of multiplication has been presented and children are comfortable with the manipulation and language associated with the concept is it time to introduce the symbolism of multiplication. The transition from concrete to abstract requires that the models and symbols be used simultaneously.

ACTIVITIES

Primary (K–3)

Objective: *to use Cuisenaire rods to illustrate the concept of multiplication.*

1. Cross two rods that are to be multiplied, one on top of the other as illustrated in Figure 8–5a. Here, we have a purple rod on top of a dark green rod. It is read from the bottom up, "dark green times purple."

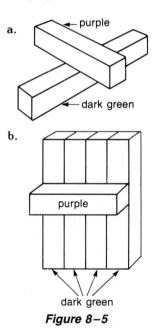

Figure 8–5

To actually perform the multiplication, ask the children to place dark green rods side by side until the purple rod reaches from one edge of the dark green rods to the other. This is shown in Figure 8–5b. Then, place the dark green rods end to end in a train (Figure 8–6). Finally, place two orange rods and a purple rod beside this train to show its value. Thus, dark green times purple equals two orange plus purple.

Objective: *to introduce the symbolism of multiplication.*

2. With the Cuisenaire rods, establish a value for the white rod (again, be careful not to declare one the permanent value of

white). If white equals 1, then a tower of yellow and light green rods is 5 times 3 (yellow times light green) and is equal to 15 (orange plus yellow). One effective way to determine the product is to put the rods side by side on centimeter grid paper, outline the rods, and count the number of centimeter squares that are contained in the outline. This technique uses the array model discussed earlier. Additional time and experience with the rods will be needed to allow children to be at ease in symbolizing the rod values.

The concept of multiplication has been presented using several embodiments. Groups containing the same number of objects were joined in some way. Numbers were then attached to the models to assist children in discovering the relationship between joining sets and multiplying numbers. We continue by introducing the concept of division in a similar manner.

Division

Teaching the concept of division is similar to teaching the concept of multiplication. The materials are the same, as is the approach. The concept, however, is different. We believe that if children are taught the concepts of multiplication and division at nearly the same time, learning division becomes easier; that one operation is the inverse of the other can more readily be seen. The operations can be directly compared and the differences noted.

The prerequisites for division are the same as those for multiplication. For the children who cannot yet sort objects by two or more attributes simultaneously and for those who have not achieved reversibility of thought, you must provide additional materials, time, and encouragement.

Division problems typically fall into two categories: measurement problems and partition problems. In a **measurement problem,** the total number of objects is provided along with the number of objects to be put into each group. It is then necessary to find the number of groups that can be made. For example, if there are fifteen pieces of paper and each child is given three pieces, how many children receive paper? Five children receive paper.

In a **partition problem,** the total number of objects is provided along with the number of groups that are to be made. It is then necessary to find how many objects will go into each group. For example, if there are fifteen pieces of paper and three children, how many pieces are given each child? The paper is partitioned into three sets of five pieces. Children should be encouraged to make up problems of their own using real-life situations to help them better understand measurement and partition division.

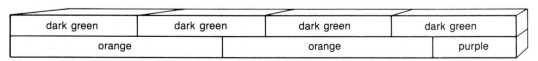

Figure 8–6

ACTIVITIES

Primary (K–3)

Objective: *to use objects as a model for division.*

1. Have the children place brown Cuisenaire rods in front of them. Ask, "How many red rods does it take to make a brown rod?" Let the children experiment to discover that four red rods are contained in a brown rod. The result will be similar to the rods in Figure 8–7a.

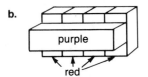

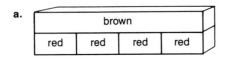

Figure 8–7

The division question we have just posed is "brown divided by red equals what?" We found that the answer is four. If we think of division as the inverse operation to multiplication, we are asking the question "Red times what equals brown?" Here, we do just the opposite of what we did in performing multiplication; we put the brown rod down, lay red rods end to end along the brown rod (Figure 8–7a), and then place the red rods side by side, searching for the rod that equals the width of the red rods, as in Figure 8–7b. It is the purple rod, so n ÷ r = p. If the white rod has the value of 1, then the purple rod has the value of 4.

2. The problem (σ + σ + k) ÷ e (orange plus orange plus black divided by blue) is solved by placing the rods together in a train and then finding the number of blue rods it takes to make that same length. It takes three blue rods as shown in Figure 8–8.

Had we divided by yellow, we would have found the result to be five yellow rods plus a red rod (Figure 8–9). The red rod represents the remainder. It is evaluated by comparing it to the white rod, which has the value of 1. Thus, (σ + σ + k) ÷ y = 5 remainder 2. Children who have used Cuisenaire rods while learning other number operations will find that division fits well with their other work.

Objective: *to practice measurement and partition division using objects.*

3. Provide children with a handful of beans. Have them count out twenty-four beans. Tell the students, "Make as many groups of six beans as you can." They will make four groups. This is a measurement problem.

Next, request that the children make six equal groups using the twenty-four beans. Ask, "How many beans are there in each group?" There will be four beans in each group. This is a partition problem.

Give the children plenty of practice using both measurement and partition problems. They should be able to recognize the difference between the two kinds of division and to use materials to illustrate the meaning of each.

Once the language describing division has been introduced, practiced, and discussed, the operation can be symbolized. The transition to the abstract symbols should take place as soon as the children are competent using objects and using the language that describes the operation they are performing.

Activities 1 and 2 above employed the Cuisenaire rods. The language the students are using is "brown divided by red equals purple, or 4." The change to number occurs quickly. If the white rod equals 1, brown has the value of 8, red has the value of 2, and purple has the value of 4. Thus, the number sentence 8 ÷ 2 = 4 results directly from the sentence using colors. Children need additional experiences with the rods to become proficient using the numerical descriptions of the rods.

In the bean activity above, students had twenty-four beans and were asked to make as many groups of six beans as they could. The children counted them and indicated there were twenty-four altogether. Repeat the activity, saying, "We are going to find out how many groups of what size?" We are going to find out how many groups of 6. Ask, "How many groups of 6 are there?" There are 4. Explain, "We say this using numbers by writing 24 ÷ 6 = 4." The symbols should be written to show the operation.

Earlier, we mentioned that you should take the opportunity to include examples of important properties of the operations being presented. There are some that are important to know when working with multiplication and division. Give the students ample opportunities to discover these properties using manipulative materials. The first of these is the **identity element for multiplication and division.** That is, any number times 1 or divided by 1 results in the number you start with. With this information, 6 × 1 = 6, 1 × 6 = 6, and 6 ÷ 1 = 6 are easy

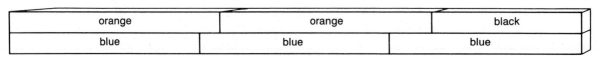

Figure 8–8

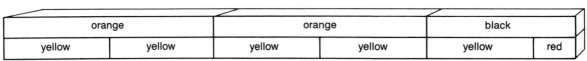

Figure 8–9

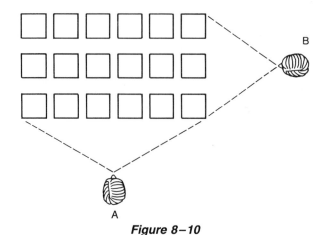

Figure 8–10

basic facts to remember. Children should discover that $1 \div 6$ (left-hand identity) does not hold for division.

A second property worth knowing is the **commutative property of multiplication.** This can be shown using any of the models presented above. For example, with the array of tiles, ask a child to stand in position A as shown in Figure 8–10 and describe the array. In that case, the array would be read three by six, or 3 times 6. There are 18 tiles. Then ask the child to move to position B and describe the array. In this second case, the array should be read six by three, or 6 times 3. The number of tiles stays the same. Students should realize that $3 \times 6 = 18$ and $6 \times 3 = 18$.

Knowing that the order has no effect on the product reduces the number of multiplication facts to be remembered. Children should have the opportunity to discover that the commutative property does not hold for division; that is, $8 \div 4 = 2$, but $4 \div 8$ does not have a whole-number solution.

A third useful pattern multiplies a number by zero or divides zero by a number. While the results are similar for both of these examples, the ideas are quite different.

Ask the children to make four rows of tiles with three tiles in each row. Have them describe their work as 4 times 3 equals 12. Next, have the children make four rows of tiles with two tiles in each row and describe the result (4 times 2 equals 8). Then, have them make four rows of tiles with one tile in each row and describe the result (4 times 1 equals 4). Finally, ask the children to make four rows with zero tiles in each row and describe the result (4 times 0 equals 0).

Have the students experiment with this activity and discuss the results anytime zero tiles are put into rows. See if the children can make a generalization regarding the result any time there are zero rows.

Encourage the children to act out these situations and to create situations of their own that involve zero. The children will quickly discover that any number times zero is zero. For example, $8 \times 0 = 0$ and $0 \times 5 = 0$.

In division, zero cannot be the divisor. To find how many groups of zero are contained in twenty-four makes no sense and is undefined in mathematics. On the other hand,

dividing zero by any number is possible. Have the children act out this situation: If there are zero pieces of clay and you wish to give each child three pieces, how many children will receive clay? Discuss how this situation is described in division (0 divided by 3 equals 0).

Similarly, describe this situation: If there are zero pieces of clay and three children, how many pieces can be given to each child? Again the children will see that 0 divided by 3 equals 0. After a number of examples, the children realize that zero divided by any number is zero. Time spent discovering and discussing these properties will make remembering the multiplication and division facts easier.

Developing and Practicing Multiplication and Division Skills

Children need to learn the skills associated with knowing how to multiply and divide. These skills are called the basic facts and algorithms. The **basic multiplication facts,** of which there are one hundred, are those ranging from 0×0 to 9×9. The **basic division facts,** of which there are ninety, are those ranging from $81 \div 9$ to $0 \div 1$.

As mentioned earlier, an algorithm is any method used to solve a problem; mathematics includes numerous algorithms for multiplication and division. As children learn multiplication and division, they can make use of paper and pencil procedures, mental arithmetic, calculators, and computers. We begin our discussion of skills with basic facts and follow with algorithms.

Basic Multiplication and Division Facts

For quick recall, children should be expected to visualize and/or memorize the basic multiplication and division facts. Memorization should begin when children understand the concepts of multiplication and division. Helping children to associate meaning with these operations is rewarded as the facts are learned. With a background in manipulative materials, children will be able to reconstruct a basic fact that they have forgotten. They can use a variety of objects as counters. They may use Cuisenaire rods, intersecting lines, arrays, or calculators.

The multiplication and division facts may be presented almost simultaneously. Typically, elementary mathematics textbooks present multiplication followed by division, with some illustration of how they are related. You may wish to start with the multiplication concept models, followed shortly by the division concept models. As children understand the concepts, they will work comfortably with both multiplication and division at the same time. Because the multiplication and division facts are closely related, they can be learned together effectively. A family of related facts includes several facts related by the numbers being multiplied and divided. One such family follows:

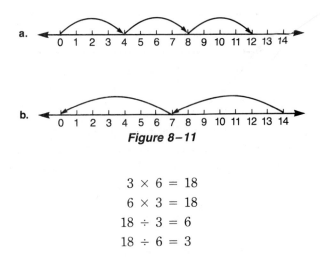

Figure 8–11

$$3 \times 6 = 18$$
$$6 \times 3 = 18$$
$$18 \div 3 = 6$$
$$18 \div 6 = 3$$

The classroom should contain tools with which to perform simple calculations. The most basic tools for calculating are the counters mentioned throughout earlier chapters. In the initial stages of learning the multiplication facts, encourage children to use cubes, beans, bottle caps, or tiles to find the answer by forming the specified number of groups, each containing a given number of objects, such as three groups of five tiles. As the work becomes more abstract and paper and pencil or calculators are used, marks on paper, including intersecting lines, help reinforce basic multiplication facts.

The number line, introduced earlier for addition and subtraction, is useful for multiplication and division. For children with little number line experience, the walk-on number line is a good place to start. Another alternative is the number line calibrated in centimeters, beside which Cuisenaire rods can be placed. Illustrate multiplication and division using a number line on which you take jumps using your finger or a pencil. When children wish to multiply 3×4, begin at 0 on the number line and take 3 jumps of 4 as in Figure 8–11a. Where you stop, 12, is the product of 3 and 4. The number line serves as an example of the repeated addition model for multiplication.

When students wish to divide 14 by 7, explain, "We are going to find how many 7s are contained in 14. Put your finger on your number line at 14." Continue, "We are going to find how many jumps of 7 we can make as we move from 14 to 0. Let's all make one jump of 7 toward 0. Are we at 0 yet? No, so let's make another jump of 7. Are we at 0 yet? Yes, how many jumps did we make?" There were 2 jumps (see Figure 8–11b). Ask, "How many 7s are contained in 14?" There are 2. Summarize, "We can say $14 \div 7 = 2$. Now let's try $12 \div 3$." This is an example of measurement division.

The calculator can play an important role during the learning of multiplication and division facts. It should be used to quickly produce facts that are forgotten, particularly during games and activities intended to help children memorize basic facts. Using the calculator should not replace the memorization of facts. It can, however, assist children by providing immediate feedback and allowing them to continue their activity.

Factor

X	0	1	2	3	4	5	6	7	8	9
0	0	0	0	0	0	0	0	0	0	0
1	0	1	2	3	4	5	6	7	8	9
2	0	2	4	6	8	10	12	14	16	18
3	0	3	6	9	12	15	18	21	24	27
4	0	4	8	12	16	20	24	28	32	36
5	0	5	10	15	20	25	30	35	40	45
6	0	6	12	18	24	30	36	42	48	54
7	0	7	14	21	28	35	42	49	56	63
8	0	8	16	24	32	40	(48)	56	64	72
9	0	9	18	27	36	45	54	63	72	81

Figure 8–12

You can help children prepare for learning the multiplication and division facts. For example, skip counting forward and backward using various numbers helps children become familiar with the multiples of these numbers. The first ten multiples of the numbers from 0 to 9 are the basic multiplication facts.

Another way to prepare children for the basic multiplication facts is to have them collect objects and record how many objects there are. For example, put groups of four beans in portion cups. Record the number of beans in one cup, then two cups, three cups, and so on, up to ten cups.

The multiplication table can help children learn the multiplication facts. The table, shown in Figure 8–12, has ten rows and ten columns. The multiplication table is read by selecting the first factor from the left side and the second factor from the top. The product of the two factors is found in the table where that row and that column meet. The fact 8×6 is shown in the figure. Use the properties discussed earlier, the identity, zero, and commutative properties, and known facts such as the twos and fives to eliminate all the facts students already know. They will be pleasantly surprised to see how few facts remain to be memorized.

The activities that follow are specifically designed for multiplication. Some are modifications of the activities presented in Chapter 7 for addition. Each activity may be used just as effectively for division by making simple changes in the materials.

ACTIVITIES

Primary (K–3)

Objective: *to help children memorize the basic multiplication facts.*

1. Commercially prepared printed flash cards have been used successfully for many years to help children reinforce basic

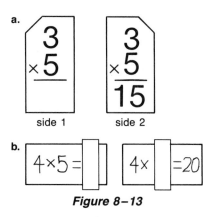

side 1   side 2

Figure 8–13

2 x 9	3 x 5	2 x 1	1 x 7	9 x 3	3 x 2	4 x 2	5 x 4	4 x 6
8 x 3	1 x 3	3 x 7	4 x 9	5 x 2	4 x 8	2 x 6	7 x 3	3 x 4
7 x 4	5 x 5	4 x 4	4 x 1	2 x 2	3 x 9	8 x 2	4 x 3	2 x 5
2 x 3	2 x 7	9 x 4	3 x 6	6 x 4	2 x 4	9 x 2	7 x 2	3 x 3
4 x 7	4 x 5	1 x 5	8 x 4	6 x 3	3 x 8	6 x 2	5 x 3	2 x 8

Figure 8–14

multiplication facts. These cards have a multiplication problem on one side and the same problem with the answer on the other side. One card from a set is shown in Figure 8–13a.

Another useful type of flash card contains the entire multiplication expression and has a sliding paper sleeve. The sleeve may be positioned over any part of the expression for special practice. This type of card helps reinforce the connection between multiplication and division. An example of this card is shown in Figure 8–13b. Show children the cards one at a time and ask them to recall the product. Put aside the products they recalled quickly, but repeat the activity with the products they did not recall. The whole class, small groups, or individuals may be involved, with cards being flashed by the teacher, an aide, or children.

Electronic flash cards resemble printed flash cards. They are displayed on preprogrammed calculators or computers. The computer records correct responses, reinforces learning with appealing graphics, provides time limits, and summarizes the results of the drill. A child may practice individually or in friendly competition with another student while the teacher attends to other important tasks.

Ideally, both printed and electronic flash cards should be used so that children can practice on problems presented in several ways. As with addition and subtraction, be careful not to frustrate children with an overemphasis on flash cards.

2. Provide practice in mental arithmetic. Practice periods should be limited to 5 or 10 minutes. Orally present the multiplication facts to children and have the children write on a piece of paper only the answer to the problem.

An alternative approach is to provide children with numeral cards (0–9) or marking boards and have them hold up the appropriate answer for you to see. Determine the difficulty of the problems based on the age and experience of the children. Initially, speedy responses are not necessary. Children need an opportunity to practice using their minds.

Ask the children to number from 1 to 5 or 10 on their papers. Explain that you are going to give them a problem and they should think of the answer and then write it down. At first, they may need to use their pencils to determine the answer; after a few sessions, they should not. Examples of mental exercises follow: What is 3 times 2? What is 2 times 4 times 3? Yes or no, 4 times 6 is more than 20? Three times 4 equals 12; what else equals 12?

As the children become more proficient, insist that they not use pencils to determine the answer. Increase the number of

questions. Expand the questions to include mixed operations. For example, begin with 5, multiply by 3, now add 5.

3. *Times up* is a game for two players that helps children learn the multiplication facts. Times up requires construction of activity materials. First, construct a board with forty-five regions, each containing a multiplication expression (see Figure 8–14).

Make two sets of twenty-four markers, each of a different color—for example, yellow and blue. In each set of markers, there should be one marker with each of the following numerals: 2, 3, 4, 5, 6, 7, 8, 9, 12, 14, 15, 16, 18, 20, 24, 25, 27, 28, 32, and 36. There should be two markers with each of the numerals 10 and 21.

To play this game, each player selects one set of markers and turns them numeral side down, mixing them together. Each player then turns over one marker to determine who will begin the game. The player with the highest number begins.

On each turn, a player turns over one marker, notes the number, and places it on a region of the playing board that corresponds to the number. For example, if a marker with the numeral 6 is drawn, it should be placed on a region containing 2 × 3 or 3 × 2. Only one marker may be placed on a region. If the marker cannot be played, it is laid aside face up. Play continues until all of the regions are covered or until no more markers can be placed on the board. The player who has placed the most markers is the winner.

This activity is unusual in that children are given the answer and are expected to find the problem. This allows children to establish a different set of associations with the basic multiplication facts. Competition in this activity is limited to the "luck of the draw." Thus, the opportunity to win is accorded all who play.

4. The game *four-in-a-row multiplication* is a variation of tic-tac-toe. Two to four students may play. This game requires a game board like the one shown in Figure 8–15 and a pair of dice with the numerals 1 to 6. Blank dice are available on which numerals may be written, or standard dice may be used for this game. Each player should be supplied with a set of unique markers, eighteen each for two players, twelve each for three players, and nine each for four players.

To start play, each player rolls the pair of dice; the numbers that show are used as factors to form a product. For example, if 3 and 4 are showing, the product is 12. The player with the largest product takes the first turn.

Each player rolls the pair of dice in turn, multiplies the two numbers together, and places a marker on the playing board

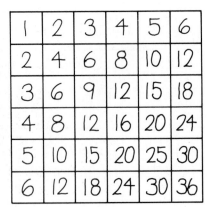

Figure 8–15

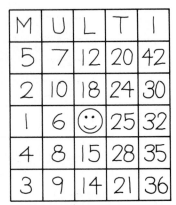

Figure 8–16

to cover the product. If the product is already covered, the turn is lost. The winner is the first player to get four markers in a row horizontally, vertically, or diagonally.

To extend this game, dice with the numerals 2 to 7, 3 to 8, or 4 to 9 may be used. For each new pair of dice, a new game board is necessary. For dice with the numerals 2 to 7, the game board contains the numerals 4 to 49. The game board is the section from the multiplication table corresponding to the numbers on the dice.

5. This is a bingo activity called *multi-bingo*. This game may be played with a small group or the entire class. Each player will need a MULTI bingo card and a number of markers. A sample card is shown in Figure 8–16. Each bingo card has one happy face region (free space) and twenty-four regions with numerals. The numerals are randomly placed in each column. Following are the numerals for each column on the card:

        M: 1, 2, 3, 4, 5
        U: 6, 7, 8, 9, 10
        L: 12, 14, 15, 16, 18
        T: 20, 21, 24, 25, 28
        I: 30, 32, 35, 36, 42

The leader draws a calling card from a shuffled deck and calls out the letter and multiplication expression on the card. The calling cards have the following letters and expressions:

    M: 1 × 1, 2 × 1, 3, × 1, 4 × 1, 5 × 1
    U: 2 × 3, 7 × 1, 2 × 4, 3 × 3, 2 × 5
    L: 2 × 6, 2 × 7, 5 × 3, 4 × 4, 9 × 2
    T: 4 × 5, 3 × 7, 8 × 3, 5 × 5, 4 × 7
    I: 6 × 5, 4 × 8, 5 × 7, 6 × 6, 7 × 6

Players complete the multiplication and place a marker on the corresponding answer in the appropriate column. The first player to get five markers in a row horizontally, vertically, or diagonally wins the game.

6. This activity, *top it*, was introduced earlier to help practice the addition and subtraction facts. Figure 8–17 shows a deck of easy and a deck of hard multiplication cards designed to accommodate the children's skill levels. The decks may be combined into a single deck representing nearly all of the basic multiplication facts.

Begin play with the stack of cards face down. The first player turns over a card and places it face up, giving the product. The second, third, and fourth players do the same.

The player with the highest product claims all of the cards played. If any players are tied with high cards, the cards remain in front of the players and those who are tied each draw another card. The player with the highest product takes all of the cards. The winner is the player with the most cards at the end of the game.

___

This chapter presents a few of the numerous activities that can be used to help children practice the basic multiplication facts. Each activity can be adapted to include division practice. The goal is for the children to be able to accurately and quickly respond to all basic multiplication and division facts. Sources of other activities and games have been included in the bibliography at the end of this chapter.

Although the properties of multiplying by one, the commutative property of multiplication, and multiplying by zero can be helpful in learning the basic multiplication facts, the **associative property of multiplication** can also be helpful for children as they multiply numbers like 4 × 16. If children rename the larger factor and apply the associative property, the problem can be made easier. For example, 4 × 16 may be renamed as 4 × (2 × 8). Using the associative property, the problem may be restated as (4 × 2) × 8, then 8 × 8, which is a basic fact and equals 64.

The associative property is shown in 4 × (2 × 8) = (4 × 2) × 8; how the numbers are grouped to multiply does not affect the product. As children learn this property, give them repeated simple examples that they can calculate from memory or with a calculator. These examples should allow them to discover that the product is unchanged regardless of the order in which the numbers are multiplied.

A property that is helpful when mentally multiplying numbers is the **distributive property of multiplication over addition.** For example, when multiplying 7 × 8, we can think of 8 as 5 + 3, multiply 7 × 5 and 7 × 3, and then add the products. Thus, 7 × 8 = 7 × (5 + 3), which then equals (7 × 5) + (7 × 3), or 35 + 21 = 56.

An illustration using arrays helps children visualize an application of the distributive property. Figure 8–18 shows

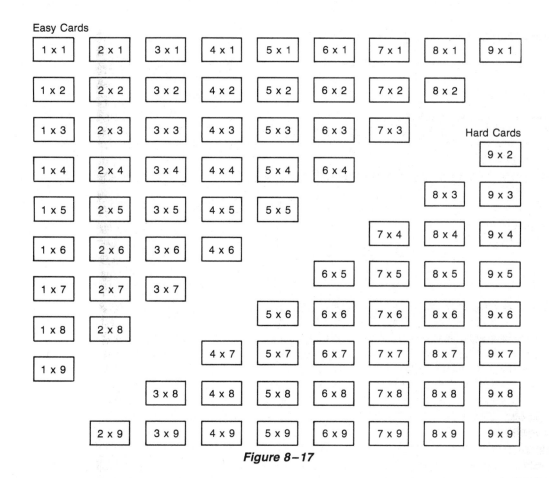

Figure 8–17

7 × (5 + 3). Children can move the dividing line to see that they could use 7 × (4 + 4) or 7 × (6 + 2) as well.

While not very efficient with paper and pencil, the distributive property helps us recall basic facts and simplify the larger problems we do in our heads. To mentally multiply 8 × 74, think 8 × (70 + 4) is 560 + 32, or 592. Find the product of a larger multiplication problem by add-

ing the products of two smaller problems. There are other variations of the distributive property, such as distributing over subtraction, but the property just described is the one most widely used. Materials for teaching multiplication, including base ten squares and an addition/multiplication grid, may be found in Appendix B.

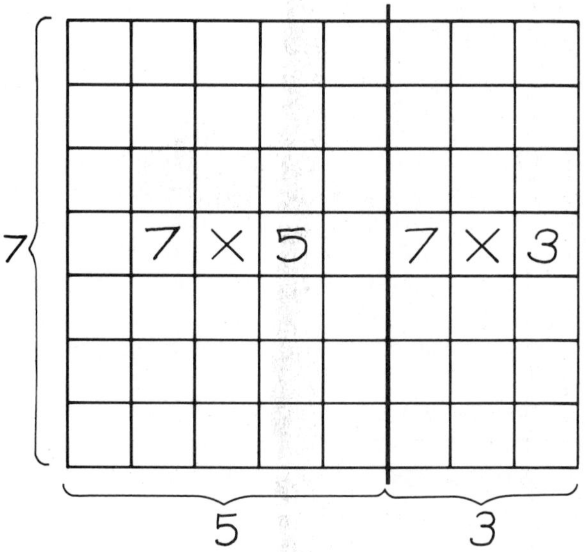

Multiplication and Division Algorithms

Children learn the multiplication and division algorithms more rapidly if they already know the basic multiplication and division facts. Further, a firm grasp of addition and subtraction is needed to perform the algorithms efficiently. As with addition and subtraction, modeling multiplication and division using the base ten blocks or other manipulative materials prepares students for learning the algorithms and gives them the opportunity to use real-life situations in posing problems, to discuss their work with one another, and to write about their work. The students should be encouraged to construct their own algorithms for multiplication and division before you present those in the book. This gives students the opportunity to explore and invent algorithms. Let the students decide which algorithm they prefer and when they no longer need to use the manipulatives. It is likely that the students will devise some algorithms similar to the standard multiplication and division algorithms, but be ready for algorithms that are

Figure 8–18

different. The process of inventing algorithms is an important part of learning how and why algorithms work.

In mathematics textbooks the multiplication and division algorithms are usually taught sequentially, beginning with the simplest problems, those requiring no regrouping (carrying). The difficulty and complexity increases until multi-digit problems with regrouping are presented. However, when students begin multidigit multiplication and division using manipulative materials, there is no need to separate problems that require regrouping from those that do not.

We believe children should master the fundamentals of multiplying and dividing, but caution you about assigning page after page of laborious multiplication and division problems. When children can demonstrate paper and pencil algorithms for multiplication and division, it is time to let them use calculators to speed up computations and allow time for higher mental processes and content that could not otherwise be covered. Using the calculator is the most efficient procedure for performing an arithmetic operation.

Paper and pencil algorithms for multiplication and division are presented in turn. We begin with the standard, and generally most efficient, paper and pencil algorithm for each operation. Alternative algorithms are then presented. Alternative procedures serve as teaching algorithms and help bridge the concrete and abstract. It was mentioned above that students sometimes invent algorithms similar to the standard multiplication and division algorithms. Likewise, they may invent algorithms similar to the alternative algorithms presented below. While in this chapter the standard algorithm is presented first followed by alternative algorithms, it is not suggested that children will invent algorithms in this order. You should not introduce standard algorithms until students have explored, invented, and discussed their own algorithms.

Multiplication.   The standard paper and pencil multiplication algorithm is applied to the six problems below, each of which has at least one two-digit factor.

$$
\begin{array}{cccccc}
12 & 62 & {}^{2}25 & 34 & 43 & {}^{34}38 \\
\times\,3 & \times\,4 & \times\,5 & \times\,10 & \times\,12 & \times\,45 \\
\hline
36 & 248 & 125 & 340 & 86 & {}^{1}190 \\
 & & & & {}^{1}43 & 152 \\
\cline{5-6}
 & & & & 516 & 1710 \\
\end{array}
$$

The first example shows multiplying by a one-digit number with no regrouping. The second problem involves multiplying by a one-digit number with regrouping from the tens to the hundreds place. The third example shows multiplying by a one-digit number with regrouping from the ones to the tens place. Typically, when a single-digit number and a multiple-digit number are multiplied, the multiple-digit number is placed on the top in the standard algorithm, as shown in the first three examples.

The last three examples show multiplying by two-digit numbers. The fourth example shows multiplying by a two-digit number that is a multiple of 10 with no regrouping. The fifth example shows multiplying by a two-digit number with no regrouping. The final example shows multiplying by a two-digit number with regrouping.

In the examples with regrouping from the ones to the tens place the regrouping numeral has been shown. Most who use this algorithm include it. It is important that children understand what happens when regrouping occurs. This is an extension of their work with place value.

Typically, it is at the fourth grade level that children are introduced to the multiplication and division algorithms. At this time, with a sound foundation of preparation, children can construct the procedures necessary for performing successfully their own or the standard algorithms. Using manipulative materials such as multibase blocks or similar models assists in the effort. Most math texts illustrate the algorithms as they are presented. Teacher's guides suggest that teachers use manipulative materials as models for a textbook algorithm.

We illustrate using an example stated as a word problem. Miss Jensen's class has completed work on a language arts project. Each student has written two poems and made an illustration of one of the poems. The 26 students believe they have turned in all their work. Just to check, Miss Jensen asks Shelley to count the papers. Because each student used a separate sheet for each part of the project, there should be 3 sheets for each person. Shelley counts 76 papers. Miss Jensen asks the class if 76 is how many sheets there should be.

Figure 8–19 shows a method the class might use to solve the problem 3 × 26 with multibase blocks representing the papers and a place value board.

In this example, the repeated addition model is used first. What the student sees is 26 + 26 + 26 (see Figure 8–19a). The unit cubes are combined into a group of ten with eight remaining. The ten cubes are exchanged for one long (see Figure 8–19b). Then the longs are joined. The result, 7 tens and 8 ones, or 78, is the product of 3 and 26 (see Figure 8–19c).

A second procedure, shown in Figure 8–20, follows the standard paper and pencil algorithm. The first step is to multiply 3 × 6. This is illustrated by three groups of six small cubes. Ten cubes are then exchanged for one long (see Figure 8–20a). The next step is to multiply 3 × 20. This is shown by three groups of two longs. The long from the exchange above is joined to these six longs (see Figure 8–20b). The final result is 7 tens and 8 ones, or 78 (see Figure 8–20c).

Another useful model using the multibase blocks is to organize the blocks into a rectangular arrangement, with each dimension representing one of the factors being multiplied. For example, to multiply 12 × 14, we would construct a rectangular region showing 12 in one dimension and 14 in the other dimension as in Figure 8–21. The

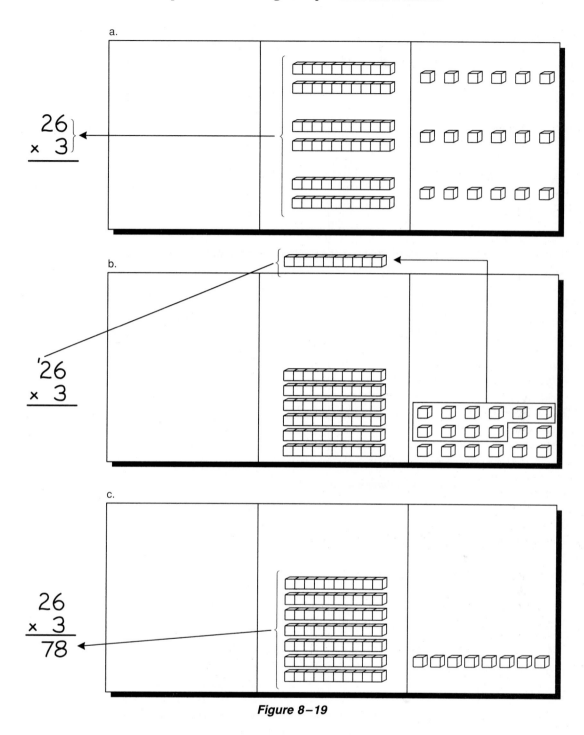

Figure 8-19

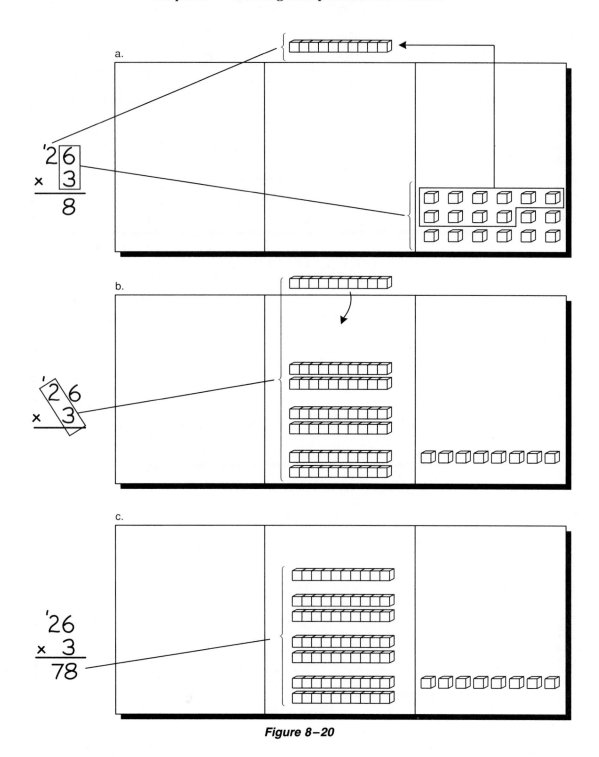

Figure 8–20

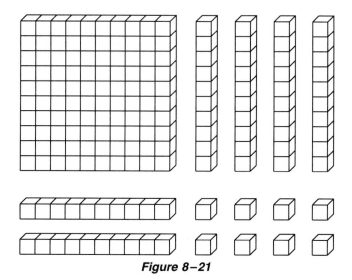

Figure 8–21

product of 12 and 14 is found by describing the pieces that make up the rectangle, one hundred plus six tens plus eight ones (168). For the problem illustrated earlier, 3 × 26, we would construct a rectangle showing 3 in one dimension and 26 in the other dimension as in Figure 8–22. The final product is found by describing the pieces, six tens plus 18 ones (60 + 18, or 78). Whether using the repeated addition approach or the standard paper and pencil algorithm, the multibase blocks help to show how the algorithm works. A similar material, the algebra blocks, can be used in place of the multibase blocks. The algebra blocks are introduced in some detail in Chapter 14.

Regardless of which procedure is used, the class finds that not all the language arts papers have been turned in. The papers are soon located.

After students have constructed their own algorithms and compared them to those of other students, you may want to introduce the algorithms in the textbook. The multiplication algorithms presented in various math texts tend to be similar and quite easy to follow. The texts provide slow, step-by-step progressions from the easiest to the more difficult multiplication problems. As well, texts provide plenty of practice exercises. Children should practice for several days after they are able to perform the algorithm without assistance. It is unnecessary for children to be assigned all of the problems on a textbook page. If children continue to have difficulty accurately completing algorithms, diagnose the difficulty. Assigning additional problems before children's error patterns are corrected is counterproductive. At this point, children need special assistance and perhaps a new approach to help them learn

the algorithms. Once children have learned the algorithms, they need periodic practice to reinforce their skill.

The activities that follow differ from those presented earlier. Their primary focus is not manipulative activities but alternative ways to present multiplication algorithms.

ACTIVITIES

Intermediate (4–6)

Objective: *to use a variety of multiplication algorithms.*

1. The **expanded notation algorithm** is an outgrowth of the work with place value. Two solutions using expanded notation are shown below. In problem a, no regrouping is necessary to solve 3 × 23.

   The problem is rewritten from standard into expanded notation. The 3 ones and the 2 tens are each multiplied by 3. Then the expanded notation is rewritten into standard notation.

   In problem b, 4 × 27 is figured in a similar manner, multiplying the 7 ones and the 2 tens by 4. As the answer is rewritten into standard form, regrouping must occur. The 28 must be expanded to 20 + 8 and the 20 added to the 80, resulting in 100 + 8, or 108. The expanded notation algorithm is similar to the standard multiplication algorithm except that regrouping occurs at a different time. The expanded notation algorithm is advantageous because children never lose sight of the numbers they are regrouping.

a.
$$\begin{array}{r} 23 \\ \times\ 3 \end{array} \qquad \begin{array}{r} 20+3 \\ \times\quad 3 \\ \hline 60+9=69 \end{array}$$

b.
$$\begin{array}{r} 27 \\ \times\ 4 \end{array} \qquad \begin{array}{r} 20+7 \\ \times\quad 4 \\ \hline 80+28= \end{array}$$
$$= (80+20)+8 = 100+8 = 108$$

2. Children should have had experience working with the loop abacus. They should know that anytime ten counters appear on a given loop, the counters must be exchanged for one counter on the loop immediately to the left. This is demonstrated as we solve 3 × 35 on the abacus. In the first frame of Figure 8–23, an abacus shows 3 × 35.

   The clips separate three representations of 35. In the second frame, we remove the holding clips in the ones column and exchange ten counters on the ones loop for one counter on the tens loop. Five counters are left in the ones column.

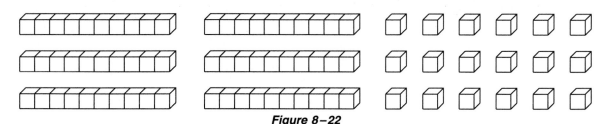

Figure 8–22

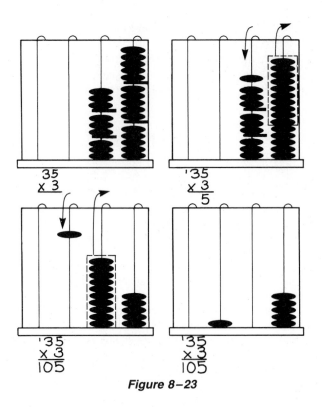

Figure 8–23

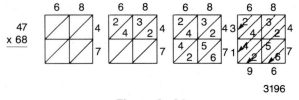

Figure 8–24

In the third frame, we remove the holding clips in the tens column and exchange ten counters on the tens loop for one counter on the hundreds loop. No counters are left in the tens column.

The fourth frame illustrates the abacus after all the exchanges have been made. The standard algorithm shows the progression from start to finish.

3. The **partial products algorithm** is another procedure that helps children through the process of regrouping. It may be used whether or not regrouping is necessary. The problem below illustrates the partial products algorithm by multiplying 8 × 57.

$$\begin{array}{r} 57 \\ \times\,8 \\ \hline \end{array} \qquad \begin{array}{r} 57 \\ \times\,8 \\ \hline 56 \end{array} (8\times7)$$

$$\begin{array}{r} 57 \\ \times\,8 \\ \hline 56 \\ 400 \end{array} (8\times50) \qquad \begin{array}{r} 57 \\ \times\,8 \\ \hline 56 \\ 400 \\ \hline 456 \end{array}$$

The first step in the solution is to multiply the number in the ones column, 7, by the multiplier, 8. The answer, 56, is the first partial product and is placed beneath the problem.

The next step is to multiply the number in the tens column, 5, by the multiplier, 8. The result, 40 tens, or 400, is placed beneath the 56.

Finally, the partial products are added to arrive at the answer, 456. Take care with this algorithm to assure that the partial products are written in the correct place value positions.

The partial products algorithm may be illustrated using manipulatives. In Figure 8–21, the multibase blocks were used to show 12 × 14. In describing the pieces that make up the rectangular region in Figure 8–21, we found that 12 × 14 was equal to one hundred plus six tens plus eight ones. The partial products algorithm for 12 × 14 is shown below.

$$\begin{array}{r} 14 \\ \times 12 \\ \hline 8 \\ 20 \\ 40 \\ 100 \\ \hline 168 \end{array}$$

In this algorithm, we find the same product that we found with the multibase blocks—one hundred plus six tens plus eight ones. Making such connections is important for students as they learn mathematics.

Incidentally, the same partial products are found when the horizontal *FOIL* algorithm is used to solve (10 + 2) (10 + 4). This algorithm is presented in Chapter 14, Introducing Algebra.

4. The **lattice algorithm** for multiplication has been around for several hundred years. For each pair of numbers multiplied, a lattice is constructed. The lattice shown first in Figure 8–24 illustrates the problem 68 × 47.

For each place value position in the factors, a region divided diagonally is provided. Begin by multiplying 4 times 8 and writing the product, 32, in the regions where 4 and 8 intersect. Then multiply 4 times 6 and write the product, 24, where 4 and 6 intersect. Continue by multiplying 7 times 8 and 7 times 6, recording each answer in the appropriate regions.

Finally, beginning at the lower right, add diagonally from right to left (see arrows) to determine the digits in the product of 68 and 47. The answer is read beginning at the upper left, down the left side of the lattice and across the bottom—the answer is 3196.

This very different algorithm provides a clever approach to long multiplication problems. It can be worked easily with factors of any size.

John Napier, a Scottish mathematician who lived from 1550 until 1617, created a unique multiplication tool for peasant workers who had little education and little knowledge of the basic multiplication combinations. The tool, Napier's rods, consisted of a series of rods on which the multiplication tables had been written. A person could carry the rods in a pocket.

The technique of lattice multiplication was used with the rods. Patterns for constructing the rods are shown in Figure 8–25. The first rod is the index rod and contains a vertical listing of

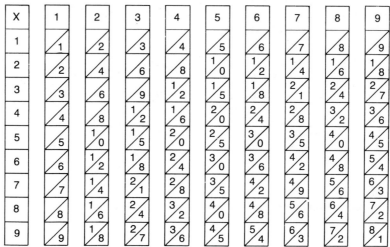

Figure 8–25

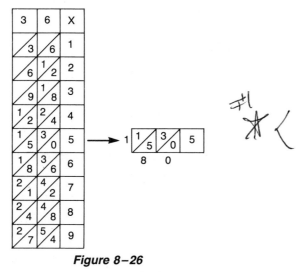

Figure 8–26

factors. The first digit at the top of each rod is another index factor.

Napier's rods resembled the orange Cuisenaire rod in size and shape. We find it easier to use the front and back of strips of oaktag to construct the rods.

When multiplying two numbers, lay the appropriate rods side by side next to the index rod. Then add along the diagonals. For example, Figure 8–26 shows how to multiply 5 × 36. The 3 rod and the 6 rod are placed side by side next to the index rod. Move down the index rod to 5. Then add diagonally as we did with lattice multiplication. The result, 180, is read in the same way as a result is read from the lattice.

Children enjoy taking a break to construct and use Napier's rods. See if they can discover how to perform multiplication such as 26 × 365.

Division.   The procedure used to perform division is quite different from that used to perform multiplication. Division does, however, rely heavily on knowledge of multiplication as well as subtraction. Skill with the basic subtraction, multiplication, and division facts is a key to success in performing division algorithms.

Recall that the basic division facts are related to the basic multiplication facts. Thus, for the multiplication fact 3 × 7 = 21, we have the corresponding division facts 21 ÷ 3 = 7 and 21 ÷ 7 = 3. Techniques for helping children memorize these facts were discussed earlier, in the section on basic facts.

When introducing division algorithms, once again be sure to let students use real-life situations and manipulatives to construct their own algorithms. Allow plenty of time for students to discuss their results with each other.

Five examples of the standard paper and pencil division algorithm are presented below.

$$\begin{array}{r} 13 \\ 3\overline{)39} \end{array}$$

$$\begin{array}{r} 44 \\ 3\overline{)132} \\ \underline{12} \\ 12 \\ \underline{12} \\ 0 \end{array}$$

$$\begin{array}{r} 44\ r2 \\ 3\overline{)134} \\ \underline{12} \\ 14 \\ \underline{12} \\ 2 \end{array}$$

$$\begin{array}{r} 12 \\ 15\overline{)180} \\ \underline{15} \\ 30 \\ \underline{30} \\ 0 \end{array}$$

$$\begin{array}{r} 16\ r13 \\ 27\overline{)445} \\ \underline{27} \\ 175 \\ \underline{162} \\ 13 \end{array}$$

The first three examples have one-digit divisors. The first of these has no regrouping, the second has regrouping, and the third has regrouping and a remainder. The last two examples have two-digit divisors; both have regrouping, and the last one has a remainder. The only markings that appear are in the last example, where regrouping was necessary during the first subtraction.

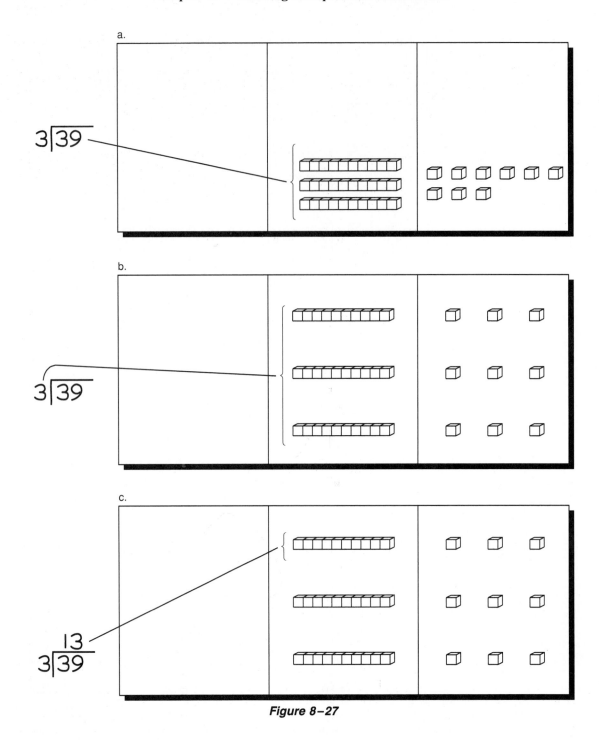

Figure 8–27

When the division algorithm is taught, it is important to initially accompany the instruction with manipulative materials. We demonstrate with multibase blocks. The first example is 39 ÷ 3. How the problem is presented determines whether children will solve it with measurement or partition division.

In a measurement problem, the total number of objects is provided along with the number of objects to be put into each group. It is then necessary to find the number of groups that can be made. In a partition problem, the total number of objects is provided along with the number of

groups to be made. It is then necessary to find how many objects will go into each group.

The first example is stated as a partition problem: We have 39 tomato seeds and 3 planting groups for a science project. How many seeds will each planting group receive? Figure 8–27 shows how to go about solving this problem.

Using the multibase blocks and the place value board, lay out 3 longs and 9 cubes to represent 39, as in Figure 8–27a. Next, separate the pieces into three groups, each of which contains the same number of pieces, as in Figure 8–27b. We find we have 1 long and 3 cubes in each group,

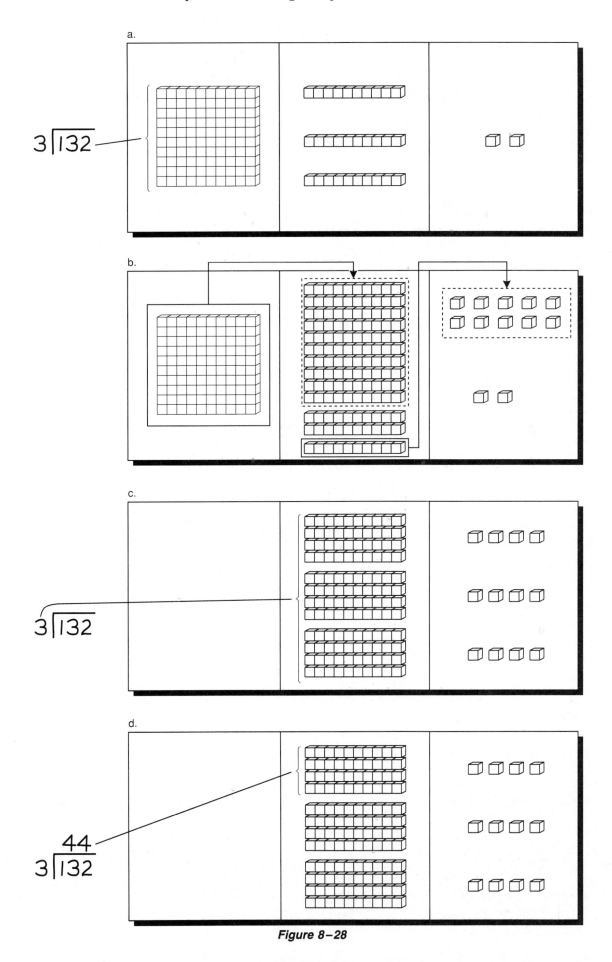

Figure 8–28

or 13 cubes, as in Figure 8–27c. Each planting group will receive 13 seeds.

The example is now stated as a measurement problem: We have 39 tomato seeds. Each student will get 3 seeds for a science project. How many students will get seeds? The solution to this problem requires that the children form groups of 3 cubes and see how many such groups can be made. It is necessary to exchange each long for 10 cubes so the grouping by 3 can be completed. In the end, 13 groups of 3 will be made. Thus, 13 students will get seeds.

While the manipulation is different for each of these problems, the algorithm is the same. How the problem is presented, however, determines how to interpret the result. In the first case, the answer is 13 seeds; in the second case, 13 students.

When regrouping is necessary in the algorithm, as in $132 \div 3$, the partition problem is most appropriate. Suppose a collection of 132 stamps is being divided among 3 children, how many stamps will each child receive? Using the multibase blocks, show 132 by displaying 1 flat, 3 longs, and 2 cubes as in Figure 8–28a.

In order to separate the pieces into 3 groups, exchange the flat for 10 longs and exchange 1 long for 10 cubes as in Figure 8–28b. Then complete the operation. Figure 8–28d shows the final result. Thus, each child will receive 44 stamps. The exchanges, from hundreds to tens and from tens to ones, represent the regrouping in this division problem.

This work with manipulative materials should not last long. Once the students have constructed algorithms and can perform the standard algorithm with manipulatives, encourage them to practice at the abstract level. As in the case of multiplication, the activities presented include alternative algorithms and approaches for division. Some children find that an alternative is superior for them and adopt it as their standard algorithm. It is appropriate for children to create their own algorithms; encourage them to do so.

ACTIVITIES

Intermediate (4–6)

Objective: *to use various algorithms for division.*

1. When the **expanded notation algorithm** is used, children are better able to see what happens in each place value position of the dividend. The problem below shows $96 \div 4$.

$$4\overline{)96} \quad 4\overline{)90+6} \quad 4\overline{)80+16}$$

$$\frac{20+4}{4\overline{)80+16}} \quad \frac{24}{4\overline{)96}}$$

The number in the dividend, 96, is rewritten in expanded form. Because 4 does not evenly divide 90, regrouping is

required from the tens to the ones place. Thus, $90 + 6$ is regrouped to $80 + 16$. Then the 80 is divided by 4 and the 16 is divided by 4, resulting in $20 + 4$. Then $20 + 4$ is rewritten into standard form, 24.

You may find it necessary to help children work through several examples of the expanded notation algorithm with concrete objects such as multibase blocks, beans, and Cuisenaire rods while performing the algorithm. Work first with problems that do not require regrouping, such as $48 \div 4$. With some practice, children should gain understanding of the division process.

2. **Repeated subtraction** may be used as an algorithm. The division problem $48 \div 12$ may be thought of as asking, "how many 12s are contained in 48?" This is the measurement concept of division. We may solve the problem by seeing how many times 12 may be subtracted from 48. This is illustrated by the problem below. Here, we see that there are four 12s in 48 because it took four subtractions of 12 to reach zero.

$$12\overline{)48} \qquad \begin{array}{r} 48 \\ -12 \ (1) \\ \hline 36 \\ -12 \ (2) \\ \hline 24 \\ -12 \ (3) \\ \hline 12 \\ -12 \ (4) \\ \hline 0 \end{array}$$

It takes many separate subtractions to solve a problem like $48 \div 4$. It is much more efficient to subtract multiples of 4. The problem below shows that the first subtraction is 10×4 and that the second subtraction is 2×4. The final result is that $10 + 2$, or twelve, 4s are contained in 48. If children had to perform twelve or more subtractions, it is questionable they would maintain much interest. Subtracting using multiples of the divisor helps eliminate this difficulty.

$$4\overline{)48} \qquad \begin{array}{r} 48 \\ -40 \ (10 \times 4) \\ \hline 8 \\ -8 \ (2 \times 4) \\ \hline 0 \end{array}$$

3. The **Greenwood,** or **down the side,** algorithm is particularly useful because it helps avoid one of the pitfalls of long division, wrongly estimating quotient figures. The Greenwood algorithm incorporates the repeated subtraction idea presented above. The problem below illustrates this algorithm for the division $597 \div 27$.

The first step is to estimate a multiple of the divisor, 27. In this case, we have estimated 10×27. The product 10×27, or 270, is subtracted from 597, leaving 327. We again estimate the multiple 10×27 in the second step. After subtracting 270 again, 57 remains. Our final estimate is

2 × 27, or 54. Subtracting 54 results in 3. Because 3 is less than the divisor, 3 is the remainder for this problem. We then add 10 + 10 + 2, which results in the quotient figure, 22. Thus, 597 ÷ 27 = 22 r3.

$$27\overline{)597}$$

$$27\overline{)597} \quad | \quad 10$$
$$\underline{-270}$$
$$327$$

$$\begin{array}{r} 27\overline{)597} \\ -270 \\ \hline 327 \\ -270 \\ \hline 57 \end{array} \begin{array}{l} 10 \\ \\ 10 \end{array}$$

$$\begin{array}{r} 22\ r3 \\ 27\overline{)597} \\ -270 \\ \hline 327 \\ -270 \\ \hline 57 \\ -54 \\ \hline 3 \end{array} \begin{array}{l} 10 \\ \\ 10 \\ \\ 2 \\ \hline 22 \end{array}$$

When the partial quotient figures were estimated, the multiples were 10 times the divisor. Using 10 or 100 times the divisor simplifies the estimation. As children gain experience with estimating, they can estimate partial quotients that are close to but less than the dividend.

A variation of the Greenwood algorithm is one in which children put their partial quotients above the dividend in the proper place value position. When all estimates have been made, the final quotient is determined by adding up. The problem below shows how this **pyramid method** works for 597 ÷ 27.

$$\begin{array}{r} 22\ r3 \\ 2 \\ 10 \\ 10 \\ 27\overline{)597} \\ -270 \\ \hline 327 \\ -270 \\ \hline 57 \\ 54 \\ \hline 3 \end{array}$$

The first estimate is 10, which is placed above the dividend, 597. Then 10 × 27, or 270, is subtracted from 597. The second estimate also is 10, which is placed above the first estimate. Then 270 is subtracted from 327, leaving 57. The final estimate, 2, is placed above the other two estimates. Then 54 is subtracted from 57, leaving 3, the remainder. The sum of the partial products, 10 + 10 + 2, is found, with the final result of 22 r3.

When performing the standard division algorithm, four basic steps must be completed, sometimes several times for a given problem. These steps are: (1) estimating the quotient figure, (2) multiplying the partial quotient times the divisor, (3) subtracting the product from the dividend, and (4) checking the difference to make sure it is less than the divisor and, if it is not, revising the quotient figure. Each of these steps provides opportunities for mistakes and frustration.

Hallmarks of the division algorithm are erasure marks on students' papers. Calculators can help students estimate quotient figures. Accord time and patience to children as they develop skill with the division algorithm. Encourage them and provide them with careful teaching and reteaching. Reward their efforts.

In this section, we have discussed developing two sets of skills. The first skills were the basic multiplication and division facts. Children should have a good command of these facts. The second skills were the algorithms for multiplication and division. Current math texts present thorough instruction to develop the standard algorithms and contain an ample number of problems for practice. You are encouraged to employ these instructional sequences. The next section provides a look at an elementary textbook example of division.

Estimating and Mental Calculating

With both paper and pencil and calculator computation involving multiplication and division, estimation and mental arithmetic play an important role. Students should know if their answers are reasonable. This is done by estimating products or quotients just before or just after working the algorithm or using the calculator, and comparing the estimated results to the calculated results. If the results vary considerably, the students should recalculate the results. The procedure takes just a few moments, and it should develop into a lifelong habit.

Estimating the quotient figure in the standard division algorithm is a skill that challenges many children. For example, below are two division problems that for most fifth graders require estimation. An older student may be able to look at the divisor, realize that 2 × 64 = 128, and estimate the quotient figure in the first example at 1 and the quotient figure in the second example at 2.

$$64\overline{)1279} \qquad 64\overline{)1289}$$

We examine two ways to estimate the quotient figure. The first is to round the divisor up or down, depending on the units digit. The numbers 61 to 64 would be rounded down to 60, while 65 to 69 would be rounded up to 70. In

PROBLEM SOLVING
Strategy: Interpreting Answers

1. Understand
2. Plan
3. Work
4. Answer/Check

The division for these problems is the same, but each problem has a different answer.

Some vans are needed to drive 37 people. Each van can carry 8 people. How many vans are needed?

$$\begin{array}{r} 4 \\ 8\overline{)37} \\ -32 \\ \hline 5 \end{array}$$

Five vans are needed.

Eight people can fill each of 4 vans. A fifth van is needed for 5 more people.

Think: What will you do to the quotient?

A box has 37 cups of unshelled peanuts. The peanuts are stored in eight-cup buckets. How many buckets can be filled from the box?

$$\begin{array}{r} 4 \\ 8\overline{)37} \\ -32 \\ \hline 5 \end{array}$$

Four buckets can be filled.

Five cups of peanuts will be left in the box.

Think: What will you do to the quotient?

CLASS EXERCISES

Solve. Then tell why you increased the quotient by one or ignored the remainder.

1. The minibus seats two people to a seat. How many seats are needed for thirteen people?

2. The theater changes shows every 3 weeks. How many shows does it complete in 52 weeks?

3. Bill has 48 free passes to the movies. He wants to give each of his 9 friends an equal number. How many passes will each friend get?

4. The rows in the theater have 9 seats per row. How many rows are needed to seat 174 people?

Figure 8–29

Ernest R. Duncan et al. Houghton Mifflin Mathematics *(Grade 5). Boston: Houghton Mifflin, 1991, p. 116. Reprinted by permission.*

THE MATH BOOK

Figure 8–29 shows a fifth grade textbook page on problem solving. Notice that in this lesson, students are learning to read a word problem carefully and to interpret their answer in terms of whether it makes sense for the problem. After students find a numerical answer to a problem, they are asked to analyze that answer and to explain what should be done with the remainder. This increases the students' ability to communicate with each other and to justify their answers.

The teacher's manual suggests that the students use counters and a table to demonstrate the solution to the first problem before working the problems abstractly. As we have seen in so many situations, the students are asked to construct their ideas concretely and to discuss them with other students before going on to more abstract work. This is as true of this fifth grade program as it was of programs for the primary grades.

Follow-up activities suggested in the teacher's manual include a tie-in to social studies with a problem on finding the number of presidential elections in the United States since 1789. Students are then asked to work in groups of three or four to research a particular time period in history and to ask three mathematical questions based upon facts they find. This helps students make a connection between mathematics and other subject areas and helps them build skills in problem creation and cooperative learning.

On preceding pages, the textbook contains several long division problems that students are asked to solve with paper and pencil after having demonstrated them with base ten blocks. That lesson focuses on common mistakes made when there is a zero in the quotient. Earlier problems ask students to decide whether to use a calculator or paper and pencil to work a problem and to look at a word problem and choose the correct operation. The chapter test on the next page has long division computation exercises as well as word problems that ask the students to choose the correct operation and to decide what should be done with the remainder. Keep in mind that it is important to use the teacher's manual in conjunction with the student text to avoid asking students to simply memorize abstract procedures.

our problem, 64 would be rounded to 60. The guide number, 6, is used to determine the first quotient figure in 1279 ÷ 64. We ask, "How many 6s are contained in 12?" Our estimate is 2. We soon discover that 2 is too large and revise the estimate to 1, which is correct. Using the same procedure for 1289 ÷ 64, we estimate a quotient figure of 2; this one is correct. With this rounding procedure, children should be willing to revise their estimates.

The second way to estimate the quotient figure is to construct a table of multiples of the divisor. Using a calculator, this takes a few moments. The multiples of 64 are:

$$1 \times 64 = 64 \quad 4 \times 64 = 256 \quad 7 \times 64 = 448$$
$$2 \times 64 = 128 \quad 5 \times 64 = 320 \quad 8 \times 64 = 512$$
$$3 \times 64 = 192 \quad 6 \times 64 = 384 \quad 9 \times 64 = 576$$

By examining the dividend of the first problem, the student will realize that 2 × 64 in the table of multiples is too large, and therefore the estimate must be 1. For the second problem, the estimate 2 is correct because 2 × 64 = 128. Not only is the first quotient figure easy to estimate, the other quotient figures in the problem are also easy to estimate.

A variation of this approach is to estimate the entire quotient using the multiples of 64 in combination with the powers of 10. Thus, the first estimated quotient figure is 10, to which you add 9 as you continue the problem. This is similar to the approach used in the Greenwood algorithm discussed earlier.

No single method of estimating quotient figures has been shown to be superior. Encourage the children to construct their own algorithms and let the children select the one they find most comfortable.

ACTIVITIES

Intermediate (4–6)

Objective: to practice estimating multiplication and division.

1. Use a set of double-nine dominoes. The large-format dominoes work well if this activity is done with the whole class. Hold up a domino for about three seconds. Ask the children if they believe the product of the two sides of the domino is more than, less than, or equal to 40. Let them signal to you that the product is 40 by putting a hand flat on the table or floor. A hand with thumb up means more than 40. A hand with thumb down means less than 40.

   For division estimation, put a numeral such as 50 on the chalkboard. Hold up a domino for about three seconds, and have the students find the sum of the two sides and then divide 50 by the sum. Have them indicate whether the quotient is more than, less than, or equal to 5. The students can use the same hand signs as before.

2. On occasion, we need to multiply large numbers or estimate their product in our heads. Sometimes these numbers are powers of 10 or multiples of powers of 10, as in 50 × 70 or 30 × 600. Other times, the numbers are not even decades, as

in 38 × 19 or 53 × 691. In the first case, numbers like 50 and 70 may be multiplied by counting the zeros (there are two) and then multiplying the remaining numbers, 5 and 7. When the result, 35, is rejoined with the two zeros, we have 50 × 70 = 3500.

With 30 × 600, follow the same procedure. Count the zeros (there are 3) and multiply the remaining numbers, 3 and 6, to get 18. Rejoin the 3 zeros to produce the answer, 18000.

In the case where the numbers are not powers of 10 or multiples of powers of 10, use the rounding strategy described in Chapter 4. The numbers are rounded up or down, then multiplied, and then the product is adjusted. For example, with 38 × 19, round 38 to 40, and round 19 to 20. Then multiply 40 × 20. The result, 800, is an estimation of the product of 38 × 19. Because both numbers were rounded up, the estimation is somewhat higher than the actual product of 38 × 19, which is 722.

For the problem 53 × 691, round 53 down to 50 and round 691 up to 700. Then compute the product of 50 × 700, which is 35000. Because one number was rounded up and one was rounded down, it is more difficult to tell whether our estimation is too high or too low. The actual product of 53 × 691 is 36623 and shows us the estimate was low. If both numbers had been rounded down, the estimate would have been below the actual product.

Another type of problem requiring mental calculation is sometimes presented. If 6 × 49 or 6 × 51 requires an exact answer, remember that 6 × 49 is 6 less than 6 × 50. Without difficulty, we know 6 × 50 = 300; six less is 294. Likewise, 6 × 51 is 6 greater than 6 × 50, or 306. It is helpful for children to practice multiplying numbers that are even decades, one less, or one greater.

Objective: to use historical material to motivate estimation and mental arithmetic.

3. Chapter 7 presented several mental exercises that appeared in an 1859 text by Greenleaf. Here are mental exercises prescribed by Fish in *Arithmetical Problems, Oral and Written,* copyrighted in 1874. Just before he presents 73 mental exercises and 120 written exercises, Fish notes:

   Any class or pupil that has gone over the Elementary Rules, in regular course of any textbook on these subjects, should be prepared for test, drill, and review in the examples of this chapter. (p. 47)

   Examples of the exercises Fish included in his book follow. The page on which each problem can be found is provided in parentheses.

   ● 8 × 9 ÷ 12 + 3 × 5 + 10 ÷ 11 × 8 + 20 ÷ 12 + 4 × 7 − 3 = how many? (p. 49)
   ● A number multiplied by 8, divided by 6, multiplied by 10, and the product increased by 5 equals 45; what is the number? (p.49)
   ● A farmer sold a grocer 15 bushels of potatoes at $1 a bushel, and bought 20 pounds of sugar at 15 cents a pound, and 10 pounds of coffee at 30 cents a pound; how many pounds of tea at 75 cents a pound could he buy for what was still due him? (p. 51)
   ● If it costs $56 for bricks to build a cistern when bricks are worth $8 a thousand, what will it cost for bricks to build it when they are worth $10 a thousand? (p. 52)

Objective: to use the calculator to improve multiplication and division estimation skills.

4. Four to six players and one calculator are needed. One player is *it* for the first round. That player generates a number on the calculator by entering and multiplying four numbers equal to or less than 12, such as 11 × 8 × 6 × 9 = 4752. The object of the game is to find numbers that will divide the number generated and then to divide by those numbers, generating new numbers. Play continues until a player correctly announces "prime," indicating that only two factors are left, 1 and the number itself.

Play begins as soon as the initial number is generated. The calculator is passed to the player to the left of the person who is it and continues to the left. The player who is it does not play in that round. The first player divides 4752 by a number he or she thinks will divide evenly by pressing the divide sign, entering the number, and pressing the equals sign. Neither 1 nor the number itself may be used as a divisor. If the number is correct, the calculator is passed to the next player. If it is incorrect, that is, if the answer is not a whole number, the player restores the previous number, by multiplying by the number used to divide, and passes the calculator to the next player. It may be necessary to round the product to restore the previous number.

For each correct division, a point is scored. The player whose division results in a prime number and who declares "prime" is awarded another point. Each player must either divide or declare "prime." The person correctly declaring "prime" generates a new number. After three rounds, the player with the most points wins.

Problem Creating and Solving

Multiplication and division are important skills in problem solving. These operations are often necessary to help solve problems. For example, in Mr. Edwin's class the students are constructing a cardboard geodesic dome. The dome radius is to be 125 centimeters. The dome is constructed using triangles of two different sizes; some triangles are equilateral with sides 0.6180 times the dome radius, and the other triangles have one side 0.6180 times the dome radius and two sides 0.5465 times the dome radius. Appliance cartons are available at the local dealer. It takes fifteen of the equilateral triangles and forty-five of the other triangles to make the dome. How many dishwasher cartons should be requested? Generate discussion about the size of each triangle and the additional information needed in order to request the cartons. Following are more problem situations that require the use of multiplication or division.

ACTIVITIES

Intermediate (4–6)

Objective: *to construct and solve problems involving multiplication and division.*

1. Provide children with a worksheet that contains several frames, each with nine numerals and an empty box at the top. Figure 8–30a illustrates one of these frames.

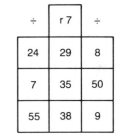

Figure 8–30

Ask the children to pick any two numbers in the frame, multiply them together, and put the answer in the box at the top. Have them do this for each frame on the worksheet. When this is done, the children will have made problem boxes for other children to solve.

Have the children exchange the worksheets and see if they can find the pair whose product equals the number on top. Instruct them to circle the two numbers. The children may check their solutions with the calculator.

Figure 8–30b shows one solution. With the same set of numbers in the frame, many different problems can be made. It is also possible to have more than one solution for a single problem.

To extend this activity to division, have the children divide any number in the box by a smaller number in the box and put the remainder in the box at the top. Figure 8–30c shows an example. We solve the problem below:

- *Understanding the problem.* The object here is to divide a larger number by a smaller number to produce a quotient with the remainder 7. The divisor must be greater than 7 because a divisor equal to or smaller than 7 cannot have a remainder of 7.
- *Devising a plan.* We should not use 7 as a divisor, as it cannot result in a remainder of 7. Thus, we begin with the smallest possible divisor, 8, and the smallest dividend, 9, and work up through each of the dividends to be sure to try

all combinations. After 9, we try these dividends: 24, 29, 35, 38, 50, and then 55 (numbers from the problem). Next, we use 9 as a divisor and work up through the dividends beginning with 24. We make a systematic list, continuing this way until we find the number with a remainder of 7.

Is it possible that none of the divisions has a remainder of 7? Is it possible that more than one of the divisions have a remainder of 7?

- *Carrying out the plan.* We begin by dividing 24 by 8, 29 by 8, and 35 by 8, finding remainders of 0, 5, and 3. It occurs to us that if we can think of the greatest multiple of the divisor less than the dividend, we can just subtract that multiple from the dividend to get the remainder. For example, we subtract $24 - 24 = 0$, $29 - 24 = 5$, and $35 - 32 = 3$ to find the first three remainders. Going on, we subtract $38 - 32 = 6$, $50 - 48 = 2$, and $55 - 48 = 7$. There is an answer!

Is it the only answer? We continue working and find another solution, 55 divided by 24.

- *Looking back.* We take the two number pairs we found, $55 \div 8$ and $55 \div 24$, and divide again to make sure the remainder is 7. It is in both cases. We have satisfied the problem.

Further extend the activity by having the children determine all the numbers for the frame as well as the solution number and operation. The problems will become difficult and challenging. Make a display of problems and solutions.

Objective: to develop a winning strategy.

2. This activity is called *high to low.* Two players may play. The players mix up double-six dominoes and place them face down on the floor or a table. Each player turns over a domino. The player with the higher product goes first.

Each player draws ten dominoes, keeping them face down. The first player turns over two dominoes as in Figure 8–31a.

Using each domino to represent a two-digit factor, the player decides which factor each domino will represent. For example, the four-two domino may be used as 42 or, reversed, 24. The three-zero domino may be used as 30 or 3. Once the two factors have been chosen, the player multiplies them together, using paper and pencil or a calculator. The player then puts the product on a scoring sheet similar to the one in Figure 8–31b.

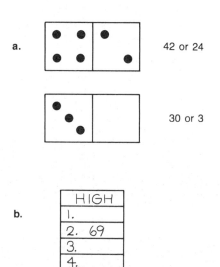

Figure 8–31

a.

×					
	15	35	10	25	40
	6	14	4	10	16
	12	28	8	20	32
	18	42	12	30	48
	27	63	18	45	72

b.

×					
	15		10	25	
		14		10	
		28			32
	18		12		
	27				72

Figure 8–32

The object is to put the product in one of the spaces so the other products will eventually be in order from highest to lowest. Suppose our first player, Harry, selected 23 and 3 as his two factors. He finds the product is 69 and places it in space number 2 on his scoring sheet as in Figure 8–31b.

The second player, Roberta, turns over two dominoes, selects two factors, multiplies them, and records the answer on her scoring sheet. Because five pairs of dominoes are selected, each player may discard one pair during play. However, the pair must be discarded before the player records the product on the scoring sheet. The player declares that turn a pass. Once a player passes, he or she must play all of the remaining dominoes.

Play continues until all the spaces on the scoring cards of both players are filled. The players calculate their final scores by marking out any products not in order and adding the remaining products. If all four numbers are in order from highest to lowest, the player doubles the score. The player with the higher score wins.

For division, have the players proceed as above except the two dominoes represent a dividend and a divisor. Have them divide the larger number (dividend) by the smaller (divisor) and put the remainder on the scoring sheet. The other rules, including those for scoring, remain the same.

Objective: to make and use problem tables.

3. This activity serves, in part, as a review of the multiplication and division facts. Begin by putting on the chalkboard, an overhead projector, or worksheets, a mixed-up multiplication table, with no factors along the left side or across the top. An example is shown in Figure 8–32a.

Explain that all the numbers in the table are in the right place. The students must find the factor that belongs to the left of each row and at the top of each column to make the table correct.

To make this activity more challenging, present a table missing some of the numbers as well as the factors, as shown in Figure 8–32b. Have the children both complete the table and insert the factors.

Notice that the examples use single-digit factors. Extend this activity by introducing numbers each of which is the product of a single-digit factor and a two-digit factor or the product of two two-digit factors. An interesting computer program from Sunburst, "Teasers by Tobbs," provides a similar activity.

Encourage the children to make up tables of their own to see if they can stump other students. See who can construct the most difficult table puzzle. Display the results on a bulletin board.

Grouping Students for Learning Multiplication and Division

When students are learning the concepts of the basic operations, the activities in which they are engaged lend themselves to either a whole-class setting or cooperative learning groups. You may demonstrate to the whole class, for example, using base ten blocks on the overhead projector, how 13 × 16 may be illustrated. This provides children the chance to ask questions of the teacher and to clarify their thinking regarding the models being used. When using Cuisenaire rods or objects to illustrate the concept of an operation or the concept of a particular operation, having the students work in groups may be most appropriate. By working in cooperative groups, students will be able to share sometimes scarce materials. As well, a variety of materials may be used simultaneously, in keeping with the principle of multiple embodiment. The conversations among group members as they work together help individuals clarify their thinking and the thinking of others as they construct new concepts and algorithms.

As students learn the skills of the basic operations, the activities in which they are engaged lend themselves, by and large, to cooperative learning group, pair, or individual learning. A cooperative group goal may be to assure that all group members have facility with some or all of the basic facts in addition, subtraction, multiplication, or division. Considerable energy will likely be put forth, and the resulting pride in achieving the "basic facts" goal will belong lasting. When students engage in games and activities, they often are engaged as pairs, threes, or fours. When practicing with the computer or calculator, students will likely work individually or in pairs.

Students who are participating in estimating and mental calculating activities may be organized in any fashion. You may find whole-group instruction efficient for practicing mental arithmetic as you can give instruction, provide examples, and present practice problems. When individuals

need or desire special work in estimating and mental calculating, then small group or individual participation may be most appropriate.

Because considerable time is spent developing the basic facts skills in elementary school, you are advised to provide instruction in a variety of group settings to help maintain student interest. The activities suggested in Chapters 7 and 8 provide you with activities that can be adapted to various modes of grouping.

Communicating in Learning Multiplication and Division

Oral and written communication provide ways for children to learn mathematics and thinking processes. Concepts and skills associated with the basic operations can be clarified, as well as illustrated, by children communicating with one another as well as with the teacher. Number stories were mentioned early in the process of developing the concepts of all the basic operations. Oral communication that includes the necessity for children to decide what is being asked, if information is missing, or if a passage or story makes sense, as well as accompanying mental calculation, should be a regular part of instruction. Stories such as those presented in the Open Court Thinking Story™ books—for example, *Measuring Bowser* (1985)—are particularly rich in calculations along with practice in logical thinking and problem solving. Opportunities for students to listen and respond to stories involving calculating should be plentiful.

The constructivist approach to teaching mathematics discussed in Chapter 3 stresses that students should be asked to share their theories about concepts, that students should be engaged in dialogue with the teacher and with one another, that you should have students elaborate on their initial responses to questions, and that you should ask thoughtful, open-ended questions and encourage students to ask questions of each other. All of these suggestions are intended to develop students' abilities to communicate mathematically.

Students are sometimes asked to keep journals as part of their language arts program. It is just as appropriate to keep a journal in mathematics. Topics for writing include algorithms that students use to perform various calculations, whether these methods involve paper and pencil, mental calculation, estimation, or the calculator or computer. These journal entries may include the standard algorithms or those devised by individuals. Further, attitudes about mathematics or learning mathematics are appropriate topics for mathematics journals. Creative stories about number may be episodes written as part of composition practice or as a journal entry. Reading Juster's delightful book *The Phantom Tollbooth* (1961) will surely inspire students in grades 5–8 to write some wonderfully imaginative stories.

Cooperative and individual writing are appropriate activities for cooperative learning groups. As group members explore a mathematical problem, the group recorder provides a chronicle of the thinking process. These records are then shared with other groups during the "debriefing" time at the end of the activity. Discussion is then invited. Excellent examples of written descriptions of calculations and thinking are presented by Marilyn Burns in her books *A Collection of Math Lessons from Grades 1 Through 3* (1988) and *A Collection of Math Lessons from Grades 3 Through 6* (1987).

Evaluating Multiplication and Division Learning

Learning the basic multiplication and division facts begins in the primary grades, generally in grades 2 and 3. Learning the algorithms for multiplication and division generally begins in earnest in grade 4. That is why the algorithm activities have been focused at the intermediate level.

Not all children who reach or even leave grade 6 can multiply or divide proficiently. What should you as a teacher do when students are not performing at your level of expectation? Use individual or small-group instruction. Determine why there is an inability to multiply or divide. Is it because the student does not know what multiplication or division is (concept)? Does the student not know the basic facts (skill)? Can it be that the student cannot perform the algorithm (skill)? Does the student lack interest because of past failure with multiplication and division (affect)? Are assignments with too many problems causing discouragement (affect)?

There are many ways multiplication and division learning are evaluated. Math texts include carefully designed testing programs. Virtually every step of the learning process is tested and retested. Children having particular problems are often referred to specific chapter sections for remediation. Part of the teacher's responsibility is to assure learning takes place at a pre-established level.

In Chapter 7, we discussed several reasons children make mistakes. Here we look at one of those reasons, using incorrect or incomplete algorithms. Regardless of how carefully you or a math text teaches multiplication and division algorithms, some children will not be able to perform them. Your diagnostic skills will be challenged soon after you enter the classroom. From the time of your first assignment and thereafter, there will be children's hands in the air or children lining up by your desk requesting help.

The first comment you may hear is "I don't get it." This means the student does not know what the assignment is, did not listen to your explanation, wants your attention, does not want to try, or is unable to do the algorithm. You must decide what caused the student to say, "I don't get it."

The diagnosis has begun. Generally, you must individually reteach part or all of an algorithm or look at a partially or fully completed algorithm and decide where the student made an error. Most teacher's guides list only the answers to the problems from the student texts. They do not show each step in the algorithm. Thus, with only the answer at hand, you must work through the algorithm with which the student is having difficulty to see where the error is.

In a multiplication problem like 24×54, you may find one or more of these errors: (a) problem miscopied from the text, (b) basic fact error, (c) carrying was not done, (d) carrying took place when not needed, (e) student forgot to add the carry after multiplying, (f) problem not completed, (g) partial products not lined up, and (h) error in adding partial products. Each of these errors is illustrated in problems a–h below. Problem i shows the correct standard algorithm.

a.	b.	c.	d.
$^{12}45$	$^{1}54$	54	$^{1}54$
$\times 24$	$\times 24$	$\times 24$	$\times 24$
180	218	206	216
90	108	108	118
1080	1298	1286	1396

e.	f.	g.	h.
$^{1}54$	$^{1}54$	$^{1}54$	$^{1}54$
$\times 24$	$\times 24$	$\times 24$	$\times 24$
206	216	216	216
108		108	108
1286		324	1276

i.
$^{1}54$
$\times 24$
216
108
1296

All of this assumes that you and the children are using the same algorithm. For children who use an alternative algorithm, there may be other types of errors.

Coping with the "I don't get it" children demands teaching time. Besides diagnosing and correcting children's written errors, children may need to return to manipulative materials to illustrate whatever is not clear from the abstract or pictorial approach. Above all, you must be supportive and encouraging. Let the children know you are working with them for the same ends.

Skillful diagnosis of the cause or causes of children's difficulties requires you to be proficient in mathematics and knowledgeable about children and the learning process. Take advantage of resource people in your school who may be able to assist in diagnosing. Sometimes students need extra time or a new approach for learning basic facts. Sometimes algorithms must be carefully retaught, perhaps using grid paper to emphasize proper alignment of digits.

For children to gain the most from them, activities must seem worth doing. A sixth grader should be convinced that she or he is not doing just fourth- or fifth-grade work. Allow children to use calculators so they may continue to learn other or more advanced mathematics and not always be frustrated by their weaknesses with basic facts. Finally, be willing to set aside certain textbook assignments; assignments can pile up and turn students away from mathematics.

Something for Everyone

Teaching operations with whole numbers involves considerable work with abstract symbols. The activities in Chapters 7 and 8 are intended to supplement and enrich the math text. In the mathematics textbook, operations are usually carefully presented using pictorial and abstract modes. Teacher's manuals recommend that teachers use concrete approaches, as well. All of these approaches, the manipulative, pictorial, and abstract, are presented because we know that children tend to learn from the concrete to the abstract.

There are other modes of learning, however, that should be attended to as children learn whole-number operations. Children who learn most effectively in the visual mode should be provided with pictures and encouraged to draw pictures and diagrams as they develop concepts and skills involving operations with whole numbers. Representing operations on the number line and drawing arrays to illustrate multiplication facts may be especially useful for visual learners. The math text can be helpful for these learners, as well, when attractive illustrations are accompanied by careful instructions, but be aware that children do not always interpret diagrams the way textbook authors intend them to. Ask the children to tell you what they think the drawings mean.

Many computer programs, such as "Math Ideas with Base Ten Blocks" by Lund, from Cuisenaire Company, have effective graphic displays and may be used to reinforce skills in operations with whole numbers as a follow-up to work with the actual manipulative materials. Seeing the graphic display next to the abstract exercise is helpful to visual learners.

Tactile/kinesthetic learners find materials such as colored cubes, base ten blocks, Cuisenaire rods, and abacuses useful in learning operations with whole numbers. Tactile learners should see and manipulate actual models of the operations as they are simultaneously writing down the abstract symbols. This manipulation helps them to connect what they are doing to the abstract algorithms they are learning in the books. Be sure that these children understand that the particular material they are using is not important. Encourage them to use several different materials to model the same exercise.

Challenge tactile learners to create their own algorithms for the operations and then test them using concrete models to see if their methods will work for all kinds of problems. Let these students teach their algorithms to other students in the class and keep a record of the favorite methods.

Children who learn most effectively in the auditory mode should find it fairly easy to learn basic facts and algorithms by oral means. These children will likely be able to understand oral instructions and suggestions for correcting faulty algorithms more easily than other children. They can sometimes help in explaining adult instructions to other students. They may enjoy listening to tapes or records of songs and poems designed to help children memorize basic facts. They may even make up their own mnemonic devices to remember facts or the steps in an algorithm.

Children who process information linearly and sequentially benefit from an emphasis on following each step in an algorithm. These children can often work comfortably at an abstract level by fifth or sixth grade and may be able to follow the written instructions in the book if accompanied by careful oral directions from the teacher. These children often do well with a traditional approach to instruction, but even these children can benefit from some work with concrete materials. Although such children may appear to be doing well, they may not have a concrete understanding of what they are doing. They should be encouraged to illustrate each step of an algorithm with manipulative materials such as chips for trading, an abacus, or bundling sticks.

Children who process information holistically or spatially are more apt to visualize an answer without going through a step-by-step process. Even though these children often have a superior number sense, they may not do well in a traditional mathematics program because they have difficulty in memorizing isolated facts or in following an algorithm with several steps.

Such children should be encouraged to explore the relationships among facts both concretely and abstractly. For example, they may be able to find the answer to $6 + 7$ because they know the doubles and they know that $6 + 7$ is 1 more than $6 + 6$ or 1 less than $7 + 7$. Some children even hide these abilities because they believe they are cheating if they use the answer to a previous fact to find a new answer instead of memorizing each fact in isolation. Be sure to let such children know that it is good to relate facts. The best mathematicians are those who can find the greatest number of relationships among known ideas and use them to discover new relationships. Children who process information visually should be able to work well with spatial materials such as base ten blocks and Cuisenaire rods.

Be sure to have children with different learning styles share their methods with each other. All students can benefit by learning to work in different modes and can strengthen their own abilities to learn by being able to choose among a number of different learning strategies.

Talented children should be provided with enriching experiences that extend their thinking abilities. Challenge them with problem-solving activities that use operations with whole numbers (the ones suggested in Chapters 7 and 8 may provide a starting point). Let gifted children make up problems for each other and the rest of the class or explore patterns in the addition and multiplication tables or in multidigit algorithms.

Use the pre- or post-tests provided in textbooks to test gifted children before you assign work with a given unit. The tests can tell you if there are any gaps in the student's knowledge of the information you are about to present. If the children have already mastered the concepts, provide them with new challenges. Never force the talented children to work every problem on a page when the other children are working only the even exercises; talented children soon learn to hide their talents to avoid boring busywork.

Challenge talented children by exploring algorithms that have been used historically for operations with whole numbers. Children sometimes believe that there is only one right way to work a problem, and they are amazed to find that people in other times or even today in other parts of the world use algorithms quite different from the ones found in American textbooks. Ask the children to explain why these algorithms work. Encourage them to make up other algorithms of their own. After they master operations with whole numbers in base 10, introduce operations in other bases.

Children who have difficulty with operations in base 10 may need a slower, more individualized approach. If children do not seem to understand the standard algorithm, try a different method. Children who have trouble with the traditional multiplication algorithm may have more success with a lattice. Children having difficulty dividing may be more successful using the Greenwood method.

Children having difficulty may need more time to work with manipulative materials. Do not rush them to learn abstract algorithms. Encourage them to write down the algorithm as they manipulate the material. When the materials are no longer needed, most children give up the materials by themselves because it is faster to write the algorithm without manipulating the materials.

Children's learning styles and abilities should help determine how you present mathematics. Be sensitive to the individual needs of all children with whom you work. Be sure, however, to include enriching and pleasurable activities for all students.

KEY IDEAS

Multiplication and division play important roles in our daily lives. Chapter 8 presents models to help conceptualize multiplication and division. Children exchange sets for objects, count and add, construct arrays, exchange objects for sets, and partition sets in learning the concepts of multiplication and division. Objects such as cubes, rods, beans, and multibase blocks help children learn.

The skills of multiplication and division begin with learning basic facts and continue as children learn various algorithms. A number of activities present ways to help children memorize the basic facts.

A variety of paper and pencil algorithms are presented. As well, children are encouraged to construct their own algorithms. Calculators and computers play an increasingly important role in multiplication and division. Estimation and mental calculation become more important as we shift to electronic computation. Children should learn to decide which calculation technique is most appropriate. Continued work on communication skills will improve students' ability to think about multiplication and division.

Carefully diagnose and check the progress of children as they learn multiplication and division. Design instruction to meet the variety of children's learning styles.

REFERENCES

BURNS, MARILYN. *A Collection of Math Lessons from Grades 3 Through 6*. New Rochelle, N.Y.: Cuisenaire Co. of America, 1987.

BURNS, MARILYN, AND TANK, BARBARA. *A Collection of Math Lessons from Grades 1 Through 3*. New Rochelle, N.Y.: Cuisenaire Co. of America, 1988.

DIENES, Z. P. *The Elements of Mathematics*. New York: Herder & Herder, 1971.

FISH, DANIEL W. *Arithmetical Problems, Oral and Written; with Numerous Tables of Money, Weights, Measures, Etc.* New York: Ivison, Blakeman, Taylor & Co., 1874.

JUSTER, NORTON. *The Phantom Tollbooth*. New York: Random House, Inc., 1961.

REYS, ROBERT E., ET AL. *Keystrokes: Multiplication and Division*. Palo Alto, Ca.: Creative Publications, 1979.

SCHOEN, HAROLD L., AND ZWENG, MARILYN J., EDS. *Estimation and Mental Computation*. Reston, Va.: National Council of Teachers of Mathematics, 1986.

SEYMOUR, DALE. *Developing Skills in Estimation, Book A*. Palo Alto, Ca.: Dale Seymour Publications, 1981.

SHEFFIELD, LINDA JENSEN. *Problem Solving in Math, Book D*. New York: Scholastic Book Services, 1982.

WAHL, JOHN AND STACEY. *I Can Count the Petals of a Flower*. Reston, Va.: National Council of Teachers of Mathematics, 1977.

NCTM *Standards*

GRADES K–4

Standard 12: Fractions and Decimals

In grades K–4, the mathematics curriculum should include fractions and decimals so that students can—

- develop concepts of fractions, mixed numbers, and decimals;
- develop number sense for fractions and decimals;
- use models to explore operations on fractions and decimals;
- apply fractions and decimals to problem situations.

GRADES 5–8

Standard 5: Number and Number Relationships

In grades 5–8, the mathematics curriculum should include the continued development of number and number relationships so that students can—

- understand, represent, and use numbers in a variety of equivalent forms (integer, fraction, decimal, percent, exponential, and scientific notation) in real-world and mathematical problem situations;
- develop number sense for whole numbers, fractions, decimals, integers, and rational numbers;
- understand and apply ratios, proportions, and percents in a wide variety of situations;
- investigate relationships among fractions, decimals, and percents;
- represent numerical relationships in one- and two-dimensional graphs.

Standard 6: Number Systems and Number Theory

In grades 5–8, the mathematics curriculum should include the study of number systems and number theory so that students can—

- understand and appreciate the need for numbers beyond the whole numbers;
- develop and use order relations for whole numbers, fractions, decimals, integers, and rational numbers;
- extend their understanding of whole number operations to fractions, decimals, integers, and rational numbers;
- understand how the basic arithmetic operations are related to one another;
- develop and apply number theory concepts (e.g., primes, factors, and multiples) in real-world and mathematical problem situations.

(NCTM, 1989, pp. 57, 87, 91. Reprinted by permission.)

Concepts about rational numbers begin to develop long before children enter school. When children are asked to share a granola bar fairly with a brother or sister, they begin to intuitively grasp the idea of $\frac{1}{2}$. In kindergarten, these intuitive ideas may be introduced more formally but the emphasis should continue to be on situations from the child's life that utilize a variety of concrete materials. Children should realize that rational numbers are very much a part of their everyday lives, and they will need a thorough understanding in order to function as intelligent adult consumers.

Ask the children to keep a record of all the times rational numbers are used in their everyday lives. You may be surprised at the large number of uses you find. Create a bulletin board with the uses the children find at home, in newspapers and magazines, and from interviewing people about the uses in their careers. Interviews may reveal such things as the baker using common fractions when preparing recipes or formulas, the bus driver using decimal fractions when buying gas and figuring mileage, the store manager using percents when planning a sale, the car salesperson using percents to figure the commission earned, the nurse using decimal fractions to measure out the medicine to give a patient, the teacher using percents to figure students' grades, and the government official using percents to determine budgets.

Children themselves must use rational numbers when they cook a meal, sew an apron, measure a garden, build a birdhouse, tip a waiter, or figure the amount they earn for $2\frac{1}{3}$ hours of baby-sitting. The United States system of money and the metric system of measurement are based on decimal fractions. Conventional measures of length, area, weight, volume, and time make extensive use of common fractions. Sales and sales tax commonly use percents. It is important, therefore, for children to have a solid understanding of all types of rational numbers.

Some people have argued that the proliferation of calculators and the move toward using the metric rather than the conventional system of measure may make it unnecessary to use common fractions, but common fractions will continue to be used to describe such everyday occurrences as eating $\frac{1}{2}$ of an apple, and operations with common fractions must be understood for later work with algebraic fractions, so we present rational numbers in common fraction as well as decimal fraction form.

It is important to remember that common fraction notation and decimal fraction notation are ways of naming the same rational number. The concept of the number is the same regardless of the form in which it is written. Before continuing, we give a formal definition of a rational number.

A **rational number** is one that can be expressed as $\frac{p}{q}$ where p and q are both integers and $q \neq 0$. Rational numbers can be expressed in different ways; they may be written as **common fractions** ($\frac{1}{2}$, $\frac{3}{4}$, . . .); as **decimal fractions,** commonly called decimals (0.5, 0.75, . . .); or as **percents** (50%, 75%, . . .). Any rational number may be represented by an infinite number of numerals. For example, $\frac{1}{2} = 0.5 = 50\% = \frac{2}{4} = \frac{3}{6} = \frac{4}{8} = \frac{5}{10} = \ldots$. Common fractions may also have several different meanings. They may represent:

1. the part-whole model, where the whole is a unit of measure, a geometric shape, or a set of objects.
2. a ratio between two subsets.
3. division.

Given the wide range of ways to represent rational numbers and the variety of meanings, it is not surprising that children often are confused when dealing with rational numbers in any form.

Children should use concrete materials when learning new concepts, and fortunately there are many good materials. Both commercial and teacher-made materials can aid learning of rational number concepts. These materials should be used in the primary grades as children begin to formalize fraction concepts and should also be used in the intermediate and middle school grades, when students learn to operate with common fractions, decimal fractions, and percents. The materials described in this chapter include Fraction Tiles, Fraction Factory, rectangular and circular fraction regions, Decimal Squares, base ten blocks, rulers, number lines, fraction strips, Fraction Bars, colored chips, Cuisenaire rods and arrays. Computer programs, such as the Fraction Bar programs and Gears, and calculators are also useful in the development of skills with rational numbers. As children use these materials, ask them to explore, question, discover relationships, and discuss their findings with each other. Learning should be both active and related to the child's world.

Developing Rational Number Concepts

As mentioned in the introduction, rational numbers can be represented in a variety of ways, and the representations may have a variety of meanings. This section begins with a description of several meanings for common fractions and several materials that may be used to develop those meanings. We then compare common fractions to decimal fractions and percents. The section ends with a discussion of equivalent fractions and ordering fractions.

Part-Whole Model for a Common Fraction

The first model for a fraction that children typically encounter in school is the **part-whole model.** In this model, the **denominator** represents the number of parts the whole or unit has been divided into and the **numerator** represents the number of parts currently under consideration. If the unit is a unit of measure such as length, area, or volume, each of the parts must be of equal size, even though they need not be congruent. If the unit consists of discrete objects such as chips or children, the objects need not be the same size. However, when children are first introduced to fractions, it is common to use congruent parts or discrete objects of the same size. The following are examples of activities and materials that can be used to help children develop the part-whole concept of a common fraction using area, length, volume, and discrete objects as the unit.

ACTIVITIES

Primary (K–3)

Objective: *to develop the concept of a fraction as part of a region divided into equal-sized pieces.*

1. Give each child a number of squares cut out of paper. Ask the children to fold a square in two sections. Unfold the square, and examine the sections. Are they the same size? If the square represents a candy bar to be split among two children, would each child receive the same amount? Compare squares that were folded in half to those that were not and discuss the differences.

   Using another of the squares, ask the children to fold it in half another way. Tell them to color one of the halves and to cut it out to compare the sections. These will probably be similar to the pieces shown in Figures 9–1a and b.

   Fold a square as shown in Figure 9–2. Then unfold it and color the two end pieces. Ask the children if the colored pieces are still $\frac{1}{2}$. Let the children cut out the pieces to prove that the two halves do indeed cover the same area.

   Challenge the children to find other ways to color $\frac{1}{2}$ of the square. Have them prove the answer by cutting the pieces out and showing that the colored pieces do fill the same area as the noncolored pieces. Figure 9–3 shows other responses that the children may give.

   After children have shown $\frac{1}{2}$ in many different ways, ask them to repeat the process for other fractions.

2. Use plastic or cardboard regions that have been divided into halves, thirds, and fourths. You may use commercial sets

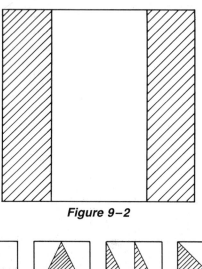

Figure 9–2

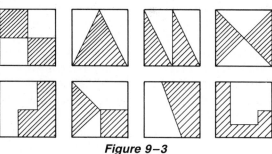

Figure 9–3

such as Fraction Tiles or Fraction Factory, which are sets made of plastic. A whole square region is one color, and congruent parts are other colors to represent various fractions (available from Creative Publications). Or, using the masters included in the appendix, make your own sets out of colored railroad board or run off copies of the regions onto colored paper and let the children cut out their own sets. It is helpful to have several units of the same color and the halves, thirds, and fourths each of a different color. Your pieces may look like those in Figure 9–4 or Figure 9–5. You may also use the square or circular fraction dies on the Ellison lettering machine to make a set of fraction pieces.

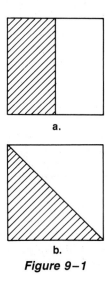

a.

b.

Figure 9–1

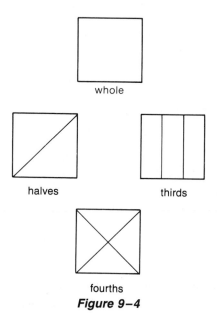

whole

halves

thirds

fourths

Figure 9–4

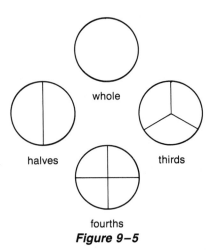

Figure 9–5

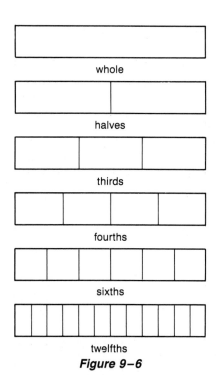

Figure 9–6

Let the children explore with the materials. Challenge the children to make a complete region using pieces all the same color. Ask what each piece is called. Ask how many thirds, fourths, or halves it takes to make a whole region. Count $\frac{1}{3}$, $\frac{2}{3}$, $\frac{3}{3}$ or $\frac{1}{4}$, $\frac{2}{4}$, $\frac{3}{4}$, $\frac{4}{4}$ as you make a whole. Let the children predict the number of sixths, eighths, and tenths it will take to make the whole region. Encourage the children to make up their own questions using the pieces. Add pieces of other sizes to the group. Sixths, eighths, and twelfths are good sizes to work with. Notice that you should ask students to make the whole when given a fractional part as well as to find the fraction when given a whole.

Make up word problems using the pieces. Use problems such as: "Mrs. Jensen has a granola bar that she wants to split evenly among her three grandchildren. Show the pieces you would use so that each child gets a piece the same size. What is each piece called?" Let the children make up their own word problems.

3. The fraction one tenth should receive special attention because of the frequent use of decimal fractions with calculators, computers, and metric measures. For young children, introduce the fractions such as halves, thirds, and fourths first because it is easier for such children to divide a unit region into two, three, or four pieces than it is to divide the region into ten pieces. The reason is similar to that for teaching children to group by threes and fours before grouping by tens when teaching the concept of place value. When tenths are introduced, you may use commercial materials such as Decimal Squares or base ten blocks or you may make your own materials, similar to those described in the last activity. The decimal paper in Appendix B can be used for this activity.

The activities described in the previous example may be repeated using tenths. Introduce tenths as common fractions before you introduce them as decimal fractions, since children will be more familiar with the common fraction form after their work with halves, thirds, and fourths. You may also introduce fifths at this time, since children are probably discovering a number of equivalent fractions as they work with the pieces.

Objective: to develop the concept of a common fraction as part of a whole unit of length.

4. Rulers, number lines, fraction strips, and Fraction Bars show fractions based on a unit of length. In each of these, a unit is chosen and then subdivided into equal-sized parts (see Figure

9–6). Fraction Bars (available from Scott Resources) are vinyl strips divided into units, halves, thirds, fourths, sixths, and twelfths, with each division printed on a different color of vinyl. The Fraction Bars are all the same length but have different amounts shaded to represent the various fractions. Teachers and students may make similar sets of fraction strips by copying the masters in the appendix onto colored construction paper or railroad board. Again, it is useful to use different colors for each fraction piece. You may also use the masters in the appendix to make fraction number lines. With either the teacher-made fraction strips or the Fraction Bars, the children may repeat the activities described above for the regions.

Fraction Bars come with sets of cards with numerals written on them and bingo cards with pictures of fractions shown as parts of circular regions. Children can play several games in which they match the numerals to the bars or the circular regions. The teacher's guides that come with the bars describe a wide variety of activities, and the children can make up others of their own. When numerals for fractions are first introduced, children may have difficulty relating the symbols to the spoken words and concepts with which they are familiar. The written symbols should be introduced only after the children have the concepts and oral language necessary to understand them. Be sure to give students plenty of experiences to understand the written symbols.

After the children have had a number of experiences with the Fraction Bars or the fraction strips, they can draw the corresponding number lines such as the ones shown in Figures 9–7a and b. Later the children will be asked to transfer this skill to using the fraction number lines without the fraction strips or Fraction Bars. Often, textbooks show only work on the number line, and children should be able to use number lines with rational numbers as well as whole numbers.

When the children are comfortable finding fractional parts of various unit lengths, they should study the marks on a

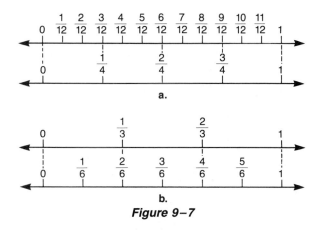

Figure 9-7

conventional ruler and discuss the meaning of the various parts. They may find that some rulers show $\frac{1}{4}$ inches, others show $\frac{1}{8}$ inches, and still others may show $\frac{1}{16}$ inches. Then have children compare English rulers to metric rulers, which show tenths of a centimeter.

Objective: *to develop the concept of a fraction as part of a whole unit of volume.*

5. Common fractions are often used in everyday life to refer to parts of units other than those for area or length. These include units of time, weight, mass, money, capacity, and volume. Children should have experience using fractions in discussions of all of these measures.

   The following are a few ideas for using common fractions to describe parts of a unit of volume. Measuring cups are good for exploring fraction concepts in connection with volume. Children should be given the opportunity to explore pouring water, sand, beans, or rice from one measuring cup into another. Ask the children to predict how many $\frac{1}{2}$ cups it will take to fill a whole cup. Will four $\frac{1}{3}$ cups be more or less than one cup? Is $\frac{1}{2}$ cup more or less than $\frac{1}{3}$ cup? After the prediction, let the children pour to see if they were right. Let the children pose questions to each other. Find simple recipes and let the children do the measuring. Discuss their observations about the fractional parts they have measured.

Objective: *to develop the concept of common fractions as part of a set of discrete objects.*

6. Often, a common fraction is used to refer to part of a set of discrete objects. A mother may refer to half a dozen eggs or the teacher may say that $\frac{1}{4}$ of the children may go to the learning center. When a common fraction is used in this way, the denominator refers to the number of equal-sized groups into which the set is divided, and the numerator refers to the number of groups currently under consideration.

   Colored cubes, bingo chips, or even the children themselves are good manipulative materials for this type of fractional representation. A child may start with 12 chips to represent a dozen eggs and then discuss what must be done in order to find $\frac{1}{2}$. The separation of the set into two equal parts should be related to earlier work separating regions, lengths, and volumes into two equal parts. After children work with halves, they may find thirds, fourths, sixths, and twelfths of the dozen eggs. Children may then use other units of discrete objects, such as finding $\frac{1}{4}$ of the children in the class or $\frac{1}{3}$ of the books in their desks.

Children may have some difficulty with the fact that fractional parts do not always contain the same number of objects. Half a dozen eggs is not the same number as half the children in the class. Have children compare this to the fact that half of a large circle is not the same size as half of a small circle. Children should realize that half of a dozen eggs is the same number as the other half of the dozen, though.

Ratio Model for a Common Fraction

The activities just described involve using common fractions to describe a part-whole relationship. Common fractions may also be used to describe a ratio between two sets. In the **ratio model,** the denominator and the numerator each represent the number of parts under consideration. The denominator does not represent the parts of the whole, as in the part-whole model. The numerator and denominator represent subsets that are being compared.

In fact, some young children have difficulty with the part-whole concept because they have difficulty with the class inclusion concept. They want to compare one subset to another subset rather than compare a subset to a whole. They may identify each of the pictures in Figure 9–8 as $\frac{1}{2}$ because they are comparing the one shaded section to the two non-shaded sections.

This is a correct concept of the fraction $\frac{1}{2}$, but only if the fraction is used in a ratio sense. It is true that the shaded section in each picture is $\frac{1}{2}$ the size of the nonshaded section, but because they are looking for a part-whole response and not a ratio response, most textbooks and most standardized tests say the child was wrong in giving the response of $\frac{1}{2}$. Teachers should encourage children to compare the two.

The following activities suggest ways that ratio concepts may be introduced and compared to part-whole models. The activities are suggested for the intermediate grades because that is when ratios are commonly introduced in textbooks. Often, only a few concrete models for ratios are included in the textbooks, and you will probably need to supplement children's work in the books. Many younger children may benefit from activities comparing ratio models and part-whole models because the ratio model fits their own intuitive concepts of common fractions.

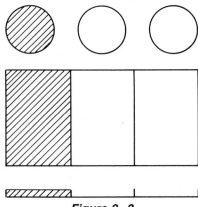

Figure 9-8

ACTIVITIES

Intermediate (4–6)

Objective: *to develop the ratio concept of a fraction using objects.*

1. Children encounter ratio ideas in everyday life, especially ratios used to describe relationships between discrete objects. The teacher may say that there are two girls for every three boys in a group or there are two cookies for each child at a birthday party.

   Use chips or colored cubes to represent the situations. The chips shown in Figure 9–9 may be used to show that there are two girls for every three boys in the group. Let green chips represent the girls and blue chips represent the boys. Ask the children what part of the whole class is girls. How is the ratio concept of a fraction related to the part-whole concept?

2. Another common use of ratio is with gears. Children may observe gears. In Figure 9–10, note that there are five teeth on the small gear and ten teeth on the large gear.

   You may buy gears from a hardware store, obtain gears from factories (which may give you some of their old ones), or use a commercial set such as TECHNIC from Lego. Mark each gear so the children can count the number of turns of each gear. Ask the children to turn the large gear one complete turn and count the number of turns of the small gear. Compare the number of turns to the ratio of the teeth on the large gear to the teeth on the small gear.

   The children should note that the small gear in Figure 9–10 will go around twice while the large gear goes around once. The ratio of the teeth on the large gear to the teeth on the small gear is 10:5, while the ratio of the number of turns of the large gear to the number of turns of the small gear is 1:2.

   Let the children make hypotheses about what will happen with gears of other sizes. Get these gears and test the hypotheses.

Objective: *to develop the concept of ratio using length.*

3. Cuisenaire rods are a good material to use to represent a common fraction as a ratio between two lengths. Use them

Figure 9–9

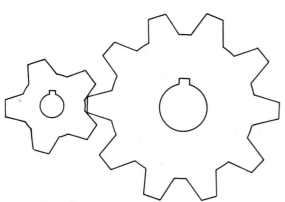

Figure 9–10

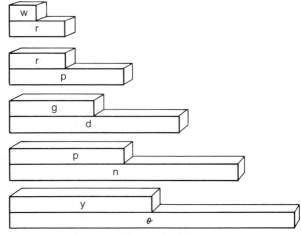

Figure 9–11

with children who have had previous experience using the rods. Ask the children to find several pairs of rods for which one rod is half the length of the other. Possibilities are shown in Figure 9–11.

Note that children are comparing the length of one rod to the length of the other and not comparing one rod to the total length of the two. Let the children suggest other fractions to display with the rods such as $\frac{2}{3}$ and $\frac{3}{4}$. Encourage comparing the other rods to the orange rod for the concept of tenths. Being comfortable with finding tenths helps children make the transition to writing decimal fractions.

Objective: *to develop the concept of ratio using volume.*

4. Ask children to try to find examples of ratios used at home. They may notice that salad dressing calls for one part vinegar for five parts oil or that ice tea mix calls for one teaspoon of ice tea mix for one cup of water. Bring in ingredients to let the children make the solutions or mixtures at school. Ask questions about the ratios as you mix. If the mixture is one part vinegar for five parts oil, what part of the total mixture is vinegar? Note that if the ratio is 1:5, the vinegar is $\frac{1}{6}$ of the total. Have the children compare this work to earlier work with ratios with discrete objects.

Using Common Fractions to Indicate Division

So far, we have discussed two different concepts that may be represented by common fractions, part-whole and ratio. Common fractions may also be used to indicate a division problem. When fractions are used to indicate division, the division problem $5 \div 6$ is shown as $\frac{5}{6}$. The models used may be similar to those used for the partitive division problems with whole numbers in Chapter 8. Therefore, introduce activities using common fractions to indicate division after children have learned to divide with whole numbers and after they understand the part-whole concept of a fraction. The following activities focus on the division concept of common fractions.

ACTIVITIES

Intermediate (3–6)

Objective: *to develop the concept of a common fraction representing division using discrete objects.*

1. Begin with problems, such as $6 \div 2$, which have whole-number answers. Tell the children you have six cookies and wish to put them into two equal groups. The children may wish to use chips to show that $6 \div 2 = \frac{6}{2} = 3$.

   Ask the children what they would do if you had seven cookies and wished to put them into two equal groups. Again, let the children model this with the chips. They will find that they have one chip left over. Some of the children may decide that they can divide the last cookie in half so that each group will have $3\frac{1}{2}$ cookies. Therefore, $7 \div 2 = \frac{7}{2} = 3\frac{1}{2}$.

   These experiences with the division model for common fractions are good for explaining the renaming of improper fractions as mixed numerals. Ask the children to use materials to explain why $\frac{8}{3} = 2\frac{2}{3}$.

Objective: *to develop the division concept for common fractions using length.*

2. Chips are not a good material to use for many of the division problems involving fractions because they cannot be broken into fractional pieces. Models involving measures of length, area, or volume are often preferable because they can be broken into smaller parts. Using these materials, start with a story situation and let the children discover the answer on their own. Give each of the children some blank 3×5 index cards and tell them the 3-inch and the 5-inch sides of the cards represent 3 yards and 5 yards, respectively.

   Tell the children that you have 3 yards of material from which to make puppets for a play. You need to make four puppets and wish to use the same amount of material for each one. What part of a yard of material can you use for each puppet? Ask the children to fold the 3×5 index cards and measure to find the answer. Many children will be surprised to find that you will have $\frac{3}{4}$ yard of material for each puppet, $3 \div 4 = \frac{3}{4}$ (see Figure 9–12).

   Ask the children to make up a story situation for the problem $5 \div 3$ and to again fold a 3×5 card and measure to find the result (see Figure 9–13). Repeat the activity with several different measures until the children can generalize that $a \div b = \frac{a}{b}$.

3. After children have had several experiences folding paper to show the division model for a fraction, they can transfer to a number line. Again, begin with the problem $3 \div 4$. Give each child a number line 3 inches long. Ask the children to divide the number line into 4 equal parts. They will probably find that this is difficult to do unless each unit on the number line is broken into smaller parts.

   Suggest that the children break each unit into 4 equal parts. Ask the children how many small parts there are in 3 units. There are 12. Now ask the children to break the number line into 4 equal parts. The children should discover that each of the parts will be $\frac{3}{4}$ inch long (see Figure 9–14).

   Ask the children to compare this work to the work folding the index cards. Repeat the activity with number lines of different sizes and with different fractions. Ask the children to make up story situations illustrated by their number lines.

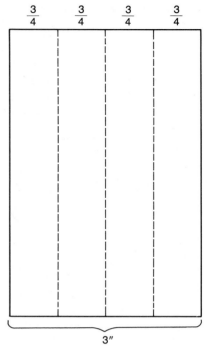

Figure 9–12

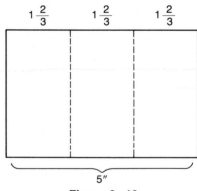

Figure 9–13

Figure 9–14

Objective: *to reinforce the division concept of a common fraction using a variety of measures.*

4. Discrete objects and length are not the only models used to show the division concept of a common fraction. Many other units of measure such as time, money, volume, area, and mass may also be used. Make up story situations for other units of measure and let children choose their own methods and materials for solving the problems. Here are a few suggestions:

   - *Money.* Mrs. Jackson has $5 to buy favors for a birthday party. She needs 10 favors and wishes to spend the same amount on each one. What part of a dollar should she spend on each favor?

   - *Time.* The Moyer relay team wishes to run the 1-mile relay in 5 minutes. If the 4 girls on the relay team each run the same distance in the same amount of time, how fast should each girl run her $\frac{1}{4}$ mile?

- *Volume.* The Coleys have a 2-liter bottle of soda. If Amy wants to split the soda evenly into 10 glasses, what part of a liter should she pour into each glass?
- *Area.* The Sheffields have a 1-acre plot of ground that they wish to plant with corn, tomatoes, and peppers. If they use the same amount of land for each vegetable, how much land will be used for each?

Encourage the children to make up their own situations and trade with each other to solve them.

Decimal Fractions

Depending on the textbook series you use, decimal fractions may be introduced before, along with, or after common fractions. You will probably wish to follow the textbook guidelines, but there are advantages to introducing decimals earlier because of the use of the decimal in place value notation, in our money system, in the metric system of measurement, and on calculators and computers, as well as the relative ease of computation with decimal fractions. Conceptually, however, it is probably easier to understand the meaning of halves, thirds, and fourths than it is to understand tenths and hundredths because there are fewer partitions involved. Whenever you decide to introduce common and decimal fractions, be sure the children realize they are simply different notation systems for the same concepts.

We focus here on the use of decimal fractions and percents to represent parts of whole units. Again, a unit may be a set of discrete objects or any unit of measure (length, area, volume, mass, money, or time). Children should have a good concept of tenths before they begin work with written decimal fractions. Work with decimals may begin as soon as children understand the concept of dividing a unit into ten equal-sized parts.

Because the money system in the United States is based on decimals, children also may be introduced to hundredths at a fairly young age. Most six or seven year olds can understand that there are one hundred pennies in a dollar and that one penny may be written as $.01. This represents $\frac{1}{100}$ of a dollar.

When decimal fractions are introduced, place value for whole numbers should be reviewed. Let the children study the following place value chart and tell you what happens as you move one place to the right on the chart.

thousands hundreds tens ones . ___ ___

Children should notice that the value of the position on the right is one tenth of the value of the position directly to the left of it. Ask the children to predict the value of the place to the right of the ones place, that is, the first place after the decimal point. (Note that in countries outside the United States, a comma is used in place of a decimal point,

for example, $1,7 = 1\frac{7}{10}$.) Children should realize that this place will have one tenth the value of the ones place and therefore is the tenths place.

The activities for teaching the concept of a common fraction as part of a whole may be repeated for decimal fractions, beginning with tenths and later expanding to hundredths, thousandths, and so on. The following activities suggest other ways in which decimal fractions may be introduced.

ACTIVITIES

Intermediate (3–6)

Objective: *to develop the concept of a decimal fraction using a length, area, or volume model.*

1. Take out the white and orange Cuisenaire rods and tell the children that the orange rod represents 1. Ask the children to tell you the value of one of the white rods. Use the decimal notation to represent the value 0.1.

   After the children understand the concept of tenths using the rods, add the orange flats from the Cuisenaire metric blocks and tell the children that the orange flat now represents 1. What is the new value of the orange rod (long)? What is the value of each white rod? What is the value shown by the rods in Figure 9–15 if the flat is 1?

   Ask the children to show 0.45, 0.89, and so on. Let the children make up problems for each other and discuss how they found the answers.

   After the children are proficient with both tenths and hundredths, add the orange cube to the set. If the cube is 1, the flat becomes 0.1, the long becomes 0.01, and the white rod becomes 0.001. Repeat the activities with thousandths.

2. After the children have worked with the Cuisenaire metric blocks, they may do the same activities using paper or cardboard models. You may use a commercial set such as Decimal Squares, or make a set out of paper. If you make your own set, a convenient size is 1 square decimeter for a unit. Mark squares with ten columns for tenths, and mark columns with ten rows for hundredths (see Figure 9–16). (A master for this square is included in Appendix B.)

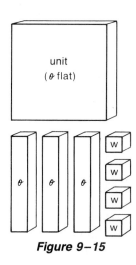

Figure 9–15

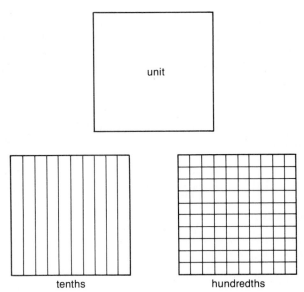

tenths   hundredths

Figure 9–16

Laminate the decimal squares so that the children may write on them with erasable markers. Make a separate set of cards with various decimal fractions written on them, and ask the children to shade the squares to match the decimal fractions written on the cards. The children may then use the cards to play various games that involve matching the numerals to the pictures, such as go fish or old maid.

3. Because the metric system is a decimal system, it provides good examples of decimal fractions. Children learning the concept of 0.1 may take strips of paper each 1 decimeter long (or they may use orange Cuisenaire rods) and line them up on a meter stick. Ask, "How many strips of paper or rods does it take to make one meter?" Ten. Ask, "How could you express three strips?" 0.3 meter.

Extend the activity to hundredths by using strips each 1 centimeter long or by using the white Cuisenaire rods. Ask, "What part of a meter is represented by twenty-three white rods? What part of a meter is represented by five orange rods and two white rods?" Let the children make up their own questions for each other and discuss their methods of solution. Thousandths may be introduced by using millimeters or by using grams and kilograms or milliliters and liters.

Percent

After children are comfortable using decimal fractions as well as common fractions, introduce them to percents. Percent simply means per hundred, or out of 100. A percent may be thought of as a ratio between some number and 100. For example, 18% is 18 out of 100. Have children practice writing percents as ratios, common fractions, and decimal fractions. Eighteen out of 100 may be written as $\frac{18}{100}$, 0.18, or 18%. Different notations are useful at different times; you might use $\frac{18}{100}$ when writing out the fractional part of a dollar on a bank check, $\frac{9}{50}$ to tell the probability of winning a game, 0.18 to give the price of an apple, or 18% to note the discount on a recent purchase.

Note that percent and percentage are not the same thing. Children (and teachers) often confuse the two. A **percent** indicates a rate, while a **percentage** indicates an amount. For example, if you take out a loan of $500 for one year at 10%, the rate is 10 percent, the base is $500, and the $50 you pay in interest is the percentage. The percentage and the base are amounts, and the percent is the ratio between these two amounts.

Materials and activities for introducing percent are similar to those for introducing decimal fractions. You may repeat any of the activities involving hundredths using a percent representation. The following activities give a few additional ideas.

ACTIVITIES

Middle Grades (5–8)

Objective: *to develop the concept of percent.*

1. The Decimal Squares with ten rows of ten squares each may be used again for this activity. Make an overhead transparency of one of the Decimal Squares, and on the overhead projector, let the children watch you shade part of the square red. Ask the children what percent of the square is shaded red.

   A hundreds board without the numbers is also useful for developing the concept of percent. Hang different colored markers on the hooks. For example, if you have fifteen green markers on the pegs and eighty-five red markers, ask the children what percent of the board has green markers and what percent has red markers.

   Let children make up different examples for each other. Ask them to compare their answers in common and decimal fractions to their answers in percents.

2. Money is good for introducing percent since the United States currency is based on the decimal system. Ask the children to express 18 cents as a percent of a dollar.

   Set up a chart such as the following:

Amount	Percent of a Dollar
$0.18	18
$0.26	—
$0.35	—
$0.92	—
$2.96	—

   Ask the children to make up other examples for each other and discuss how they found their answers.

Equivalent Fractions

After children understand the concept of a common fraction as part of a unit, they may begin to explore the concept of **equivalent fractions.** Unlike whole numbers, each fraction has an infinite number of symbolic representations. The common fractions that name the same number are called equivalent fractions.

State-operator-state machines may be used to demonstrate this idea. Fraction pieces representing a common fraction in one form are the input and fraction pieces representing the fraction in a different but equivalent form are the output. The following activities include a variety of examples of finding equivalent fractions.

ACTIVITIES

Intermediate (3–6)

Objective: *to develop the concept of equivalent fractions.*

1. Choose one of the sets of fraction pieces for this activity. The examples here are given for the Fraction Bars. The operator in the state-operator-state machine is "give an equivalent." Use the machines you used earlier for operations with whole numbers.

   Demonstrate the idea first to the whole group before asking the children to work with individual machines. Give the children who will be supplying the input one set of Fraction Bars and the children doing the operation another set of bars. Ask one child to choose a bar such as $\frac{1}{2}$ and place it in the machine. The operators in the machine should then find a bar with an equivalent amount shaded and place the bar in the output. Ask the children to name the amount in the output. Do this for several amounts for which equivalent fractions can be found with the bars.

   After the children understand these exchanges, ask them what they would do with a bar such as $\frac{7}{12}$. They will not be able to find a different equivalent amount with the bars. Does that mean that no equivalent amount exists? Lead the children to discuss how they can create new bars to show an equivalent amount such as $\frac{14}{24}$.

   After the children have worked with the machine as a whole group, let them work in pairs or individually to find other equivalent fractions. Ask the children to write down all the sets of equivalent fractions they find.

   When the children get a list such as $\frac{1}{2} = \frac{2}{4} = \frac{3}{6} = \frac{6}{12}$, ask them if they notice any patterns in the list. They should realize that to convert from one fraction to an equivalent fraction, they can either multiply or divide the numerator and denominator by the same number. Let the children repeat the activity with other materials such as the Fraction Tiles or the fraction strips to see that the same rules are true regardless of the material used.

2. Give each child a strip of adding machine tape 1 foot long. Have the children label the left end of the tape with a 0 and the right end with a 1.

   Ask the children to fold the tape in half. Have them now label the left end $\frac{0}{2}$, the middle $\frac{1}{2}$, and the right end $\frac{2}{2}$.

   Ask the children to fold the tape in half again and to relabel it. They should now have $\frac{0}{4}$, $\frac{1}{4}$, $\frac{2}{4}$, $\frac{3}{4}$, and $\frac{4}{4}$.

   Unfold the tape and label the thirds, and then fold the tape in half again and label the sixths. Fold it in half once more and label the twelfths. The final tape should look like the one in Figure 9–17.

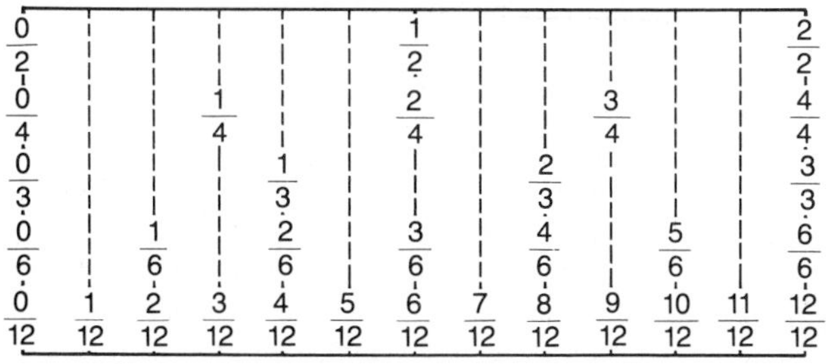

Figure 9–17

Ask the children to list all the fractions that name the same fold on the tape. Ask them to compare this list to their lists of equivalent fractions from the work with the state-operator-state machines. Ask them how the lists may be extended by continuing to fold the paper. Ask if there is ever an end to the number of equivalent fractions.

Ordering Fractions

After the children have worked with equivalent fractions, have them use the same materials to order fractions. Following are a few activities for developing the concept of ordering fractions.

ACTIVITIES

Intermediate (4–6)

Objective: *to develop the concept of ordering fractions.*

1. Use one of the sets of fractional regions such as the pie pieces or the rectangular regions. Ask the children to name two common fractions and to find the corresponding regions. By placing the pieces on top of each other, compare the areas of the two regions to determine which of the fractions is larger.
2. Use the folded adding machine tape that the children have labeled with common fractions. Ask the children to locate $\frac{1}{4}$ and $\frac{1}{2}$ on the tape. Ask the children which is smaller and have them write the answer using the symbol for less than. They should write $\frac{1}{4} < \frac{1}{2}$.

   Let the children suggest several other pairs of numbers from the number line and write comparisons using $<$, $=$, or $>$. Are there any pairs of common fractions that cannot be compared using one of these three symbols? Can any pairs be compared using more than one of these symbols? This is called the **trichotomy principle:** Any rational number is greater than, less than, or equal to any other rational number.

   After the children have compared pairs of numbers, ask them to compare several numbers at once using $<$. They may write $\frac{1}{12} < \frac{1}{6} < \frac{1}{4} < \frac{1}{3} < \frac{5}{12} < \frac{1}{2} < \frac{7}{12} < \frac{3}{4} < \frac{5}{6} < \frac{11}{12} < 1$. Is it possible to make a longer chain of numbers? Are there any fractions between $\frac{1}{12}$ and $\frac{2}{12}$? The fact that there is always another rational number between any two rational numbers is called the **density property.**

After children have a good concrete understanding of the rational number concepts, develop more abstract skills with the numbers in different forms. Be sure you do not rush the children into this abstract work before they have developed the ideas concretely. The following section focuses on developing skills with rational numbers written as common fractions, decimal fractions, and percents.

Developing and Practicing Rational Number Skills

The skills associated with rational numbers are renaming equivalent common fractions, reading and writing decimal fractions, converting common fractions to decimal fractions and percents and vice versa, ordering both common and decimal fractions, and using proportions. These skills involve making the transition from work with manipulative materials to abstract work using either mental calculation, paper and pencil, or calculators or computers. We begin with procedures for renaming common fractions.

Renaming Common Fractions

After children can find equivalent fractions using a variety of concrete models, they are ready to discover more abstract procedures for finding equivalent fractions. Children should have made lists of the equivalent fractions that they found using materials and should have generalized the fact that equivalent fractions may be found by multiplying or dividing the numerator and the denominator by the same number. Ask the children to make a list of at least ten ways to name 1 ($\frac{1}{1}$, $\frac{2}{2}$, $\frac{3}{3}$, . . .). Ask the children what they notice about the numerator and the denominator of each fraction that is equivalent to 1.

Notice that when you have $\frac{a}{b} \times \frac{c}{c}$ or $\frac{a}{b} \div \frac{c}{c}$, you are multiplying by the identity or dividing by the right-hand identity 1. Multiplying or dividing by the identity does not change the value of the original number. Let the children make up tables with several names for different common fractions, such as $\frac{1}{2}$, $\frac{1}{3}$, $\frac{1}{4}$, $\frac{2}{3}$, and $\frac{3}{4}$.

After children can list several equivalent fractions for any given fraction, have them practice writing fractions in **simplest terms.** We prefer using *simplest terms* or *simplest form* rather than *reduced,* or *lowest, terms,* because some children will think a fraction has gotten smaller when it is reduced even if they have been working with equivalent fractions.

To find fractions in simplest terms, return to the concrete materials you worked with for finding several ways to name a fraction. Tell the children that when you have a list of equivalent fractions, the one in simplest terms has the smallest number in the denominator, and it is not possible to divide both the numerator and the denominator evenly by a whole number greater than 1. It is in simplest terms because it uses the fewest parts.

To simplify a fraction abstractly, ask the children if there are any numbers that will divide evenly into both the numerator and the denominator. Continue dividing until no more numbers will divide into both. If you are using a calculator such as the Math Explorer, experiment by pushing the "simp" button and the "equals" button when the calculator is displaying a common fraction. What happens if you push the "simp" button more than once? Try it for several different fractions and discuss the results.

Children should find the **greatest common factor** (greatest common divisor) of the numerator and denominator to simplify the fraction in one step. If c is the greatest common factor, then dividing both the numerator and the denominator by c gives you a fraction in simplest form.

After children are comfortable with finding equivalent fractions for a given fraction, have them find equivalent fractions with a common denominator for two or more common fractions. Have children reverse the process used to simplify fractions, that is, have them multiply a fraction by $\frac{a}{a}$ in order to find equivalent fractions. This skill is used in working with addition, subtraction, and division of common fractions, when it is often necessary to write fractions in a form with a common denominator before any operation can be performed. There are several ways to find common fractions with a common denominator, and three of them are discussed here.

ACTIVITIES

Intermediate (3–6)

Objective: *to find a common denominator for two common fractions.*

1. Sometimes a common denominator is the denominator of one of the common fractions. This is the case for two fractions such as $\frac{1}{2}$ and $\frac{3}{10}$ or $\frac{3}{4}$ and $\frac{5}{16}$. To rename $\frac{1}{2}$ as tenths, the children should first realize that 10 is a multiple of 2. Ask the children what they must multiply by 2 in order to get 10. Since $2 \times 5 = 10$, the numerator must also be multiplied by 5. Therefore, $\frac{1}{2} = \frac{1 \times 5}{2 \times 5} = \frac{5}{10}$. Let the children practice with several pairs of numbers in which one denominator is a multiple of the other before you introduce pairs of common fractions in which this is not the case.

2. Ask the children to find a common denominator for the fractions $\frac{2}{3}$ and $\frac{3}{4}$. First, ask the children to get out their lists of the equivalence classes of these two common fractions. Find examples of equivalent fractions for $\frac{2}{3}$ that have denominators that also appear on the list of equivalent fractions for $\frac{3}{4}$. The children may find denominators of 12, 24, 36, and 48.

   Tell the children that twelfths are the **least common denominator** since 12 is the smallest number that appears as a denominator in both equivalence classes. Therefore, to write $\frac{2}{3}$ and $\frac{3}{4}$ with the least common denominator, the children would write that $\frac{2}{3} = \frac{8}{12}$ and $\frac{3}{4} = \frac{9}{12}$. Let the children find common denominators for other pairs of fractions and discuss their methods.

3. Sometimes it is necessary to find a common denominator for two fractions when one denominator is not a multiple of the other and it is laborious to list equivalence classes until a common denominator appears. Thus, it is often most efficient to use prime factorization to find the **least common multiple** of the two denominators. The least common multiple is also the least common denominator.

   For example, if you wish to add $\frac{5}{12}$ and $\frac{7}{30}$, first find the prime factorization of the two denominators, 12 and 30. $12 = 2 \times 2 \times 3$ and $30 = 2 \times 3 \times 5$. Put these factors into a Venn diagram such as the one in Figure 9–18.

   The union of the two sets in the least common denominator, $2 \times 2 \times 3 \times 5$, or 60. (Notice that the intersection of the two sets is the greatest common factor.) Each of the fractions must be written in sixtieths, the least common multiple. Each numerator is multiplied by the number or numbers in the Venn diagram that do not appear in the denominator of that fraction. The numerator of $\frac{5}{12}$ is multiplied by 5, and

factors of 12   factors of 30

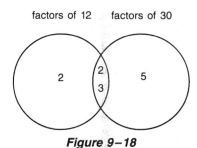

Figure 9–18

the numerator of $\frac{7}{30}$ is multiplied by 2. Thus, $\frac{5}{12} = \frac{5}{2 \times 2 \times 3} =$ $\frac{5 \times 5}{2 \times 2 \times 3 \times 5} = \frac{25}{60}$; and $\frac{7}{30} = \frac{7}{2 \times 3 \times 5} = \frac{7 \times 2}{2 \times 3 \times 5 \times 2} = \frac{14}{60}$. Let the children make up other problems for each other and discuss their solutions.

4. For a fraction with a large number in the denominator, children will probably need to use a factor tree to find the prime factors. To make a factor tree, place the number to be factored at the top and choose two factors that when multiplied give you the original number. Continue until all factors are prime numbers. For example, for 48, begin with 6 × 8 as shown:

$$48 = 2 \times 2 \times 2 \times 2 \times 3$$
$$= 2^4 \times 3$$

Notice that the order in which the factors are found does not matter. You still end up with the same prime factors. This is called the **Fundamental Law of Arithmetic.**

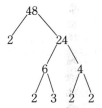

Most textbooks include a chapter on number theory that precedes the unit on operations with common fractions and includes lessons in finding least common multiples and greatest common factors. We also discuss the necessity for finding common denominators in the next chapter when we talk about operations with common fractions.

After children understand renaming fractions and can rename fractions in simplest terms and rename two fractions with a common denominator, let them play games that reinforce those skills. Card games, board games, and bingo are good for this. The following activities are designed to give children practice in renaming common fractions.

ACTIVITIES

Intermediate (3–6)

Objective: *to practice finding equivalent fractions.*

1. Make a set of cards with fractions on them in both pictorial and numerical form. Make four cards for each of the following

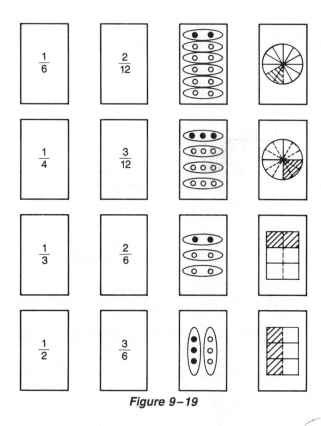

Figure 9–19

fractions: $\frac{1}{6}, \frac{1}{4}, \frac{1}{3}, \frac{1}{2}, \frac{3}{4}, \frac{2}{3}, \frac{5}{6}$, 1. For each fraction, have two pictures depicting the fraction, the fraction itself, and an equivalent fraction. Four sets are shown in Figure 9–19.

Have the children play a game like go fish. They should try to get four cards with equivalent fractions. A child who gets four equivalent fractions should lay the cards down.

On each turn, a child should ask, "Do you have any cards showing —?" If the child asked has any of the designated cards, he or she must give them to the asker. If not, the child says, "Go fish." The child who asked then draws a card from the pile. If this card shows the requested fraction, the child may continue asking. If it does not, it is the next child's turn. When all sets of four have been laid down, the child with the most sets is the winner.

After the children become proficient at playing with pictures and numerals, make another set of cards which has just numerals, or encourage the children to make their own cards for the game.

2. Make up bingo cards with equivalent fractions. The caller's cards may have the following fractions:

B:	$\frac{1}{2}$,	$\frac{4}{5}$,	$\frac{5}{12}$,	$\frac{2}{3}$,	$\frac{3}{4}$,	$\frac{5}{8}$,	$\frac{2}{6}$
I:	$\frac{3}{6}$,	$\frac{4}{5}$,	$\frac{1}{3}$,	$\frac{5}{8}$,	$\frac{5}{6}$,	$\frac{4}{6}$,	$\frac{2}{8}$
N:	$\frac{2}{3}$,	$\frac{2}{9}$,	$\frac{4}{8}$,	$\frac{2}{5}$,	$\frac{3}{5}$,	$\frac{4}{12}$,	$\frac{5}{6}$
G:	$\frac{2}{4}$,	$\frac{10}{12}$,	$\frac{10}{12}$,	$\frac{1}{5}$,	$\frac{2}{3}$,	$\frac{4}{12}$	
O:	$\frac{5}{10}$,	$\frac{5}{15}$,	$\frac{2}{3}$,	$\frac{5}{12}$,	$\frac{1}{4}$,	$\frac{3}{5}$,	$\frac{3}{8}$

B	I	N	G	O
$\frac{3}{6}$	$\frac{8}{10}$	$\frac{6}{10}$	$\frac{1}{4}$	$\frac{6}{16}$
$\frac{9}{12}$	$\frac{3}{9}$	$\frac{1}{3}$	$\frac{1}{2}$	$\frac{10}{24}$
$\frac{4}{6}$	$\frac{1}{2}$	FREE	$\frac{1}{3}$	$\frac{4}{6}$
$\frac{15}{24}$	$\frac{2}{3}$	$\frac{1}{2}$	$\frac{5}{6}$	$\frac{1}{2}$
$\frac{1}{3}$	$\frac{1}{4}$	$\frac{4}{10}$	$\frac{4}{6}$	$\frac{1}{3}$

Figure 9–20

One such card is shown in Figure 9–20. Be sure you do not make two cards identical.

When the cards are ready, call the fractions and let the children put markers on the equivalent fractions. The first child to get five in a row wins. Be sure to check the card to make sure the child has covered equivalent fractions.

3. The Fraction Bars Computer Programs by Albert Bennett, Jr., and Albert Bennett, III, available from Scott Resources, are designed to reinforce the concrete work with Fraction Bars. There are seven programs in the series, which begins with basic concepts and goes on to operations with fractions. All are good for practice after the children have developed the concepts with the concrete materials. Children can practice their skills with equivalent fractions by working with the program on equivalent fractions.

Objective: to determine if two fractions are equivalent.

4. After children are competent at finding equivalent fractions, ask them how they would determine if two fractions such as $\frac{84}{126}$ and $\frac{104}{156}$ are equal. The children may try to find a common denominator, which is very time-consuming; they may try simplifying both fractions, which in this case would work very well; or they may get out their calculators and divide to determine if both common fractions are equivalent to the same decimal fraction.

Another method that children enjoy is cross multiplying. Ask the children to list several pairs of common fractions that they know are equivalent. They should have fractions in the form $\frac{a}{b} = \frac{c}{d}$. Ask the children to multiply $a \times d$ and $b \times c$. Encourage the use of calculators so the children may check several pairs without being bogged down by the computations.

After the children have cross multiplied with several pairs of equivalent fractions, ask them to compare the findings. Does $a \times d$ always equal $b \times c$? What happens if the original fractions are not equivalent? Gifted children may wish to find out why this works.

Reading and Writing Decimal Fractions

After children understand the concepts of decimal fractions and can extend the place value into hundredths and thousandths, have them practice both reading and writing decimal fractions. Be sure this abstract work is based on a solid concrete foundation. The following activities are designed to give children practice reading and writing decimal fractions.

ACTIVITIES

Intermediate (4–6)

Objective: *to practice reading and displaying decimal fractions.*

1. Make a set of cards with decimal fractions written in both numerals and words. On one card, write a decimal fraction in words and on another write the same decimal fraction in numerals, such as three hundredths and 0.03. Make about fifteen such pairs.

   Let the children use the cards to play the match game. Turn all the cards face down on the table in front of the children. The children should take turns turning over a pair of cards. The goal is to get one numeral card and one word card that show the same decimal fraction. If the two cards match, the child gets to keep the cards and takes another turn. If the cards do not match, they are placed face down on the table in their original positions and it is the next child's turn. Play continues until all the cards have been matched. The child with the most matches at the end is the winner.

   Children who still need work on matching the decimal fractions to pictorial representations can play this same game matching pictures and numerals or words. It is helpful in both variations of the game to have an answer key with which to check any matches that the players are not sure of.

2. Ask each child to make a decimal place value chart. Have each child fold a piece of tagboard or make a pocket chart like the one in Figure 9–21 by folding and stapling a piece of oaktag.

   Give each child two sets of cards with the numerals 0–9 on them. The caller should then read numbers such as thirty-five and four hundred twenty-six thousandths. The children should place the numeral cards in the correct positions and hold the chart up for the caller to check. The teacher or student caller may then quickly check to see which students are correctly displaying the number.

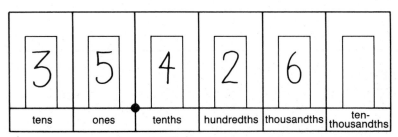

Figure 9–21

This activity may be done by the whole class, small groups, or a pair of children working alone. For another variation of this activity, furnish each child with a calculator and ask the children to put the numbers in the calculator as they are read. You or a caller may then check each calculator display, or the children may compare their displays.

Renaming Common Fractions, Decimal Fractions, and Percents

After children understand the concepts of fractions in common, decimal, and percent form and are proficient at naming equivalent fractions and reading and writing decimal fractions, they are ready to learn to rename fractions as decimals or percents and vice versa. Results from the Fourth National Assessment of Educational Progress show that students have difficulty with this. Only 30 percent of seventh graders could express 0.9 as a percent and 8 percent as a decimal (Kouba, Brown, Carpenter, Lindquist, Silver, and Swafford, 1988). Children who can read a decimal fraction should also be able to write it in common fraction form. Children who can tell you that 0.23 is twenty-three hundredths should also be able to write it as $\frac{23}{100}$. When they realize that percent simply means per hundred, they can also write it as 23%. Some fractions must be simplified once they are written in common fraction form, and children should be reminded of that.

Converting a common fraction to a decimal may not be as easy as converting a decimal to a common fraction because just reading a common fraction may not tell you its decimal form. Reminding the children that one meaning for a common fraction is division can help them make the conversion. The common fraction $\frac{5}{6}$ is the same as 5 divided by 6. The division may be done very quickly on the calculator or slightly more laboriously by hand.

Dividing fractions may be the children's first introduction to nonterminating decimals. Children will realize as they divide 5 by 6 that they could go on forever and never reach a point where there is no remainder. Their answer of .8333333 . . . will end in 3 no matter how long they continue the division. Other fractions such as $\frac{1}{2}$, $\frac{1}{4}$, $\frac{3}{5}$, and $\frac{7}{10}$ terminate in a few places.

Have children explore converting common fractions to decimal fractions to see if they can predict which common fractions will terminate and which will not. They may need the hint that it will help to first write each fraction in simplest terms and then to factor the denominator into prime factors. Compare the prime factors to the factors of the powers of ten. Using a calculator will help the children explore a greater number of conversions. Lead children to the generalization that whenever the denominator has only 2s and/or 5s as factors, the decimal equivalent will terminate. Otherwise, the common fraction will be equivalent to a repeating decimal. Children who are learning to write computer programs may wish to write a program to convert common fractions to decimals and percents and vice versa.

After children are comfortable converting decimal fractions to common fractions and common fractions to decimal fractions, they may learn to convert decimal fractions to percents. Begin with decimal fractions in hundredths. These are very easy to convert to percents, since percent means per hundred. For example, 0.16 is simply 16%.

After children can convert hundredths to percent, ask them to try fractions that are in tenths. The children may notice that 0.3 = 0.30 and is therefore equal to 30%. (You may need to ask them how many hundredths are equal to 0.3 if they do not notice this on their own.)

After children can convert tenths and hundredths to percents, try other decimal fractions, such as 0.005. Ask the children if this is more or less than 1%. Children may need to return to the decimal squares or other pictorial materials to determine that this is $\frac{1}{2}$ of 0.01 and is therefore $\frac{1}{2}$% or 0.5%.

After they successfully convert several decimal fractions to percents, ask the children what they notice about the decimal point as they move from a decimal fraction to a percent. Yes, it moves two places to the right. Ask, "How do you convert a percent back to a decimal?" Move the decimal point two places back to the left. If the children have a % key on their calculators, have them explore its use at this point.

After children can easily convert decimal fractions to percents and vice versa, ask them how they would convert a common fraction to a percent. For most children, the easiest method is probably to convert the common fraction to a decimal fraction and then to convert the decimal fraction to a percent. Calculators come in handy when using this method.

The following activities are designed to give children practice in converting decimal fractions, common fractions, and percents.

ACTIVITIES

Middle Grades (5–8)

Objective: *to practice converting decimal fractions, common fractions, and percents.*

1. The games described earlier for practice with renaming common fractions and with reading and writing decimal fractions may be adapted for practice with renaming decimal fractions, common fractions, and percents. Bingo cards can be made with common fractions on the calling cards and decimal fractions and/or percents on the bingo cards. Children can play a match game in which they match a card with a common fraction to a card with an equivalent decimal fraction to a card with an equivalent percent. Card games such as old maid can also be adapted so that children match these three types of cards.

   Encourage the children to create other games of their own to practice making conversions. Keep calculators handy as the children play the games so that the children may check their work.

2. Ask the children to bring in the stock page out of the local paper. Stock prices are typically quoted as mixed numerals and common fractions. Ask the children to tell you the price of a stock that is listed at $28\frac{1}{2}$. This stands for $28.50. Children should be able to list the price for any stock in standard dollars and cents notation.

Have each child invest an imaginary $1000 in the stock market and keep track of how the stocks are doing. Let the children discuss their methods for recording their investments and their earnings. If any of the parents are accountants or stockbrokers, ask them to come in to talk to the children about their careers and the mathematics they use every day. Women working in these fields are good role models and may give some of the girls in your class an incentive.

Ordering Common and Decimal Fractions

After children can find equivalent common fractions and read and write decimal fractions, they are ready to practice ordering both common and decimal fractions. Often, children must determine such things as which of two fractions is greater or whether a given fraction is greater, less than, equal to, or between other fractions. Many children do not have accurate strategies for determining the relative sizes of rational numbers. They may have difficulty because of the infinite number of ways of writing equivalent fractions or because of the denseness of rational numbers. It is difficult for children to realize that there is always another rational number between any two given rational numbers, which is not true for whole numbers.

Children may need to return to some of the work with concrete materials before they begin to practice ordering fractions on an abstract level. The following activities are designed to give the children practice in ordering both common and decimal fractions. Be sure to discuss strategies with the children as they work. Note that it is not always possible to determine incorrect strategies by simply looking at mistakes on the children's papers.

ACTIVITIES

Intermediate (4–6)

Objective: *to practice ordering common and decimal fractions.*

1. Ask the children to predict, and then use a concrete material such as the circular regions to determine, which fraction in each of the following pairs is the greater:

$$\frac{1}{2} \qquad \frac{3}{4}$$

$$\frac{2}{3} \qquad \frac{5}{6}$$

$$\frac{3}{8} \qquad \frac{3}{5}$$

$$\frac{3}{8} \qquad \frac{2}{5}$$

$$\frac{1}{3} \qquad \frac{3}{8}$$

Have the children explain why they made the predictions that they did and tell whether or not they were right. Ask children who were wrong to explain why they were wrong. Can you tell which is larger by looking only at the numerator or only at the denominator?

When children can make fairly accurate estimates of the sizes of the fractions, ask them to cross multiply each pair of fractions the way they did to determine if the two fractions were equivalent and to observe the findings. Ask, "For each pair of common fractions $\frac{a}{b}$ and $\frac{c}{d}$, what is the relationship of the fractions if $ad > bc$? What is the relationship if $ad < bc$?" Cross multiply with several pairs of fractions and check the results by using concrete materials, converting both fractions to a common denominator, or using a calculator to convert both common fractions to decimal fractions. Let the children come up with a generalization. Does the generalization fit their intuitive ideas about ordering fractions?

2. At least one computer game is designed to give children practice in working with decimals, not only in ordering them but also in performing operations with them. *Get to the Point* by Judah Schwartz, available from Sunburst, consists of three games, one of which is designed to help children order decimals. Children can practice their skills and have fun at the same time.

3. The game *pyramid* can be played by two or more children. The first child must name a common fraction between 0 and 1 and write it on the bottom line of a sheet of paper. This child should then use the calculator to convert the common fraction to a decimal fraction and write the decimal fraction on the same line next to the common fraction. The next child must then name another common fraction that is greater than the first but still less than 1. This common fraction and its decimal fraction equivalent are then written on the next line of the pyramid. Play continues with larger and larger common fractions until the fraction named is not larger than the previous one, until the fraction named is larger than 1, until a time limit is reached, or until the children realize that the game could go on for an infinite number of moves and they tire of it. The beginning of a game is shown below:

$$\frac{1}{2} = 0.5$$

$$\frac{2}{5} = 0.4$$

$$\frac{3}{8} = 0.375$$

$$\frac{1}{3} = 0.333\ldots$$

$$\frac{1}{4} = 0.25$$

$$\frac{2}{9} = 0.222\ldots$$

$$\frac{1}{6} = 0.1666\ldots$$

$$\frac{1}{8} = 0.125$$

$$0$$

Children may make the game harder by not allowing denominators that are powers of ten or not allowing unit fractions. To play this game well, some children may need to return to the activities for building the concepts of ordering common and decimal fractions.

Children may explore what happens to a repeating decimal when it is displayed on the calculator, since the calculator obviously cannot show an infinite number of decimal places. Calculators either round or truncate the decimals. When the calculator **truncates** a decimal, it merely cuts off all the decimal places beyond those that it can display. When the calculator **rounds** a decimal, if the first digit beyond the last digit on the display is less than five, the last digit is not changed. If the first digit beyond the last digit on the display is greater than five, the last digit is rounded up. If the first digit beyond the last digit on the display is five, most calculators round up, although the rules vary.

Let the children try a fraction such as $\frac{2}{3}$ to see the response on the calculator. Some calculators will display 0.6666666 and some will display 0.6666667. Ask the children which is more accurate. The calculator function may be compared to a computer that continues to divide until told to stop when converting common fractions to repeating decimals.

Proportions

One of the uses of common fractions is to show a ratio of one part to another part. A **proportion** consists of two equal ratios. If you know that there are two girls for every three boys in the class and you know the whole class has ten girls, you may use a proportion to find the total number of boys in the class. Set up the proportion

$$\frac{2 \text{ girls}}{3 \text{ boys}} = \frac{10 \text{ girls}}{n \text{ boys}}$$

The technique of cross multiplying works in solving a proportion. In this example, when you cross multiply you get $2 \times n = 3 \times 10$. If $2n = 30$, then $n = 15$. The total number of boys in the class is 15.

Children may use proportions to solve problems involving percents and inverse proportions to solve problems involving more complex relationships such as comparing the number of teeth in a gear to the number of rotations it will make. Textbooks for the middle grades give several more examples. However, don't assume that children understand proportions after working these examples abstractly. Proportional reasoning is one of the signs of formal operational thinking, and many adults have difficulty with formal operational thinking. Try some of the following questions yourself (assume letters not mentioned will remain the same):

$\frac{a}{b} = \frac{c}{d}$ What will happen to c if a gets larger?

$\frac{e}{f} = \frac{g}{h}$ What will happen to h if e gets larger?

$\frac{j}{k} = \frac{l}{m}$ What will happen to l if k gets smaller?

Did you have difficulty with any of these? Try to explain your reasoning to a classmate. Do you feel that you could explain proportions well to children? Explaining on an abstract level is difficult. Try returning to concrete examples for your explanations. You may need to substitute numerals for the letters. Now try this one:

The red string is 3 paper clips long. The blue string is 5 paper clips long. If you measure the red string in bingo chips, you find it is 6 chips long. How many chips will it take to measure the blue string?

Was this any easier? Did you draw a picture or visualize the answer mentally? How does this question compare to the earlier ones? Keep these activities in mind as you introduce proportions to the children. Be sure to include concrete examples to help children develop the concepts. The following activities give a very brief idea of some of the uses for proportions. Be sure to add other ideas of your own.

ACTIVITIES

Middle Grades (5–8)

Objective: *to practice using proportions.*

1. Proportions are used frequently in children's everyday lives. Give the children some examples of their uses and ask them to make up their own problems using proportions or to collect examples of problems they encounter, say, in the grocery store or while reading the newspaper. The following are a few examples:

- José is baking a birthday cake. The recipe feeds 12 and calls for 3 eggs. José is planning a large party and wants to make a cake for 36 people. How many eggs does he need? Use the ratio $\frac{3}{12} = \frac{n}{36}$ or $\frac{3}{n} = \frac{12}{36}$. Discuss with the children how to set up the proportion. There are two other proportions the children may set up that would be equivalent to these. Can you find them?

- Mrs. Montoya is buying prizes for José's party. The prizes are 3 for $.50. Mrs. Montoya wishes to buy 36 prizes. How much will they cost? How do you decide on the proportions $\frac{3}{\$.50} = \frac{36}{n}$ or $\frac{3}{36} = \frac{\$.50}{n}$?

- Juanita is building a scale model of a car to give José for his birthday. The scale is 5 centimeters per meter. If the model is 30 centimeters long, how long is the car?

$$\frac{5 \text{ centimeters}}{1 \text{ meter}} = \frac{30 \text{ centimeters}}{n \text{ meters}}$$

or

$$\frac{5}{30} = \frac{1}{n}$$

Children may wish to use concrete materials to model each of these problems. Be sure to connect the work they are doing concretely to the abstract solution of the equations.

Objective: *to practice using proportions to solve percent problems.*

2. One method of solving problems involving percents is through the use of the following proportion:

$$\frac{\text{rate}}{100} = \frac{\text{percentage}}{\text{base}}$$

This proportion works regardless of whether the children need to find the rate, the percentage, or the base in a given problem. For example, a store may be having a sale on all its jeans. In one case, the jeans originally cost $20 (base). They have been reduced by $5 (percentage) and you wish to find the percent of the original cost that the reduction represents. Use the proportion $\frac{n}{100} = \frac{\$5}{\$20}$. By cross multiplying, you find that $\$20 \times n = 100 \times \$5$, or $20n = 500$, or $n = 25\%$. The jeans are marked down by 25% of the original cost.

In another problem, you see a sign that says 20% off everything on this rack. You wish to buy a pair of jeans that were originally $30 and want to find out how much you will save buying them on sale. Again, you can use the proportion to solve the problem. This time you know the percent and the base and wish to find the percentage. The proportion is $\frac{20}{100} = \frac{\$n}{\$30}$. By cross multiplying, you find that $20 \times \$30 = \$n \times 100$, or $100n = 600$, or $n = \$6$. You will save $6 on the jeans.

In the third case, the sign says that the jeans have been marked down 40% and that you will save $10 on each pair. What was the original cost of the jeans? This time you know the percent and the percentage and you wish to find the base. Use the same proportion. The proportion is $\frac{40}{100} = \frac{\$10}{\$n}$. By cross multiplying you find that $40 \times \$n = 100 \times \$10$, or $40n = 1000$, or $n = \$25$. The jeans originally cost $25.

The children will find many more examples of ways proportions can be used, not only in their textbooks but also in their own shopping. You may wish to introduce a unit on becoming a wise consumer and ask children to bring in examples of percent problems from their own lives. The newspaper is a rich source of problems. The children may be surprised to find the number of examples of store sales that do not report the percentages correctly. Challenge the children to find examples of misleading or incorrect ads in the paper. They may even wish to inform the store managers of their mistakes.

Ask the children if it makes a difference if the percents are reported as a percent of the original cost or as a percent of the discounted cost. Let the children work several examples to see that it does indeed make a difference. Encourage children to ask questions and explore other aspects of percents. They may even be able to teach their parents and other adults some aspects of becoming wise consumers.

Objective: to practice using proportions with inverse relationships.

3. If children have worked with the gears mentioned earlier in the chapter, they should have noticed that the larger gears with more teeth went around fewer times. If a gear with 10 teeth turned a gear with 5 teeth, the larger gear went around only $\frac{1}{2}$ the number of times that the smaller gear went around. The proportion for the gears is

$$\frac{\text{the number of teeth in the first gear}}{\text{the number of teeth in the second gear}} =$$

$$\frac{\text{the number of turns of the second gear}}{\text{the number of turns of the first gear}}$$

If the larger gear turns three times, the proportion is $\frac{10}{5} = \frac{n}{3}$, where n is the number of turns for the smaller gear.

By cross multiplying, you get $10 \times 3 = 5 \times n$, or $30 = 5n$, or $n = 6$. The smaller gear turns 6 times as the larger gear turns 3 times.

The computer program Gears by Robert Kinbal and Dave Donoghue, available from Sunburst, gives children the opportunity to collect data on gear rotation through simulation and then to practice solving problems involving this inverse proportion by asking children to select the number of gears and their sizes to match a given problem. Children may also create challenges for each other with the program.

Estimating and Mental Calculating

Test results from the Fourth National Assessment of Educational Progress (Kouba, Brown, Carpenter, Lindquist, Silver, and Swafford, 1988) show that students generally are poor at estimations that involve common or decimal fractions. As we rely more on calculators and computers, estimation and mental calculation skills become increasingly important. Children should be encouraged to estimate or calculate mentally in activities involving rational numbers just as they were in the whole-number activities.

Some of the activities mentioned earlier, such as the pyramid game, require that children be able to estimate whether one fraction is larger than another.

There are also a number of times when children should be able to accurately calculate an answer mentally. For example, they should know the common fraction equivalents for percents that they encounter every day. Children should have several estimation and mental calculation experiences as they learn rational number skills. The activities presented below give a few more ideas for practicing estimation and mental calculation skills. The activities are listed for a wide range of grade levels because although, ideally, children should learn to estimate when they first learn fractions, many children do not. Therefore, the activities may be used with students in the middle grades as well.

ACTIVITIES

Elementary (2–8)

Objective: to develop estimation skills with common fractions.

1. Children should be able to use certain model fractions to compare various amounts. Give children experiences with $\frac{1}{2}$, $\frac{1}{4}$, and $\frac{3}{4}$ so that they may estimate whether other amounts are more or less than these. Begin by letting children estimate when something is $\frac{1}{2}$ full. Use a glass container and let children pour beans into it until they feel it is half full. After they fill it to what they think is half full, have them empty the beans into another container and again fill the container half full with different beans. Leave these beans in the container and then pour the first amount of beans back into the container. If both estimates were good, the container should be full. Or, mea-

sure the total container and the amount poured into it to see how close the amount is to $\frac{1}{2}$ of the container.

After the children can estimate $\frac{1}{2}$ fairly well, pour varying amounts into the container and ask the children if the amounts are more or less than $\frac{1}{2}$. When the children can accurately compare amounts to $\frac{1}{2}$, then repeat the activity with $\frac{1}{4}$ and $\frac{3}{4}$. Older children with more estimation experience may use a liter and estimate amounts in tenths of a liter (deciliters).

2. Ask children to keep a list of times outside of school that they need to use common or decimal fractions to estimate. They may include some of the following:

- part of an hour to eat breakfast, take a bath, do homework, etc.
- part of a mile or kilometer to walk to a friend's house, the store, school, etc.
- part of a quart or liter of milk drunk for dinner
- part of a cake or pie eaten for dessert
- part of a pound or kilogram of meat or cheese to buy for lunch

Children can think of many other examples themselves. Encourage them to share the techniques they use to estimate.

Objective: to practice mental conversion of percents to common fractions.

3. Older children should have experience in converting percents to common fractions mentally because of the widespread use of percents in everyday life. Bring in ads from department store sales and look for the percents that are used. Ask the children to tally how often different percents are used in the ads. They will probably find frequent use of percents that convert to tenths, thirds, fourths, and halves.

Ask the children the benefits of knowing the common fraction equivalents. Discuss how to use proportions to find discounts if you know the common fraction equivalents of the commonly used percents. Ask the children how they convert 20%, 30%, 40%, . . . to common fractions if they know that 10% is $\frac{1}{10}$. If 33.$\overline{3}$ % is $\frac{1}{3}$, what is $\frac{2}{3}$? If 25% is $\frac{1}{4}$, what is $\frac{3}{4}$? If an item is marked 25% off, how can you use the common fraction equivalent to find the discount? Let the children ask similar questions of their own and share their strategies with each other.

Problem Creating and Solving

Throughout this chapter, we have mentioned a number of opportunities for children to solve problems and to create problems of their own, which they will then solve or which their classmates will solve. Because of the importance of this topic, more ideas for problem solving and problem creating are included here. As with any other topic, problem creating and solving should become the child's natural approach to work with rational numbers.

ACTIVITIES

Intermediate (3–8)

Objective: to explore relationships among common fractions, recognizing that each fraction may be represented in a number of different ways.

1. Pattern blocks are designed so that the areas of many of the blocks are multiples of the areas of other blocks. This makes them well-designed for exploring fractional relationships. The book *Fractions with Pattern Blocks* by Mathew E. Zullie, available from Creative Publications, suggests many activities for the exploration of these relationships. Many of the activities require that the children use problem solving and higher-level thinking strategies.

Have the children use orange squares, uncolored parallelograms, green triangles, blue rhombuses, red trapezoids, and yellow hexagons for these activities. Ask, "If the yellow hexagon represents 1, what are the values of each of the other pieces?" If the children have difficulty determining the other values, ask, "How many green triangles does it take to cover the yellow hexagon? How many red trapezoids does it take to cover the hexagon? How many triangles does it take to cover the parallelogram?"

Next, challenge the children to find as many examples as they can that show $\frac{1}{3}$ using any number of pattern blocks.

- *Understanding the problem.* I will need to find ways to show $\frac{1}{3}$ using pattern blocks. This means that I should look for single blocks or groups of blocks that can represent the value of 1 (unit). Then I will try to find blocks to represent $\frac{1}{3}$. I will try also to find other ways to make a unit.
- *Devising a plan.* One way I can solve this problem is to take three blocks of one shape, for example, three orange squares. Then I place them together to form a row. This rectangular row will have the value of 1. Any orange square is $\frac{1}{3}$ of the row. Another way to solve the problem is to look for combinations of blocks that can be placed together into shapes to represent 1. For example, a blue rhombus combined with a green triangle can have the value of 1. Then a green triangle has the value of $\frac{1}{3}$. I will try some of these ways.
- *Carrying out the plan.* I can place three of the uncolored parallelograms together to make a shape with the value of 1. One of the parallelograms represents $\frac{1}{3}$ of the unit shape. That works also with three green triangles, orange squares, blue rhombuses, red trapezoids, and yellow hexagons.

I can also use a combination such as a green triangle and a blue rhombus as the unit. Then a green triangle represents $\frac{1}{3}$. Another combination representing the value of 1 is two red trapezoids. Then a blue rhombus represents $\frac{1}{3}$. If a yellow hexagon and a red trapezoid are combined into a unit, a red trapezoid represents $\frac{1}{3}$. Still another solution is to use six green triangles as a unit; then two green triangles represent $\frac{1}{3}$. I have found at least ten different ways to show $\frac{1}{3}$ using the pattern blocks.

- *Looking back.* I need to check to make sure each of the fraction models I made shows the fraction $\frac{1}{3}$. I will then try to generalize my solutions. I can continue to use more blocks to show the unit and try to find other examples in which more than one block is used to show the fraction $\frac{1}{3}$.

Encourage the children to make up questions of their own. Children can make designs to represent the unit and trace the outline. Have them trade outlines and challenge each other to find various fractional amounts of the unit designs. Is there more than one solution for each design? Let the children find out.

Exploring Algebra

Suppose you write the numerators and denominators of a set of equivalent fractions as number pairs. Then you graph them. What do you notice about the points?

1. Copy the table below. Write numbers for △ and □ .

$$\frac{2}{3} = \frac{\triangle}{\square}$$

△	2	4	6			
□	3	6	9			

2. Graph the number pairs (△ , □) from the table above on a graph like this.

3. Connect the points on the graph. What do you notice?

4. Repeat using the fraction $\frac{2}{5}$. What do you notice about this graph?

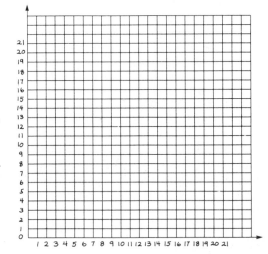

Figure 9–22

Robert E. Eicholz et al. Addison-Wesley Mathematics Series *(Grade 4). Menlo Park, Ca:* Addison-Wesley, 1991, p. 342. Reprinted by permission.

THE MATH BOOK

The fourth grade textbook page shown in Figure 9–22 is a lesson that reinforces ideas about equivalent fractions that were introduced earlier in the chapter and that extends those ideas to include some algebra concepts. Earlier in the chapter, the students explored equivalent fractions using such materials as fraction strips, two-color counters, egg cartons, and geoboards. The teacher's manual suggested that they use graph paper and calculators in cooperative learning groups to explore patterns and to make generalizations about equivalent fractions such as you can multiply or divide the numerator and the denominator by the same number to find an equivalent fraction and you can use "crisscross patterns" to determine if two fractions are equivalent.

In this lesson, the students convert fractions to ordered pairs, complete a table of equivalent fractions, and then graph the ordered pairs on a Cartesian coordinate graph. Students are asked to look for patterns in the table as well as on the graph. The teacher's manual suggests that students make graphs for several sets of equivalent fractions and then compare the graphs. Students then discuss the fact that on all the graphs the points form a line. As a challenge, students are asked to compare the steepnesses of the lines for different sets of equivalent fractions and to relate the steepness of the line to the size of the fraction. This is an excellent activity to prepare fourth grade students for later algebra activities on graphing in which students compare the slopes of lines for different equations.

This lesson is followed by a midchapter review/quiz and then a mental math lesson on finding a fraction of a number. Later lessons in the chapter include using a calculator to find fractions of a number, and applying fraction concepts to money, time, and measurement. Most lessons begin with suggestions in the teacher's manual for the use of concrete manipulatives in cooperative learning groups and connections using problems of the day, life skills, computers, and data collection. Ideas for having students write about their reasoning are also included.

2. Earlier in this chapter, we suggested that you have the children explore many ways to show $\frac{1}{2}$ using paper folding and Cuisenaire rods. A similar activity should be repeated using other materials and other fractions.

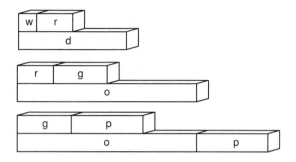

Ask the children to make two trains of Cuisenaire rods such that one train is $\frac{1}{2}$ the length of the other train. This will help the children later when they learn to use the rods to show addition of fractions. Ask the children to use the rods to show other fractional relationships, such as $\frac{2}{3}$, $\frac{3}{4}$, and 0.6.

After the children can find these relationships, tell them that the dark green rod represents 1 and ask them to find the rod for $\frac{1}{3}$. If the dark green rod represents 2, which rod is $\frac{1}{3}$? Let the children challenge each other with problems of their own. Is it always possible to find a rod to represent a given fraction? Why not?

Grouping Students for Learning Rational Number Concepts

Because the concepts of rational numbers and the rules for operations with them frequently are different from the concepts and rules that students have for whole numbers, it is important that students be encouraged to explore these ideas concretely in small groups to encourage open discussion of ideas. When younger students are first discovering the concepts of fractions, they can work in groups of two, three, or more to find ways to divide different amounts of materials equally among the members of the group. For example, three students may be given a large sheet of paper and asked to fold it and then cut along the folds so that each person in the group has the same amount of paper, or four students may be given a bottle of juice or a bag of peanuts to share equally. In each case, discuss the fractional parts that are found and compare the methods used by one group to those used by another.

After students have developed the concepts of rational numbers, they may work in groups to expand upon these ideas. Ask a group of four to brainstorm everything they know about the fraction $\frac{3}{4}$ or to write equivalent names for $\frac{3}{4}$ in as many ways as possible. They may also list all the ways that they see rational numbers being used every day.

Communicating Rational Number Concepts

Payne and Towsley (1990) state that work with fractions and decimals should be mostly oral in grades K–4 and that it should be only oral in grades K–2, using models and realistic problems. Children of this age are not ready for the abstract symbolic notation of fractions. Even in grades 5–8, they recommend that all work begin with concrete models, realistic problems, and oral language. This oral communication of ideas is essential to the development of the concepts of rational numbers. Mack (1990) notes that while students bring a rich informal background to the study of rational numbers through real-life situations, this knowledge is not connected to the abstract symbols and procedures. Indeed, knowledge of rote procedures may actually hinder students in building on prior knowledge. If students try to learn abstract symbols and procedures before they are given a chance to orally discuss their understanding of rational number concepts, they may never develop a true understanding of rational numbers.

Evaluating Rational Number Learning

As with other concepts, children should be tested informally as well as formally on their rational number concepts. As children work with concrete materials, keep track of which children demonstrate competence with the rational number concepts and skills and which children are still having difficulty. For additional instruction, group together children having difficulty with the same concepts or skills. Let the other students explore some of the applications of rational numbers.

As you test children, present problems in a variety of ways, since some children may be able to perform satisfactorily using one model or one type of presentation but not another. Differences are found in performance when children are asked to create their own models for a given rational number rather than to select a model from a number of choices, or when they are asked to write the correct common fraction to match a given picture rather than to select or create a picture to match the fraction. Some children can select the correct model from concrete materials but not from pictures, or can draw a picture but not create a concrete model such as paper folding. Children may select or create the correct model using areas but not using a number line. It is therefore important to present rational number tasks to children using a variety of models and a variety of methods of presentation.

Some tasks may be paper and pencil or calculator tests but others should be individual interviews in order to get

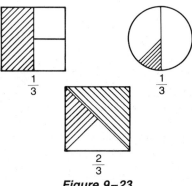

Figure 9-23

the clearest picture of the child's conceptions or misconceptions about rational numbers. Following are some of the common problems that arise as children work with rational numbers.

1. Some children do not realize that in the part-whole model of a fraction, the parts must be of equal size. They may give some of the responses shown in Figure 9–23. Children having difficulty with the part-whole concept should return to work with concrete materials such as paper folding. Have them cut out the regions to show that all halves, thirds, fourths, etc., take up the same amount of area even if they are not congruent.

2. Some children give a ratio answer when a part-whole answer is expected. They may give the responses shown in Figure 9–24.

   This problem is related to difficulty with the Piagetian tasks of class inclusion and reversibility. Children may have difficulty realizing that in the part-whole model, the fractional part must be compared back to the whole that includes it; they compare one part to the other parts. Or they may not realize that something divided into two halves will again become one whole when you put the two halves back together. Such children need more experience manipulating concrete models and discussing their meanings before they continue with abstract work with fractions.

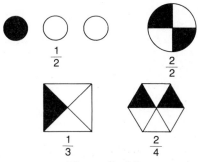

Figure 9-24

3. Some children have difficulty writing a common fraction as a division problem. They think that $\frac{2}{3}$ means $3 \div 2$ rather than the other way around. This misunderstanding is particularly troublesome as children attempt to convert common fractions to decimal fractions.

   Children will also need instruction on which number to put into the calculator first when they divide. Remind them that the numerator is the dividend and the denominator is the divisor. You may also need to return to the concrete models and remind them that in the division model, the numerator tells you the number you begin with and the denominator tells the number of parts it is being divided into.

4. Some children do not realize that a common fraction is a single number. They think of it as two numbers, one sitting on top of the other. This causes them to work with the numbers separately when they perform operations with common fractions; they may add together numerators and denominators when they attempt to add two common fractions. Work with concrete materials and with renaming common fractions as decimal fractions helps children realize that common fractions represent only one number.

5. Some children have difficulty ordering rational numbers, even unit fractions. They think that $\frac{1}{3}$ is bigger than $\frac{1}{2}$ because 3 is bigger than 2. These children may need to return to work with concrete materials to discover that there is an inverse relationship between the size of the denominator in a unit fraction and the size of the piece that the fraction represents. The activities suggested for estimation should help these children.

   Vance (1986) has found that similar problems arise when children order decimal fractions, especially when the numbers do not contain the same number of decimal places. When ordering .5, .34, and .257, some children decide that .257 is the largest because it uses the most digits or that it is the smallest because it is written in thousandths, which are smaller than tenths or hundredths. Such reasoning would work for the problem above but not for .5 and .678.

   Children having difficulty ordering decimals should return to work with concrete materials and with estimation. They should work with rewriting decimals so that all the decimal fractions contain the same number of decimal places.

   Calculators can be useful here. Vance suggests a calculator game in which the students try to "wipe out" a digit in the calculator display. For example, if the calculator shows 23.9876, wipe out the 7 by subtracting .007. Ask students to see if the same thing happens when they subtract .0007 or .0070.

6. Some children have difficulty understanding that rational numbers are dense. It is hard to understand

that with rational numbers there is no next number, as there is with whole numbers; there is always another rational number between any two rational numbers. Children who find it hard to give a rational number between $\frac{1}{4}$ and $\frac{2}{4}$ or between 0.1 and 0.2 may need to return to work with concrete materials and with renaming rational numbers. If $\frac{1}{4}$ and $\frac{2}{4}$ are renamed as $\frac{2}{8}$ and $\frac{4}{8}$, or if 0.1 and 0.2 are renamed as 0.10 and 0.20, it is easier for children to find another number between them. The pyramid game described earlier also will help children practice finding a rational number between two other rational numbers.

7. Some children understand unit fraction concepts but not concepts of fractions with numerators greater than 1 or concepts of mixed numbers. Children who have a concrete understanding of these concepts may still have difficulty with improper fractions. Such children should return to a variety of models, including concrete regions and the number line, to develop a better understanding of improper fractions.

8. Some children have difficulty converting decimals to percents. They remember that they should move the decimal point two places, but they move it the wrong direction. This is especially true for percents that are less than 1 or greater than 100.

   This problem with the decimal point also arises in writing dollars and cents. A number of adults seem not to realize that when used with a cent sign, the decimal point represents hundredths of a cent. It is tempting to offer the grocer a penny for four oranges when the sign above the oranges reads .25¢ for one orange. Children enjoy looking for mistakes such as this.

   Similar mistakes involving the decimal point arise when children are using a calculator. Be sure children use their estimation skills to determine whether or not their answers make sense. Examples from real life can be useful here. Misplacing the decimal point when solving a real-life problem involving money can be very costly.

9. Some children have difficulty with proportional reasoning. Tasks involving proportions are often difficult on the abstract level, even for adults. Be sure to include concrete work on these concepts and encourage children to discuss their reasoning with each other.

Something for Everyone

Again, we recognize that while children learn in a variety of modes, some children may learn more comfortably in a particular mode, such as visually or auditorily. Because both Chapters 9 and 10 are directed at teaching rational numbers, the specific learning modes associated with rational numbers are discussed at the end of Chapter 10.

KEY IDEAS

Rational numbers in all forms are commonly used in our everyday lives, but research shows that they are generally poorly understood by elementary and middle school students. Chapter 9 presents many ideas for helping students better understand these numbers. Several models for presenting rational numbers in common and in decimal fraction form are shown, including the part-whole, ratio, and division models.

Materials described in this chapter include Fraction Tiles, rectangular and circular fraction regions, Decimal Squares, base ten blocks, number lines, fraction strips, Fraction Bars, colored chips, Cuisenaire rods, and arrays. These concrete materials, as well as pictures, computer programs, calculators, and reinforcement activities, are important to the development and retention of good rational number concepts. Children should explore the meaning of rational numbers written in any form and should be able to read, rename, and order fractions, decimals, and percents.

They should practice estimating and mentally calculating with rational numbers in every form. Problem solving and problem creating should encourage the children to delve even deeper into the meanings of rational numbers.

Diagnosis should include looking at children's understanding of the meanings of rational numbers using a variety of concrete and semi-concrete models, not just looking at a child's ability to rotely memorize some poorly understood rules such as how to write equivalent fractions.

REFERENCES

ALLINGER, GLENN D., AND PAYNE, JOSEPH N. "Estimation and Mental Arithmetic with Percent." *Estimation and Mental Computation*. National Council of Teachers of Mathematics, 1986 Yearbook. Reston, Va.: NCTM, 1986.

BEHR, MERLYN J.; POST, THOMAS R.; AND WACHSMUTH, IPKE. "Estimation and Children's Concept of Rational Number Size." *Estimation and Mental Computation*. National Council of Teachers of Mathematics, 1986 Yearbook. Reston, Va.: NCTM, 1986.

BEHR, MERLYN J., ET AL. "Order and Equivalence of Rational Numbers: A Clinical Teaching Experiment." *Journal for Research in Mathematics Education*. Vol. 15, No. 4 (November 1984), pp. 323–341.

BENNETT, ALBERT B., JR. *Decimal Squares*. Fort Collins, Col.: Scott Resources, 1982.

BENNETT, ALBERT B., JR., AND DAVIDSON, PATRICIA A. *Fraction Bars*. Fort Collins, Col.: Scott Resources, 1973.

BEZUK, NADINE, AND CRAMER, KATHLEEN. "Teaching About Fractions: What, When, and How?" In Trafton, Paul R., and Shulte, Albert P., eds. *New Directions for Elementary School Mathematics*. Reston, Va.: National Council of Teachers of Mathematics, 1989, pp. 156–167.

BRADFORD, JOHN. *Everything's Coming Up Fractions with Cuisenaire Rods*. New Rochelle, N.Y.: Cuisenaire Company of America, 1981.

CARPENTER, THOMAS P., ET AL. "Decimals: Results and Implications from National Assessment." *Arithmetic Teacher*. Vol. 28, No. 8 (April 1981), pp. 34–37.

DOSSEY, JOHN A.; MULLIS, INA V.S.; LINDQUIST, MARY M.; AND CHAMBERS, DONALD L. *The Mathematics Report Card: Are We Measuring Up? Trends and Achievement Based on the 1986 National Assessment.* Princeton, N.J.: Educational Testing Service, 1988.

EICHOLZ, ROBERT E., ET AL. *Addison-Wesley Mathematics, Book 7.* Menlo Park, Ca.: Addison-Wesley Publishing Co., 1987, p. 204.

JENKINS, LEE, AND McLEAN, PEGGY. *Fraction Tiles: A Manipulative Fraction Program.* Hayward, Ca.: Activity Resources Co., 1972.

KOUBA, VICKY L.; BROWN, CATHERINE A.; CARPENTER, THOMAS P.; LINDQUIST, MARY M.; SILVER, EDWARD, A.; AND SWAFFORD, JANE O. "Results of the Fourth NAEP Assessment of Mathematics: Number, Operations, and Word Problems." *Arithmetic Teacher.* Vol. 35, No. 8 (April 1988), pp. 14–19.

LICHTENBERG, BETTY K., AND LICHTENBERG, DONOVAN R. "Decimals Deserve Distinction." *Mathematics for the Middle Grades (5–9).* National Council of Teachers of Mathematics, 1982 Yearbook. Reston, Va.: NCTM, 1982.

MACK, NANCY K. "Learning Fractions with Understanding: Building on Informal Knowledge." *Journal for Research in Mathematics Education.* Vol. 21, No. 1 (January 1990), pp. 16–32.

NATIONAL COUNCIL OF TEACHERS OF MATHEMATICS. *Curriculum and Evaluation Standards for School Mathematics.* Reston, Va.: NCTM, 1989.

PAYNE, JOSEPH N., AND TOWSLEY, ANN E. "Implications of NCTM's *Standards* for Teaching Fractions and Decimals." *Arithmetic Teacher.* Vol. 37, No. 8 (April 1990), pp. 23–26.

POST, THOMAS R. "Fractions: Results and Implications from National Assessment." *Arithmetic Teacher.* Vol. 28, No. 8 (May 1981), pp. 26–31.

POST, THOMAS R.; BEHR, MERLYN J.; AND LESH, RICHARD. "Research-Based Observations About Children's Learning of Rational Number Concepts." *Focus on Learning Problems in Mathematics.* Vol. 8, No. 1 (Winter 1986), pp. 39–48.

POST, THOMAS R., ET AL. "Order and Equivalence of Rational Numbers: A Cognitive Analysis." *Journal for Research in Mathematics Education.* Vol. 16, No. 1 (January 1985), pp. 18–36.

"Rational Numbers (Focus Issue)." *Arithmetic Teacher.* Vol. 31, No. 6 (February 1984).

VANCE, JAMES. "Ordering Decimals and Fractions: A Diagnostic Study." *Focus on Learning Problems in Mathematics.* Vol. 8, No. 2 (Spring 1986), pp. 51–59.

ZULLIE, MATHEW E. *Fractions with Pattern Blocks.* Palo Alto, Ca.: Creative Publications, 1975.

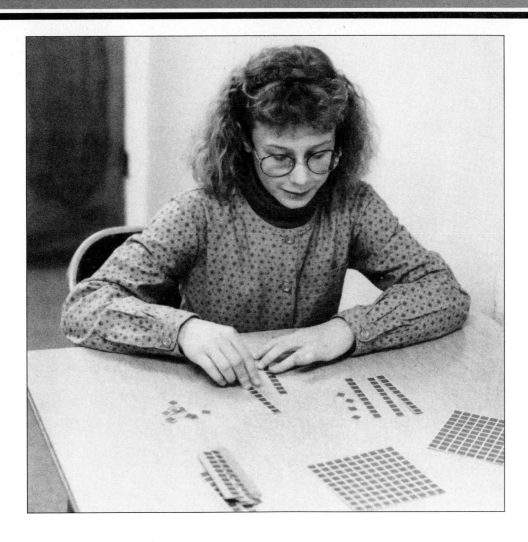

Operations with rational numbers are generally emphasized in the upper elementary and middle school grades. Operations with decimal fractions may be introduced earlier along with the decimal notation for money or perhaps with measurements in the metric system, but generally textbooks wait until fourth or fifth grade to teach rational number operations.

There is some controversy over whether operations with common fractions should be delayed until seventh or eighth grade because of the increased use of decimal fractions and the difficulties many children experience with common fractions, but most textbooks do not delay the instruction that long. Many textbooks introduce operations with decimal fractions the same year they introduce operations with common fractions. That is one of the reasons we do not have separate chapters for common and decimal fractions.

Perhaps more importantly, we do not have these operations in separate chapters because conceptually the two represent the same situations. Children should be able to recognize that if they use three tenths of a meter of material for a doll's blouse and seven tenths of a meter of material for a doll's coat, it does not matter if they find the total amount of material by adding $\frac{3}{10} + \frac{7}{10}$ or by adding $0.3 + 0.7$. No matter how it is written, they have used a meter of material.

It is important that operations with rational numbers arise from examples in the "real world." Have children make a list of occasions when rational numbers are used in their homes or everyday lives. The following are a few suggestions for events or places where rational numbers are used. Ask children to discuss situations in which rational numbers might be used and to make up story prob-

lems with these settings. Have the children add to the list as situations arise.

- recipes
- sewing
- gardening
- medicine
- building
- unit pricing
- odometer
- hourly wages
- track and field events
- kilowatt hours of electricity
- stock market
- scale drawings (architecture, engineering, drafting, surveying, mapmaking, etc.)
- measurements of all types (time, length, area, volume, money, etc.)
- probability and statistics
- graphing

As we discuss the concepts of operations with rational numbers, we frequently draw examples from daily events. You and the children should do the same in your classroom.

Working with concrete materials continues to be important as children learn to operate with rational numbers. Even children in the seventh and eighth grades are generally concretely operational and cannot fully understand new concepts on an abstract level. Work with manipulative materials helps children construct their knowledge about rational number operations. Continue to use the manipulative materials discussed in Chapter 9 for rational number concepts; these include Fraction Bars, Fraction Tiles, circular and rectangular regions, pattern blocks, Cuisenaire rods, base ten blocks, Decimal Squares, number lines, fraction strips, paper for folding, colored chips, and arrays. Continue to stress calculator skills, especially with decimal fraction operations, and use appropriate computer programs to help the children develop skills with common and decimal fractions and percents.

As children develop understanding and skill in their work with rational numbers, they begin to explore some of the early concepts of operations with these numbers. When they found like denominators for two common fractions, they may have noticed that when two fractions have like denominators, it is very easy to add them together, to find how much bigger one is than the other, or even to find how many times bigger one is than the other. Encourage and expand upon this intuitive understanding of operations with rational numbers as you introduce the operations more formally. Let children continue to question, explore, and discover relationships and algorithms of their own as they expand their knowledge of rational numbers to include operations.

Developing Concepts of Operations with Rational Numbers

After children understand the meaning of rational numbers written as both common and decimal fractions and can rename equivalent fractions, they are ready for operations with rational numbers. Operations with rational numbers should be understood before they are practiced abstractly. Too many children memorize rules such as "invert and multiply" but are not able to explain when or why this should be done. In this section, we concentrate on the meaning of operations with all types of rational numbers.

In the next section, we discuss developing and practicing skills with the operations. Be sure to allow children plenty of time for concept development before you move on to practicing skills. The more time children spend on the concepts, the less time they will need to practice the skills.

Addition of Common Fractions

When you first introduce children to addition of common fractions, continue to use the manipulative materials they used for learning the concepts of common fractions. Introduce addition using problems from the children's lives for which the children are able to discover solutions by exploring with familiar materials.

Textbooks generally begin with problems involving common fractions with like denominators and later move to problems with unlike denominators. Children working with concrete materials may not need to separate problems with like from those with unlike denominators, however. If they have worked with renaming fractions concretely in developing concepts of common fractions, the addition and subtraction of common fractions should follow naturally. Following are some examples of situations that may be used to introduce addition of common fractions.

ACTIVITIES

Intermediate (4–6)

Objective: *to develop the concept of addition of common fractions with like denominators.*

1. Give the children an example of a situation that would require the addition of common fractions and let them use concrete materials such as the circular or rectangular regions (see Appendix B) to find the answer. A beginning example follows: Danny baked a pie and left it to cool. His sister Maureen came along and cut it into 6 equal pieces. She ate $\frac{1}{6}$ of the pie and gave her friend Suzanna $\frac{1}{6}$ of the pie. How much of the pie did they take altogether?

    Let the children manipulate the pieces to show the answer of $\frac{2}{6}$. Then, have the children make up and illustrate other problems of their own. After children have worked several problems using the materials, ask them to write down what they have done. They may initially write 1 sixth + 1 sixth = 2 sixths. After comparing this to earlier work with whole numbers, they may write the equation as $\frac{1}{6} + \frac{1}{6} = \frac{2}{6}$ or

$$\begin{array}{r} \frac{1}{6} \\ + \frac{1}{6} \\ \hline \frac{2}{6} \end{array}$$

Ask the children to write equations for all their work with the manipulatives. They may have some of the following:

$$\frac{1}{6} + \frac{1}{6} = \frac{2}{6}$$

$$\frac{1}{3} + \frac{1}{3} = \frac{2}{3}$$

$$\frac{1}{4} + \frac{2}{4} = \frac{3}{4}$$

$$\frac{2}{5} + \frac{3}{5} = \frac{5}{5}$$

Ask the children if they notice anything consistent about the problems. Lead them to discuss the fact that in each problem, the denominator remains the same and the numerators are added together. After the children have worked several problems using regions and have recorded the equations, ask them to show the same equations using another material such as the Cuisenaire rods, the number line, or the Fraction Bars. Figure 10–1 shows a few possibilities.

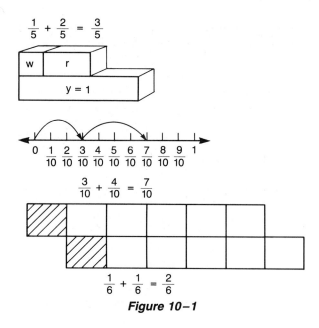

Figure 10-1

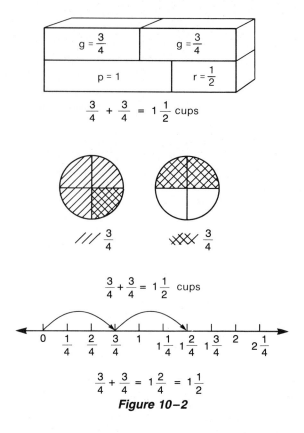

Figure 10-2

Discuss with the children the fact that the sum remains the same no matter what material is used to illustrate the equations. Ask the children to analyze what is wrong with examples such as $\frac{2}{3} + \frac{1}{3} = \frac{3}{6}$ and $\frac{2}{5} + \frac{1}{5} = \frac{3}{10}$. Ask how they can use the materials to help another child understand why those examples are incorrect.

2. For this activity, use the state-operator-state machine that you used to explore addition of whole numbers. Begin with a machine in which the operation is join. Use fraction pieces such as the circular regions cut into eighths. Have one child pick any number of eighths and put the regions into the machine. Ask the children watching to write down the input. Have another child choose another number of eighths to put in the other input. The children observing should then write down that amount. Have another child join the two amounts, and have the observers write + between the numerals they have just written.

   Ask the observers to predict the output of the machine. Put the joined amounts into the output and check to see if the prediction was correct. Let the observers finish writing down the equation.

   Try this with a number of different inputs, asking the children to write the equation and predict the output each time. After several examples using the operator join with the circular regions, switch to other types of fraction pieces. Let the children suggest their own equations and show each with the pieces as the others write the corresponding equations.

3. Children sometimes think that whenever you add two fractions, the answer will be less than 1 because that is often the case with the early examples. They need early exposure to problems where the sum is greater than 1. Again, use examples from the children's everyday lives. For example, for Danny's pie, he used $\frac{3}{4}$ cup of sugar for the crust and $\frac{3}{4}$ cup of sugar for the filling. How much sugar did he use altogether? Again, let the children use any of the familiar manipulatives to find the sum.

   Some children may have difficulty when they realize that the answer is larger than 1 cup. Ask them how much larger

than a cup the sum is; yes, you have one whole cup and $\frac{2}{4}$, or $\frac{1}{2}$, of another cup. Figure 10–2 shows some of the possible solutions with the materials.

Again, ask the children to write the equation for the problem. Discuss the fact that $\frac{6}{4}$ and $1\frac{2}{4}$ name the same amount (this should be familiar from the work with equivalent fractions). Let the children suggest other word problems, solve them with the manipulatives, and write the equations. Encourage the children to predict whether the answer to each example will be larger or smaller than 1. After the children have had some practice, let them work the problems mentally, without using the materials or writing the equations.

Some calculators show fractions in common as well as decimal form. Allow the students to use one of these calculators to explore the addition and subtraction of common fractions. What happens when you simplify a fraction or change from a mixed numeral to an improper fraction?

Objective: to develop the concept of addition of common fractions with unlike denominators.

4. After the children have had experience adding common fractions with like denominators, suggest a problem that involves adding fractions with unlike denominators. Again, let the children choose a concrete material with which to solve the problem. Following is one example: After Danny's pie was eaten, he decided to bake a cake. His father said the cake looked so good, he would like a huge piece. Danny's mom was on a diet, so she wanted only a small piece. Danny cut a piece $\frac{1}{4}$ of the whole cake for his dad and a piece $\frac{1}{8}$ of the cake for his mom. How much of the whole cake did Danny give his parents?

Figure 10–3a shows how the children may demonstrate this with the fraction pieces. Ask the children what they

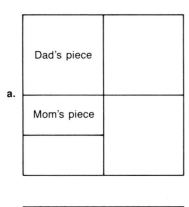

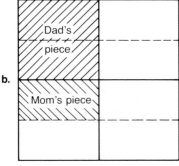

Figure 10–3

2 wholes + $\frac{7}{6}$ = $3\frac{1}{6}$ yards

$1\frac{2}{3}$ = $1\frac{4}{6}$

$1\frac{1}{2}$ = $1\frac{3}{6}$

$2\frac{7}{6}$ = $3\frac{1}{6}$ yards

Figure 10–4

would call the answer. It is difficult to name the answer unless the pieces are all the same size. In Figure 10–3a, you have two pieces, but they are not two fourths or two eighths. It is not convenient to say you have one fourth and one eighth of the whole cake. Encourage children to draw upon their experience with renaming common fractions to suggest that the fourth may be renamed as eighths. Let the children exchange the fraction pieces as shown in Figure 10–3b.

Now ask the children to tell you the sum. The two pieces are divided into portions of the same size, eighths, so the sum is three eighths. Encourage the children to show their work in equation form. At this point, it is easiest to work with equations in vertical form, so show the conversion as:

$$\begin{array}{r} \frac{1}{4} = \frac{2}{8} \\ +\frac{1}{8} = +\frac{1}{8} \\ \hline \frac{3}{8} \end{array}$$

Let the children suggest other word problems and solve them using other concrete or semi-concrete aids. Some of the problems should involve common fractions that must be renamed with a common denominator. Remind children of their earlier work with renaming fractions. Ask children to write up word problems for each other to leave in the learning center or for you to use on worksheets for the whole class.

Encourage children to discuss their methods of solution with each other. If disagreements about the solutions arise, ask questions to lead the children to discover which solution is correct. They may find that several methods (or none)

work. Compare the results to answers obtained using a calculator that shows common fractions.

Objective: *to develop the concept of addition of mixed numerals.*

5. With mixed numerals as well as with proper fractions, students should develop addition concepts beginning with "real-life" situations. An example follows: Maria needs $1\frac{2}{3}$ yards of material to make a skirt and $1\frac{1}{2}$ yards of material to make a matching jacket. How much material should Maria buy for the outfit?

Again, the children should use the familiar fraction materials to work out the problem. Let the children discuss what to do with the fractions with unlike denominators. Encourage the children to begin by putting the whole sections together and then trading in the parts for amounts shown in sixths, the common denominator. Ask the children what to do with the sixths, since they make more than another whole yard. Encourage the children to write down the equations as they work with the materials to record what they are doing. The children's work with fraction strips may look like that in Figure 10–4.

After the children have added mixed numerals with the aid of one manipulative, encourage them to show the same problem with other manipulatives or pictures. Figure 10–5 shows other possibilities for $1\frac{2}{3} + 1\frac{1}{2}$.

Addition of Decimal Fractions

Depending on the textbook series you use, you may decide to introduce addition with decimal fractions either before or after addition with common fractions. Many children find addition with decimal fractions easier than addition with common fractions because of their familiarity with adding amounts of money and the ease of adding decimal fractions on the calculator. As the metric system becomes more popular, even more examples of decimal fractions will be familiar from everyday life. Whether decimal fraction addition is introduced before or after common fraction addition, have students use the same manipulative materials they used to understand the basic concepts of decimal fractions; these include Cuisenaire rods, Decimal Squares, number lines, arrays, and base ten blocks.

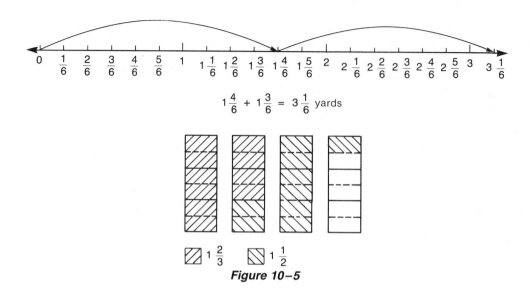

$$1\frac{4}{6} + 1\frac{3}{6} = 3\frac{1}{6} \text{ yards}$$

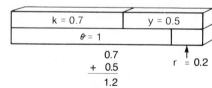

Figure 10-5

If addition of common fractions precedes addition of decimal fractions, have children write equations in both common and decimal fraction form and compare the results. If work with common fractions follows addition of decimal fractions, have children compare the two forms of addition when they learn to compute with common fractions. The activities described for common fraction addition can be repeated with decimal fractions. Following are a few other suggestions.

ACTIVITIES

Intermediate (3–6)

Objective: *to develop the concept of addition of decimal fractions in tenths using Cuisenaire rods and number lines.*

1. As with common fractions, children should begin work with decimal fractions using a situation from everyday life. The following is one example: Amy bought a new odometer for her bike because she wanted to find out how far it was to her friends' homes. She put the odometer on her bike and saw that it was set at zero. After she rode to Maureen's house, the odometer read 0.7 kilometers. Amy rode from there to Ahmad's house and told Ahmad that he lives 0.5 kilometers from Maureen. What did Amy's odometer read when she arrived at Ahmad's?

   Let the children use Cuisenaire rods to show the addition. Let the orange rod represent 1. Which rod shows 0.7? Which rod shows 0.5? Notice that once the children select the black and yellow rods, the addition is shown in the same manner that addition of whole numbers was illustrated.

   Since the sum is more than the orange rod, we know that the odometer will show more than 1 kilometer. Ask the children how they can tell how much more than a kilometer will be shown. They should fill in the space next to the orange rod to match the total of the black and yellow rods. The total is an orange rod and a red rod. The red rod is 0.2 of the orange rod, so the odometer must show 1.2 kilometers (see Figure 10–6).

   Let the children suggest other problems of their own and show them with the Cuisenaire rods.

2. After the children can use the Cuisenaire rods to show problems, they should transfer to the number line. Use a number line with 1 decimeter (10 centimeters) representing one unit. Each centimeter is then 0.1 of the unit. The children can lay the Cuisenaire rods on the number line to show the addition in the same way that they showed addition of whole numbers. When the children use the number line, they can simply read the answer off the number line, as shown in Figure 10–7.

3. After the children work several problems using Cuisenaire rods on the number line, they can show the addition on the number line by simply using arrows, as shown in Figure 10–8.

   As children work the problems with any of the models, encourage them to write the algorithms in vertical form near the pictures. After the children have worked several problems, ask them what they notice about the addition algorithm. They should notice that the algorithm for adding decimal frac-

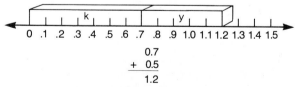

$$\begin{array}{r} 0.7 \\ + 0.5 \\ \hline 1.2 \end{array}$$

Figure 10-6

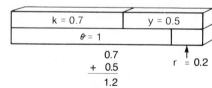

$$\begin{array}{r} 0.7 \\ + 0.5 \\ \hline 1.2 \end{array}$$

Figure 10-7

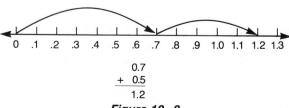

$$\begin{array}{r} 0.7 \\ + 0.5 \\ \hline 1.2 \end{array}$$

Figure 10-8

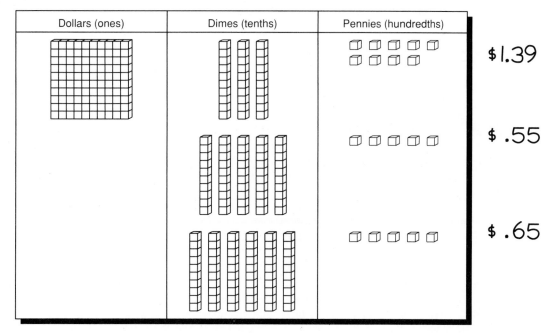

Dollars (ones)	Dimes (tenths)	Pennies (hundredths)	
			$1.39
			$ .55
			$ .65

Figure 10–9

Dollars (ones)	Dimes (tenths)	Pennies (hundredths)	
			$ 1.39
			.55
			.65
			$2.59

Figure 10–10

tions is the same as that for adding whole numbers if the decimal points are lined up.

If the children have previously worked with adding common fractions, ask them to write the number sentence for the same exercise in common fraction form next to the number sentence in decimal fraction form, as shown below:

$$
\begin{array}{cc}
\dfrac{7}{10} & 0.7 \\[2mm]
+\dfrac{5}{10} & +\,0.5 \\[2mm]
\hline
\dfrac{12}{10} = 1\dfrac{2}{10} & 1.2
\end{array}
$$

Ask them to compare the two examples and discuss their findings.

Objective: *to develop the concept of addition of decimal fractions in hundredths using base ten blocks, Decimal Squares, and arrays.*

4. Children should be familiar with addition of hundredths because of previous work adding money. They may not relate this to a concrete model, however. Use the flat 10 × 10 base ten block to represent one dollar. Ask the children how they would show a dime and a penny. They should show you the long and the small cube, respectively. Ask the children to show amounts such as $.23, $.96, and $1.48.

Ask the children to use the blocks to determine Bill's total bill if he orders a hamburger for $1.39, a cola for $.55, and french fries for $.65. The blocks should initially look like those in Figure 10–9.

Have the children make all the trades possible to end up with the least number of separate blocks. After all the trades, the response should appear as in Figure 10–10.

Have the children write the number sentence in vertical form near the work with the blocks and again discuss the algorithm for addition of decimal fractions. The children should realize that the algorithm remains the same as that for addition of whole numbers as long as the decimal points remain in line.

Let the children suggest other problems of their own and work them with the blocks. After they have worked several problems with the blocks, have them work a few problems

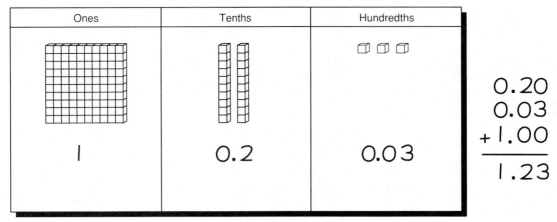

Figure 10–11

with other materials, such as the Decimal Squares or arrays using 10 × 10 grid paper. Encourage the children to discuss their methods of solution. They should compare these to other methods, such as calculating mentally, working the example on a calculator, or computing with paper and pencil.

Objective: to develop the concept of addition of decimal fractions with thousandths.

5. Once children are comfortable working problems involving tenths and hundredths, introduce problems involving thousandths. Following is one example: José is measuring chemicals for an experiment. He has measured 0.125 liter of water and 0.375 liter of oil into the same beaker. How much mixture is in the beaker?

Children may use actual beakers to work out the problem or they may model it using the base ten blocks or the Decimal Squares. If they use base ten blocks, the large cube should represent 1 and the smallest cube should be 0.001. Again, ask the children to write the number sentence in vertical form below the work and to compare this algorithm to the whole-number algorithms.

Children should compare their work with decimal fractions to the algorithm for common fractions if they have had previous work with common fractions. Have children use calculators to compare the work with concrete objects. If you are using calculators that show both decimal and common forms, ask the students to work the problems in both forms and compare the results. Encourage the children to discuss their methods with each other and to estimate results without actually computing them.

Objective: to develop the concept of addition with decimal fractions when some decimals are written in tenths and others are in hundredths or thousandths.

6. Children's main difficulties in adding decimal fractions come when the decimals are not all written to the same decimal place. Children may not realize that it is important to keep the decimal points lined up and they may try to write the numerals as shown below:

$$\begin{array}{r} 0.2 \\ 0.03 \\ 1 \\ \hline .06 \end{array}$$

Ask the children to use one of the concrete models to show the exercise. Using the base ten blocks, with the flat as 1, the materials would look like those in Figure 10–11. Have the children write the number sentence below their work with the blocks. Be sure that they realize the importance of lining up the decimal points and writing all decimal fractions in hundredths.

This is a good time to use a calculator to compare the results to paper and pencil computation. Ask the children to try examples such as .5 + .35 + .456 and .500 + .350 + .456. Do they get the same result? Ask the children to explain why this is true. If the children have previously added common fractions, ask them to compare this process to converting all common fractions to like denominators before adding.

Subtraction of Common Fractions

When children begin work with subtraction of common fractions, again start with realistic situations. As with whole-number subtraction, problems may involve taking away one amount from another, comparing two amounts, or figuring out how much more is needed. Again, have children use manipulative materials to work out the examples when you first introduce subtraction of common fractions. Build upon previous work with converting common fractions to equivalent fractions with like denominators and work with addition of common fractions.

Because addition and subtraction are inverse operations, you may introduce addition and subtraction of common fractions almost simultaneously. Students in third or fourth grade generally can reverse one operation to develop the other. Therefore, soon after children learn addition of common fractions, they may learn subtraction. Subtraction of mixed numerals may quickly follow addition of mixed numerals.

Following are a few suggestions for teaching the concepts of subtraction of common fractions. You and the children should develop other ideas related to the earlier work with addition and the children's experiences outside of school.

ACTIVITIES

Intermediate (4–7)

Objective: *to develop the concept of subtraction of common fractions with like denominators.*

1. Suggest a problem such as the following and ask the children to work it using a manipulative such as the circular regions: Emilio had $\frac{3}{4}$ of a pound of cheese. He used $\frac{1}{4}$ of a pound to make macaroni and cheese. How much cheese does Emilio have left? The children's regions should be similar to those in Figure 10–12.

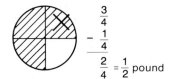

$$\begin{array}{r} \frac{3}{4} \\ - \frac{1}{4} \\ \hline \frac{2}{4} = \frac{1}{2} \text{ pound} \end{array}$$

Figure 10–12

Ask the children to write the equation next to the pieces as they work the problem. Let the children suggest other word problems themselves and use other types of materials to work them. They should discuss their methods as they work.

One common mistake that children make when they write word problems for subtraction of common fractions is that they want to subtract a part of the first fraction rather than a part of a whole amount. For example, a child may say, "Jeff had $\frac{1}{2}$ of an apple. Amy ate $\frac{1}{4}$ of what Jeff had. How much does Jeff have left?"

Discuss with the children that $\frac{1}{4}$ of $\frac{1}{2}$ of an apple is only $\frac{1}{8}$ of the whole apple. This is less than $\frac{1}{2}$ of a whole apple. The equation for this problem would be $\frac{1}{2} - (\frac{1}{4} \times \frac{1}{2}) = n$. This is not $\frac{1}{2} - \frac{1}{4}$. Encourage children to show both problems with the manipulatives and to explain their solutions (more is said about problems involving multiplication in the section on multiplication of common fractions).

Objective: *to introduce subtraction of fractions with unlike denominators.*

2. Use a word problem to introduce subtraction of fractions with unlike denominators. For example, Andreas has $\frac{3}{4}$ of a bushel of apples. Marlo has $\frac{2}{3}$ of a bushel of apples. What part of a bushel more does Andreas have than Marlo?

Notice that this is a comparison type of subtraction problem. Children can use number lines or Fraction Bars to compare the two amounts. If children do not convert the fractions to like denominators, they can tell that $\frac{3}{4}$ is greater than $\frac{2}{3}$ but they will have trouble determining how much greater.

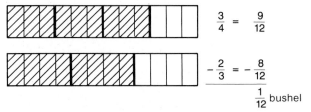

$$\frac{3}{4} = \frac{9}{12}$$
$$-\frac{2}{3} = -\frac{8}{12}$$
$$\overline{\frac{1}{12}} \text{ bushel}$$

Figure 10–13

Ask the children to always estimate which amount is greater before they begin to subtract and to estimate *about* how much greater that amount is. Compare this to adding $\frac{3}{4}$ and $\frac{2}{3}$. For addition, they first changed both fractions to like denominators. They should do the same when subtracting fractions with unlike denominators. Have the children write the algorithms in vertical form beside the work with the number line or manipulatives as shown in Figure 10–13 and discuss their strategies of solution.

Objective: *to develop the concept of subtraction of mixed numerals.*

3. Again, subtraction of mixed numerals should begin with an example from everyday life, such as: Cara has knitted $2\frac{3}{4}$ yards of a scarf. She wants the scarf to be $3\frac{1}{2}$ yards long. How many more yards does she need to knit?

This type of problem, which requires renaming one unit, often is difficult for children. Let the children use materials or a number line. On the number line, have the children show the $2\frac{3}{4}$ yards already completed and add on until they reach the $3\frac{1}{2}$ yards, as shown in Figure 10–14.

Have the children record the algorithm in vertical form below the materials as they work. Work several problems with mixed numerals with the children, some that require renaming the whole number and some that do not. Have the children use other manipulative materials to demonstrate exchanging the whole number for equivalent parts. Discuss with them the fact that the mathematics remains the same regardless of the concrete materials chosen to illustrate the equations. Figure 10–15 shows an example with pie pieces that some students may use. Other students may begin with $2\frac{3}{4}$ circular pieces and add on fourths until reaching $3\frac{1}{2}$.

Again, encourage the children to estimate or mentally compute several of the examples. Paper and pencil computations are not always necessary. Children should compare their various strategies as they work.

Subtraction of Decimal Fractions

Subtraction of decimal fractions may be introduced immediately following addition of decimal fractions. If subtrac-

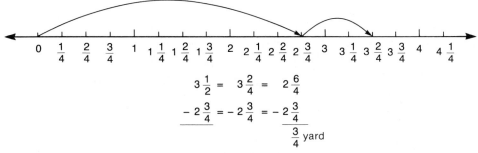

$$3\frac{1}{2} = 3\frac{2}{4} = 2\frac{6}{4}$$
$$-2\frac{3}{4} = -2\frac{3}{4} = -2\frac{3}{4}$$
$$\overline{\frac{3}{4}} \text{ yard}$$

Figure 10–14

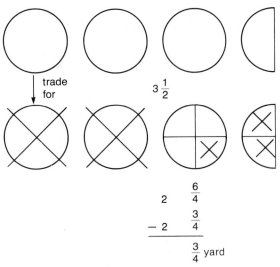

$$3\frac{1}{2}$$

$$
\begin{array}{r}
2\ \dfrac{6}{4} \\[4pt]
-\ 2\ \dfrac{3}{4} \\[2pt]
\hline
\dfrac{3}{4}\ \text{yard}
\end{array}
$$

Figure 10–15

tion of decimal fractions follows subtraction of common fractions, have the children compare the new work with decimal fractions to the previous work with common fractions. Have them write algorithms in both forms and compare the processes and answers. Let them also use calculators for comparison.

If the work with decimal fractions comes first, compare addition and subtraction of decimal fractions with the new work when you introduce addition and subtraction of common fractions. Use the same manipulative materials for both addition and subtraction of decimal fractions. Compare the subtraction algorithms for decimal fractions to the addition algorithms for decimal fractions and the subtrac-

tion algorithms for whole numbers. Use the same types of word problems for decimal fractions as you did for subtraction of whole numbers and for subtraction of common fractions. Money and metric measures are good sources of word problems because of the frequent use of decimals in these examples.

The activities for subtraction of common fractions and for addition of decimal fractions can be adapted to subtraction of decimal fractions. Following are a few other ideas. You and your students should suggest others.

ACTIVITIES

Intermediate (3–6)

Objective: *to develop the concept of subtraction of decimal fractions.*

1. Because dealing with money gives children a number of chances to add and subtract using decimals, have the children set up a store or restaurant. Initially, stock the cash register only with dollar bills, dimes, and pennies to reinforce the regrouping the children must do to perform the subtraction algorithm. For example, Ray is told that his bill for lunch comes to $2.57. He hands the cashier $3.00. Have the children make the change either by using the traditional decomposition algorithm or by counting on from $2.57. Figure 10–16 shows the trades the cashier must make to give Ray his change if the decomposition algorithm is used.

   Counting on from $2.57 has the advantage of helping the children learn to count back change (this seems to be a lost art in many stores). Encourage children to write down an algorithm for the method used. Let children create other algorithms of their own. Because so many stores now use computerized cash registers, encourage children to use cal-

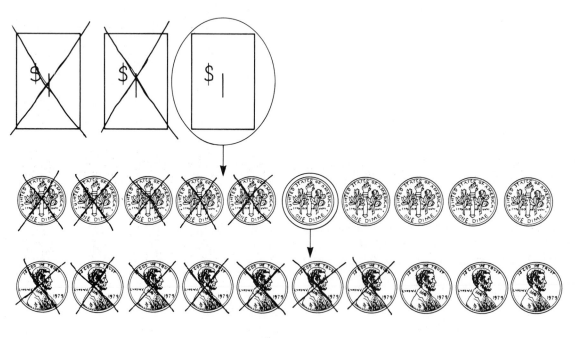

$$
\begin{array}{r}
2\overset{\scriptscriptstyle 9}{\cancel{}}\overset{\scriptscriptstyle 10}{\cancel{}}10 \\
\$\cancel{3}.\cancel{0}\cancel{0} \\
-\ 2.57 \\
\hline
\$.43
\end{array}
$$

Figure 10–16

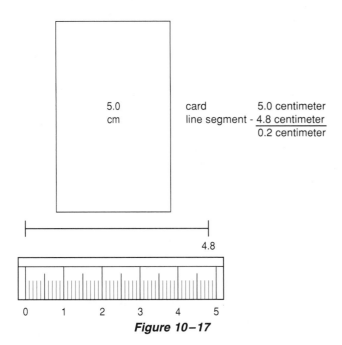

5.0
cm

card   5.0 centimeter
line segment - 4.8 centimeter
              0.2 centimeter

4.8

0   1   2   3   4   5

Figure 10–17

culators to compute the change. A parent or other adult in the community may be willing to come in to explain the use of various machines that aid clerks in making change.

2. Have children play an estimation game that also involves measurement and subtraction of decimal fractions. Make a set of cards that show amounts such as 5 centimeters, 6.2 centimeters, 8.4 centimeters, and 3 centimeters. Place the cards face down on the table.

   Turn over one card and ask a child to draw a line segment approximately the length designated by the card. Then have the child measure the line segment to the nearest tenth of a centimeter (nearest millimeter) and subtract to find how close the length of the line segment is to the length specified on the card. The difference is the child's score for that round. Then turn over a card for the next child and repeat the activity.

   After several rounds, the child with the lowest total is the winner. Children may find the difference on the ruler by comparing the length of the line segment to the length indicated by the card, or they may subtract as shown in Figure 10–17.

   After children have had experience using the rulers, ask them to find the difference mentally, with a calculator, or by writing the subtraction algorithm.

Multiplication of Common Fractions

Too often, children memorize rules for multiplying and dividing common fractions but have very little concept of the meaning behind the rules. If the rule is forgotten, the children cannot reconstruct the algorithm because they have no understanding of the operation. Nor do they know which operation to choose when faced with a word problem involving fractions because they do not associate any real meaning with the operations. They can use a calculator to do the computation for them only if they know what computation is needed.

Unfortunately, this is also a problem for many adults. Ask anyone to give you a word problem that can be solved by $\frac{2}{3} \times \frac{3}{4}$ or $\frac{2}{3} \div \frac{3}{4}$. Perhaps this is a problem for you as

well. After reading this section, you should not only be able to write word problems involving multiplication and division of common fractions but should also be able to demonstrate the solutions with a number of manipulatives.

Children usually begin by multiplying a common fraction by a whole number when they are introduced to multiplication of common fractions. This is perhaps the easiest type of problem to understand and to model with the manipulatives. Again, as with the other operations, the children should start with familiar examples and manipulatives or number lines.

The following activities begin with multiplying a common fraction by a whole number and then multiplying a whole number by a common fraction. Finally, we give ideas for multiplying two proper fractions and for multiplying two mixed numerals. If multiplication of decimal fractions has preceded multiplication of common fractions, have the children compare the processes and word problems to note the similarities. The concepts are not new, although the algorithms may be in a different form.

ACTIVITIES

Middle Grades (4–8)

Objective: *to develop the concept of multiplying a common fraction by a whole number.*

1. Suggest a situation such as the following to the children when they are first learning to multiply common fractions. Ask the children to find the answer using any of the manipulative materials with which they are familiar from earlier work with addition and subtraction of common fractions. The following example lends itself to being solved by repeated addition, so children who can add common fractions should be able to solve it. The school track is $\frac{3}{8}$ of a mile around. Sandy is on the track team, and she has run around the track three times in practice. How far has Sandy run?

   Figure 10–18 shows some of the ways in which the children may work the problem. Ask them to write the multiplication sentence below their work and to write the answer as a mixed numeral.

   Ask the children to suggest other examples of problems that can be solved by multiplying a common fraction by a whole number and to show the solution with the manipulatives. Have them exchange problems with each other as well as work their own problems. Encourage them to discuss their methods of solution with the author of the problem. If they have solved the problem in different ways, lead them to discover whether or not each solution works.

Objective: *to develop the concept of multiplying a whole number by a common fraction.*

2. This concept is one with which children should be familiar from earlier work with the concept of a common fraction as part of a set of discrete objects, but review it at this time. Suggest familiar examples such as finding $\frac{2}{3}$ of a dozen eggs, the number of ounces in $\frac{1}{2}$ of a pound, or the number of inches in $\frac{1}{4}$ of a foot. Point out that the word *of* in these examples indicates that the children should multiply. To find $\frac{2}{3}$ of a dozen, have the children multiply $\frac{2}{3} \times 12$. Since there are 16

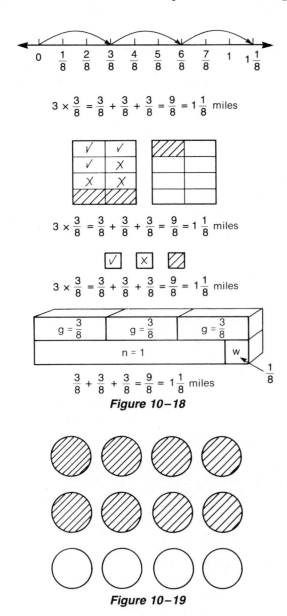

$$3 \times \frac{3}{8} = \frac{3}{8} + \frac{3}{8} + \frac{3}{8} = \frac{9}{8} = 1\frac{1}{8} \text{ miles}$$

$$3 \times \frac{3}{8} = \frac{3}{8} + \frac{3}{8} + \frac{3}{8} = \frac{9}{8} = 1\frac{1}{8} \text{ miles}$$

$$3 \times \frac{3}{8} = \frac{3}{8} + \frac{3}{8} + \frac{3}{8} = \frac{9}{8} = 1\frac{1}{8} \text{ miles}$$

$$\frac{3}{8} + \frac{3}{8} + \frac{3}{8} = \frac{9}{8} = 1\frac{1}{8} \text{ miles}$$

Figure 10–18

Figure 10–19

ounces in a pound, $\frac{1}{2}$ of a pound is $\frac{1}{2} \times 16$, or 8, ounces. Let the children work these exercises using an array of concrete materials such as chips, if necessary, and to discuss their thought processes with each other. Figure 10–19 shows $\frac{2}{3} \times 12$.

Objective: *to develop the concept of multiplying two proper fractions.*

3. Just as arrays can illustrate multiplication of whole numbers and multiplication of a whole number by a common fraction, they can also illustrate multiplication of two common fractions. Again, introduce this multiplication with a "real" situation such as the following: Steve is planting a rectangular garden. He wants $\frac{1}{3}$ of his garden to be flowers. Steve likes roses, so $\frac{1}{2}$ of his flowers will be roses. What part of the total garden will be roses?

The array in Figure 10–20 represents the garden with $\frac{1}{3}$ planted in flowers and $\frac{1}{2}$ of that part planted in roses. Ask the children to write the corresponding number sentence $\frac{1}{2} \times \frac{1}{3} = \frac{1}{6}$ below the drawing.

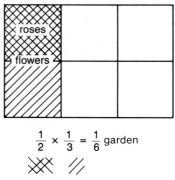

$$\frac{1}{2} \times \frac{1}{3} = \frac{1}{6} \text{ garden}$$

Figure 10–20

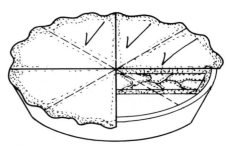

Figure 10–21

Use several other examples, including ones that do not use unit fractions, and ask the children to show the solutions with arrays. Situations involving measurement, such as cooking, sewing, and building, are good sources of word problems. After the children have worked several examples and written the corresponding number sentences, ask them to compare the sentences to develop a rule for multiplying two common fractions.

After the children realize that they can multiply the numerators and the denominators, ask them if their rule works for multiplying a whole number by a common fraction. Remind children to write the whole number over 1 so that it is in fraction form before they multiply. Encourage the children to solve equations by computing mentally and then comparing their results to the results with the concrete models. They may also use a calculator that displays common fractions to compare results.

4. After the children have used arrays to illustrate multiplication of common fractions, have them use other manipulatives. Following are a few examples.

Fred sees $\frac{3}{4}$ of a pie on the counter. He is starving, so he eats $\frac{1}{2}$ of the $\frac{3}{4}$ of the pie. How much of a whole pie does Fred eat?

Notice that Fred is eating $\frac{1}{2}$ of the $\frac{3}{4}$, not $\frac{1}{2}$ of the whole. Figure 10–21 shows the solution. Notice that Fred first cuts each of the fourths into halves; now he has 6 eighths. Then he takes $\frac{1}{2}$ of the 6 remaining pieces, or 3 eighths.

Rachel is building a scale model of a toy tower. The original tower is $\frac{1}{2}$ of a foot tall. Rachel wants her tower to be $\frac{1}{2}$ of that height. How tall should Rachel build her tower?

Use a number line to represent the original tower. Split the number line into halves and mark the height of the original tower. Since the height is one space and you cannot divide one into halves, divide each space into halves. Look at the

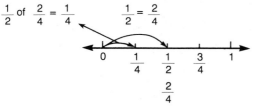

$\frac{1}{2}$ of $\frac{2}{4}$ = $\frac{1}{4}$ $\frac{1}{2}$ = $\frac{2}{4}$

$\frac{2}{4}$

Figure 10–22

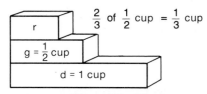

$\frac{2}{3}$ of $\frac{1}{2}$ cup = $\frac{1}{3}$ cup

Figure 10–23

number line and realize that the unit is now split into four pieces, so each piece is $\frac{1}{4}$. The original tower takes up $\frac{2}{4}$ of the unit (check to make sure $\frac{1}{2} = \frac{2}{4}$). One-half of two is one. The scale model should be $\frac{1}{4}$ of a foot tall (see Figure 10–22).

Brian is making punch. His recipe will serve three people, but Brian needs to serve only two, so he has decided to make $\frac{2}{3}$ of the recipe. The recipe calls for $\frac{1}{2}$ cup of sugar. Brian needs to know how much sugar to use for $\frac{2}{3}$ of the recipe.

Use Cuisenaire rods to solve the problem (see Figure 10–23). Use the dark green rod to represent 1 cup of sugar. Why? Can you use any other color rod to represent 1 cup? Try others and see what happens?

If the dark green rod is 1 cup, which rod represents the $\frac{1}{2}$ cup of sugar needed for the original recipe? The light green rod is $\frac{1}{2}$ of the dark green rod.

Which rod is $\frac{1}{3}$ of the light green rod? Which rod(s) represent $\frac{2}{3}$ of the light green rod? Either the red rod or two white rods can be used to represent $\frac{2}{3}$ of the light green rod, but each representation is only $\frac{1}{3}$ of the dark green rod, which represents one cup. Therefore, Brian will need $\frac{1}{3}$ cup of sugar to make $\frac{2}{3}$ of the recipe.

Encourage the children to make up other problems and to discuss their methods of solution. They should realize that the concrete model used does not change the answer to the problem. The mathematics remains the same as the materials change.

Objective: to develop the concept of multiplication of mixed numerals.

5. After children are comfortable multiplying two proper fractions, let them encounter situations involving the multiplica-

tion of a proper fraction by a mixed numeral or the multiplication of two mixed numerals. Following is one example. You and the children can make up other examples of your own.

Raul is looking at a scale drawing of a flower. The scale says the actual flower is $2\frac{1}{2}$ times as large as the drawing. Raul measures the flower in the drawing and sees it is $1\frac{3}{4}$ inches long. How long is the actual flower?

Use a number line to solve the problem, as in Figure 10–24. Measure off $1\frac{3}{4}$ inches twice to show that the actual flower is 2 times as large. Then figure that $\frac{1}{2}$ of $1\frac{3}{4}$ inches is $\frac{7}{8}$ inch and go that much farther on the number line. The real flower must be $4\frac{3}{8}$ inches long.

Again, encourage the children to write the number sentence beneath the materials as they work. Lead the children to discover that they should convert mixed numerals to improper fractions before they multiply, then multiply as they did with proper fractions, and finally convert the answer back to a mixed numeral. Let the children create other problems for each other and solve them using a variety of models, comparing their methods of solution with each other.

Multiplication of Decimal Fractions

If multiplication of decimal fractions follows multiplication of common fractions, children should compare the two by writing number sentences in both forms and comparing the answers. Actual situations that give rise to multiplication of decimal fractions are very similar to those for common fractions. The main difference is that decimal fractions are always tenths, hundredths, or another power of ten. Because multiplying tenths by tenths results in an answer in the hundredths, we do not go beyond multiplying tenths using concrete models. A picture showing 0.52×0.63 must include ten thousandths and is fairly difficult to draw.

Children multiplying decimal fractions should begin as they did when multiplying common fractions. Give examples of multiplying a whole number times a decimal and then of multiplying a decimal times a whole number. Multiply two decimal fractions later. Adapt the examples given for common fractions to decimal fractions. Have the children again use familiar manipulatives and add other ideas of their own. Following are a few other suggestions.

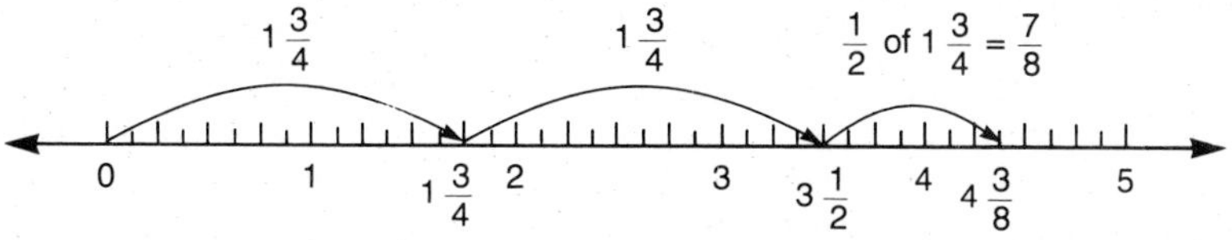

$1\frac{3}{4}$ $1\frac{3}{4}$ $\frac{1}{2}$ of $1\frac{3}{4}$ = $\frac{7}{8}$

$$1\frac{3}{4} + 1\frac{3}{4} + \frac{7}{8} = 4\frac{3}{8} \text{ inches}$$

$$2\frac{1}{2} \times 1\frac{3}{4} = \frac{5}{2} \times \frac{7}{4} = \frac{35}{8} = 4\frac{3}{8} \text{ inches}$$

Figure 10–24

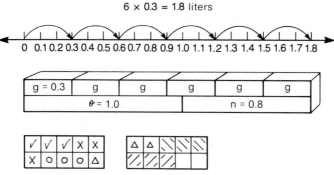

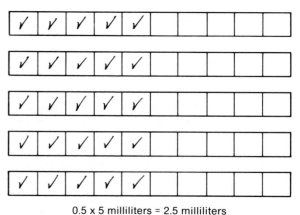

Figure 10–25

ACTIVITIES

Middle Grades (4–8)

Objective: to develop the concept of multiplying a decimal fraction by a whole number.

1. Money and metric measurement are good contexts for problems involving the multiplication of decimal fractions. Children may begin with the following situation: Jane has 5 friends coming to her birthday party. She wants to give each friend and herself 0.3 liter of orange juice. How much orange juice will Jane need so that the 6 children will each get 0.3 liter?

   Work this problem as a repeated addition problem. Children who have had previous practice with addition of decimal fractions should have no difficulty in finding the product. Figure 10–25 shows some possible methods of solving the number sentence 6×0.3.

   Have the children record the solution near the picture or materials as they work. Ask the children to compare their work to the addition problem $0.3 + 0.3 + 0.3 + 0.3 + 0.3 + 0.3 = n$. If children have previously multiplied common fractions, ask them also to compare their work to $6 \times \frac{3}{10} = n$.

   This is a good time to use a calculator. Ask the children to compare entering $+ .3 = = = = = =$ to entering $.3 + .3 + .3 + .3 + .3 + .3$ and to entering $6 \times .3$. Have them try several examples to see if they always get the same answer with the three methods. Ask them to explain the results using concrete materials.

   Encourage the children to suggest and work other word problems, discussing their results with you and each other.

Objective: to develop the concept of multiplying a whole number by a decimal fraction.

2. The following is an example of a problem involving this concept: Jerry has a recipe calling for 5 milliliters of salt. He wants to make only 0.5 of the recipe. How much salt should Jerry use?

   Figure 10–26 shows one possible method of solution. Notice that each of the 5 units representing the milliliters is split into 10 equal parts, and in each case, Jerry takes 5 of the parts.

   Again, ask the children to record the number sentence as they work. Let the children suggest and work similar word problems. Encourage them to estimate their answers and to use calculators to compare the results to their concrete work.

Ask them to decide if the answer will be larger or smaller than the beginning amount and to explain why this is true.

Objective: to develop the concept of multiplying two decimal fractions.

3. The following is one possible example to use to develop the concept of multiplying two decimal fractions: Marie is planting a small garden. It is 0.7 meter by 0.8 meter. How many square meters are there in Marie's garden?

   Figure 10–27 shows a rectangular array for working this problem. Notice that the garden is less than 1 square meter. Ask the children to predict before working problems such as

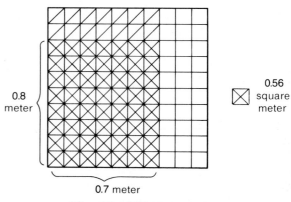

0.5 x 5 milliliters = 2.5 milliliters

Figure 10–26

0.7 × 0.8 = 0.56 square meter

Figure 10–27

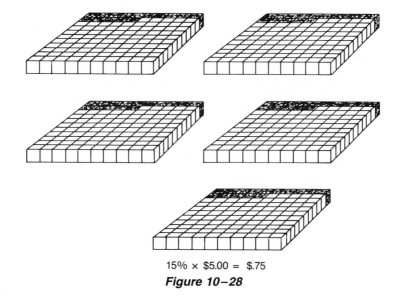

$$15\% \times \$5.00 = \$.75$$

Figure 10–28

this whether the answer will be more or less than 1 and to explain their reasoning. In this case, the square meter is broken into hundredths, and the garden takes up fifty-six of those hundredths. Therefore, the garden covers 0.56 of a square meter.

Use other examples such as the following, which can be shown using string and a meter stick: Amy has a board 0.8 of a meter long. She needs 0.3 of the board for a project. How many meters of board will the project take?

Measure 0.8 meter of string and fold it into 10 equal parts. Take 3 of the parts, and measure that amount.

Ask children to work the following using dimes and pennies: Jeff has $.70. He is feeling generous and wants to give 0.4 of his money to his sister. How much should he give his sister?

Jeff can trade in all his money for pennies and then split the pennies into 10 equal piles. He can then give four of the piles to his sister.

Work with the children to make up other word problems. Notice that in each case you need to take one part of the other part and want to know how the final answer relates to a unit. Children will need a great deal of practice with word problems and concrete materials before they become proficient at writing the word problems, but the time spent will pay off well in understanding. If the children have previously worked with multiplication of common fractions, they may also write the problem with common fractions and compare the results, for example, $0.4 \times 0.7 = \frac{4}{10} \times \frac{7}{10} = \frac{28}{100} = 0.28$. After students have worked several problems concretely with decimal fractions, ask them if they can make a generalization about the number of decimal places in the product.

Multiplication with Percent

Many percent problems can be solved as proportions. Another method of solving some problems involving percent is by using the formula $p = r \times b$. This stands for percentage = rate × base. The method is often useful for estimating or computing amounts mentally, such as figuring a 15% tip at a restaurant or deciding if you have enough money to buy pants that are 30% off.

The following activities give a few ideas for introducing the concept of multiplying a whole number or decimal fraction by a percent. Because a percent is simply another way of writing a common or decimal fraction, any of the activities described earlier for those forms may be adapted to percents.

ACTIVITIES

Middle Grades (5–8)

Objective: *to develop the concept of multiplying a whole number or decimal fraction by a percent.*

1. Because percent means per hundred, children should first practice finding percents of multiples of 100. Use the Decimal Squares or base ten blocks to introduce finding percents. Tell the children that you want to leave a 15% tip for your dinner. Your bill was $5.00.

   Use the 10 × 10 square from the Decimal Squares or the 10 × 10 flat from the base ten blocks to represent $1.00. You now want to show 15% × $5.00. Ask the children to show 15% × $1.00. Do this for each of the dollars. How much do you have? Figure 10–28 shows one response.

   Ask the children to write a number sentence below the work with the squares or the blocks. Encourage the children to suggest and work other problems of their own. After the children have worked several problems with the materials, ask them to rewrite the number sentences using decimals instead of percents and to compare the answers. Use calculators to compute the same problems. If you have a % key on the calculator, compare its use to performing the calculation with the amount written in decimal form. Ask the children to practice finding the answers with mental calculation. This is certainly a skill adults frequently need.

2. When amounts are not multiples of 100, the task may not be so easy. Ask the children how they would find 25% of 16. Let the children suggest different solutions. They may need to be reminded that a percent is another way of writing a common or decimal fraction. Discuss the benefits of converting 25% to 0.25 and multiplying 0.25 × 16, or converting 25% to $\frac{1}{4}$ and

multiplying $\frac{1}{4} \times 16$. Let the children suggest other problems of their own and decide whether to convert to a common fraction or a decimal fraction to find a solution.

In these examples, we were looking for the percentage each time. In the section on developing and practicing skills, we discuss problems involving percentage where the base or the rate needs to be found.

Division of Common Fractions

As mentioned earlier, children and adults often cannot give examples from "real life" of occasions when division of rational numbers is called for. They may vaguely recall that you invert one of the numbers and multiply, but they cannot tell you why this algorithm works or even which number should be inverted. This section focuses on developing the meaning of division of rational numbers, and understanding the "invert and multiply" algorithm. It introduces an algorithm for division of common fractions that builds on the measurement concept of division and previous work with finding a common denominator. We explore division of common fractions using both measurement and partition division situations, building on previous work with whole numbers. We then use similar situations to explore division of decimal fractions. If children divide decimal fractions before they divide common fractions, have them use this work as a basis for comparison when they begin dividing common fractions.

The following activities begin with the measurement concept of division of common fractions with a whole-number answer and move to fractional or mixed-numeral answers. The partition concept of division of common fractions follows the measurement idea, beginning with whole-number divisors and moving to fractional divisors. Appropriate algorithms are discussed for each type of division.

ACTIVITIES

Middle Grades (5–8)

Objective: to develop the measurement concept of division of common fractions with a whole-number quotient.

1. As with all other operations, begin division of common fractions with an example from the children's lives. The following is one possibility: Ernie has $\frac{1}{2}$ of a quart of orange juice. He wants to pour $\frac{1}{8}$ of a quart into each juice glass. How many juice glasses can Ernie fill?

   Ask the children to compare this problem to earlier measurement division problems with whole numbers. A similar whole-number problem might have been: Ernie has 12 quarts of orange juice. He wants to put 3 quarts into each pitcher. How many pitchers can he fill?

   Ask the children to first show the problem with whole numbers using materials. Then ask the children to work the problem with fractions. Figure 10–29 shows some possible solutions, using a variety of different familiar manipulatives or pictures.

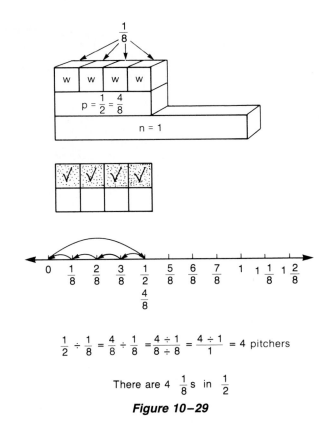

$$\frac{1}{2} \div \frac{1}{8} = \frac{4}{8} \div \frac{1}{8} = \frac{4 \div 1}{8 \div 8} = \frac{4 \div 1}{1} = 4 \text{ pitchers}$$

There are 4 $\frac{1}{8}$s in $\frac{1}{2}$

Figure 10–29

Notice that in each case, the $\frac{1}{2}$ is exchanged for $\frac{4}{8}$ and the number of $\frac{1}{8}$s in $\frac{1}{2}$ is figured. The algorithm for this may be written as $\frac{1}{2} \div \frac{1}{8} = \frac{4}{8} \div \frac{1}{8} = \frac{4 \div 1}{8 \div 8} = \frac{4 \div 1}{1} = 4 \div 1 = 4$. When the fractions are written with like denominators and the numerators and denominators are divided, the old denominators cancel out and the new denominator is 1. Then divide the numerators the same way you divide whole numbers.

This algorithm appears in an example from a textbook that refers to dividing numerators and denominators as an incorrect algorithm. Such division is actually quite proper and describes well the procedure for working measurement division problems with common fractions. For children familiar with renaming common fractions with like denominators, the method of dividing numerators and denominators is quite similar to the algorithm of multiplying numerators and denominators.

Even though textbooks may not mention this algorithm, children may create it on their own from work with materials. Children easily understand and remember this algorithm after they master algorithms for addition, subtraction, and multiplication of common fractions. We encourage you to use this common-denominator algorithm with your students, and if you decide to teach it instead of the more traditional invert and multiply algorithm, be sure to send a note home to the parents to explain what you are doing and the reason for it.

Following are other suggestions for measurement situations involving the division of common fractions. Discuss the solutions with the children as they work the problems.

Emma works for a cheese packing company. She puts $\frac{3}{4}$ of a pound of Swiss cheese into each party package. Emma has 6 pounds of Swiss cheese. How many party packages can Emma fill? (See Figure 10–30.)

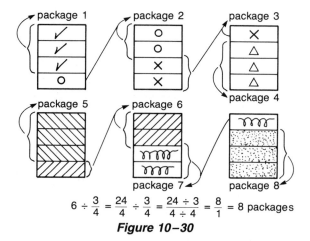

$$6 \div \frac{3}{4} = \frac{24}{4} \div \frac{3}{4} = \frac{24 \div 3}{4 \div 4} = \frac{8}{1} = 8 \text{ packages}$$

Figure 10–30

Erica is making doll clothes. She needs $\frac{2}{3}$ of a yard of material for each outfit. She has $2\frac{2}{3}$ yards of material. How many outfits can she make? (See Figure 10–31).

Encourage children to make up other problems for each other. Ask the children to write the corresponding number sentences as they work.

Objective: to develop the measurement concept of division of common fractions with a fractional or mixed-numeral quotient.

2. In the examples given earlier, the quotient was a whole number. Children are often confused when they must divide common fractions that do not have a whole-number quotient. They do not know what to do with the remainder. They may need to be instructed about the algorithm for division of whole numbers that allows them to write the remainder as part of the divisor. For example, the solution to $5 \div 2$ may be written as either 2 with a remainder of 1 or $2\frac{1}{2}$. When written as $2\frac{1}{2}$, the remainder 1 becomes the numerator of a fraction with the divisor 2 as the denominator. Thus, when we say $5 \div 2 = 2\frac{1}{2}$, the remainder 1 is expressed as part of the divisor. In terms of fractions, the problem $5 \div 2$ may be thought of as $\frac{5}{2} = 2\frac{1}{2}$, which is consistent with the division concept of common fractions presented earlier in this chapter. This also is the process used when children divide common fractions and the answer is not a whole number, as in the following example.

Danny has to take $\frac{2}{3}$ of a teaspoon of medicine in each dose. The doctor gave him 3 teaspoons of medicine. How many doses can Danny take? Figure 10–32 shows that Danny can take four full doses of medicine, and he will then have $\frac{1}{3}$ of a teaspoon of medicine left, that is, $\frac{1}{2}$ of another $\frac{2}{3}$ teaspoon dose of medicine.

Again, we show the common-denominator algorithm for division. Danny has $\frac{9}{3}$ teaspoons of medicine from the doctor

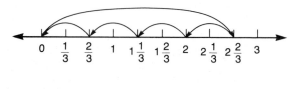

$$2\frac{2}{3} \div \frac{2}{3} = \frac{8}{3} \div \frac{2}{3} = \frac{8 \div 2}{3 \div 3} = \frac{4}{1} = 4 \text{ outfits}$$

Figure 10–31

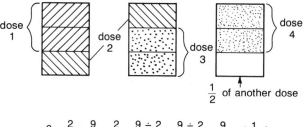

$\frac{1}{2}$ of another dose

$$3 \div \frac{2}{3} = \frac{9}{3} \div \frac{2}{3} = \frac{9 \div 2}{3 \div 3} = \frac{9 \div 2}{1} = \frac{9}{2} = 4\frac{1}{2} \text{ doses}$$

Figure 10–32

and he wants to know how many doses of $\frac{2}{3}$ of a teaspoon each he can make. He can make 4 full doses and $\frac{1}{2}$ of another dose.

Children may wish to try to use a division algorithm similar to that for whole numbers. Study the following:

$$
\begin{array}{r}
4r\frac{1}{3} \\
\frac{2}{3} \overline{\smash)\dfrac{9}{3}} \\
-\frac{8}{3} \\
\hline
\frac{1}{3}
\end{array}
\quad \text{or} \quad
4\frac{\frac{1}{3}}{\frac{2}{3}} = 4\frac{1}{2}
$$

This algorithm will also work.

Encourage the children to suggest other examples and to solve them using the materials and the algorithm of their choice. Ask the children to predict whether the answers will be more or less than 1 before they compute the results. Ask them to discuss their methods of solution. They may use a calculator showing common fractions to compare results.

Objective: to develop the partition concept of division of fractions with whole-number divisors.

3. Recall that division may be shown as either a measurement or a partition concept. In a partition problem the total and the number of groups are known, and you are asked to find the amount in each group. Children should work a partition problem with whole numbers before they work a corresponding problem with fractions. Following is one suggestion: Neil plans to spend 4 hours on his homework this weekend. He plans to spend equal amounts of time on Saturday and Sunday. How many hours should Neil spend studying each day? Children should recognize this as a partition division problem; they divide the 4 into 2 equal parts to decide that Neil should spend 2 hours studying each day.

After working a problem with whole numbers, try the following: Jack plans to spend $\frac{1}{2}$ hour on his homework this weekend and wants to spend the same amount of time on Saturday and Sunday. How much time should Jack spend on his homework each day?

Figure 10–33 shows different materials used to solve this problem. Ask the children to compare this solution to $\frac{1}{2} \times \frac{1}{2}$. Jack will do $\frac{1}{2} \div 2$ of his homework on each day, or spend $\frac{1}{2}$ of $\frac{1}{2}$ hour.

Have children make up word problems for other number sentences and compare them to the corresponding multiplication

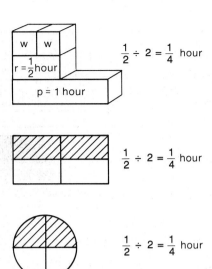

$\frac{1}{2} \div 2 = \frac{1}{4}$ hour

$\frac{1}{2} \div 2 = \frac{1}{4}$ hour

$\frac{1}{2} \div 2 = \frac{1}{4}$ hour

Figure 10–33

problems. Each problem should be worked with concrete materials, with the number sentences written next to the work.

Following are a few other examples: Compare $\frac{2}{3} \div 3$ to $\frac{1}{3} \times \frac{2}{3}$. Susan has $\frac{2}{3}$ of a pie. She wants to divide it equally among her two friends and herself. How much of the whole pie should each person get? Each will get $\frac{1}{3}$ of the $\frac{2}{3}$ pie, or $\frac{2}{9}$ of the whole pie.

Compare $\frac{3}{4} \div 5$ to $\frac{1}{5} \times \frac{3}{4}$. Antonio has $\frac{3}{4}$ of a pound of cheese. He wants to make 5 sandwiches and he wants to put the same amount of cheese on each sandwich. How much cheese should he put on each sandwich? Each sandwich should get $\frac{1}{5}$ of $\frac{3}{4}$ of a pound, or $\frac{3}{20}$ of a pound.

After you compare several multiplication and division problems, introduce the children to the invert and multiply algorithm for division of common fractions. Discuss why dividing by 3 gives the same answer as multiplying by $\frac{1}{3}$. Compare the results using the invert and multiply algorithm to the results using the common-denominator algorithm.

Let the children discuss which method they prefer. Do all children prefer the same algorithm? Does the preferred algorithm depend upon the type of problem being worked? As the teacher, you should determine whether the algorithm you prefer is based on the ease of understanding, the ease of use, or simply past familiarity.

The partition concept of division of fractions is not difficult to understand as long as the divisor is a whole number, but it can be fairly difficult to explain if the divisor is less than 1. We recommend that you use measurement examples for most of the problems with fractional divisors. Some of the children may suggest partition word problems with fractional divisors, however, so we present them here. Encourage elementary gifted children and middle school students to explore these problems further.

ACTIVITIES

Middle Grades (5–8)

Objective: *to develop the partition concept of division of fractions with fractional divisors.*

1. Following is a partition word problem for the number sentence $\frac{2}{3} \div \frac{1}{2} = n$: Mr. Jensen has $\frac{2}{3}$ of a ton of grain. It is enough to feed $\frac{1}{2}$ of his herd of cattle. How much grain will it take to feed the whole herd?

   Ask the children to compare this to a similar problem with whole numbers, such as: Mr. Ali has 6 tons of grain. This is enough to feed 3 herds of cattle. How much grain will it take to feed one herd of cattle?

   Have the children demonstrate how they would solve the problem with whole numbers using manipulatives. Then ask the children to show the fraction problem with the materials. Figure 10–34 shows two possible solutions.

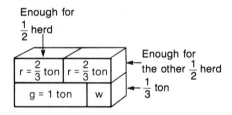

Need $1\frac{1}{3}$ tons for 1 herd

$\frac{2}{3} \div \frac{1}{2} = 2 \times \frac{2}{3} = 1\frac{1}{3}$ tons

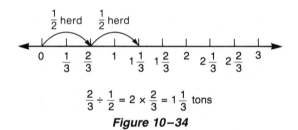

$\frac{2}{3} \div \frac{1}{2} = 2 \times \frac{2}{3} = 1\frac{1}{3}$ tons

Figure 10–34

Have the children write the number sentence below the work. Discuss with the children the similarity between solving this problem with the materials and showing $2 \times \frac{2}{3}$ with the materials. Repeat this process with other word problems and other manipulatives. Compare the results to those obtained using a calculator that works with common fractions.

2. Problems in which the divisor is not a unit fraction may pose more difficulties. Challenge the students who are ready to solve a problem such as the following using materials: Evelyn has $\frac{3}{4}$ of a pound of nuts. This is $\frac{2}{3}$ of the original package of nuts. How much did the original package weigh? Write the number sentence $\frac{2}{3} \times n = \frac{3}{4}$ or the sentence $\frac{3}{4} \div \frac{2}{3} = n$.

   Children who think of this as a partition division problem may reason this way: "If $\frac{3}{4}$ of a pound is $\frac{2}{3}$ of the package, I should first find $\frac{1}{3}$ of the package and then multiply by 3 to find the whole package. To find $\frac{1}{3}$ of the package, I need to divide $\frac{3}{4}$ by 2. This tells me that $\frac{1}{3}$ of the original package was $\frac{3}{8}$ of a pound. I should now multiply this by 3, and this tells me that

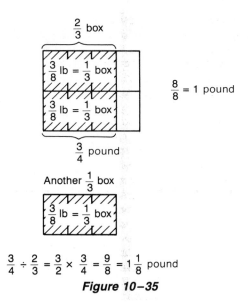

$$\frac{3}{4} \div \frac{2}{3} = \frac{3}{2} \times \frac{3}{4} = \frac{9}{8} = 1\frac{1}{8} \text{ pound}$$

Figure 10–35

there must have been $\frac{9}{8}$, or $1\frac{1}{8}$, pounds of nuts in the package originally." (See Figure 10–35.)

Note that by dividing by 2 and then multiplying by 3, the children have performed the same operation as inverting the $\frac{2}{3}$ and then multiplying by the inverse ($\frac{3}{2}$). The invert and multiply algorithm is discussed further in the section on skills. At this point, let the children work several examples using various materials, and encourage them to discuss their solutions.

Division of Decimal Fractions

If division of decimal fractions follows division of common fractions, the children should compare the two algorithms by solving number sentences in both common fraction and decimal fraction form and comparing the two quotients. Word problems and models for the two types of number sentences are essentially the same. Any of the suggestions given earlier for common fractions may be adapted for work with decimal fractions.

Review with children the algorithm for division of whole numbers before you attempt division of decimal fractions. Introduce division of decimal fractions with whole-number divisors first, and then move to divisors with one or two decimal places. Following are a few suggestions for introducing division with decimal fractions.

ACTIVITIES

Intermediate (4–7)

Objective: to develop the concept of division of a decimal fraction by a whole number.

1. Metric measures and money are good sources of problems. The following is one suggestion. Encourage children to create other word problems of their own and to discuss their methods of solution.

Sasha has a board that is 9.3 meters long. She wishes to make a bookshelf with three shelves of equal length from the board. How long should each shelf be?

Notice that this is a partition type of division problem. Model the problem. Cut a piece of string 9.3 meters long and then fold it into 3 equal parts. Ask the children to estimate how long each of the parts is. Will each part be longer than 1 meter? Will each one be longer than 10 meters? About how many meters long is each part? Is each longer or shorter than 3 meters? Ask the children to measure the parts to find the answer to the nearest 0.1 of a meter.

Show the children the division algorithm in the following form:

$$
\begin{array}{r}
3.1 \\
3\overline{\smash{)}9.3} \\
\underline{9} \\
3 \\
\underline{3}
\end{array}
$$

Ask the children what should be done with the decimal point? After the children predict that the division algorithm is performed in the same manner as the algorithm for whole numbers, with the decimal point in the quotient directly above the decimal point in the dividend, ask them to work other problems with whole-number divisors using that algorithm, and to check the answer by measuring and folding string. Have them use a calculator to see if they get the same results.

Objective: to develop the concept of division of decimal fractions with non-whole-number divisors.

2. After children have had experience dividing with whole-number divisors, give them a problem such as the following: Maureen has a lemonade stand. She has made 2.4 liters of lemonade. She wants to sell glasses with 0.3 liter of lemonade in each one. How many glasses can she fill?

This measurement type of division problem can be shown using any of the materials familiar from the work with decimal fractions. Figure 10–37 shows several possibilities. Ask the children to write the number sentence below each example. They may also use a repeated subtraction algorithm to show the measurement idea of division.

If children have had previous experience with dividing common fractions, ask them to write the number sentence in common fraction form. They may show the problem as $2\frac{4}{10} \div \frac{3}{10} = \frac{24}{10} \div \frac{3}{10} = n$. Using the common-denominator algorithm, this becomes $\frac{24 \div 3}{10 \div 10}$. Since the denominators cancel, the problem is simply $24 \div 3 = 8$. Discuss with the children that this has the same effect as moving the decimal point in both the divisor and the dividend one place to the right.

The children may also use the division concept of common fractions to show the problem as $\frac{2.4}{0.3}$. In this form, both the numerator and the denominator may be multiplied by 10 so the divisor will be a whole number. Thus, $\frac{2.4 \times 10}{0.3 \times 10} = \frac{24}{3}$.

This is a good time to use a calculator. Try the problem as $2.4 - .3 - .3 - .3 - .3 - .3 - .3 - .3 - .3$ or $2.4 - .3 = = = = = = = =$. How many glasses of 0.3 liter each can you make? Compare this result to $2.4 \div 0.3$.

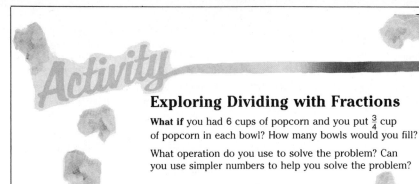

Exploring Dividing with Fractions

What if you had 6 cups of popcorn and you put $\frac{3}{4}$ cup of popcorn in each bowl? How many bowls would you fill?

What operation do you use to solve the problem? Can you use simpler numbers to help you solve the problem?

Working together

Materials: Workmat 12 or grid paper

A. What if you had 6 cups of popcorn and you put 3 cups in each bowl? How many bowls would you fill? Write a division sentence.

B. Write a division sentence for putting 2 cups in each bowl; for putting 1 cup in each bowl.

C. What operation do you think you would use to solve the problem at the top of the page?

D. Write a division sentence for putting $\frac{3}{4}$ cup in each bowl. Use a variable for the quotient.

E. This is a model for putting 3 cups in each bowl.

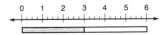

F. Use Workmat 12 or grid paper to draw a model for 2 cups in each bowl; 1 cup in each bowl. Does each model show the correct quotient?

G. Draw a model for $\frac{3}{4}$ cup in each bowl. What is the quotient?

Sharing Your Results

1. If you had 6 cups of popcorn and you put $\frac{3}{4}$ cup of popcorn in each bowl, how many bowls would you fill?

2. Why is the quotient in **G** greater than the quotients in **E** and **F**?

Figure 10–36

Ruth I. Champagne et al. Mathematics: Exploring Your World (Grade 6). Morristown, N.J.: Silver Burdett & Ginn, 1991, p. 286. Reprinted by permission.

THE MATH BOOK

The activity shown in Figure 10–36 is from a sixth grade textbook and is an introduction to division of fractions. Students are asked to work in groups of three to explore the concept of dividing fractions. They are asked to use either the workmat provided with the book or grid paper to answer the questions. This is similar to an activity students did earlier in the chapter that involved multiplying fractions. Notice that they begin by dividing a whole number by a fraction using the measurement concept of division and then they extend the activity to division of a whole number by a fraction. In the section for sharing results, students are asked to explain why the quotients get larger as the divisors get smaller. On the following page, students are asked to write division sentences for models that are pictured and then to draw models for given sentences. They then have to explain the meaning of each model. To sum up, money examples are given and discussed that use division of fractions.

The teacher's manual describes ideas for using fraction strips and base ten blocks to explore division of fractions concretely in small groups before using the pages in the book. Students are asked to challenge other groups to model division sentences they have written after they have done some initial exploration.

In the following lesson, students draw models, complete charts, and analyze results to show connections between number sentences such as $\frac{3}{4} \div \frac{1}{4} = 3$ and $\frac{3}{4} \times 4 = 3$. They are then led to discover the rule for multiplying by the reciprocal. Again, the lesson is set up for cooperative learning groups and the teacher's manual suggests different roles for the students in the group.

The teacher's manual has a number of suggestions for working with students with different modality strengths and for connections to the "real world" and other subject areas. It also has suggestions for using calculators with fraction capabilities. There are also several interesting problems and simulations involving fractions both in the student textbook and in the teacher's manual.

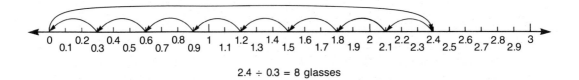

2.4 ÷ 0.3 = 8 glasses

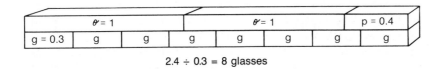

2.4 ÷ 0.3 = 8 glasses

2.4 ÷ 0.3 = 8 glasses

Figure 10–37

After the children have worked several examples, you may show them the division algorithm that makes use of the carat to show the movement of the decimal point in both the divisor and the dividend. Ask the children to give you a rule for the number of places that the decimal point should be moved. Be sure the children have experience with divisors and dividends with varying numbers of decimal places so they can generalize that the number of places the decimal point moves in both the divisor and the dividend is equal to the number of decimal places in the divisor. They may need to add decimal places in the dividend to make this work.

Let the children use calculators to check their predictions about moving the decimal points. Using calculators allows them to make predictions and check a much larger number of examples than doing all the calculations with paper and pencil.

Developing and Practicing Skills with Operations with Rational Numbers

After children understand the concepts behind the algorithms for operations with both common and decimal fractions, they can practice their skills with these operations. Do not rush the students into this practice. In many textbooks, one page explains why the algorithms work and then the next page includes thirty or forty exercises for the children to practice the algorithm. It is more beneficial to spend more time on the development of understanding through the use of concrete materials than it is to rush to the abstract algorithms. Children who understand what

they are doing often become much more proficient with the algorithms than children who have had a great deal more practice with the algorithms but did not understand the processes they were using.

Because of the proliferation of inexpensive calculators, it is not necessary to spend a great deal of time dividing by a divisor with five or six decimal places or adding two fractions with denominators such as 57 and 34. Very few of us as adults would do those problems with paper and pencil even if the occasion arose for us to work the problems at all. We would work the problems on a calculator. Even fractions with nasty denominators can be easily converted to decimals and the problem worked on a calculator, or the problem can be worked on a calculator that computes with common fractions.

Children should practice simpler problems with paper and pencil, however, and this section is devoted to developing those skills. It is also important that children learn to estimate and calculate mentally with rational numbers and the following section focuses on these skills.

Addition and Subtraction of Common Fractions

After children understand the algorithms for adding and subtracting common fractions, they may practice the two skills together. The activities for reinforcing operations with whole numbers may be adapted to reinforcing the same operations with rational numbers. Children enjoy inventing and playing their own card games, bingo games, and board games, and they can make the playing cards with number sentences involving addition and subtraction of common fractions. Following are a few other suggestions.

ACTIVITIES

Intermediate (4–7)

Objective: *to reinforce the skills of adding and subtracting common fractions.*

1. Magic squares, which have the same sum for each row, column, and diagonal, give the children a great deal of practice in both addition and subtraction of fractions. You may fill in as many or as few of the positions of the magic square as you wish, depending upon the level of the children. For a real challenge, give the children a blank square and the list of fractions and ask the children to fill in all the spaces so that each row, column, and diagonal has the same sum. Figure 10–38 shows a partially completed square. Discuss with the children the best place to begin and strategies for completing the square. Fraction calculators may be useful here.

2. Children often enjoy using codes and solving puzzles. To solve the riddle in Figure 10–39, children must break the

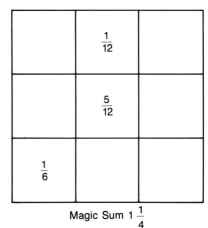

Magic Sum $1\frac{1}{4}$

Figure 10–38

Use these values to solve the following riddle.

$a - \frac{1}{3}$	$b - \frac{2}{3}$	$c - 1$	$d - \frac{1}{4}$
$e - \frac{3}{4}$	$f - 0$	$g - \frac{1}{5}$	$h - \frac{2}{5}$
$i - \frac{3}{5}$	$j - \frac{4}{5}$	$k - \frac{1}{8}$	$l - \frac{3}{8}$
$m - \frac{5}{8}$	$n - \frac{7}{8}$	$o - \frac{1}{10}$	$p - \frac{3}{10}$
$q - \frac{7}{10}$	$r - \frac{9}{10}$	$s - \frac{1}{2}$	$t - 1\frac{1}{2}$
$u - 1\frac{1}{4}$	$v - 1\frac{3}{4}$	$w - 1\frac{1}{3}$	$x - 1\frac{2}{3}$
$y - 1\frac{1}{5}$	$z - 1\frac{3}{5}$		

Why was the elephant wearing blue sneakers?

$$\frac{1}{3} + \frac{1}{3} \qquad 1 - \frac{1}{4} \qquad \frac{1}{4} + \frac{3}{4} \qquad \frac{1}{6} + \frac{1}{6} \qquad \frac{3}{4} + \frac{1}{2} \qquad \frac{1}{4} + \frac{1}{4} \qquad \frac{1}{2} + \frac{1}{4} \qquad \frac{1}{5} + \frac{1}{5} \qquad 1\frac{1}{5} - \frac{3}{5} \qquad \frac{3}{4} - \frac{1}{4}$$

$$\frac{1}{2} + \frac{2}{5} \qquad \frac{7}{8} - \frac{1}{8} \qquad \frac{1}{2} - \frac{1}{4} \qquad \frac{3}{5} - \frac{1}{2} \qquad 1\frac{1}{8} - \frac{1}{4} \qquad \frac{1}{4} + \frac{1}{2} \qquad \frac{3}{8} + \frac{1}{8}$$

$$\frac{2}{3} + \frac{2}{3} \qquad \frac{1}{2} + \frac{1}{4} \qquad 1\frac{1}{10} - \frac{1}{5} \qquad 1\frac{1}{2} - \frac{3}{4} \qquad 1\frac{1}{2} - \frac{1}{6} \qquad 1\frac{1}{4} - \frac{1}{2} \qquad \frac{3}{4} + \frac{3}{4}$$

Figure 10–39

code by completing the addition and subtraction number sentences. Ask the children to make up other codes for each other and trade them.

3. Because computers can individualize work to the level of the children, many computer programs give children practice on the skills they need. The Fraction Bars Computer Program series has disks specifically for adding and subtracting common fractions. These disks are good to use as a follow-up to the actual work with the Fraction Bars. Other programs, such as Microcomputer Software Mixed Numbers by Media Materials, are designed to give the children more practice with the abstract algorithms. Look for programs that individualize instruction to the needs of the children and explain, not just repeat, the algorithms that the children have missed.

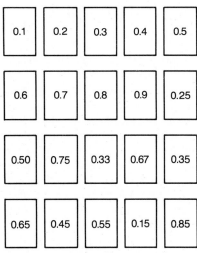

Figure 10–40

Addition and Subtraction of Decimal Fractions

After children understand the algorithms for addition and subtraction of decimal fractions using the concrete materials, have them begin to practice the algorithms more abstractly. Be sure to make full use of calculators and computers as the children are practicing these skills. Emphasize estimation and mental calculation, because once they leave school, the children will probably be performing more operations either mentally or with a calculator than they will with paper and pencil. Following are a few activities for paper and pencil as well as for calculators and computers. Other ideas for estimating and mental calculating are described in that section.

ACTIVITIES

Intermediate (3–6)

Objective: *to practice addition and subtraction of decimal fractions.*

1. Make up a set of cards for the game *sum of one,* as shown in Figure 10–40.

    Children should shuffle the cards and deal them all out to the players. From two to six children may play. At a given signal, each player turns the top card on the dealt stack face up. If any player sees two or more cards that add to one, that player says, "Sum of one." If the player is correct, he or she gets to keep the cards that add to 1. If players disagree over whether the sum is correct, the children should use a calculator to check. Play continues with every player turning over his or her top card on each turn. The game ends when the designated time period ends or when one player has all the cards. The player with the most cards at the end of the game is the winner. The game may be played with other sums or with differences.

2. Bring in newspaper ads or catalogs from different stores. Make up problems that require the children to add and subtract various amounts using money written as decimal fractions. Ask how much cheaper one item is than another or how much several items would cost all together. Challenge the children to find exactly three items that would have a total

cost of $10.00. See if the children can find two items with a difference in price of $.57. Ask the children to make up other problems of their own.

Encourage mental computation but allow children to use a calculator as they work, if necessary. Let them discuss their strategies for choosing the items and for performing the mental computations.

3. The target game allows children to use calculators but requires them to estimate and compute mentally in order to be successful. To play this game, one child begins by displaying a rational number on the calculator. A second child names another rational number as the target and the third child tries to hit the target by adding to or subtracting from the number currently in the calculator. If the target is hit, the child hitting the target scores one point and begins a new game by leaving that number in the display and announcing a new target. The calculator is then passed to the next child, who attempts to hit the new target with addition or subtraction.

Each time the target is hit, a point is scored and a new target named. If the target is not hit, the calculator is passed on, and the next player adds to or subtracts from whatever number is currently in the calculator display. A sample game is described below.

Maureen, Esteban, and Marlo are playing. Maureen puts 8.6 into the calculator and Esteban announces a target of 17.2. Marlo is given the calculator and thinks, "If I add 9, I will have 17.6; 17.2 is 0.4 less than that. I will add 0.4 less than 9, or 8.6." Marlo pushes + 8.6 and hits the target. She scores 1 point, announces a new target of 12.5, and hands the calculator to Maureen.

Maureen thinks, "12 is 5 less than 17 and 0.2 is 0.3 less than 0.5." She subtracts 5.3, and the calculator shows 11.9. Maureen realizes she did not figure the 0.3 correctly. It is now Esteban's turn, and the target is still 12.5.

Encourage the children to discuss their strategies aloud as they play the game. Strategies should be perfected as play continues. This game may also be played with common fractions, if you have a calculator that can handle them.

4. Programs such as the Decimal Squares Computer Games by Albert Bennett, Jr., and Albert Bennett III, available from Creative Publications, build on the children's earlier work with concrete materials. These are good for expand-

ing their understanding as well as for practicing skills. Other programs, such as the Microcomputer Software Decimal Skills from Media Materials and Get to the Point by Judah I. Schwartz, available from Sunburst, give more practice with the abstract algorithms. Review software before giving it to the children to be sure it meets their needs.

Multiplication and Division of Common Fractions

After children have a good concrete understanding of multiplication and division of common fractions and can relate the algorithms to both concrete and semi-concrete algorithms, it is time to practice the algorithms more abstractly. As with addition and subtraction, emphasize understanding, not memorization of algorithms. Do not assign thirty or forty exercises to be worked rotely each night in hopes of building greater proficiency. Be sure children understand any algorithm before they practice it because it is very difficult to "unteach" an incorrectly practiced algorithm. Encourage use of estimation and mental calculation, and teach the use of calculators to solve more difficult problems.

The reinforcement ideas described earlier for whole-number operations or for addition and subtraction of rational numbers may be adapted to multiplication and division of rational numbers. Following are a few additional ideas for extending the algorithms and reinforcing the learning of multiplication and division of common fractions.

ACTIVITIES

Middle Grades (5–8)

Objective: *to extend the algorithm for multiplying or dividing common fractions by using cancellation.*

1. After children have discovered and practiced the multiplication algorithm and the invert and multiply algorithm for common fractions, you can help them discover a shortcut for multiplying some fractions. Ask the children to complete the following number sentence and simplify the answer when they are finished:

$$\frac{3}{4} \times \frac{4}{5} = n$$

$$\frac{3}{4} \times \frac{4}{5} = \frac{3 \times 4}{4 \times 5} = \frac{12}{20} = \frac{12 \div 4}{20 \div 4} = \frac{3}{5}$$

Ask the children if there is an easier way to work the problem than to first multiply the 3 and the 5 by 4 and then to divide the 12 and the 20 by 4 to get the 3 and the 5 again. Ask the children to suggest a way to avoid having to multiply both the numerator and the denominator by the same number and then having to simplify the answer.

Have the children try their suggestions in solving the following problems:

$$\frac{8}{9} \times \frac{9}{15} = n$$

$$\frac{5}{6} \times \frac{7}{5} = n$$

Does the process work for problems such as:

$$\frac{4}{5} \times \frac{3}{8} = n$$

$$\frac{2}{3} \times \frac{6}{7} = n$$

Suggest factoring the numerators and denominators before attempting the cancellation. Try to solve the following problem by factoring first:

$$\frac{24}{35} \times \frac{7}{36} = n$$

Let the children suggest other problems and use cancellation before they multiply. Let children use calculators to assist with the larger multiplication problems. In this way, children can explore more examples. (This skill will be very useful in algebra later on.)

Ask the children if they can also use cancellation for division when using the invert and multiply algorithm. Encourage them to explore what happens if they do not invert the divisor before they attempt to cancel.

Objective: *to develop the concept of reciprocals.*

2. Ask the children to multiply several pairs of common fractions in the form $\frac{a}{b} \times \frac{b}{a}$. They may solve the following problems:

$$\frac{1}{2} \times \frac{2}{1} = n$$

$$\frac{2}{3} \times \frac{3}{2} = n$$

$$\frac{3}{4} \times \frac{4}{3} = n$$

Give the definition of a reciprocal. The **reciprocal** of any number *a* is the number that you must multiply *a* by to get 1, the multiplicative identity. Ask the children to find the missing number in each of the following:

$$\frac{5}{6} \times b = \frac{30}{30}$$

$$\frac{7}{8} \times c = \frac{56}{56}$$

$$1\frac{2}{3} \times d = \frac{15}{15}$$

You may need to suggest that the children first write $1\frac{2}{3}$ as an improper fraction. Ask them to give you a rule for finding the reciprocal of any common fraction.

Objective: *to develop the meaning of the invert and multiply algorithm.*

3. After children understand the concept of reciprocals, ask those who seem to grasp the concept well to work a problem

involving division of common fractions using a new algorithm. This will reinforce the idea that there are several ways to compute and the children may choose the method easiest for them. Remind the children that common fractions can be used to represent division: $3 \div 4$ can be shown as $\frac{3}{4}$.

Use this idea to show that $\frac{2}{3} \div \frac{4}{5}$ can be written as

$$\frac{\frac{2}{3}}{\frac{4}{5}}$$

How can children solve the number sentence written in this form? Suggest to the children that they multiply the numerator and the denominator by the same number to get a denominator of 1. This renames the fraction with an equivalent fraction and eliminates the messy denominator. (This skill is useful for later work with algebraic fractions.) Notice that it is necessary to multiply both the denominator and the numerator by the reciprocal of the denominator. This has the same effect as inverting the divisor and multiplying the inverted number by the dividend. See below:

$$\frac{2}{3} \div \frac{4}{5} = \frac{\frac{2}{3}}{\frac{4}{5}} = \frac{\frac{2}{3} \times \frac{5}{4}}{\frac{4}{5} \times \frac{5}{4}} = \frac{\frac{2}{3} \times \frac{5}{4}}{1} = \frac{2}{3} \times \frac{5}{4}$$

Encourage the children to try other examples of their own. Ask them to explain in their own words why the invert and multiply algorithm works.

Multiplication and Division of Decimal Fractions

After children understand the concepts of multiplying and dividing decimal fractions and can relate the abstract algorithms to both concrete and semi-concrete models, they can begin to practice the algorithms more abstractly. Again we caution you against an overreliance on abstract practice without understanding. Be sure that children understand the concepts and are correctly working the algorithms before they take any work home to practice.

We remind you that most adults work problems involving decimal fractions either mentally or with a calculator the majority of the time. This practice will certainly not lessen as our children enter adulthood. Use estimation, mental calculation, and calculators fully as children practice their skills. Following are a few suggestions for practice.

ACTIVITIES

Middle Grades (4–8)

Objective: *to practice operations with decimal fractions.*

1. Ask the children to predict which of the following will have an answer of 2.4. Let them check by using either a calculator or paper and pencil.

$$0.2 \times 12 = n$$
$$3.0 \times 0.08 = n$$
$$60 \times 4.0 = n$$
$$0.1 \times 240 = n$$
$$0.24 \times 10 = n$$
$$30.0 \times 8.0 = n$$
$$24.0 \div 10 = n$$
$$240 \div 0.1 = n$$
$$0.24 \div 0.1 = n$$

Ask the children to tell you why they made the predictions they did and if they were correct. Let the children make up other problems for each other.

2. Ask the children to describe the average person in each group of three or four. Tell the children to make several measurements (in meters) of each person in their group. They may measure each person's height; the length of each person's arms, legs, feet, and fingers; the circumference of each person's head, etc. Avoid any measurements that may embarrass some of the children. After all the measurements are made, ask the children to find the mean of each measurement by totaling the measurements and dividing the total by the number of people in the group. Let the children compare groups to see which group has the smallest or largest mean on certain measures. Use the means to describe an average student. Can you find a student who matches this description? Let the children suggest other problems that involve any of the operations with decimal fractions. Use computer programs to help children practice their skills. The programs mentioned earlier for practicing addition and subtraction of decimal fractions also provide practice for multiplication and division and should be used for these skills as well.

Operations with Percents

Because percents are used so frequently in everyday life, children should develop their skills in this area. After the children understand the meaning of percent using concrete materials and can connect the algorithms to the models, they should practice their skills mentally and with calculators as well as with paper and pencil. Following are a few activities for developing this skill.

ACTIVITIES

Middle Grades (5–8)

Objective: *to practice using percents.*

1. Earlier, we mentioned using percent in the formula $p = r \times b$. In the earlier discussion, the problems involved knowing the rate and the base and finding the percentage. After mastering those problems, children should also work with problems in which the base or the rate is unknown. Allow children to use a calculator to solve problems such as the following: Juanita bought a coat that was marked $10.00 off. The coat was originally $50.00. What percent was taken

off the original price? What percent is the discount of the selling price? Does the discount stay the same? Are the two percents the same? Why not?

Fred bought a jacket that was $20.00 off. Fred said he saved 25% off the original price. What was the original price?

Ask the children to make up other problems and to trade the problems with each other. They can use ads from the newspaper as a resource. Children usually prefer solving each other's problems over solving the problems in the book. Discuss what to do with problems that have too much or too little information and the methods used to solve the problems that could be solved.

2. Another common use of percent is to figure the interest on a loan or savings account. Simple interest is figured using the formula $I = prt$, that is, interest = principal × rate × time. Ask the children to figure the amount of interest a savings account will pay if it pays simple interest at 5% for 2 years on $500.00. Children should take $500.00 × 5% × 2 to find the interest.

After children have worked several problems, some of them may wish to explore problems with compound interest or problems with a discount rate rather than simple interest. Children may wish to visit a bank or savings and loan to discuss ways to find the best interest rate on a savings account or a car loan.

Estimating and Mental Calculating with Rational Number Operations

Throughout the chapter, we have discussed the importance of children using estimation and mental calculation skills. The Second National Assessment of Educational Progress found that only 24 percent of the 13 year olds tested could correctly choose the number closest to the sum of $\frac{12}{13} + \frac{7}{8}$ from the choices 1, 2, 19, and 21 (Post, 1981). Both 19 and 21 were more popular choices. This demonstrates that most children do not realize that both $\frac{12}{13}$ and $\frac{7}{8}$ are fairly close to 1 and that, therefore, the sum must be almost 2. They are apparently trying to use some misunderstood algorithm that tells them to add either the numerators or the denominators. They probably do not realize that a fraction represents one number to be operated on as one entity and not two numbers, one on top of the other, to be operated on separately. They do not have the necessary understanding of the size of rational numbers. The Fourth National Assessment of Educational Progress found that fewer than 40% of the seventh graders tested could identify the largest and smallest of four fractions and fewer than 50% recognized that $5\frac{1}{4}$ was the same as $5 + \frac{1}{4}$ (Kouba, Brown, Carpenter, Lindquist, Silver, and Swafford 1988). This basic understanding of the size of rational numbers is crucial to understanding problems that require the use of rational numbers.

Reys and Bestgen (1981) found that students do not perform any better with decimals. Fewer than 30 percent

of 13 year olds could correctly estimate the sum of $95.0 + 865.2 + 1.583$ to the nearest power of 10.

It is therefore essential that we help children learn to both estimate and calculate exactly with rational numbers. We cannot assume that children who can compute accurately with paper and pencil also can estimate. Many children who can correctly compute the product of two rational numbers cannot estimate whether the product will be larger or smaller than 1. The skills of estimation and mental calculation need to be specifically taught. Most children do not automatically transfer skills with paper and pencil calculation to mental estimation and calculation. The following activities are designed to help children develop these skills further.

ACTIVITIES

Intermediate–Middle Grades (3–8)

Objective: *to develop estimation skills for addition and subtraction of rational numbers.*

1. In Chapter 9, we mentioned activities for helping children estimate values of rational numbers by relating them to wholes, halves, and fourths. These skills help the children as they learn to add and subtract fractions. For example, if the children have an exercise such as $1\frac{15}{16} + \frac{12}{23}$, they may think the following way in order to estimate the sum: "$1\frac{15}{16}$ is almost 2. $\frac{12}{23}$ is about $\frac{1}{2}$. Therefore, the sum is about $2\frac{1}{2}$."

A similar activity that relates the problem to money is useful for decimal addition and subtraction because of the children's familiarity with the decimal notation of money. When given a problem such as $2.4876 - 1.98$, the children may think, "2.4876 is about $2.50 and 1.98 is about $2.00. Therefore, since $2.50 - $2.00 is $.50, the answer is around 0.5."

Children may wish to use a calculator to see how close the estimates are. Estimates are also a good check on calculator work. Let children suggest other problems for each other to work. After they have worked several examples, children may suggest strategies for refining their estimates. Let them share strategies with each other and decide which ones work best for them.

Objective: *to develop estimation skills for multiplication and division of rational numbers.*

2. Children are often surprised when they multiply two rational numbers and see that the product is smaller than either of the two factors or when they divide two rational numbers and get a quotient larger than either the dividend or the divisor. These experiences may go against generalizations they have made for whole numbers about the product being larger than either factor or the quotient being smaller than the dividend. Have the children return to the meaning of multiplication and division of rational numbers in order to help them predict the size of the answer to a number sentence involving multiplication or division of rational numbers.

Give the children a word problem such as the following and ask them to tell you whether the scale drawing will be larger or smaller than the original: Jacque is making a scale drawing of his room. He wants the drawing to be 0.1 the size of his room. His room is 3.0 meters by 2.5 meters. How large should the drawing be?

The children should realize that the drawing will be smaller than the room. Ask the children to give you a number sentence for the word problem. Discuss with the children what happens when you take 0.1×3.0. Why is the answer less than 3? Ask the children to tell you the answer without using paper and pencil.

Have the children make up other word problems involving multiplication and predict the answer without using paper and pencil. Let them check using paper and pencil or a calculator if they wish. Encourage children to round mixed numerals to the nearest whole number to estimate their product or to mentally calculate the product of a whole number and a mixed numeral by multiplying the two whole numbers and then multiplying the first whole number times the fraction. For example, $3 \times 2\frac{1}{2} = (3 \times 2) + (3 \times \frac{1}{2})$.

Use word problems for estimation involving division of rational numbers also. Use a problem such as the following, and ask the children if the answer is more than one apple: Dan has $\frac{3}{4}$ of a pound of apples. Each apple weighs $\frac{1}{4}$ pound. How many apples does Dan have? Ask the children to write a number sentence for the word problem and to compute the answer mentally. Encourage the children to make up other word problems for each other and to predict the answers before using paper and pencil algorithms or a calculator.

Objective: to use estimation skills to correctly place the decimal point in any problem involving operations with decimal fractions.

3. As the use of calculators becomes widespread, children need to become even better at estimating to determine whether or not the answer shown on the calculator is reasonable. One way to practice this skill is to give the children a list of number sentences along with their solutions, only without the decimal points. Ask the children to put the decimal point in the correct place without using paper and pencil or a calculator. Use a list such as the following:

$$6.9 \times 45.0 = 3105$$
$$8.14 \div 0.2 = 407$$
$$0.5 \times 26 = 130$$
$$0.18 \div 0.5 = 36$$
$$3.62 \times 54.789 = 19833618$$

Let children make up other problems on their own.

Objective: to use mental calculations to answer problems involving percents.

4. Because we encounter percents frequently as consumers and often want to figure an amount exactly without using a calculator or paper and pencil, this area is an especially crucial one. Give children experience finding 1% and 10% of an amount. Ask the children to give you a rule for multiplying by these percents. Let them use calculators to see if the rule always works.

After the children can easily move the decimal point to find 1% or 10%, ask them for a way to find 5% or 15%. Good mental calculators might tell you that 5% is half of 10% and

that to get 15% you should add 5% and 10%. Let the children practice finding the exact amount they should leave for a 15% tip for meals of varying amounts.

After children can easily find 1%, 5%, 10%, and 15%, ask them for an easy way to find 25%, 50%, and $33\frac{1}{3}$%. The children should know the fraction equivalents of these percents and practice finding $\frac{1}{4}$, $\frac{1}{2}$, and $\frac{1}{3}$ of varying amounts. Bring in sale ads from the paper and ask the children to figure how much they will save on sales of 25%, 50%, or $33\frac{1}{3}$% off.

After children are able to find these percents, ask them for 75%, $66\frac{2}{3}$%, and 40%. Good mental calculators will use 25% and/or 50% to find 75%, $33\frac{1}{3}$% to find $66\frac{2}{3}$%, and 10% and 50% to find 40%. Given these abilities, the children should be able to either find the exact amount or give a good estimate for any percents they encounter as consumers.

Problem Creating and Solving

Throughout the book, we have emphasized the importance of children both creating their own problems and solving the problems created by others. In this chapter, we have given suggestions for using problem creating and problem solving in teaching operations with rational numbers. Problem creating and solving should be the approaches to learning. Children should learn by asking questions, making mistakes, and finding the answers. Here are a few other suggestions for problems involving rational numbers.

ACTIVITIES

Intermediate–Middle Grades (3–8)

Objective: to use common fractions to describe data collected in an experiment.

1. Ask the children to suggest experiments they can conduct that involve rational numbers. They can use common fractions to describe the number of heads that come up when flipping a coin, the number of spades drawn from a deck of cards, or the number of times a sum of seven is shown when rolling two dice. Conduct one of the experiments five times, ten times, and one hundred times. For example, write fractions that represent the number of heads that come up when you flip a coin.

- *Understanding the problem.* If I flip a coin once, the fraction that represents the number of heads is either $\frac{1}{1}$ (one head, no tail out of one flip) or $\frac{0}{1}$ (no head, one tail out of one flip). Now, what I want to do is to flip a coin and see what I get. Each time I flip the coin I can write a new fraction based on one flip, two flips, three flips, and so on. Let's see, if I flip a coin five times I can get 5, 4, 3, 2, 1, or 0 heads. That is one of six fractions. I will need to try.

- *Devising a plan.* I will flip a coin and write a fraction describing the number of heads compared to the number of flips. The denominator will be 1. Then I will flip a coin again and write a different fraction describing the number of heads compared to the number of flips. The denominator

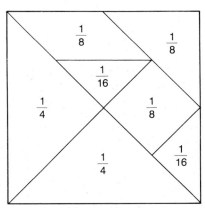

Figure 10-41

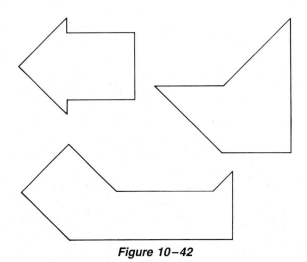

Figure 10-42

will be 2. I will continue to flip the coin, recording a new fraction each time. The denominator will increase by 1 each time I flip the coin (make and use a drawing or model).

- *Carrying out the plan.* On the first flip, I have a head; my fraction is $\frac{1}{1}$. On the second flip, I have a head; my fraction is $\frac{2}{2}$. Next, a tail comes up; my fraction is $\frac{2}{3}$. I continue flipping the coin, and the fractions I get are $\frac{3}{4}$, $\frac{3}{5}$, $\frac{4}{6}$, $\frac{4}{7}$, $\frac{4}{8}$, $\frac{5}{9}$, $\frac{6}{10}$.

  Now, I want to flip the coin ten more times to see if I get the same fractions. After that, I will attempt one hundred flips.

- *Looking back.* Are all of my fractions written correctly? Do the denominators increase from 1 to 10? (I could write all of the denominators before I ever flip the coin.) Is there any pattern to the numerators? Can I predict the final numerator? Could I write a computer program to simulate the flips so I would not have to do all the work myself? If so, I can repeat the experiment as many times as I want and see if there is a pattern to the results.

Ask the children to describe the fractions they got each time. Which experiment would be the most accurate predictor in the long run? What would happen if they performed the task a thousand or a million times? These activities are a good introduction to teaching probability, which is described in the next chapter.

Objective: to solve problems involving spatial visualization and rational numbers.

2. Use the tangrams in Appendix B for the following activity. Tell the children that all seven pieces together have a value of 1. Ask them to find the value of each shape.

   After the children have determined that the pieces have the values shown in Figure 10-41, ask the children to find the total value of each shape in Figure 10-42. Encourage the children to make other shapes of their own and to challenge each other to find the total values. How many different shapes can they make with the same value? Determine how much more one shape is than another.

Objective: to use decimals and percents to describe data and make predictions.

3. Ask the children to design a survey and collect data on some area of interest to them, such as the food offerings in the cafeteria. Let them list available offerings and suggest others and then ask students to select their favorites. When the

survey is complete, have the children report the percent of children who chose each offering.

Have the children suggest uses for their data. They may wish to discuss the results with the head of the cafeteria service. Discuss with the children things that must be considered in planning menus, such as cost and nutrition. Have the children interview people from companies that are in charge of planning. How do they make decisions regarding the products or services they offer? The children may even wish to set up a company of their own and to keep records and collect data to plan for the future.

Grouping Students for Learning Rational Number Operations

Several researchers recommend that more time be spent developing the concepts of rational number operations (Hiebert, 1987; Payne and Towsley, 1990; and Mack, 1990). Students need to work in groups and discuss the concepts that they are constructing using concrete materials. Otherwise, students often have very incomplete or incorrect concepts of operations with rational numbers. They try to apply memorized algorithms to operations with rational numbers, and these algorithms are frequently remembered incorrectly, if at all. Symbols are poorly understood and introduced too early. If the symbols are not understood, and students have attempted to memorize rules rotely, they have nothing to fall back on when their memories fail them. They are not able to judge when they have an incorrect answer from mental estimation, written computation, or a calculator.

To construct correct algorithms, students need to work on real problems using the manipulatives and to discuss their findings with each other. Much of the early work on the development of algorithms should be spent in small cooperative learning groups working on real problems and constructing meaning for each of the operations. For example, to introduce multiplication and division of fractions, you might begin by giving a group a problem such as determining the number of pizzas you need to serve the whole class, if each student wants $\frac{1}{4}$ of a pizza. Let them work with circular fraction pieces to determine the answer. Later determine the number of servings you can get from $3\frac{3}{4}$ pizzas if each serving is $\frac{1}{4}$ of a pizza. Which operations were involved in these problems? Ask the students to write their findings and share them with the entire class. Did each group work the problems in the same way? Which method did they prefer? In this way, the grouping may move from a whole class with the teacher introducing the problem and the students discussing the background knowledge needed to get started, to small groups working cooperatively on the problems, back to the whole class listening to students in different groups explaining their methods of solution.

After the concepts have been constructed, other types of grouping may be used. Practice on the operations may be done in larger groups as students play a game to reinforce the computation skills, or individually as students work on puzzles and problems that require the application of these operations. At times, the teacher may need to work with a small group to remediate any difficulties they may be having, or the teacher may work with the talented students to challenge them to go beyond the explorations of the rest of the class.

Communicating Understanding of Rational Number Operations

Like the work with rational number concepts, the work with operations with rational numbers must include a heavy emphasis on verbal communication. The overreliance on rote memorization of symbol manipulation hinders students as they attempt to understand operations with rational numbers. As students work in groups solving problems that involve rational number operations, they should constantly be communicating with each other their understanding of the meaning of the problems. This oral communication frequently helps students make the necessary connections to the concepts they have already learned, such as operations with whole numbers and how they relate to operations with decimal fractions, or the connections between decimal fraction operations and operations with common fractions. When you circulate among the groups as they work on a problem, ask them how this problem relates to earlier ones.

In addition to the oral communication, ask the students to record what they are learning in daily logs and in work for their portfolios. Ask them to be on the lookout both for connections to earlier learning and for models that demonstrate what they have learned. Students should make use of concrete materials to exemplify meanings and use pictures in the logs and portfolios to illustrate the written messages.

Evaluating Learning of Rational Number Operations

Use a number of means to evaluate the children as they add, subtract, multiply, and divide rational numbers. The textbook you use may contain both pre- and post-tests for the rational number chapters. Test the children before beginning a new unit to see which children already understand some of the concepts and which children need to begin at the beginning. If you use a test that pictures concrete materials and uses word problems, you will get a better idea of the children's abilities than if the test has only abstract number sentences for the children to complete. To get a better understanding of the children's levels, add a few questions on the concrete level.

As the children work through the unit, keep anecdotal records of their progress and note any areas of difficulty. Ask the children to explain their thinking processes as they work algorithms or manipulate concrete materials. Tell the children to record the algorithms they create and their solutions to problems in their math journals. They should keep examples of some of their best work in a portfolio.

Check the children's written work for any patterns of errors. Following are some of the common ones. If you see children making any of these mistakes, talk to them about what they were thinking. You will often need to ask the children to return to the concrete models to correct any misunderstandings.

1. In addition and subtraction of common fractions, children add or subtract numerators and denominators.
2. To find like denominators, children add or subtract the same amount from the numerator and the denominator.
3. To multiply common fractions, children cross multiply; that is, they multiply the numerator of one fraction by the denominator of the other.
4. To cancel when multiplying common fractions, children cancel two numbers from the numerators or two from the denominators.
5. Children cancel when dividing common fractions before they invert and multiply.
6. Children forget to invert when dividing common fractions, or they invert the dividend rather than the divisor.
7. When multiplying mixed numerals, children multiply the whole numbers, multiply the fractions, and then add the products together.
8. When adding or subtracting decimal fractions, children do not line up the decimal points.
9. When multiplying or dividing decimal fractions, children keep the same number of decimal places in the answer as appeared in one of the original numbers.

10. When multiplying or dividing a decimal fraction by a power of ten or when converting a decimal to a percent or vice versa, children move the decimal point in the wrong direction.

If you see any of these errors, be sure to discuss them with the children. Do not just mark a problem incorrect without any feedback; that does not help the children make the necessary corrections. You will often find in talking to the children that they do not have a clear understanding of rational numbers or their usage in operations. If this is the case, return to the section on concept development in this chapter and use concrete materials with the children. Use word problems and materials that will help guide the children to correct the algorithms for themselves.

Something for Everyone

The manipulative materials discussed in Chapters 9 and 10 should be used by all students initially when learning concepts and operations with rational numbers. Many teachers in the upper elementary grades seem to think that manipulative materials are for only the kindergartners and maybe the first graders. They do not believe that older children need these "crutches." In fact, many of the students in these grades also believe that the manipulatives are baby stuff.

This is definitely not the case, however. Research has shown repeatedly that manipulatives are helpful for students of any age in learning mathematics and specifically in learning rational number concepts (Suydam, 1986). Visual, tactile, and kinesthetic learners need materials to develop a concrete understanding of the concepts. Students with tactile/kinesthetic strengths need to manipulate materials such as region models, base ten blocks, Decimal Squares, Cuisenaire rods, and pattern blocks in order to internalize the concepts discussed in this chapter. They may have difficulty transferring this understanding to paper.

Allow children to devise their own methods of recording what they have learned with the materials. When children are presented with abstract rules or algorithms for such things as finding equivalent fractions, ordering common fractions, or converting a common fraction to a decimal fraction or a percent, be sure they have first modeled the rule with the concrete materials.

If the child is an auditory learner, make sure the child is allowed to explain the rule to you after you explain it. For children who are visual learners, drawing models of the rational numbers is helpful. Make sure children have a variety of models, since some children may understand the concepts for a region model but not for a number line model. Let these children draw pictures to explain why certain rules work, such as why you can multiply or divide the numerator and the denominator of a common fraction by the same number and not change the value of the frac-

tion. Encourage children to write the algorithm or rule being demonstrated below the picture.

Some children who are able to abstractly manipulate symbols and correctly perform the algorithms may still not understand the meaning of the operations. Therefore, even children who seem to be performing well abstractly benefit from being introduced to the materials.

Make sure to assist students in making the transition from the concrete materials to the abstract algorithms. They will probably not make this transfer automatically. Ask the children to write the algorithms as they manipulate the materials. Do not use the materials one day and expect the children to write the algorithms the next without the materials present. Let each child decide when he or she no longer needs to use the manipulatives. Children are usually good judges as to when the concrete knowledge has become internalized and the manipulation simply slows them down.

Depending on the students' maturational levels and abilities, a few children may only need to see the concepts demonstrated a few times to abstract the necessary information, while others will need to work individually with the materials for a long time as they work through the algorithms. Do not worry about covering all the pages in the textbook as the children learn. You are trying to teach children concepts, not a textbook.

If you find that some children are having difficulty, you may need to use a different concrete model. Some children may not be able to understand a length model such as the Cuisenaire rods or the number line but may be successful with an area model. Some children may not understand the concept of class inclusion or reversibility and may have difficulty with the part-whole model for common fractions. You must have a variety of models available for the children, and you must interview the children individually to determine the best method for teaching each one.

Children who can quickly understand and perform the operations with rational numbers will need more challenging work. Let them demonstrate that they can perform the algorithms correctly, but do not punish them by assigning more exercises from the book.

Many of the problem-solving and problem-creating activities described in this chapter are especially useful for gifted children. Be sure to encourage them to design problems and experiments of their own, but do not always expect them to work alone. They need interaction with their peers and with you. Following are a few other suggestions for avenues of exploration for the gifted students and any others who enjoy a challenge.

ACTIVITIES

Intermediate–Middle Grades (3–8)

Objective: *to develop the concept of scientific notation.*

1. In scientific notation, a number is written as a value from 1 to 10 multiplied by a power of ten. For example, 3 245 654 can

be written as $3.245\ 654 \times 10^6$ and 0.000078 can be written as 7.8×10^{-5}.

Children should explore what happens to the exponent when you multiply and divide powers of ten. For example, $10^5 \times 10^6 = 10^{11}$ and $10^6 \div 10^5 = 10^1$. Ask the children to explain why these statements are true. They may wish to write out all the powers of ten in expanded form and use the rules they know for multiplying and dividing by powers of ten.

Ask the children how this knowledge can help them estimate the product or quotient of two numbers written in scientific notation. What happens if you try to add or subtract two numbers written in scientific notation if the powers of ten are not the same?

As children explore scientific notation, they can also explore the topic of significant digits. If you have one measurement accurate to four decimal places and another accurate to one decimal place, how many decimal places should you have in the sum, difference, product, or quotient of the two numbers?

Objective: to understand why it works to invert and multiply when dividing common fractions.

2. Children should constantly be challenged to discover why rules work, not just how they work. The rule for dividing common fractions by inverting and multiplying is just one such example. Earlier in this chapter, we demonstrated this algorithm with numbers using the division concept of a fraction and the multiplicative inverse. Using the following, gifted children could show that this algorithm would work for all rational numbers:

$$\frac{a}{b} \div \frac{c}{d} = \frac{\dfrac{a}{b}}{\dfrac{c}{d}} = \frac{\dfrac{a}{b} \times \dfrac{d}{c}}{\dfrac{c}{d} \times \dfrac{d}{c}}$$

$$= \frac{\dfrac{a}{b} \times \dfrac{d}{c}}{1} = \frac{a}{b} \times \frac{d}{c}$$

Another way to demonstrate the reason for invert and multiply is to use the fact that multiplication and division are inverse operations:

$$\frac{a}{b} \div \frac{c}{d} = n$$

Therefore,

$$\frac{c}{d} \times n = \frac{a}{b}$$

To solve for n, multiply both sides of the equation by the reciprocal of $\frac{c}{d}$:

$$\frac{d}{c} \times \frac{c}{d} n = \frac{d}{c} \times \frac{a}{b}$$

Because $\frac{d}{c} \times \frac{c}{d} = 1$, and 1 is the multiplicative identity,

$$n = \frac{d}{c} \times \frac{a}{b}$$

Use the commutative property of multiplication to show that this equals

$$\frac{a}{b} \times \frac{d}{c}$$

Ask the children to explain other algorithms for rational numbers. For example, they could use the distributive property to show why, when multiplying two mixed numerals, you cannot just multiply the whole numbers together, multiply the proper fractions together, and then add the two products. Ask them to explain why cancellation works when multiplying or dividing common fractions. Encourage the children to discover and prove other rules for themselves.

Other topics gifted children may be interested in exploring include operations with common fractions in other numeration systems, such as the Egyptian. Reference books on the history of mathematics and mathematics activity books will give you and the children other ideas.

Many computer programs are designed to individualize instruction based on a child's knowledge and ability. If you have computers available, you should explore the possibilities. In any case, do not try to teach the whole class together from the book. That level of instruction will be inappropriate for most of your students.

KEY IDEAS

Operations with rational numbers are generally introduced in the upper elementary grades and continue to be a major topic of study throughout the rest of elementary and middle school. In spite of their emphasis in the curriculum, operations with rational numbers in any form are generally not well understood by students. In this chapter, we have presented a number of ways to help children understand operations with rational numbers. Concrete materials, computer programs, and calculators are again emphasized for their usefulness in helping students to gain understanding.

Much emphasis is placed on the use of everyday examples and word problems to help children understand when to add, subtract, multiply, or divide common and decimal fractions and percents. The meaning behind rules such as invert and multiply is described, and alternative algorithms are discussed.

Because so often we need to use common and decimal fractions and percents when we estimate or mentally calculate—for example, in finding a discount at a department store sale or in calculating a tip for a waiter or waitress—these skills are emphasized. Problem creating and problem solving with rational numbers are discussed, including the uses of rational numbers in experiments, data collection, and prediction.

Diagnosis of rational number operations should include data other than just paper and pencil tests; include a search for common error patterns. Individual skills and learning styles should be analyzed when you plan instruction. Challenge each child accordingly.

REFERENCES

BEHR, MERLYN J.; WACHSMUTH, IPKE; AND POST, THOMAS R. "Construct a Sum: A Measure of Children's Understanding of Fraction Size." *Journal for Research in Mathematics Education.* Vol. 16, No. 2 (March 1985), pp. 120–131.

BENNETT, ALBERT B., JR. *Decimal Squares.* Fort Collins, Col.: Scott Resources, 1982.

BENNETT, ALBERT B., JR., AND DAVIDSON, PATRICIA A. *Fraction Bars.* Fort Collins, Col.: Scott Resources, 1973.

BEZUK, NADINE, AND CRAMER, KATHLEEN. "Teaching About Fractions: What, When, and How?" In Trafton, Paul R., and Shulte, Albert P., eds. *New Directions for Elementary School Mathematics.* Reston, Va.: National Council of Teachers of Mathematics, 1989, pp. 156–167.

BRADFORD, JOHN. *Everything's Coming Up Fractions with Cuisenaire Rods.* New Rochelle, N.Y.: Cuisenaire Company of America, 1981.

DOSSEY, JOHN A.; MULLIS, INA V. S.; LINDQUIST, MARY M.; and CHAMBERS, DONALD L. *The Mathematics Report Card: Are We Measuring Up? Trends and Achievement Based on the 1986 National Assessment.* Princeton, N.J.: Educational Testing Service, 1988.

HARNADEK, ANITA. *Word Problems (Fractions–1, Decimals–1, Percents–1, Percents–2, Mixed Concepts A–1, Mixed Concepts B–1, Mixed Concepts C–1).* Pacific Grove, Ca.: Midwest Publications Co., 1980.

HIEBERT, JAMES. "Research Report: Decimal Fractions." *Arithmetic Teacher.* Vol. 34, No. 7 (March 1987), pp. 22–23.

JENKINS, LEE, AND MCLEAN, PEGGY. *Fraction Tiles: A Manipulative Fraction Program.* Hayward, Ca.: Activity Resources Co., 1972.

KINDIG, ANN C. "Using Money to Develop Estimation Skills with Decimals." *Estimation and Mental Computation.* National Council of Teachers of Mathematics, 1986 Yearbook. Reston, Va.: NCTM, 1986.

KOUBA, VICKY L.; BROWN, CATHERINE A.; CARPENTER, THOMAS P.; LINDQUIST, MARY M.; SILVER, EDWARD A.; AND SWAFFORD, JANE O. "Results of the Fourth NAEP Assessment of Mathematics: Number, Operations, and Word Problems." *Arithmetic Teacher.* Vol. 35, No. 8 (April 1988), pp. 14–19.

MACK, NANCY K. "Learning Fractions with Understanding: Building on Informal Knowledge." *Journal for Research in Mathematics Education.* Vol. 21, No. 1 (January 1990), pp. 16–32.

NATIONAL COUNCIL OF TEACHERS OF MATHEMATICS. *Curriculum and Evaluation Standards for School Mathematics.* Reston, Va.: NCTM, 1989.

PAYNE, JOSEPH N., AND TOWSLEY, ANN E. "Implications of NCTM's *Standards* for Teaching Fractions and Decimals." *Arithmetic Teacher.* Vol. 37, No. 8 (April 1990), pp. 23–26.

POST, THOMAS R. "Fractions: Results and Implications from National Assessment." *Arithmetic Teacher.* Vol. 28, No. 8 (May 1981), pp. 26–31.

POST, THOMAS R.; BEHR, MERLYN J.; AND LESH, RICHARD. "Research-Based Observations About Children's Learning of Rational Number Concepts." *Focus on Learning Problems in Mathematics.* Vol. 8, No. 1 (Winter 1986), pp. 39–48.

"RATIONAL NUMBERS (FOCUS ISSUE)." *Arithmetic Teacher.* Vol. 31, No. 6 (February 1984).

REYS, ROBERT E., AND BESTGEN, BARBARA J. "Teaching and Assessing Computational Estimation Skills." *Elementary School Journal.* Vol. 82 (November 1981), pp. 117–127.

ROBBINS, PAUL R., AND HAUZE, SHARON K. *Word Problems with Fractions.* Portland, Me.: J. Weston Walch, 1981.

SHEFFIELD, LINDA JENSEN. *Problem Solving in Math.* New York: Scholastic Book Services, 1982.

SUYDAM, MARILYN N. "Research Report: Manipulative Materials and Achievement." *Arithmetic Teacher.* Vol. 33, No. 6 (February 1986), pp. 10, 32.

THOMPSON, CHARLES. "Teaching Division of Fractions with Understanding." *Arithmetic Teacher.* Vol. 26, No. 5 (January 1979), pp. 24–27.

VANCE, JAMES. "Ordering Decimals and Fractions: A Diagnostic Study." *Focus on Learning Problems in Mathematics.* Vol. 8, No. 2 (Spring 1986), pp. 51–59.

VANCE, JAMES H. "Estimating Decimal Products: An Instructional Sequence." *Estimation and Mental Computation.* National Council of Teachers of Mathematics, 1986 Yearbook. Reston, Va.: NCTM, 1986.

WOODCOCK, GARY E. "Estimating Fractions: A Picture Is Worth a Thousand Words." *Estimation and Mental Computation.* National Council of Teachers of Mathematics, 1986 Yearbook. Reston, Va.: NCTM, 1986.

YEAGER, DAVID CLARK. *Story Problems: Fraction, Decimal, Percent.* Palo Alto, Ca.: Creative Publications, 1983.

ZULLIE, MATHEW E. *Fractions with Pattern Blocks.* Palo Alto, Ca.: Creative Publications, 1975.

It is impossible to pick up a newspaper or listen to the news on television without noticing the extensive use of statistics, probability, and graphs. Computers have made it easier than ever before to handle and report very large amounts of data. Students should be able to understand the statistical reports they will face everyday.

Statistics are frequently reported through the use of charts and graphs. Even very young children can make and read simple object, picture, and bar graphs. As children get older, they should learn to read and construct line and circle graphs. They also should learn to plot and read points on a Cartesian coordinate system.

Children should be on the lookout for uses of probability, statistics, and graphs in their everyday lives. We are all familiar with probability from listening to the weather report each day. If there is a 90 percent chance of rain, we do not want to forget to take our umbrellas. If we buy a lottery ticket, we want to know our chances of winning a million dollars. Any type of scientific experiment depends upon the laws of probability. Drugs are tested to determine if the recovery rate improves when the drug is used and to determine if there are any harmful side effects. Businesses use probability for everything from determining the optimum number of check-out lines to quality control on an assembly line. Insurance company officials use probability to set their rates. Wildlife experts tag wild animals and then use probability to determine the number of animals in an area by looking at the ratio of the number of tagged animals spotted to the number of untagged animals spotted.

In many cases, statistics and graphs are used in conjunction with probability. The statistics for a sports player are used to help determine the probability of having a winning team if that player is hired. Statistics are compiled from the results of polls taken to help businesses plan for the future, and probabilities are determined for the success of the business if certain changes are made. Politicians hire experts in the area of statistics to help them decide the best strategies for winning an election.

An intelligent consumer in today's world must be able to decipher sometimes conflicting sources of information. It is important for children to know some of the fundamental concepts of probability and statistics as well as how to interpret the charts and graphs used to describe those ideas.

Like other topics, probability, statistics, and graphing should be taught using manipulative materials. Children should have numerous chances to perform experiments and collect data to give them practical experience with these topics. Some of the materials include dice, coins, cards, colored cubes, chips, spinners, graph paper, squares, and objects for making concrete graphs.

Children also should make full use of calculators and computers when exploring these topics. Calculators and computers give children the ability to handle more data than they could handle with paper and pencil calculations alone. Graphics calculators allow students to graph various types of information and to explore the effects of changing different variables. Computer programs include those for teaching probability, statistics, and graphing, as well as simulations that give children the opportunity to experiment and make predictions.

Developing Concepts of Probability, Statistics, and Graphing

Probability, statistics, and graphing may not be covered in depth in elementary math texts, but it is important for elementary children to begin to develop a basic understanding of the ideas, and teachers may need to supplement the textbook materials. Because of the attention received by the NCTM *Standards,* however, recent textbooks do include more on these topics. In this chapter, we present a few ideas for teaching some of the elementary concepts. Children may wish to explore some of the ideas further on their own. They should find many examples of statistics and graphs in their everyday lives, and they may devise a variety of experiments and surveys to further investigate probability. Do not try to formalize the topics or to teach them on an abstract level. Let the children learn at their own rates on their own levels.

Probability Concepts

Many of the concepts of probability involve terms the children may hear frequently. These include *impossible, certain, uncertain, likely, equally likely, unlikely, random, sample space, independent, biased, unbiased,* and *random sample.* Children should be given opportunities to become familiar with these terms and their meanings. The following activities may be used to introduce young children to the terminology and to give them a chance to apply the terms to daily events.

ACTIVITIES

Primary (1–3)

Objective: *to define and apply the following terms: certain, impossible, uncertain, likely, and unlikely.*

1. Discuss with the children events they are familiar with and ask whether they can be certain that the events will happen. Are there some things you are positive will happen and some you are positive will not happen? Events that we are positive will happen are called **certain,** and events that cannot happen are called **impossible.** At other times, we are **uncertain** whether or not something will happen.

   Give the children a list of statements and ask them to sort them into three piles labeled "Certain," "Uncertain," and "Impossible." Use statements such as the following:

   ● Tomorrow, it will rain.
   ● Jane will be here every day next week.
   ● I will get 100% on my next spelling test.
   ● The teacher is older than every child in the room.
   ● Tomorrow, we shall all go to the moon.
   ● Tonight, the sun will shine brightly at midnight.
   ● If I flip a coin, it will land either heads or tails.

   As the children classify the statements, discuss with them the reasons for the classifications. When they have finished, ask them to further classify the uncertain statements as either **likely** or **unlikely.** Will all children agree on whether or not an event is likely? Is it possible for two children to have different answers and for both to be correct? As children discuss these questions, you will get an insight into their conceptions and misconceptions about early probability concepts. When misconceptions appear, lead the children to activities and experiments to help them correct those ideas. Encourage the students to brainstorm a list of synonyms for the terms *certain, impossible, likely* and *unlikely.* Ask them to make their own list of statements to classify into these categories.

Objective: *to develop the concepts of sample space, independent, equally likely, fair, random, unbiased, and biased through use of an experiment and tallies.*

2. The **sample space** for a problem consists of all the possible outcomes. The sample space for flipping a coin is heads and tails. Ask the children to predict whether the coin will land with the head or the tail showing if you flip it. Flip the coin and show the children the result. Ask the children to predict the outcomes of several flips of the coin.

trials											
	1	2	3	4	5	6	7	8	9	10	total
H											
T											

Figure 11–1

Discuss with the children whether one flip seems to have any influence on the next flip. Events are called **independent** if one event has no effect on another. Give each child a penny and ask the children to make a tally of the heads and the tails out of ten flips, using the chart in Figure 11–1.

Tell the children to record the total number of heads and the total number of tails out of the ten trials. Use calculators to find the total number of heads and the total number of tails for the whole class. Are the totals close to each other? Can you say that heads or tails are more likely?

If two events have the same probability, then we say that they are **equally likely.** We can say that the coin is a **fair** one if it does not favor either heads or tails. We can also say that heads and tails come up **randomly.** This means that heads and tails have an equal chance of coming up. We may also say that the coin in **unbiased.** It does not favor heads more than tails or vice versa. Use these terms frequently with the children as you perform experiments. Ask the children to define the terms in their own words after they have had a chance to experiment.

Ask the children if the coin would be **biased** if they dropped it flat on the table instead of flipping it. Would the coin be biased if they flipped a quarter instead of a penny? Would the coin be biased if you flipped a play coin with two heads instead of a head and a tail?

Children may wish to explore other questions, such as how many times a certain number of heads came up out of ten flips of a fair coin. Which number of heads was most likely? Which was least likely? What about tails? Let children propose other questions and experiments of their own and help them tally the results.

Children who are learning to write computer programs can write a simple program to instruct the computer to simulate the flips of a coin and record the results. In this way, the children can collect data on a far greater number of trials.

As children get older, they should begin to explore other areas of probability. They may look at events with more than two possible outcomes, and they may begin to use rational numbers to describe probability. The **probability** of an event is

the number of favorable outcomes
the total number of outcomes

If an event is certain, the probability is 1 because all possible outcomes are favorable. If an event is impossible, the probability is 0 because there are no favorable outcomes. All uncertain events have a probability somewhere between 0 and 1. An event with a larger probability is more likely than an event with a smaller probability. The following activities give a few ideas for teaching these concepts

to students in the intermediate and middle grades and for expanding on other concepts introduced earlier in the primary grades.

ACTIVITIES

Intermediate–Middle Grades (4–8)

Objective: *to develop the concept of assigning a numerical probability to equally likely events.*

1. Show the children a spinner with three colors as in Figure 11–2.

   The sample space in this experiment is red, blue, and yellow. Ask the children whether landing on one color is more likely than landing on another color. Spin the spinner a few times to demonstrate that it is a fair spinner, that is, that it does not get stuck on any of the colors. Tell the children that if the spinner is fair, and if each color is equally likely, they can write the probability of landing on any one color as

   the number of favorable outcomes
   the total number of outcomes

If you wish to know the probability of getting yellow, the number of favorable outcomes is 1 because there is only one yellow section. The total number of outcomes is 3 because there are three sections altogether. Therefore, the probability of getting yellow is $\frac{1}{3}$.

Ask the children to predict the number of times they could expect to get yellow if they spin the spinner thirty times. Spin the spinner thirty times and record the resulting color each time. How many yellows did you get? Was the actual number close to the prediction?

Ask the children to find the probability of landing on the blue section. What is the probability of landing on the red section? What should you get if you add the probability of the yellow section to the probability of the red section and the probability of the blue section?

Try the experiment using different colors and different numbers of spins. Can you find a formula for predicting the number of times a color will come up? Multiply the probability of the color by the number of spins. Will this always give you the exact number of times the color will come up in the actual experiment? What happens if you add together the probabil-

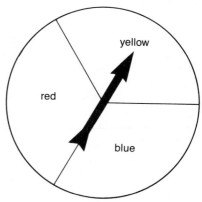

Figure 11–2

ities of all the possible outcomes? Will you ever get an answer larger than 1? Ask the children if the spinner would be biased if each child were allowed to turn the spinner so it pointed at her or his favorite color.

2. After the children feel comfortable making predictions with the spinner, try a different material, such as a deck of playing cards. Ask the children to predict the number of times they would draw a black card, a diamond, or a king out of twenty draws. What is the sample space each time? Which is more likely: a black card, a diamond, or a king? Let the children suggest other questions. Discuss with them their strategies and reasoning as they make predictions and perform experiments.

Objective: to distinguish between probable and certain events and between independent and dependent events.

3. Put the name of each child in the class on a 3 × 5 card and put all the names in a box. Predict the number of times a girl's name will be drawn out of a given number of draws when you put the name back in the box after each draw. Can you be certain that your own name will be drawn in ten draws if the name is replaced after each draw? Can you be certain that a girl's name will be drawn in ten draws? Can you be certain that a girl's name will be drawn in one hundred draws?

Even though it is very likely that a girl's name will be drawn, you cannot be certain. It is possible to draw only boys' names if the name is replaced after every draw.

Does it make a difference if the name is not replaced after each draw? Some children may wish to research the difference between dependent and independent events.

As mentioned earlier, if events are independent, the outcome of one event has no effect on the outcome of another event. If the first event does effect the next one, the events are said to be **dependent.** Drawing cards with no replacement involves dependent events, while drawing cards with replacement involves independent events. Ask the children to explain why this is true. If you were drawing names of the class members without replacing the name each time, can you be certain that you will eventually draw a girl's name?

Objective: to develop the concept of a random sample.

4. A **random sample** is a sample drawn in such a way that every member of the sample space has an equal chance of being chosen. Random sampling is often used in taking surveys when it is not feasible to survey every possible person or in designing experiments when it is not possible to use every person of interest.

Discuss with the children how they might take a random sample of marbles from an opaque bag. Would the sample be random if they were allowed to look into the bag while picking out a color? Would it be random if each child was allowed to pick his or her favorite color? Would it be random if all the red marbles were on the top and the blue ones on the bottom? How could the ideas of random sampling be expanded to taking a random sample of students in the school for a survey?

Put 5 red marbles and 45 blue marbles in an opaque bag. Tell the children you are going to take a random sample of 10 marbles from the bag and use the sample to try to predict the total numbers of red and blue marbles. Let them know that there are 50 marbles total and that there are only red and blue marbles. Draw 10 marbles from the bag and record the results. Ensure randomness by replacing the marble each time, blindfolding the person drawing, and shaking up the marbles in the bag.

If you drew 3 red marbles and 7 blue marbles, what would be your prediction for the total numbers of red and blue marbles in the bag? Would you trust your prediction based on the results of only 10 draws? What if you repeated the 10 draws several times?

You and the children can create or find many interesting probability activities as time permits. Some sources of activity ideas for the topics in this chapter are listed in the bibliography at the end of the chapter.

Statistics Concepts

Statistics are very often reported using rational numbers. Some of the concepts from statistics are quite abstract and should be reserved for higher grades, but children should be introduced to the basic concepts in the primary grades.

Statistics is the collecting and reporting of data. With all the information collected and reported in the world today, we need to have a way of organizing and describing it so that it is understandable. This collection, tabulation, organization, presentation, and interpretation of data is called **descriptive statistics.** This is the type of statistics most suited for study by elementary students.

The study of statistics and probability is a means of making predictions. The ability to draw inferences and make predictions based on the data collected is studied in **inferential statistics.** Inferential statistics is often not presented until college, but with the proliferation of data and computers in the information age, we are more likely to see inferential statistics presented in high school. Our focus here, however, is descriptive statistics and ways of presenting the subject to elementary and middle school students.

Tables and graphs are often used to describe the data collected (graphs are discussed in the next section). In addition, there are at least two ideas from the area of statistics you should present to elementary children; these are the concepts of measures of central tendency and measures of dispersion.

Measures of central tendency are ways of reporting data in the middle. Perhaps the most familiar of these measures is the arithmetic average, or the **mean.** The mean is frequently referred to simply as the **average,** the number found by adding together all the values of interest and dividing by the total number of addends. Another measure of central tendency is the median. The **median** is the value found by listing the values of interest in order from the highest to the lowest and taking the one in the middle. The **mode,** another measure of central tendency, is the most frequently reported value.

Measures of dispersion tell you how spread out scores are. The simplest measure of dispersion is the **range,** which is the difference between the highest score and the lowest score. Statisticians use several other measures, but they are generally too complex for children in elementary or middle grades.

Following are a few suggestions for teaching some elementary statistics concepts. You may find that your younger students can understand the concepts when they are presented on a concrete level.

ACTIVITIES

Intermediate (2–5)

Objective: *to develop the concepts of mode, median, and mean on a concrete level.*

1. Tell children the following story and ask them to illustrate it using blocks on a grid as shown in Figure 11–3: Sam practiced his spelling words every day. On Monday he missed 5 words, on Tuesday he missed 1, on Wednesday he missed 3, on Thursday he missed 1, and on Friday he got a perfect score.

   Ask the children if any score showed up more than once. It did; on Tuesday and on Thursday, Sam missed 1 word. Since that is the only score that shows up more than once, that is the mode.

   List the scores from the highest to the lowest: 5, 3, 1, 1, 0. The score in the middle is the median. Therefore, 1 is the median as well as the mode. Are the median and the mode always the same? The children can make up other scores to check their predictions.

   Will the mean be the same? Use the blocks on the grid and the original scores to find the mean number of words Sam missed each day—given the total number of missed words, figure out how many words would be missed each day if the same number were missed each day. Encourage the children to rearrange the blocks until there are the same number of blocks on each day. The children should be able to rearrange the blocks so that there are two blocks on each day.

   Ask the children what would happen if Sam missed 3 words on Thursday instead of 1 word. Now, can they rearrange the blocks so that there are the same number on each day? What should they do with the extra blocks? Can they use rational numbers to answer the question? Could Sam miss an average of 2.4 words per day? Did Sam actually miss 2.4 words on any day? If you read an article that says the average family has 2.2 children, does that mean that any family actually has 2.2 children?

   Let the children keep records of their own test scores or other data of interest and find the mean, median, or mode. They might wish to survey the class to determine the mean, median, and mode for the number of children in each family or the number of pets in each household. After surveys are taken, encourage the children to ask their own questions about the data and to work in groups to come up with answers.

2. Ask each child to draw around one hand on a sheet of paper and to cut out the drawing. Split the class into groups of six to eight, and ask each group to order the hand drawings from the smallest to the largest. The children may need to decide what is meant by the "smallest" hand drawing. Is it the shortest drawing or the one with the least area? Does the hand begin at the wrist and end at the tip of the longest finger?

   Once an ordering scheme is decided upon, ask the children to find the median hand. What is the median if an even number of students are in the group? It should be the average of the two middle hands. To find the average length of two middle hands without measuring, have the students tape the two hands together end to end and then fold the strip in half.

   Are there any hands that seem to be the same size? Is there a mode?

   Suggest that children find the mean hand length by taping the hands together end to end and then folding the strip. For example, if there are eight children in the group, tape the eight hands together and then fold the strip of hands in half three times to make eight equal sections. The length of one of those sections should be the mean of the lengths of the children's hands.

   Notice that the mean, median, and mode are found here without measuring. After the children have found these values concretely, ask them to measure the length of each of the hands and to find the mean, median, and mode of the measures. How do these methods compare to the concrete work? Encourage the children to discuss their methods with each other as they are working. Let the children suggest other measures for which they would like to find averages. They might find the average height, shoe size, neck size, and hat size of their group. Is anyone in the group "average. . . ?" Does that person have the average size for all the measures?

   Are mean, median, and mode always equally useful? If you have to stock dresses for a department store, would you be more interested in the mean, median or mode? Which does the government use in a census report?

Middle Grades (4–8)

Objective: *to explore differences in reporting data using the mean, the median, and the mode.*

3. The following are the salaries of five professional basketball players: $80,000, $80,000, $100,000, $120,000, and $620,000. The players are complaining about their salaries. They say that the mode of the salaries is $80,000 and that they deserve more money for all the games and practices. The owners claim that the mean salary is $200,000 and that this is plenty for any team. Which side is correct? Is anyone lying? How can you explain the differences in the reports?

   Ask the children to look in newspapers and magazines for reported averages. Are there any discrepancies in the reports? Are the statistics ever deceiving? Bring in reports for discussion in class. Encourage the children to read any reported statistics critically.

Objective: *to use a **line plot** as a quick way to organize numerical data.*

4. A line plot is a quick way to organize data to show the range and central tendencies, especially if the range is fairly small.

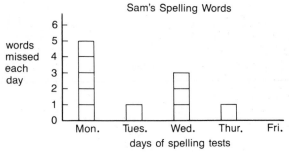

Sam's Spelling Words

words missed each day

days of spelling tests

Figure 11–3

It is not a formal graph, but rather is a working graph that can be used in initial data analysis. A line plot is simply a sketch along the horizontal axis in which X's are used to show the frequency of certain values.

Ask your students to collect data on some information with a limited number of expected values, such as the number of pets in each household. These data can easily be recorded as students are polled in the classroom. The following line plot shows how students might record the results:

```
     X
 X   X
 X   X
 X   X
 X   X   X
 X   X   X
 X   X   X   X                   X
 0   1   2   3   4   5   6   7   8   9   10
```

From this line plot, students can quickly see that most of their classmates have no pets or only one pet, while one student has nine pets. This information can then be analyzed further or graphed more formally, if the children wish.

*Objective: to use a **stem-and-leaf plot** as a quick way to organize numerical data.*

5. A stem-and-leaf plot is another quick way to represent the shape of a data set. It works best for data that span several decades rather than just a few numbers, unlike the line plot. This plot is most frequently organized by tens and can also be made quickly as data are reported orally. To make this plot, you divide each value into tens and units. The tens are the stems of the plot and the units are the leaves. Ask the children to collect data on information that may have values spanning several decades. The following plot shows the ages of the people at a family reunion:

```
0 | 2  3  2  5  8
1 | 4  9
2 | 2  0  8  9  5  8
3 | 1  6  9
4 | 5  1  0  9  0  0  6  8
5 | 6  8  9  4
6 | 2  4  2
7 |
8 | 3
```

From this plot, students can quickly see that there are five children under ten, two teenagers, and no one in their seventies. As you can see, these data are too spread out to put on a line plot, but they can be neatly organized and analyzed using the stem-and-leaf plot.

You may find several other suggestions for handling data in the Used Numbers series, which is listed in the reference section.

Graphing Concepts

Unlike probability and statistics, which can be very complex and abstract, the concepts of graphing can be learned as early as kindergarten, and children should graph data frequently in the primary grades. By their very structure, graphs are either concrete or semi-concrete and therefore quite appropriate for elementary school.

Graphs, tables, and charts are often used to display the data collected for descriptive statistics. Children need to learn to read, interpret, and create these displays in order to become intelligent, discriminating consumers. We have already discussed the importance of making a table or chart when solving problems. In this section, we present ideas for teaching graphing using concrete objects, pictures, and bar graphs in the primary grades and extending to line, circle, and Cartesian graphs in the intermediate and middle grades.

ACTIVITIES

Primary (K–3)

Objective: *to develop the concepts of graphs on a concrete and on a semi-concrete level.*

1. The math book insert shows first graders making graphs using concrete objects. Graphing can be begun as early as kindergarten. When graphs are made concretely, it is important to use a system to ensure that each object occupies the same amount of space, so that children are not misled by the differences in the sizes of the objects. The focus should be on the number of objects and not their volume. The example from the program shown in the math book insert uses a simple device, marking off congruent squares upon which to place the objects. This keeps the objects lined up, gives each object the same amount of space, and forms a good foundation for later work with bar graphs. Try the following with your class.

    Set up a two-column grid system on the floor (see Figure 11–5 on page 238). Bring in a large bowl of apples and oranges for the children's snack. Let each child choose one piece of fruit, but ask the children not to eat the fruit yet.

    While the children are holding the fruit, ask if more children have chosen apples or if more have chosen oranges. Let the children discuss ways to find the answer. They may want to walk around the room to count the children with each kind of fruit or they may want the children to sort themselves into two groups before counting.

    After some discussion, ask the children to sort themselves into two groups, one with apples and one with oranges, and then ask the children if they can tell which group is larger without counting. If the sizes of the two groups are similar, they may not be able to tell.

    One of the children may suggest that they line up instead of standing in a group and then look to see which line is longer. This is a good introduction to the squares on the floor. When the children line up, they may not all stand the same distance apart, and therefore, the longer line may not have more children.

    Have the children stand on the squares on the floor. In this way, they are all the same distance apart. Ask the children again if they can tell which group is larger.

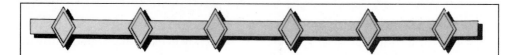

7. Solid of the Day

Gather the children together and display a model of a sphere. Invite them to find objects in the room which are the same shape. As the children identify objects, list them on a chart. Have the children bring the objects they identify to a display area. Encourage the children to bring objects shaped like a sphere from home. These objects can be displayed for the day and recorded on the chart. Also list any objects the children notice in the environment.

On following days focus on different solids so that a list and display are created for the cylinder, cube, rectangular prism, and cone. If you keep these lists displayed, the children will continue to search their environment for objects having the same shape as the solids introduced.

You may wish to store the objects collected and the materials from home in cartons under or by a table. Divide the table top into sections using pieces of yarn or masking tape. Direct the children to the materials and encourage them to sort the collection into sections on the table top. Have the children discuss and label the different sets they create.

8. Graphing Classroom Solids PS

Set up a graphing mat with a cube, a sphere, a cylinder, and a rectangular prism as concrete labels to indicate the objects to be graphed. Invite the children to find things in the classroom which have the same shape as the geometric solids and place them on the appropriate places on the graph. When the children have found as many objects as they can, have them gather around the graph. Ask questions about each object to determine if it is correctly placed. For example, if something is in the cylinder column, ask, **Does it look like a can? Does it roll and slide?**

When the children have examined all the objects, ask, **Which objects did you find the most of in the classroom? The least? How many more _____ than _____?**

When the children have finished interpreting the graph, have them return the objects to their original place.

Variations

• Use the boxes, cans, cartons, cylindrical potato chip containers, etc., which the children have brought from home to make a concrete graph. Interpret the graph. **Which geometric solids do we have most of? The least of? How many more _____ than _____?**

• Divide the children into small groups. Have each group focus on finding objects having a particular shape.

Figure 11–4
Betty Coombs and Lalie Harcourt. Explorations 1 *(Teacher's Manual). Don Mills, Ontario: Addison-Wesley of Canada, 1986, p. 183. Reprinted by permission.*

THE MATH BOOK

The page shown in Figure 11–4 is actually from a first grade teacher's manual rather than from the children's textbook. This particular program is based on children's concrete manipulation of materials rather than on a text or workbook, so there are no books for the children. This lesson focuses on creating a bar graph using actual objects and classifying them as cubes, spheres, cylinders, and rectangular prisms. This is a follow-up to a lesson on solid figures.

In earlier lessons, the children actively explored and built with the solids and then related their work to the outside world by making a playground with the figures. Children brought solid figures from home and sorted them according to a variety of characteristics using Venn diagrams. They built an inclined plane to determine if the shapes would roll or slide.

In lessons following this one, children will make a graph using glasses of juice they have chosen from either apple or orange. After the children have made a graph with the actual glasses, each child will draw a picture of his or her choice for a picture graph. Children will then move on to graphs that describe themselves and their preferences.

Throughout the experiences, the children are led to discuss their work and the uses of graphs to describe data collected. Suggestions are given for including the topic of graphs in other subject areas, such as language arts, art, science, social studies, and physical education. A project to extend the concepts is also suggested.

Because of the lack of a textbook, teachers choosing a program such as this one may have some difficulty adjusting their instruction, but the children's enthusiasm for learning and the concepts built through their active involvement should more than make up for any inconvenience. Teachers should not be tied to a text; they should be free to develop their own program to best meet the needs of the children. A program such as this one, based on the concrete development of concepts, should be very well suited to the needs of most children.

Children's Snacks

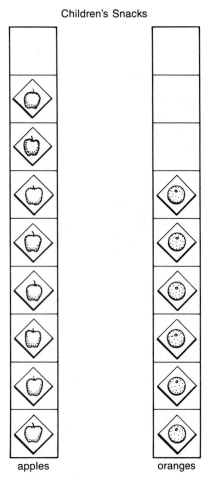

Figure 11–5

Children's Snacks

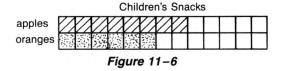

Figure 11–6

After the children have decided that the longer line represents the most popular fruit, tell them to put the fruit on a napkin on the square on which they are standing and to move away. Now ask the children which is the most popular fruit. Discuss whether it matters if they are standing on the squares or if the fruit alone can represent them. Ask if pictures of the fruit could represent them as well.

Give the children congruent sheets of square paper and ask them to draw a picture of the fruit they chose. (Square Post-It notes are good for this because they can easily be stuck to the blackboard or wall and moved around to answer your questions.) Tape the pictures of the fruit in two horizontal rows, with one row directly beneath the other. Again, discuss which fruit was chosen more often.

To go one step further, tell the children that you want to just use a red square to represent a child who chose an apple and an orange square to represent a child who chose an orange. Give each child a piece of centimeter graph paper (you may copy the master in the appendix) while you demonstrate on large graph paper on the board or on graph paper on the overhead projector. Label the graph and the rows as shown in Figure 11–6, and lead the children to color in the appropriate squares.

Ask the children whether all the graphs they made now give the same information. Which type of graph would be easiest to use if you wanted to print the information in the school paper?

The children can follow up this lesson by suggesting other things to graph, such as eye or hair color or favorite books or pets. Children will be much more involved in lessons on graphing if they are graphing information in which they have a personal interest. Use the information students collect in their graphs as a springboard for more questions. Graphs frequently raise as many questions as they answer.

Objective: *to use surveys to collect data and to display the data using graphs.*

2. Tell the children to each develop a question with a limited number of answers to ask their classmates or a group of adults. They may have questions such as "What is your favorite subject in school?" or "How many hours a night do you watch TV?" for the students and "Do you think seven year olds should get an allowance?" or "Should teachers get higher salaries?" for the adults. Or a child might ask the question of both students and adults and compare the responses. Discuss with the children the importance of selecting a random sample and ask how they would choose one.

After the children have each decided upon a question to ask, assist them in devising a tally sheet upon which to record the results. Decide how many people should be surveyed for interesting results. Children with similar questions may wish to team up to be able to question more people.

When the results have been tallied, discuss the best ways to display them. Children asking about favorite pets may make a graph using plastic animals, children asking about favorite sports may draw a picture of each sport for a picture graph, and children asking about allowances may use graph paper to make a bar graph.

Be sure every graph has a title and labels on all the pertinent parts. Encourage the children to make up questions about their graphs for other students to answer.

Even though concrete, picture, and bar graphs should be introduced in the primary grades, children in the intermediate grades should review these types of graphs before moving on to other types. For older children, each picture on a picture graph or each square on a bar graph may represent more than one object. Children in the intermediate and middle grades should have experience with these graphs as well as line, circle, and Cartesian coordinate graphs and stem-and-leaf plots. Following are several ideas for teaching graphing concepts in the intermediate and middle grades.

ACTIVITIES

Intermediate–Middle Grades (4–8)

Objective: *to create and read picture and bar graphs in which a picture or a bar represents more than one object.*

1. Use the results from one of the children's probability experiments as data to be graphed. The children may use their

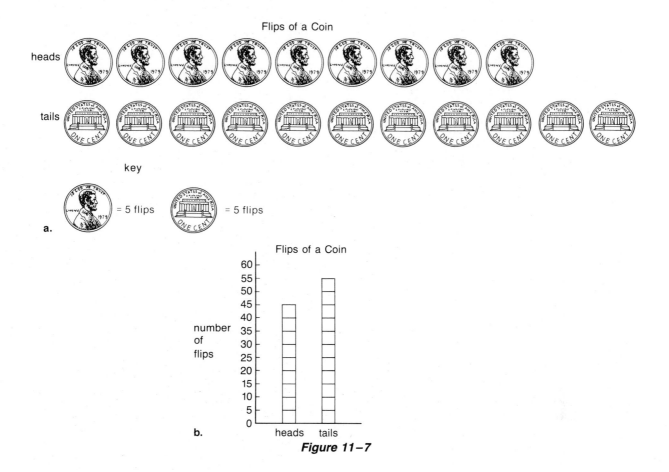

Figure 11–7

tallies from flipping a coin one hundred times. Use a picture or bar graph to show the results, but let each picture or bar represent five flips. Graphs showing forty-five heads and fifty-five tails would look like those in Figures 11–7a and b.

Ask the children how they would graph the same information if each bar or picture represented ten flips. Discuss how to use ½ of a bar or picture. Tell the children to bring to school bar or picture graphs from newspapers or magazines, and discuss what each bar or picture represents in the graph.

Objective: *to create and read line graphs (frequency polygons).*

2. When there is a continued trend from one point to the next, a line graph is useful. Line graphs should not be used unless both axes of the graph represent continuous data, such as time and temperature. It is not appropriate to use a line graph to show the numbers of people with different eye colors, for example, because the line between blue and green eyes would have no meaning.

Line graphs can be used in conjunction with subjects other than mathematics. The following activity may be used with a science lesson on plants. You and the children should find several occasions during the day to use their graphing skills.

Plant a fast-growing seed such as a bean seed and graph the plant's height each day after it has sprouted. If the plant is measured in centimeters, it is convenient to use centimeter graph paper to make the line graph. Children may grow the bean plants under several different conditions, such as in a dark closet, on the window sill, with no water, with water every day, and with water once a week. Graph the growth under each condition and discuss the results.

Follow this activity with the computer program Botanical Gardens by Robert Kimball and David Donoghue, available from Sunburst, which produces a simulation of growth under different conditions. Students must learn to interpret graphs in order to make hypotheses about the ideal growth conditions.

Find examples of line graphs in newspapers and magazines and discuss whether they are used properly. Encourage children to make up questions for each other that can be answered by looking at graphs. Leave these questions in the learning center or write them on dittos for the class to answer.

Objective: *to create and read circle graphs.*

3. A circle graph is used to show information dealing with parts of a whole. The circle represents a whole, and the information being graphed is some fractional part of that whole. When children construct their own graphs, it is helpful if they can use a compass and measure angles. This activity can be used before the children have acquired these skills if you cut out sections of a circle ahead of time.

Tell the children that you are going to make a circle graph to represent their pets. If you have twelve children with pets, cut several circular regions out of different colors of construction paper. Cut each circular region into twelve congruent sections (for a different number of children, adjust the size of the regions accordingly). Select a different color of paper to represent each type of pet (dogs, cats, etc.). Let each child select a colored region that represents her or his pet and place the region into the circle. Figure 11–8 shows a circle graph for one group of twelve children.

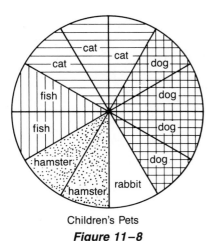

Children's Pets
Figure 11-8

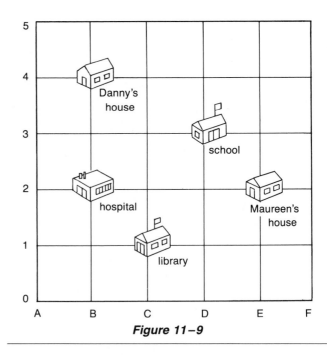

Figure 11-9

Encourage the children to ask questions about the completed graph. Fractions and percents can be used to describe the sections of the graph if children have studied these topics.

After children have learned to use a compass and protractor, discuss with them the fact that every circle has 360 degrees. Ask how many degrees would show that $\frac{1}{2}$ of the class have brown eyes. How would they show that $\frac{1}{4}$ have blue eyes? Let the children find and graph other data. Discuss circle graphs in newspapers and magazines.

Objective: to create and read Cartesian graphs.

4. Cartesian graphs are used for locating positions in two-dimensional space. Children may be introduced to Cartesian coordinates in social studies when they look up a point on a map. Maps generally use a letter and a number to give a location. Ask children to study the map in Figure 11-9 and then tell you where the school, the hospital, and the library are, giving the letter for the horizontal location first and then the number for the vertical location.

Encourage the children to make up other maps of their own. They may make up treasure maps and ask other children to find the secret location of the treasure by moving around on the map using horizontal and vertical locations.

The game Battleship is a good follow-up to this activity. Children must give the coordinates of their opponent's hidden ships in order to sink them.

After children are comfortable using letters and numbers to locate points, introduce them to ordered pairs of numbers, in which the first number gives the *x*-coordinate and the second number the *y*-coordinate. When they have some concept of negative integers, let older children use all four quadrants of the Cartesian plane.

After children have had concrete experiences, use computer programs to reinforce graphing concepts. Programs include Exploring Measurement, Time and Money, available from IBM; Interpreting Graphs, available from Conduit; Interpreting Graphs by Sharon Dugdale and David Kibbey, available from Sunburst; Graphing and Hurkle, available from MECC; and Exploring Tables and Graphs, available from Weekly Reader. Computer programs allow the children to quickly see the relationships between tables and graphs and to see the effects of adding data and changing variables.

Developing Skills with Probability and Graphing

One function of the mathematics curriculum in the elementary and middle grades is to give children a solid foundation in the basic concepts of probability, statistics, and graphing, but most of the skills practice in these areas is reserved for secondary school. However, some additional skills in probability and graphing can be developed before high school, and following are some ideas for developing these skills.

Probability Skills

Probability is a good topic with which to reinforce a student's ability to think systematically. In Chapter 5, we discussed using tree diagrams to find various combinations for attribute materials. Tree diagrams are also quite useful in probability for finding combinations and permutations. The concepts of combinations and permutations are important ones in the area of discrete mathematics and should be explored by elementary and middle school students. The following activities suggest ways to reinforce these skills.

ACTIVITIES

Intermediate–Middle Grades (4–8)

Objective: to use a tree diagram to find combinations and permutations.

1. The topics of combinations and permutations are fundamental to the subject of probability. When we want to know how

Flip 1   Flip 2   Results

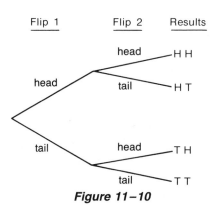

Figure 11–10

many ways we can select one number of objects from another number of objects, we are finding the number of **combinations.** The order of the selection does not matter when finding the number of combinations. When we are interested in the order, we are finding the number of **permutations.**

If we flip a coin two times, we may be interested in all the possible combinations of heads and tails we could get. We can list the combinations, as 2 heads, 1 head and 1 tail, and 2 tails. Are these combinations equally likely?

A tree diagram is useful for determining the sample space. Make a tree like the one in Figure 11–10, where the tree gives us all the possible permutations. They are: HH, HT, TH, and TT.

Do you think that the chance of getting two heads is the same as the chance of getting two tails? Is it the same as the chance of getting one head and one tail? Remind the children that the probability of getting one head and one tail is

$$\frac{\text{the number of favorable outcomes}}{\text{the total number of outcomes}}$$

Here, you have two favorable outcomes, HT and TH, out of the four possible outcomes.

Give the children a penny and a nickel to flip, and tell them to record the results of flipping the coins together twenty times. Compile the results from the whole class. Did two heads come up as often as one head and one tail? Does this result match your expectations?

Notice that you are reinforcing the skill of making a table as students record the data from the experiment. You may also ask the children to make a bar graph to show the results of flipping the coins. Does the information on the graph coincide with your expectations based on the tree diagram?

After the children can make a tree diagram to show the possible outcomes from two flips of a coin, ask them to expand it to show the outcomes from three or more flips. Encourage the children to make up other experiments and to use a tree diagram to find all the possible permutations to predict the outcome before performing the experiment. Can they show other results on a tree, such as the suit of two cards drawn or the numbers shown on the roll of two dice? How does this relate to earlier work with tree diagrams and attribute materials? How can probability be used to discuss the chance of randomly choosing a given attribute piece?

Objective: *to extend skills with combinations and permutations.*

2. If we want to know the number of different committees of three we can form from a group of five students, we are looking for the number of combinations. It does not matter who is selected first for the committee. On the other hand, if we are going to pick a chairperson, a secretary, and a treasurer for the committee, we are looking for the number of permutations, because the position of each of the three selections does matter.

Suppose we have three children in the group—Danny (D), Maureen (M), and Amy (A)—and wish to select two of them. We could select Danny and Maureen; this would be considered the same as selecting Maureen and Danny. Listing all the possibilities, we have: DM, DA, and MA.

Are there any other choices? If we were looking for permutations rather than combinations, then the selection of Maureen and then Danny is considered different from Danny and then Maureen. We must add the following to our list to include all the possible permutations: MD, AD, and AM.

What happens if you are choosing a committee of three people out of five? How many combinations do you have now? Use a tree diagram to help find all the possible permutations. (Remember that for permutations, the order is significant.) Simply list the terms in order from the tree to find all the possible permutations. How can you use a tree diagram to find all the combinations? After you have found the permutations, remove any combinations from your list that have the same elements only in a different order as another combination in the list.

Draw a tree diagram to show all the possible three-dip ice cream cones if the choices for each dip are vanilla, strawberry, chocolate chip, and rocky road. Assume that it makes a difference which dip is on the top, which is in the middle, and which is on the bottom. What if the order does not make a difference? Will it affect your tree if you do not want any two dips the same flavor?

Objective: *to distinguish between odds and probability.*

3. Children (and adults) are often confused about the difference between odds and probability. As mentioned earlier, the probability of an event is

$$\frac{\text{the number of favorable outcomes}}{\text{the total number of outcomes}}$$

The **odds** of an event is

$$\frac{\text{the number of favorable outcomes}}{\text{the number of unfavorable outcomes}}$$

For example, on a fair coin the probability of getting a head is $\frac{1}{2}$, but the odds of getting a head is $\frac{1}{1}$. The total number of outcomes in the sample space is two (head or tail) and the total number of unfavorable outcomes is one (the tail). If you randomly draw a marble from a bag containing five green marbles and three red marbles, the probability of getting a green marble is $\frac{5}{8}$, but the odds of getting a green marble is $\frac{5}{3}$. Because the odds of getting a green marble are greater than 1, does that mean you will always choose a green marble?

Ask the children to return to some of the earlier activities for probability and determine the odds as well as the probability of an event. How could you determine the odds if you knew only the probability? Could you determine the probability if you knew only the odds?

Computer programs such as Graphing from MECC can list all the possible outcomes of an experiment as well as simulate events such as coin flips thousands of times. Such listings and simulations can be useful as children practice probability skills. Children may also enjoy writing computer programs for simple probability concepts using the random number generator. Even some calculators contain random number generators.

Graphing Skills

After children have mastered the basic concepts of graphing and are able to decide on the appropriate graphs to use to display different types of data, they should use graphs in all subject areas whenever they have collected or need to interpret data. Much graphing practice may therefore take place in subject areas other than mathematics. As children construct their own graphs, be sure they clearly label the entire graph as well as all of its pertinent parts, such as the two axes on a line graph or every bar on a bar graph.

Children should practice reading as well as constructing graphs. Children should learn not only to read the data on a graph but also to use the data to solve problems.

One area of mathematics in which graphing may be practiced now is graphing equations. This will be a very useful skill later in algebra.

ACTIVITIES

Intermediate–Middle Grades (4–8)

Objective: *to reinforce the skill of plotting points on a Cartesian coordinate graph.*

1. After children have learned to use ordered pairs of numbers to locate points, they may practice by using points to draw a picture. If you put a dot on each of the following points on a Cartesian coordinate system and connect the dots in order, you will make the picture in Figure 11–11: (0,6), (3,4), (4,5), (5,4), (8,6), (7,1), (5,2), (4,1), (3,2), (1,1).

   Have the children draw their own pictures and then list the points for the other children to use.

Objective: *to graph linear equations by plotting points.*

2. Use information children may encounter as consumers or in other subjects as a basis for data to be graphed. For example, you may see an advertisement for corn at 15 cents an ear. Tell the children you want to make a graph to help you find the price of any number of ears. Show the children how to make a table that shows the number of ears of corn purchased and the total price. Figure 11–12 shows a table and the corresponding line graph for the data on the corn.

   In *n* stands for the number of ears of corn and *T* stands for the total price, the equation $T = \$.15n$ can be written to show the relationship between the price and the number of

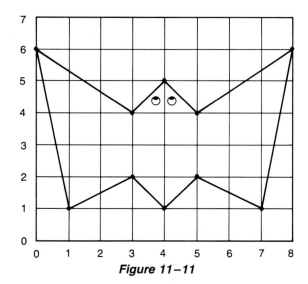

Figure 11–11

ears of corn. A spreadsheet might be used to help students make a table of cost per ear vs. total cost. Students can experiment with changing a variable such as cost per ear or adding other factors such as the prices of other fruits or vegetables. If corn is 15¢ an ear and lettuce is 90¢ a head, how many different ways could you spend exactly $3 on corn and lettuce? A spreadsheet lets you explore this and many related questions. Articles by Edwards and Bitter (1989) and Edwards, Bitter, and Hatfield (1990) give you some ideas to get started.

Suggest that the children make up other graphs of their own. Can they write the equations for each other's graphs if

Corn	Prices
ears of corn	total price
1	$.15
2	$.30
3	$.45
n	$.15 n

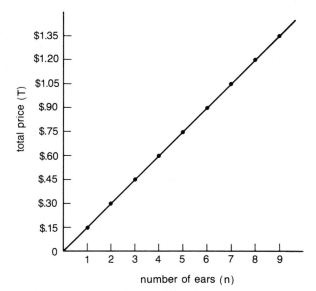

Figure 11–12

they have not seen the original data? Can they add additional pairs to the table by looking at the graph? Ask the children how this work relates to the work with proportions, which were discussed in Chapter 9. Children now have an additional method of solving proportions.

Computer programs such as Graphing Equations from Conduit, Master Grapher from Addison-Wesley, Mathgrapher from Queue, SuperPlot from EduSoft, Exploring Measurement, Time and Money from IBM, and Green Globs and Graphing Equations by Sharon Dugdale and David Kibbey, available from Sunburst, can give the students additional practice in writing equations and graphing. Graphics calculators are also an excellent tool for the exploration of graphing equations. Using a graphics calculator, you can very quickly see the effects of raising the price of an ear of corn to 25¢, or you could plot two graphs to determine the number of ears of corn you could buy for $1.35.

Estimating and Mental Calculating

One of the main uses of probability is to predict future events. In fact, the study of probability began in European gambling halls about 300 years ago in analyzing games of chance to better predict their outcomes. These predictions are based on past events and/or mathematical models and are not just random guesses. The field of probability has grown since then, however, and probability is now used to predict everything from next year's corn crop to population growth in the twenty-first century.

Statistics go hand in hand with probability in predicting future events. Statistics are used to analyze the results of polls and experiments. The results of such an analysis affect, for example, what shows are on television next season, who will be the next president of the United States, and which drugs are approved for fighting cancer. Graphs are often used to display and search for trends in these data.

As part of the activities already mentioned in this chapter, you should frequently ask the children to predict outcomes before performing an experiment. The following are a few other suggestions for reinforcing estimation skills.

ACTIVITIES

Elementary–Middle Grades (2–8)

Objective: *to use experimental probability to predict future events.*

1. For some events, such as flipping a coin or drawing a card from a standard deck, we can find a mathematical probability without ever touching a coin or a card. For other events, such as tossing a tack or a paper cup, we must actually perform the experiment before we can accurately predict what will happen. Give each child a paper cup and ask him or her to mark a spot 12 inches above a table from which to drop the cup. The cup should be dropped bottom down as in Figure 11–13.

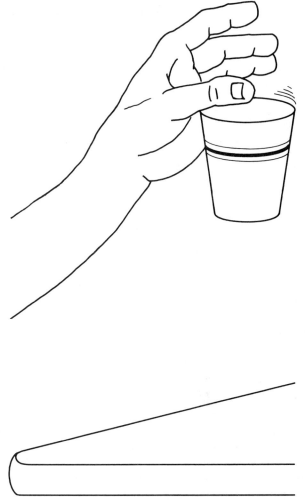

Figure 11–13

Ask the children to describe the possible outcomes in the sample space. Let each child drop the cup twenty times and record the numbers of times the cup lands bottom down, top down, and on its side. Let the children predict the number of times it will land in each of those positions for the next twenty tries, for fifty tries, and for one hundred tries.

Combine the results from the whole class and make predictions. Do the predictions become more accurate as students collect more data? How close are the predictions?

How do scientists use their experiments to make predictions? You may be able to get a parent or another adult in the community whose career involves using experimental data to make predictions to talk to the children.

Objective: *to use opinion polls to make predictions.*

2. Before the next class election, take an opinion poll to try to predict the results. Do you have to poll everyone in the school to make a prediction? What is an unbiased sample? If everyone in the school votes, will your sample be biased if you poll only sixth graders? What if you ask only girls? Do some research to learn about the prediction of the winner in the 1948 presidential race between Dewey and Truman.

Objective: *to use commercial materials to make predictions.*

3. Children enjoy using materials that they see every day to make predictions. It is interesting to determine whether

common materials such as bags of pretzels or chips or boxes of raisins always contain the same number of objects. Give each pair of children a small bag of pretzels and ask them to predict the number of pretzels in the bag before opening it. After the predictions are recorded, ask the children to open the bag and count the number of pretzels. Make a graph to show the number of pretzels in each bag. Predict the mean, the median, and the mode. Are these the same? Compute to determine each one. Which would best describe the "average" number of pretzels in a bag? Does the size of the pretzel make a difference? Repeat the activity with another material, for example, raisins or small candies such as M&Ms. With the M&Ms, you can also predict and determine the most common color. Is the most or least common color the same for every bag? Make a graph to show your findings and make a list of questions that the class would like to explore further. This activity may take place over a month or so and may include writing to the manufacturer for information on how the product is marketed. How do they use statistics and probability to ensure that every bag has at least some minimum amount of food in it?

Simulation games for the computer, such as Oh, Deer! and Lemonade Stand from MECC, can help children learn to use data to make predictions for the future.

Problem Creating and Solving

Probability, statistics, and graphing offer students a number of excellent opportunities to design and carry out their own experiments. Throughout the chapter, we have discussed the importance of having children make up their own problems and questions. Following are a few additional suggestions for teaching students to create and solve problems in these areas.

ACTIVITIES

Elementary–Middle Grades (2–8)

Objective: *to use probability to determine the most likely sum of two dice.*

1. Make a chart like the one in Figure 11–14 and ask the children to keep a tally of the number of times each total comes up when they roll two dice. Combine the tallies from the whole class. Which total came up the most often? The least often? How would you explain these results?

    Tell the children to use a tree diagram to predict the most common total. Make a tree such as the one in Figure 11–15 to find all the possible permutations of two dice.

    List the ordered pairs at the ends of the branches and find the totals. Make a table to show the number of times each total came up. How does this table compare to the actual totals found in your experiment?

    ● *Understanding the problem.* Each member of the class has already rolled a pair of dice two times, so we have fifty-two rolls altogether. And the chart based on our tree diagram

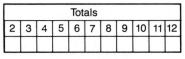

Totals										
2	3	4	5	6	7	8	9	10	11	12

Figure 11–14

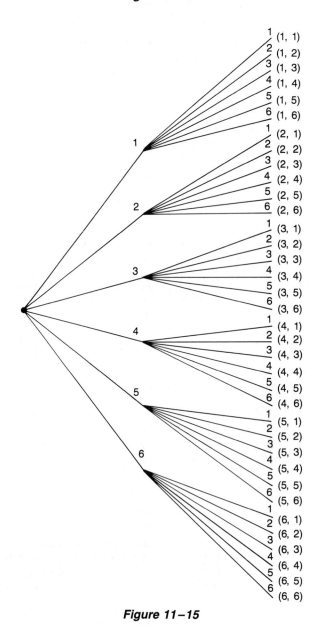

Figure 11–15

shows what would come up most often if everything were perfect. We need to compare the charts to see what is alike and what is different.

When we rolled the dice, we got fifty-two answers. Will that make a difference when we compare? We can tell which numbers come up most often and then next most often. Or we can change the numbers to fractions or percents and then compare them.

● *Devising a plan.* To compare the different numbers that came up most often, we'll make a small table with two rows. In one row, we'll put the percent of times each number came up when we threw the dice. In the next row, we'll

WHICH NUMBER CAME UP MOST OFTEN?											
	2	*3*	*4*	*5*	*6*	*7*	*8*	*9*	*10*	*11*	*12*
Dice	4%	6%	12%	10%	6%	23%	15%	8%	2%	13%	2%
Tree	3%	6%	8%	11%	14%	17%	14%	11%	8%	6%	3%

put the percent of times the number appeared on the tree. Then we can compare the percents (make a systematic list).

- *Carrying out the plan.* We count how many times each number was rolled by the class, and Darlene and Ben use their calculators to find the percent each number represents. We put that information in the table. Then Candy and Jackie calculate the percents from the tree diagram. We put the results in the table, as shown above.

    Now we can compare the results. The first thing we observe is that the number 7 was rolled most often (23%) and also appears on the tree diagram most often (17%). Now we can make several other important observations.

- *Looking back.* We were able to find that 7 was the most likely sum of two dice. We found even more information because the table was systematic.

Have the children discuss the other findings from the table. What were the least frequent totals? How many times would you expect to get a total of 7 out of one hundred rolls? Would you expect to get a total of 2 very often? Why?

Encourage the children to devise other experiments. Use two spinners, each numbered 1, 2, and 3, or keep track of cards drawn out of two hands that each have cards numbered 1, 2, 3, and 4. Tell children to draw tree diagrams to show all the possible permutations and to use these permutations to predict results for their experiments. How close are the predictions? Encourage students to discuss their strategies as they carry out the experiments.

Objective: to use statistics to persuade an audience.

2. Have the children select a problem that directly affects them or the school. This may be a hazardous street in front of the school, lack of appealing food in the cafeteria, or a need for supervised activities after school.

    Help the children develop an instrument to collect data to support their case, such as a method of clocking and counting the traffic passing the school, surveys to determine the need for some change in the cafeteria menu, or a method of counting children's after-school activities. Discuss finding a random sampling of times for the traffic count or a random sampling of students and adults to participate in the survey. Discuss the best ways to organize and display the data collected. Help the children present their data to those responsible for making changes, such as members of the traffic bureau or the head cook in the cafeteria.

    Claus (1989) reported on a statistics project undertaken by her fourth grade class. They polled the entire school on cafeteria preferences such as fruit bars vs. ice cream and hot vs. cold lunches. They presented results to the principal, the PTA, and the school nurse, who used the information to make changes. The use of data collection and statistical interpretation gives students a powerful tool for change.

Objective: to create questions for other children to answer by looking at graphs.

3. Graphs are often presented along with a ready-made list of questions to be answered. Turn the tables by asking the children to write the questions. Tell them that the only rule is that it must be possible to answer the questions using the information found on the graph. Encourage children to think of questions that no one else in the room will think of.

    Structure some questions by specifying an operation or operations that must be used in the solution. Play a game like the television show "Jeopardy," in which you give an answer and ask the children to come up with the question. Keep the questions in the learning center for the children to work on later.

Grouping Students for Learning Probability, Statistics, and Graphing

The topics of probability, statistics, and graphing offer a natural opportunity for cooperative learning groups to explore real phenomena in their environments. Russell and Friel (1989) point out that leading elementary textbook series generally present work in data analysis as exercises requiring students to find information on a table or graph that has already been constructed for them. The textbooks rarely ask students to interpret or collect data. They note that the use of real data requires that students select the problems and questions they wish to explore, collect the data, work with a range of different data representations, including those that make use of technology, and learn to analyze data that is frequently "messy," with either too much or too little information that may or may not answer the questions that the students have asked. After the data have been collected and analyzed, students learn to use data not only to answer previous questions, but also to generate new ones. In their Used Numbers Project, Russell and Friel (1989) developed a number of curriculum modules to help K–6 students learn how numbers can be used to describe, interpret, and make predictions about real phenomena. A typical activity begins with the teacher or a student presenting some information to the whole class about an area of interest, such as the average family size in the community. Students then discuss questions that they are interested in exploring and break into small groups to come up with ways to investigate the topic. The teacher circulates to give help as needed, and students can move from whole group, to small group, to individual work in the exploration of the problem. Many of the activities presented in this chapter can be done in this same manner.

Communicating in Learning Probability, Statistics, and Graphing

By the very nature of the topics, this is an excellent area for students to use concrete or visual models to demonstrate findings. Students who have taken a survey to find the answer to a question of interest to the class, can use concrete or visual graphs, tables, or charts to display the information they have found. This is an excellent example of information that should be communicated to a larger audience, not just to classmates. Surveys of interest to the children often pertain to the policy decisions of others, such as the principal, the mayor or the city council, the PTA, your congressperson, or the cafeteria manager. Students should be encouraged to write reports of their findings and conclusions and to illustrate their reports liberally with charts and graphs. They may then present the reports to the appropriate parties. Local newspapers or local cable television companies may be willing to publish or broadcast the results if they are approached by a committee of the students. This is an excellent way to integrate mathematics with other areas of the curriculum such as social studies and language arts and to empower the students at the same time.

Evaluating Learning of Probability, Statistics, and Graphing

Most of the evaluation in the areas of probability and statistics should be informal. Standardized tests for elementary children may include reading graphs and finding measures of central tendency, but they may not include questions on probability and statistics. This does not mean that probability and statistics are unimportant, however. A good foundation in the basic concepts of these topics will greatly help the children as they progress in school.

Piaget and Inhelder (1975) studied the development of the idea of chance in children and found that children's thinking in this area can be divided into three stages. In the first stage (up until about age 7 or 8), children do not understand the concept of randomness. They look for some hidden order and make predictions based on their own preferences or on the misconception that an outcome should "catch up" to the others. Very unlikely events do not surprise them.

In the second stage (from about age 7 or 8 until about age 11 or 12), a broad understanding of randomness is achieved, but children do not understand the effects of large numbers. Very unlikely events do surprise them and cause them to look for a reason. In the third stage, that of formal reasoning, children understand the effect of large numbers and can assign numerical probabilities.

In a review of a number of studies conducted after the Piaget and Inhelder research, Shulte (1987, p. 32) concluded that "the research indicates that students have some understanding of probability and related topics, that this understanding increases with age and instruction, and that probability can successfully be taught in the elementary school in carefully selected experiments." He recommends that probability be included among the topics presented in elementary school.

Teachers should be aware of some of the misconceptions that may arise in the area of probability as well as in the areas of graphing and statistics. The following list gives some of these misconceptions:

1. Children may believe that a mathematical probability should give the exact outcome of an experiment. They may believe that if the probability of getting a head is $\frac{1}{2}$, then they should always get ten heads out of twenty flips of a coin. Be sure the children have many opportunities to actually perform experiments to see the differences between mathematical and experimental probabilities.

2. Children may believe that the instrument used in a probability experiment has a memory. They may believe that if they have gotten five heads in a row with a coin, then it is more likely that a tail will come up on the next flip. Discuss whether the coin has any memory of what came up on the previous flips. The probability of getting a tail remains the same, no matter what came up previously.

3. Children may hold a bias against or in favor of a particular outcome in an experiment. They may feel that 3 should come up most often on a die because it is their favorite number or that 6 should not come up very often because they do not like it. The activities suggested in this chapter in which the children are asked to tally a large number of experimental results should be discussed with the children to see if the results match their expectations.

4. Children may expect all outcomes to be equally likely, even if they are not. They may think that a total of 2 on two dice should be just as likely as a total of 7. Children should discuss the activity relating the roll of two dice to the tree diagram described in the problem-solving section of this chapter.

5. Children may wish to overgeneralize from their own experiments or surveys. If they take a poll of all the fifth graders in their school, they may wish to say that the results would be the same all over the country. Discuss with them what is wrong with such thinking. Making correct inferences from data is a skill that takes time and experience to develop.

6. Children may confuse the different measures of central tendency. They should be given a number of chances to find the mean, the median, and the mode and to discuss which measure is used in the articles they read and which is most appropriate for a given situation.

7. Children may try to use an inappropriate graph to display data. Discuss with them why a line graph is not good for showing favorite colors. Does the line from blue to green have any meaning? What is a more appropriate type of graph?

8. Children may forget to completely label a graph. How can they interpret a graph if some of the labels are missing?

9. Children may use pictures of different sizes on a picture graph, bars of different widths on a bar graph, or spaces of different sizes between numbers on a line graph. How does this affect the interpretation?

When testing children's understanding of these topics, be sure to observe their work during experiments and while they are developing surveys. Do not rely on paper and pencil instruments. You may wish to use an instrument such as "How Many Questions?" (Jensen, 1973), which tests children's ability to ask questions involving graphs rather than their ability to answer such questions. This will give you an idea of how well the children understand what graphs may be used for and how creative the children can be in posing questions.

Encourage children to keep records of their work and to write down any questions they have as they progress. Children often are excellent judges of how well they understand the lesson.

Something for Everyone

Many of the topics discussed in this chapter are particularly appropriate for gifted students. Some gifted students may seem to prefer working on abstract problems and may not use concrete materials, even though using these materials may be of benefit to them. Probability and statistics give gifted children an opportunity to be involved with concrete models and simulations that have direct applications in many fields of interest to them, such as future careers in business and the professions.

Gifted students can further explore topics in probability and statistics, such as the counting processes for finding the number of possible outcomes and methods for finding the probability of a statement involving *and* and *or*. Use counting processes to find the number of five-digit zip codes or seven-digit phone numbers. Encourage the students to develop their own rules for finding the number of possible combinations, with or without repetition. How can you find the number of ways Event A *and* Event B can happen? Is this the same as finding the number of ways Event A *or* Event B can happen?

Compare this work to the work with Venn diagrams using union and intersection (discussed in Chapter 5). Pascal's triangle (which was introduced in Chapter 6) also has interesting applications for finding the number of possible combinations of anything with two possibilities for each move. Encourage gifted children to make connections to earlier learning whenever possible.

The topic of graphing offers opportunities for learning to suit many varied learning styles. Tactile/kinesthetic children learn well when they have the opportunity to make graphs out of the objects themselves. Visual children can learn well using any type of graph. Graphs suit their style of forming a mental image to fit the data collected. Auditory children enjoy the discussions that accompany collecting data and analyzing the graphs.

Use graphing in conjunction with other subjects, such as science, reading, and social studies, to help children learn those subjects as well. Many of the concepts from this chapter do not need to be a separate unit in mathematics. Use opportunities throughout the school day to help children learn the concepts and their uses in the world surrounding them.

KEY IDEAS

The study of mathematics is changing with the increased use of technology in the world. In this chapter, we discussed three topics that are currently receiving more attention in the elementary school—probability, statistics, and graphing. We described relationships among these three topics, as well as their integration into other areas of the curriculum.

In the elementary grades, teach these topics concretely, using materials such as dice, coins, cards, colored cubes, chips, spinners, graph paper, squares, and a variety of concrete materials for the graphs. Fully utilize calculators and computers to help children collect and organize large amounts of data. Emphasize helping children understand the concepts, not rote memorization of terms.

Much of the skill work in probability and statistics will come in the later grades, although children should become fairly skillful at making and reading graphs of all types by the end of elementary school. The topics in this chapter should help children become more skillful at predicting future events. Use these topics as children devise and plan their own experiments or surveys. Evaluate informally, and take into account developmental levels as well as individual learning styles and abilities.

REFERENCES

BARRETT, GLORIA, AND GOEBEL, JOHN. "The Impact of Graphing Calculators on the Teaching and Learning of Mathematics." In Cooney, Thomas J., and Hirsch, Christian R., eds. *Teaching and Learning Mathematics in the 1990s.* Reston, Va.: National Council of Teachers of Mathematics, 1990, pp. 205–211.

BESTGEN, BARBARA J. "Making and Interpreting Graphs and Tables: Results and Implications from National Assessment." *Arithmetic Teacher.* Vol. 28, No. 4 (December 1980), pp. 26–29.

BRUNI, JAMES V., AND SILVERMAN, HELENE J. "Developing Concepts in Probability and Statistics—and Much More." *Arithmetic Teacher.* Vol. 33, No. 6 (February 1986), pp. 34–37.

CLAUS, ALLISON. "Making Mathematics Come Alive Through a Statistics Project." In Trafton, Paul R., and Shulte, Albert P., eds. *New Directions for Elementary School Mathematics*. Reston, Va.: National Council of Teachers of Mathematics, 1989, pp. 177–182.

CONFERENCE BOARD OF THE MATHEMATICAL SCIENCES. *The Mathematical Sciences Curriculum K–12: What Is Still Fundamental and What Is Not?* Washington, D.C.: CBMS, 1983.

COOMBS, BETTY, AND HARCOURT, LALIE. *Explorations 1.* Don Mills, Ontario: Addison-Wesley Publishing Co., 1986, p. 183.

CORWIN, REBECCA B., AND FRIEL, SUSAN N. *Used Numbers: Statistics: Predicting and Sampling.* Palo Alto, Ca.: Dale Seymour, 1990.

CORWIN, REBECCA B., AND RUSSELL, SUSAN JO. *Used Numbers: Measuring: From Paces to Feet.* Palo Alto, Ca.: Dale Seymour, 1990.

DICKINSON, J. CRAIG. "Gather, Organize, Display: Mathematics for the Information Society." *Arithmetic Teacher.* Vol. 34, No. 4 (December 1986), pp. 12–15.

DIFAZIO, MARTHA HUNT. "Graphics Software Side by Side." *Mathematics Teacher.* Vol. 83, No. 6 (September 1990), pp. 436–446.

DOSSEY, JOHN A.; MULLIS, INA V. S.; LINDQUIST, MARY M.; AND CHAMBERS, DONALD L. *The Mathematics Report Card: Are We Measuring Up? Trends and Achievement Based on the 1986 National Assessment.* Princeton, N.J.: Educational Testing Service, 1988.

EDWARDS, NANCY TANNER, AND BITTER, GARY G. "Teaching Mathematics with Technology: Changing Variables Using Spreadsheet Templates." *Arithmetic Teacher.* Vol. 37, No. 2 (October 1989), pp. 40–44.

EDWARDS, NANCY TANNER; BITTER, GARY G.; AND HATFIELD, MARY M. "Teaching Mathematics with Technology: Data Base and Spreadsheet Templates with Public Domain Software." *Arithmetic Teacher.* Vol. 37, No. 8 (April 1990), pp. 52–55.

FRIEL, SUSAN N.; MOKROS, JANICE R., AND RUSSELL, SUSAN JO. *Used Numbers: Statistics: Middles, Means and In-Betweens.* Palo Alto, Ca.: Dale Seymour, 1992.

JENSEN, LINDA. "The Relationships Among Mathematical Creativity, Numerical Aptitude and Mathematical Achievement." Unpublished doctoral dissertation. Austin, Tx.: The University of Texas at Austin, 1973.

NATIONAL COUNCIL OF TEACHERS OF MATHEMATICS. *Curriculum and Evaluation Standards for School Mathematics.* Reston, Va.: NCTM, 1989.

NUFFIELD FOUNDATION. *Pictorial Representation.* New York: John Wiley & Sons, 1967.

_____ . *Probability and Statistics.* New York: John Wiley & Sons, 1969.

PIAGET, JEAN, AND INHELDER, BARBEL. *The Origin of the Idea of Chance in Children.* New York: W. W. Norton & Co., 1975.

RUSSELL, SUSAN JO, AND FRIEL, SUSAN N. "Collecting and Analyzing Real Data in the Elementary School Classroom." In Trafton, Paul R., and Shulte, Albert P., eds. *New Directions for Elementary School Mathematics.* Reston, Va.: National Council of Teachers of Mathematics, 1989, pp. 134–148.

RUSSELL, SUSAN JO, AND STONE, ANTONIA. *Used Numbers: Counting: Ourselves and Our Families.* Palo Alto, Ca.: Dale Seymour, 1990.

SANDEFUR, JAMES T., JR. "Discrete Mathematics: A Unified Approach." In *The Secondary School Mathematics Curriculum.* National Council of Teachers of Mathematics, 1985 Yearbook. Reston, Va.: NCTM, 1985.

SCHOOL MATHEMATICS STUDY GROUP. *Probability for the Intermediate Grades.* Stanford, Ca.: SMSG, Stanford University, 1966.

_____ . *Probability for the Primary Grades.* Stanford, Ca.: SMSG, Stanford University, 1966.

SHIELACK, JANE F. "Teaching Mathematics with Technology: A Graphing Tool for the Primary Grades." *Arithmetic Teacher.* Vol. 38, No. 2 (October 1990), pp. 40–43.

SHULTE, ALBERT P. "Learning Probability Concepts in Elementary School Mathematics." *Arithmetic Teacher.* Vol. 34, No. 5 (January 1987), pp. 32–33.

SHULTE, ALBERT P., ED. *Teaching Statistics and Probability.* National Council of Teachers of Mathematics, 1981 Yearbook. Reston, Va.: NCTM, 1981.

SHULTE, ALBERT, AND CHOATE, STUART. *What Are My Chances? Books A and B.* Palo Alto, Ca.: Creative Publications, 1977.

Most of our buildings and decorations are based on geometric forms. Also, much of nature can be described in geometric terms, and, this accounts, in part, for the origin of geometry. The work of Babylonian astronomers and Egyptian surveyors became the beginnings of mathematics. It is appropriate, then, to help children recognize the geometry that surrounds them.

The environments most familiar to children are those of the home, neighborhood, and school. By and large, the objects in these environments are the products of human effort. The products of nature are evident, as well, and provide rich, intriguing objects of study. Once children are made aware of various shapes and geometric forms, they will find them everywhere. The patterns and forms in nature may not be as obvious but will capture children's interests for long periods of time. Stevens noted that:

> [W]hen we see how the branching of trees resembles the branching of arteries and the branching of rivers, how crystal grains look like soap bubbles and the plates of a tortoise's shell, how the fiddleheads of ferns, stellar galaxies, and water emptying from the bathtub spiral in a similar manner, then we cannot help but wonder why nature uses only a few kindred forms in so many different contexts. Why do meandering snakes, meandering rivers, and loops of string adopt the same pattern, and why do cracks in mud and markings on a giraffe arrange themselves like films in a froth of bubbles? (1974, p. 3)

Children's awareness of geometry in the environment is heightened considerably as teachers focus their attention on various applications of geometry. This awareness also strengthens students' appreciation for and understanding of geometry and helps develop students' spatial sense.

The foundations for learning geometry lie in informal experiences in the elementary school. These experiences should be carefully planned and structured to provide youngsters with a variety of concepts and skills. These concepts and skills serve as a basis for later, more formal work in geometry. That is why it is important to provide extensive, systematic exposure to geometric ideas from kindergarten through grade 8.

Infants explore space initially by thrashing about in a crib or playpen and crawling toward objects or open doors. Children discover that some objects are close, while others are far. They discover that rooms have boundaries, and that sometimes, if a door is left open, the boundaries can be crossed. They discover that certain items belong inside boundaries—for example, father's nose belongs within the boundaries of his face, or the bathtub belongs within the confines of the bathroom.

Children also discover that events occur in a sequence or an order. Early in their lives, they learned that their own crying was often followed by the appearance of a parent, who then attended to their needs. Later, children notice that a stacking toy is put together by putting certain parts in a particular order.

These examples illustrate children's initial experiences in space. They are far removed from school experiences with geometric shapes but nonetheless help show how children discover spatial relationships. Children learn first about the common objects in their environments. Piaget (1967) found that young children view space from a topological perspective. For example, shapes are not seen as rigid; they may readily change as they are moved about. Later, children use projective viewpoints as they make the transition to a Euclidean point of view. Shadows provide an example of projective geometry. In projective geometry, distances and dimensions are not conserved, but the relative positions of parts of figures and the positions of figures relative to one another are conserved. Employing projective viewpoints helps children, by ages 5 to 7, to begin to perceive space from a Euclidean point of view, from which they see shapes as rigid—the shapes do not change as they are moved about.

Elementary school geometry plays an important developmental role as children learn geometry. Introduced in the United States by Wirszup (1976), the work of P. M. van Hiele described five levels through which individuals pass as they learn to work comfortably in the most abstract geometries. These are:

- *Level 1: Recognition.* Children learn to recognize various shapes after repeatedly seeing them as separate objects. Children do not notice the common characteristics of similar figures.
- *Level 2: Description.* Children observe and manipulate figures, thus determining the properties necessary for identifying various shapes. Measuring is one way children learn the necessary properties.
- *Level 3: Relation.* Children establish relationships between figures and their properties. Children understand that a square is also a rectangle, a parallelogram, and a quadrilateral. The importance of definitions is recognized.
- *Level 4: Deduction.* Children use deduction while working with postulates, theorems, and proof.
- *Level 5: Rigor.* Children employ rigorous applications in their study of various geometries.

Most high school geometry courses begin work at the fourth level. Burger (1985) noted, however, that many high school students are working at the levels of younger children—Levels 1, 2, and 3. Thus, teachers and students may have difficulty understanding each other. It is important, therefore, for the elementary school mathematics program to provide informal geometry experiences to help children progress through the first, second, and third levels. The activities suggested in this chapter illustrate the types of geometric experiences necessary to assist children through the early van Hiele levels.

There are many manipulative materials available that enhance the direct instruction of teachers. Among those that the authors recommend in teaching geometry are: pattern blocks, geoblocks, geoboards, mira, paper models, and Logo (the computer language of turtle graphics). These and other useful materials are described and illustrated as they are presented in this chapter.

Geometry also serves as an instructional medium in its own right. Geometric models are used to introduce and illustrate a variety of mathematical topics. For example, geometric models are used to illustrate algorithms in Chapters 7 and 8 and geometric models are used to illustrate the concept of fractions in Chapter 9. Visualizing mathematics through models is well established as a teaching method. Materials such as *Math and the Mind's Eye* (1987) have been designed for use in grades 4–9 to help students develop their visual thinking. You are encouraged to explore these and other materials that employ geometry to model other mathematical topics.

Developing Geometric Concepts

The concepts upon which geometry is built begin with the simplest figure, the point, and expand to lines, line segments, rays, curves, plane figures, and space figures. We briefly discuss each of these.

The **point**, like all geometric figures, is an abstract idea. A point has no dimensions. It may be thought of as a location in space. For example, the tip of a pencil, the corner of a table, or a dot on a sheet of paper can represent a point.

A **line** is determined by two points and consists of a set of points connecting the two points and continuing endlessly in both directions. Figure 12–1a represents the line AB, defined by the points A and B.

Line segments and **rays** are subsets of a line. Like the line, each is determined by two points. The line segment, however, has two end points and the ray has only one end point. Line segment AB in Figure 12–1b is described by the two points A and B. Ray AB in Figure 12–1c includes end point A and a set of points continuing endlessly beyond point B. The arrowhead indicates the direction of a ray.

Lines, line segments, and rays have one dimension, length. When three or more points are not on the same line, a different kind of geometric figure results. It is a **plane figure** or a figure in two dimensions. Figures such

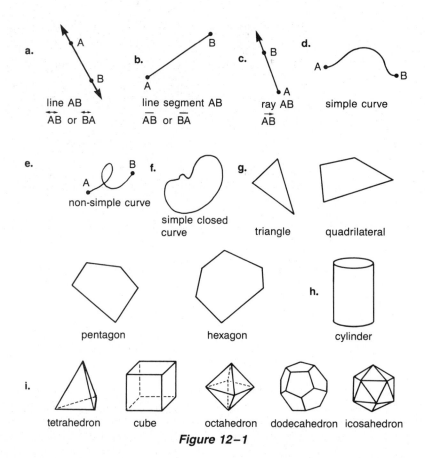

a. line AB
$\overleftrightarrow{AB}$ or $\overleftrightarrow{BA}$

b. line segment AB
$\overline{AB}$ or $\overline{BA}$

c. ray AB
$\overrightarrow{AB}$

d. simple curve

e. non-simple curve

f. simple closed curve

g. triangle   quadrilateral

pentagon   hexagon   h. cylinder

i. tetrahedron   cube   octahedron   dodecahedron   icosahedron

Figure 12–1

as angles (the union of two rays) and triangles (the union of three segments) are plane figures. We now consider curves and other plane figures.

A **curve** is a set of points that can be traced on paper without lifting the pencil. Figure 12–1d shows a simple curve between points A and B. It is **simple** because it does not cross over itself as it is drawn. The curve in Figure 12–1e is not simple because it crosses over itself as it is drawn from point A to point B. These two curves are not closed because they both have end points. When a curve has no end points, it is a **closed curve**. Figure 12–1f illustrates a simple closed curve.

Plane figures that are simple closed curves formed by joining line segments are called **polygons**. A polygon is named by the number of segments joined to make it. There are **triangles** (3 sides), **quadrilaterals** (4 sides), **pentagons** (5 sides), **hexagons** (6 sides), and so on. Figure 12–1g shows several polygons. A common simple closed curve not formed by joining line segments is the **circle.**

A polygon may have certain properties that provide a more specific description. For example, a **regular figure**, such as a square or an equilateral triangle, has sides that are the same length and angles of the same measure. Having sides that are parallel and having right angles are other descriptive characteristics of plane figures. A **square** is a quadrilateral with all sides the same length and all angles the same size. A **rectangle** is a quadrilateral

with opposite sides parallel and the same length, and all angles the same size. A **parallelogram** is a quadrilateral with opposite sides parallel and the same length. A **rhombus** is a quadrilateral with opposite sides parallel and all sides the same length. A rhombus is sometimes called a diamond.

A **space figure** is one that does not lie wholly in a plane. A soup can represents one such figure, called a **cylinder,** shown in Figure 12–1h. Other space figures include spheres, pyramids, prisms, and cones. The playground ball serves as a model of a **sphere,** the set of all points in space equidistant from a given point. A **pyramid** is a figure with a base the shape of a polygon and sloping triangular sides that meet at a common vertex. A **prism** is a figure whose ends are congruent polygons and parallel with each other, and whose sides are parallelograms. A **cone** is a figure with a circular base and a curved surface that tapers to a point.

Polyhedra are space figures that have four or more plane surfaces. **Regular polyhedra** are those in which each face is a regular polygon of the same size and shape, and in which the same number of edges join at each corner or vertex. There are only five regular polyhedra: the **tetrahedron** (4 faces), the **cube** (6 faces), the **octahedron** (8 faces), the **dodecahedron** (12 faces), and the **icosahedron** (20 faces). These are shown in Figure 12–1i.

The geometric concepts described above form a major part of the elementary mathematics textbook presentation of geometry. How these ideas are presented to children is of concern. Suydam (1985, p. 26) noted that "for many children, instruction in geometry in the elementary school revolves around only two points: recognition of shapes and development of vocabulary." She goes on to say that working with concrete materials and illustrations is valuable in learning geometric ideas. It is these activities that help develop spatial sense and expand the content of geometry. In the NCTM *Standards* it is noted that:

> Spatial sense is an intuitive feel for one's surroundings and the objects in them. To develop spatial sense, children must have many experiences that focus on geometric relationships; the direction, orientation, and perspectives of objects in space; the relative shapes and sizes of figures and objects; and how a change in shape relates to a change in size. . . . When children examine the result of combining two shapes to form a new shape, predict the effect of changing the number of sides of a shape, draw a shape after it has been rotated a quarter or half turn, or explore what happens when the dimensions of a shape are changed, they acquire a deeper understanding of shapes and their properties. (1989, p. 49)

The following development of geometric concepts expands on the elementary textbook presentation of recognition of shapes and definition of terms. We begin with a description of the views young children have of the world when they enter school, followed by activities introducing plane figures and their properties, symmetry and motion, and space figures and their properties.

Young Children's Views of the World

The perceptions of children before they are five to seven years old are topological. **Topology** is the study of space concerned with position or location, where length and shape may be altered without affecting a figure's basic property of being open or closed. For example, a five year old shown a triangle and asked to make several copies of it may draw several simple closed curves but not necessarily triangles, as in Figure 12–2a. To the child, all of the drawings are the same, because the child perceives that the triangle has only the property of being closed (younger children often draw figures that are not closed). As well, a triangle may be stretched into any closed figure, as in Figure 12–2b (Copeland, 1984, p. 216).

The study of space in which a figure or any enclosed space must remain rigid or unchanged is called **Euclidean geometry.** The historical development of geometry was Euclidean; that is, geometry developed from ideas such as points, lines, and polygons. Some of Piaget's research has implied that children do not develop geometric concepts in a Euclidean manner. Because of their topological perspectives, children need an active, exploratory period when they enter school (1953, p. 75).

In Chapters 5 and 6, relationships among objects and numbers were discussed as the concept of number was developed. Likewise, spatial relationships can be identified as the concepts associated with space are developed. Children who perceive the world from a topological point of view are developing an understanding of four basic relationships:

1. *Is close to* or *is far from*
2. *Is a part of* or *is not a part of*
3. *Comes before* or *comes after*
4. *Is inside of, is outside of,* or *is on*

During kindergarten and first grade, children develop to the point where they can understand the meaning of Euclidean space. That is, children develop their abilities to reproduce shapes without significantly altering the characteristics of those shapes. For example, in the earlier topological stage, children copy a figure but allow corners to become round and distances to change. At the stage of Euclidean understanding, corners remain corners and distances are unchanged—the figure is considered rigid.

The shift from topological to Euclidean thinking is not sudden. It may occur over a period of two years. Thus, usually between the ages of four and six, children can recognize and name the more common figures: square, triangle, rectangle, circle. Other figures are neither identified nor differentiated from these shapes. For example, the square and other rhombi may be confused, as may the rectangle and other parallelograms. Even more difficult for children is copying various shapes from blocks or drawings. Children may be able to accurately identify shapes long before they are able to produce their own examples.

During kindergarten and first grade, it is important to continue activities that relate to topological space. The following are typical activities that extend topological ideas.

ACTIVITIES

Primary (K–1)

Objective: *to develop and reinforce the concepts of near, far, on, in, under, over, inside, and outside.*

1. Developing language in concert with activities is a natural part of teaching. Have children sit in small groups at tables on which numerous objects are placed. Give directions to various children. For example, "Julia, please put the red block as far away from the plastic cup as you can" or "David, please

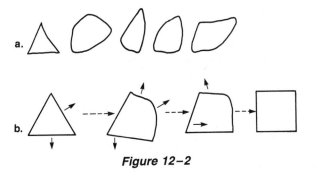

Figure 12–2

put the short pencil in the tin can." Several children may participate simultaneously. Check the understanding of the language and the concept. Encourage discussion.

2. Draw three regions on the playground or on the floor of the multi-purpose room. The regions represent a red base, a green base, and a catchers' region. Select two groups of children: those who attempt to change from the red base to the green base when a whistle is blown and those who begin at the catchers' region. As the children are changing from the red to the green base, the catchers run from their region and tag those who are changing.

The catchers may tag the changers as long as they are outside of both the red and green bases. Once the changers reach the green base, they try to return to the red base. They continue running back and forth between bases as long as possible. Children who are tagged join the catchers. The game is over whenever there are no more children to run between the red and green bases.

Children participating in this activity are concerned about being inside or outside of the various regions. Occasionally during the activity, have the children freeze. Then instruct, "Raise your hand if you are inside the green region. Raise your hand if you are outside the green region. Raise your hand if you are inside the red region. Raise your hand if you are outside the red region. Raise your hand if you are outside both the red and green regions."

3. Construct the following activity on the playground or on paper. Put drawings such as those in Figure 12–3 on the ground and invite the children to stand *inside* and to see if they can get to the outside by walking. There is one rule: you cannot step over a boundary line.

Students unable to get outside are inside a closed curve. All other students are outside the closed curve or are stand-

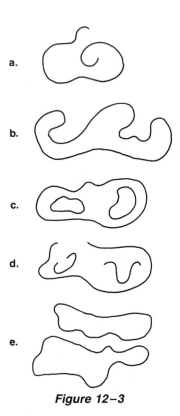

Figure 12–3

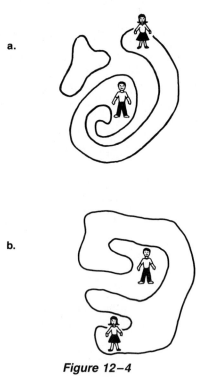

Figure 12–4

ing on the curve itself, or else the region is not closed. Have the children experiment with several curves until they can easily determine if they are inside or outside a region, or if there is a closed region at all.

A simple curve like that in Figure 12–3a does not divide the plane in which it is drawn. Thus, only one region exists, whereas in Figure 12–3b, two regions exist because the simple closed curve separates the plane into two regions. In Figure 12–3c, there are four regions and the curve itself. The region outside the figure is counted. Figure 12–3d shows one region; Figure 12–3e shows three regions.

If these activities are performed on paper, the children may benefit from coloring each region a different color. Devise variations of this sort of boundary exercise.

4. Another type of boundary activity is the maze. The object of this activity is to see if two children are in the same region. On the playground, the children attempt to walk to one another without walking on or across a boundary. On paper, have children trace the regions with their fingers. The variations and the complexity of these designs are nearly unlimited. Figure 12–4 provides two examples of simple mazes. The children in Figure 12–4a are able to walk to each other because they are in the same region. In Figure 12–4b, the children cannot reach each other because they are in different regions.

5. A third, more complex boundary activity involves having children construct maze puzzles for themselves and other children. Maze puzzles may be constructed by beginning with a simple frame with a door to go in and a door to go out, as in Figure 12–5.

To complete the maze, draw lines from any wall. The only rule is that no line can connect one wall with another wall. Steps a, b, and c in Figure 12–6 show how a maze puzzle was constructed. Children are fascinated by the construction of mazes.

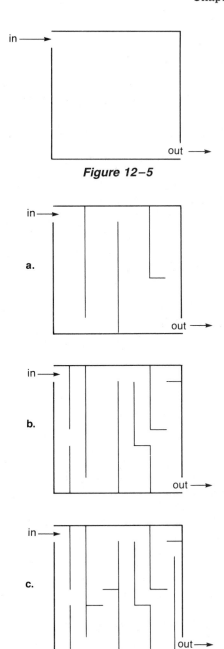

Figure 12-5

Figure 12-6

Objective: *to develop the ability to verbalize about geometric figures and patterns.*

6. Encourage children to construct and manipulate space figures. Materials may include tiles, attribute blocks, geoblocks, cubes, cans, empty milk cartons, Unifix cubes, Cuisenaire rods, pattern blocks, parquetry blocks, and clay. **Geoblocks** are pieces of unfinished hardwood, cut into a wide variety of space figures. Allow the children to talk with one another as they work. During that time, circulate and ask individuals, "Tell me what your picture shows. Can you find another shape like this one? How would you describe this piece? How are the buildings the same?"

   Children can learn to be analytical when questions are carefully phrased. For example, "Can you make another house just like the one you have made there? I would like you

to try." At the same time, the questions can serve to gather information for the teacher. Be sure to allow children to explain an answer.

Objective: *to use visual clues in matching shapes.*

7. Encourage children to construct picture jigsaw puzzles. Challenge the students with difficult puzzles, and discuss informally with an individual or a small group how they have gone about putting the puzzle together. It should be evident that strategies are developed as puzzles are completed. Edge pieces are generally put together first, followed by pieces that form distinct images or those that have easily matched colors. Pieces are added to the puzzle when their shapes fit a region that has been surrounded by other pieces. Finally, all other pieces are put into place by the process of elimination.

The preceding activities have been presented to help reinforce the topological ideas of youngsters. They serve as preparation for the following activities, which help introduce children to the Euclidean shapes.

Projective Geometry

As children investigate figures and their properties through shadow geometry, they are being assisted in the transition from a topological perspective of their world to a Euclidean perspective. Piaget (1967, p. 467) noted that "Projective concepts take account, not only of internal topological relationships, but also of the shapes of figures, their relative positions and apparent distances, though always in relation to a specific point of view." Children explore what happens to shapes held in front of a point source of light, such as a slide projector or a bright flashlight. They also explore what happens to shapes held in the sunlight when the sun's rays are nearly parallel. They discover which characteristics of the shapes are maintained under varying conditions. Children need to make observations, sketch the results of their work, and discuss their observations. As a result, children develop a viewpoint that is not part of a topological perspective.

The activities that follow are intended to provide children with experience with projective geometry.

ACTIVITIES

Primary (1–3)

Objective: *to produce and describe the shadows of squares and other shapes, using the sun as a source of light.*

1. Provide pairs of children with square regions such as wooden or plastic geoboards or regions cut from railroad board. Take the children to an area of the playground that has a flat, smooth surface such as blacktop or concrete. Have the children hold the square regions so that shadows are cast on the ground, as in Figure 12–7.

   Encourage the children to move the square regions so that the shadow changes. Be sure both members of a pair have a chance to experiment with shadow-making. After a few minutes, gather the children around you and ask them to talk

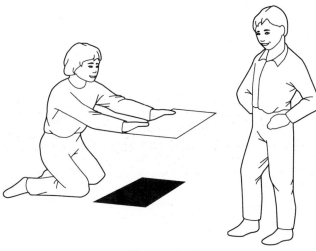

Figure 12–7

about the shadows they found as they moved their square regions. If it is difficult for a child to explain the shape of the shadow, have the child illustrate the shadow for the others. Have the children discuss how they were able to make the shapes larger and smaller. See what other observations they have made.

To make a permanent record of shapes, have one member of each pair of children put a piece of paper on the ground and let the shadow fall on the paper. Have that child draw around the outline of the shadow on the sheet of paper. When each student has had a chance to draw a favorite shape, there will be a collection of interesting drawings that can serve as a source for discussion, sorting, and display.

2. Using the square regions from Activity 1, challenge the children to make the shadow into a square. Ask the students what they had to do to produce a square shadow. Give the children square regions that have been cut from paper to put on the ground. Have the children use their square regions to make a shadow just large enough to exactly cover the paper square on the ground. Have them make a square smaller than the paper square, then one larger than the paper square. Let the children discuss how they were able to make the shadows different sizes.

Next, give each pair of children a paper diamond region to put on the ground and ask them to try to make the same shape using the square region. Have them exactly cover the diamond shape, then make diamond shapes smaller and larger than the paper diamond.

Have the children discover if they can make a triangle or a pentagon shadow using the square region. See if they can make a rectangle or another parallelogram. It will be necessary to provide paper shapes as models for the children to use. Be sure to have the children sketch their results and discuss their findings.

3. Introduce diamond, triangular, and hexagonal regions to see what kinds of shapes their shadows are. Can a diamond shadow be made with a diamond region? Can a square shadow be made? What other shadow shapes can be made? Can a triangular shadow be made with a triangular region? Can square or diamond shadows be made?

Other shapes should be available with which the children can experiment. Again, outlining the shadows will produce a

permanent record of the shadow shapes. Expect the children to make discoveries that you had not thought of, and join in the excitement of such discoveries.

4. Using the outlines that the children drew of shadows cast by square regions, see if the children can find things that are alike and things that are different in the drawings. Encourage the students to count the number of corners and the number of sides of each shadow shape and to compare those numbers. Write down the conclusions made based on these observations.

Pose problems such as: "Suppose we take one of our shadow drawings and cut it out and glue it to a piece of railroad board cut exactly like the outline. Would it be possible to use that shape to make a shadow that would just match the square region that we started with? How do you think it could be done? Why do you believe that it can't be done?" Let the children perform the experiment to see if they can do this. Have them put their square regions on the ground and see if they can exactly cover the square region with a shadow from the outline region.

Objective: to produce and describe the shadows of squares and other shapes, using a point source of light.

5. Set up a slide or filmstrip projector or use a flashlight so that the light is projected onto a screen or wall. Let the children play in the light by making shadows using their hands or by holding small objects. After this introductory activity, provide the children with square regions and encourage them to explore the different ways that shadows can be produced. Tape paper to the wall and have the children outline the shadows to provide a record of their work that can be displayed on a bulletin board and discussed.

As an extension, introduce other shapes such as triangular, rectangular, and hexagonal regions and let the students find out what their shadows look like. Let the children describe their shadow shapes and explain how various shadows were made.

6. Compare the outline drawings of the shadows of the square regions made using the sun as a source of light with those made using a point source of light. A bulletin board display can have the shadows sorted, with shadows made using the sun on one side and those made using the projector or flashlight on the other side. Can all of the same shadow shapes be made? Do the shapes look similar? For those shadows that are different, would it be possible to make that shape if we tried again using the sun or a point source of light?

To extend this activity, cut out the outline of a shadow of a square region made with a point source of light, glue it to a piece of railroad board cut to exactly the same shape, and see if it is possible to make a shadow that matches the original square region. Are the results of this activity the same as the results obtained using the sun as a source of light?

7. Take a square region and place it on a block or paper cup so that the square region is supported parallel to the floor as in Figure 12–8. Have the children hold a flashlight above the square region, moving it from side to side, and ask them to observe the shadow that is produced.

What characteristics of the shadow shape are noted? If a sheet of paper is placed beneath the block, an outline of the shadow shape can be drawn. Then direct comparisons can be made between the square region and its shadow, such as comparing the sizes of the corners and the lengths of the sides.

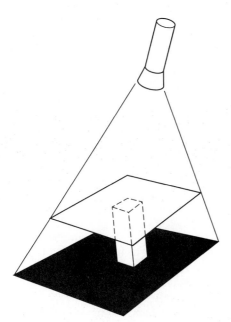

Figure 12-8

Is it possible to make diamonds or rectangles by moving the light to various positions? As the flashlight is moved higher and lower how does the size of the shadow change? Next, use triangular and rectangular regions and explore their shadows.

The preceding activities in projective geometry have been presented to help children as they make the transition from topological notions to Euclidean notions. You may also wish to examine the activities suggested by Dienes (1967) and Mansfield (1985) in the works listed in the bibliography at the end of the chapter.

Plane Figures and Their Properties

Children's abilities to learn the names and properties of common plane figures, such as triangles, squares, rectangles, circles, parallelograms, rhombuses, hexagons, and so forth, vary considerably within any group of children. Those who are able to observe a shape and then easily find another like it or those who are able to look at a figure and then draw it maintaining the characteristics essential to the figure are ready to proceed with more systematic instruction on Euclidean shapes.

Piaget and Inhelder (1967, p. 43) indicated that the learning of shapes requires two coordinated actions. The first is the physical handling of the shape, being able to run fingers along the boundaries of the shape. The second is the visual perception of the shape itself. It is insufficient for children merely to see drawings or photographs of the shapes. A variety of materials and activities can help to present plane figures to children. Some of these materials and activities are presented below.

ACTIVITIES

Primary (K–3)

Objective: *to develop tactile understanding of common plane figures.*

1. Give children flat shapes to explore. The shapes may be commercially produced, such as attribute blocks, or they may be teacher-constructed from colorful railroad board. Allow the children time for free play with little or no teacher direction. Perhaps the children will construct houses, people, cars, animals, patterns, or larger shapes.

    After having plenty of free time with the shapes, the children will be ready for the teacher to ask a few questions or to compliment them on their work. If someone has constructed a truck, ask several children to construct others just like it. Challenge the children to make an object that is the same except upside down.

    If a pattern is made, perhaps it can be extended. Encourage children and ask, "What shapes have you used to make your picture? What would happen if we changed all of the triangles to squares? What would happen if all the pieces were exchanged for larger pieces of the same shape? Let's try it."

2. Construct models of various shapes for the children to handle. One way to construct a model is to bend heavy wire in the shape of a triangle, square, rectangle, circle, parallelogram, rhombus, or hexagon. A touch of solder should hold the ends together. Another way is to glue small doweling to a piece of railroad board. The children can then develop a tactile understanding of the shapes.

    Once the children have handled the shapes, encourage them to describe the shapes. Ask, "How many corners does it have? How many sides does it have? What else do you notice?" Ask them to draw a particular shape while looking at and feeling the model. Later, ask them to draw the shapes while only feeling or seeing the models. Finally, ask the children to draw the shapes without either seeing or feeling the models.

Objective: *to make patterns using geometric shapes.*

3. Parquetry blocks (Figure 12–9) are a unique material to use to learn about plane figures. **Parquetry blocks** are geometric shapes of varying colors and sizes. The first attempt to use them should be in a free-play activity. Then, there are several ways to use the blocks to present shapes.

    ● Copy activities include holding up one of the shapes and having children find another block of the same or a different shape. Next, put three or four of the blocks together in a simple design and ask the children to copy the design. It

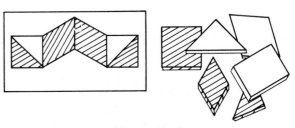

Figure 12-9

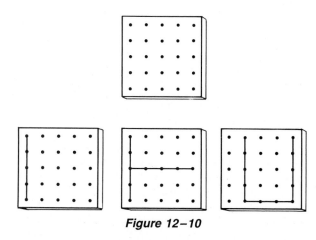

Figure 12–10

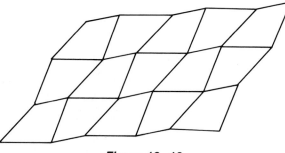

Figure 12–12

may be copied exactly or with a slight variation, such as with different colors. Finally, put the blocks into a simple design but separate them from each other. Copying this design requires the children to visualize across the separations.

• Present outlines of parquetry blocks and ask the children to find a piece the same color, shape, and size and place it on the outline. Later, have them match just shape and size. Present more complicated outlines, using designs of two or more blocks, after the children have worked with single blocks.

• Ask children to make their own outlines for others to fill in either by drawing around the various shapes or by putting all the shapes down and drawing around the entire design. The latter variation produces a challenging puzzle for children to complete.

While our discussion has centered on the parquetry blocks, another learning aid, pattern blocks, works equally well for the activities just mentioned.

Objective: to construct common geometric figures.

4. The geoboard is a dynamic aid for use in teaching geometry. It consists of a board 20 to 25 centimeters square with five rows of five escutcheon pins in each row (Figure 12–10). Stretch rubber bands around the pins to form various figures. Instructions for constructing a geoboard can be found in Appendix B.

   After a period of free play to allow the children to discover some of the patterns, shapes, and pictures that can be constructed, direct some copying activities. Construct a particular configuration or shape and show it to the children, asking them to copy it. Initially, construct a line segment, then perhaps combinations of two, three, or more line segments (Figure 12–10). Next, construct simple shapes. Gradually make the shapes more complex and challenging, as in Figure 12–11. As soon as the children understand the nature of the copying exercises, allow them to construct shapes for others to copy.

   As the children gain experience in recognizing and naming shapes, use the names to describe shapes for the children to

construct. Say, for example, "Let's make triangles on our geoboards. If we look at everyone's triangles, can we find some things that are the same? Are there any triangles that are completely different? Who has the biggest triangle? Who has the smallest? Who has the triangle with the most nails inside the rubber band? Who can make a shape that is not a triangle? Now, let's make some squares."

Objective: to discover characteristics of various shapes.

5. **Tessellating** is covering or tiling a region with many pieces of the same shape. Countertops and floors are often tessellated with square pieces. Of the regular Euclidean figures (that is, those with sides of equal length and angles of equal measure), only triangles, squares, and hexagons will completely cover a region without the need for additional pieces to fill in gaps. There are, however, many irregular shapes with which a region may be tessellated. Figure 12–12 shows a tessellation of quadrilaterals. All quadrilaterals will tessellate.

   Give children a sheet of paper to serve as a region and numerous pieces of some shape with which to tessellate. Pattern blocks are a handy and colorful material to use in tessellating. Ask the children to cover the paper with a particular shape and to decide which shapes will work. Later, ask them to try to use a combination of two or three shapes to tessellate. Before they begin, have the children estimate whether or not they can use the shapes to tessellate.

Objective: to develop the concept of an angle.

6. An **angle** may be thought of as a change in direction along a straight line. On the floor or playground, have children walk along a line that at some point changes direction, however slightly or sharply, as in Figure 12–13a. Discuss with the children that the change in direction forms an angle.

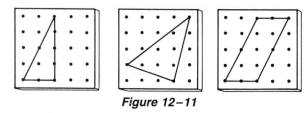

Figure 12–11

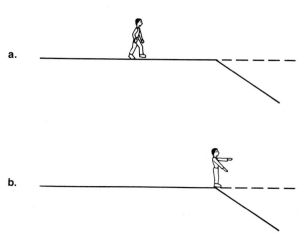

Figure 12–13

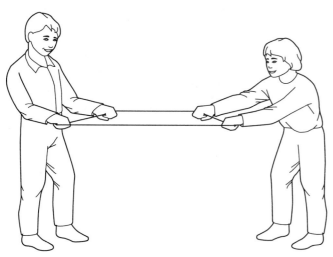

Figure 12–14

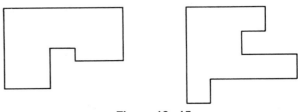

Figure 12–15

Ask the children if they can think of a figure that has an angle or corner. Children discussing the characteristics of a plane figure will mention the corners or bends that help give the figure its shape. The concept of an angle is being developed at an intuitive level.

Later, more formally define an angle as two rays sharing a common end point. Have children walk along chalk or tape lines that form a zig-zag path. By pointing one arm in the direction in which they have been walking and the other arm in the direction of change, children can form the angle of change. Figure 12–13b illustrates using the arms.

Later, have the children walk on large polygons. This activity serves as an introduction to one aspect of the computer language Logo. Logo activities follow.

Objective: *to practice making polygons and discover how their properties may change.*

7. Provide the members of a learning group with a 10-foot length of yarn that has been knotted at the ends to form a large loop. First, have two members of the group each hold it with both hands so that four vertices are formed, as in Figure 12–14. The other members of the group serve as observers and recorders. The holders pull the yarn taut, producing a quadrilateral, then they explore what happens to the shape as they change the sizes of the angles and the lengths of the sides of the figure. An observer's job is to describe what happens and a recorder's job is to sketch the shape as it changes. Will it be possible to produce a triangle? How about a pentagon? What must be done to make a square, a rectangle, and a parallelogram? When the children have had a chance to discuss their findings and to look at the sketches of the figures, ask them what they can conclude about changing the angles, changing the lengths of the sides, and making other types of geometric figures?

To extend this activity, have a third student hold the yarn so that there are six vertices. Now, what different shapes can they make and how does changing the angles and the lengths of the sides affect the appearance of the figure? Does changing the length of the yarn affect the results of this activity? Again, discussion helps children share their observations and their sketches help to verify their conclusions.

Objective: *to explore geometric figures using Logo activities.*

8. **Logo,** the computer language of turtle graphics, provides primary children with a rich source of experiences. Logo is available for the most popular microcomputers used in schools. To prepare children to use Logo, engage them in a series of noncomputer activities. Following is a brief series of introductory activities.

a. Find an activity that the children are familiar with. List the individual parts that make up the activity in a series of steps. For example, to put the cat out, we might:

- Call the cat.
- Does the cat come?
- If not, go find the cat.
- When the cat comes, pick it up.
- Carry the cat to the door.
- Open the door.
- Put the cat outside.
- Shut the door.

Next, write each step on a separate card. Mix up the cards, and challenge the children to put them back in the correct order. Once the children discover how to do this activity, present a series of cards without first showing the appropriate sequence. Have the children order the steps of the procedure by figuring out the sequence. Procedures besides putting the cat out may include making a peanut butter sandwich, preparing for and taking a bath, and getting ready for bed. Allow the children to make up sequences to challenge one another.

b. With masking tape or yarn, construct a geometric figure on the floor. It may be a square, a triangle, or a rectangle, at first. Later, make a more complex figure, such as those in Figure 12–15.

Ask the children to begin by going to any corner and facing an adjacent corner. Have them describe what they are doing as they walk around the boundary of the figure and end up where they started. Limit the descriptions to the use of forward, back, left, and right. It may be helpful to have direction cards that show what is meant by the four commands. Figure 12–16 illustrates what such cards might look like.

Later, have one child give directions to a second child that will guide the second child in walking the boundary of a figure. The second child should follow the directions exactly.

It will soon be necessary to tell a student how many steps forward or backward to take. For example, "Go

forward   back   left   right

Figure 12–16

Computer Graphics: Map Making

You need to give directions to a friend about your route home from school. Based on your verbal directions, your friend draws a map. How could a computer help your friend plot the route?

DATA

Use graph paper and pencil to draw a path. Each unit on the grid will represent one city block. Write the labels for north, east, south, and west. Begin your map in the middle of the graph paper. Then follow this route:

a. Walk two blocks south.

b. Turn east and walk three blocks.

c. Turn south and walk one block.

d. Turn east and walk three blocks.

e. Turn south and walk four blocks.

f. Turn east and walk one block.

g. Turn south and walk three blocks.

h. Turn west and walk half a block.

THINKING ABOUT COMPUTERS

1. Compare your map to those of other students. How are they the same? How are they different?

2. How would the map change if step **b** were a 90-degree turn west?

3. Why would a computer be a useful method for drawing the map?

4. ***What if*** you are an architect planning a blueprint for a house? How could a computer be useful?

5. An astronomer is drawing maps of the constellations that can be seen during different times of the year. Why would a computer be useful in drawing the maps?

Figure 12–17

Reproduced with permission of Macmillan/McGraw-Hill School Publishing Company from Mathematics in Action, Grade 5, p. 339 by Alan R. Hoffer et al. Copyright 1991.

THE MATH BOOK

The activity from a fifth grade textbook shown in Figure 12–17 asks students to work in pairs to draw routes using graph paper and a pencil and then to discuss what they have done. In the section on thinking about computers, they discuss how a computer could be used in different occupations such as architecture and astronomy. Notice that to complete this page, the students do not actually use a computer. They do an activity that may prepare them to use a program such as Logo and they talk about the uses of computers, but they do not actually use a computer. The support materials for this series do include a technology section in the Teacher's Resource Center. It is in this section that Logo is described and that students learn to draw the map on the computer.

This activity would be much enhanced if the students actually used a computer to learn Logo for map drawing. Just doing the activity with paper and pencil and talking about how map makers and architects might use a computer will not give the students a very good understanding of the uses of computers. It is very important, therefore, that teachers take advantage of all the resources available with the series or that they supplement the textbook by teaching Logo.

This activity is included in the chapter on geometry and follows lessons on topics such as angles, symmetry, visual reasoning, motion geometry, and congruent, similar, plane, and space figures. Most of these topics are covered in one two-page lesson. The teacher's manual does contain several ideas for setting up learning centers, and the brevity of each lesson in the book again points to the need for supplementing these lessons with a variety of concrete materials and lessons. Students who only did the exercises in the book without working through either the concrete activities suggested in the teacher's manual or other activities, such as those suggested in this chapter, would have a very limited concept of geometry even though the topics covered in the book are good ones.

forward twelve steps." Agree that such steps are taken by putting one foot directly in front of the other.

Next, give a child a drawing of a figure and instruct the child to give another student commands for making the figure. The teacher or a student can lead the entire class in this activity.

   c. Put a blindfold on a child and arrange the desks in a simple maze. Have children carefully give commands that will, if followed, lead the blindfolded child around the desks and out of the maze. Use particular caution to avoid any possibility of injury.

9. Introduce turtle geometry on the computer by putting a small colored sticker on the computer monitor and asking the children to see if they can find the appropriate commands to hide the turtle under the sticker. Encourage the children to estimate the commands before they actually try them. In the beginning, use RIGHT 90 and LEFT 90 to designate the turns but allow the children to experiment with other degrees of turns very soon. The activity with making angles introduced earlier should help students understand how to construct various angles. It will take a little experimentation to determine the size of the turtle steps, as well. Fairly quickly the children will become accomplished at moving the turtle freely around the screen.

To encourage moving the turtle about, put an overhead transparency on the screen with several regions drawn on it. A thin transparency will cling tightly to the screen. Have the children move the turtle from one region to another until it has entered all the regions.

Sketch a simple maze on another transparency and place the transparency on the screen with the turtle in the maze. Challenge the children to get the turtle out of the maze without crossing any boundaries.

From here on, use one or more of several well-written Logo manuals. The Logo manuals are carefully sequenced. They contain many challenging figures to test children's abilities to use Logo. Several of these resources are listed at the end of the chapter.

The activities just presented are intended to give primary children experiences with plane figures to complement work from the mathematics textbook. We now turn to activities for older students.

Most of the following activities are intended to support children as they work in the second and third of the van Hiele levels, description and relation. This means the students will continue to analyze the properties of Euclidean figures and will begin to understand the characteristics of the figures in terms of definitions.

ACTIVITIES

Intermediate (4–8)

Objective: *to discover important properties that define a variety of polygons.*

1. Periodically designate a bulletin board as a shape board. Attach a label such as "quadrilateral" and invite the children to put as many different quadrilaterals as they can on the board. After two or three days, have the children describe the ways in which the shapes are different. Thus, the children look at the defining properties of quadrilaterals. Ask the children to classify the quadrilaterals as squares, rhombuses, rectangles, parallelograms, and trapezoids. Which categories overlap? How do the shapes relate? Another time, the board theme may be triangles, then hexagons and octagons.

2. On the overhead projector or chalkboard, display a set of properties of a particular quadrilateral. Reveal the properties one at a time until a student decides a sufficient number of properties have been displayed to identify the shape. That student must then convince the rest of the class that enough characteristics have been given to identify the figure. For example, the following list may be presented.

   - It is a closed figure with 4 straight sides.
   - It has 2 long sides and 2 short sides.
   - The 2 long sides are the same length.
   - The 2 short sides are the same length.
   - One of the angles is larger than one of the other angles.
   - Two angles are the same size.
   - The other 2 angles are the same size.
   - The 2 long sides are parallel.
   - The 2 short sides are parallel.

   To vary this activity, have the children develop lists, individually or in small groups, that can be used to challenge the others in the class. They may select particular triangles, quadrilaterals other than the parallelogram described above, or various other polygons.

Objective: *to discover the numerous configurations a polygon may have.*

3. Challenge the children to find as many different triangles as possible on the geoboard. By different, we mean noncongruent, that is, not the same size and shape. Because of the variety of such figures, it is helpful to structure this activity using the problem-solving skill of simplifying the problem. For example, ask for as many different triangles as can be made using only two adjacent rows on the geoboard (we count fourteen such triangles). Before the children begin, have them estimate how many triangles they can make.

You may wish to simplify the problem even more by asking the students to make triangles on a 2×2, 2×3, or 2×4 arrangement of nails. As the children find the triangles, have them sketch the triangles on a piece of dot paper (see Appendix B) and discuss how they went about finding them. Figure 12–18 shows a few of the possible triangles.

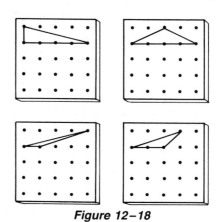

Figure 12–18

A little later, ask the children to make as many triangles as possible on a 3 × 3 nail arrangement on the geoboard. Put a rubber band on the geoboard surrounding the 3 × 3 area as a guide. Of course, you may use another arrangement as the basis for constructing triangles.

Discuss the types of triangles found. There will be **right triangles** (one angle of 90 degrees), **isosceles triangles** (a pair of congruent sides), **acute triangles** (all angles less than 90 degrees), **obtuse triangles** (one angle more than 90 degrees), and **scalene triangles** (no sides of equal length).

Extend this activity by seeing how many quadrilaterals may be made on a certain part of the geoboard. Be sure to have children estimate before they begin. We know that sixteen noncongruent quadrilaterals can be formed on a 3 × 3 geoboard. How many squares or rectangles or hexagons may be constructed?

As a further extension, have the children use the triangles they constructed on a 3 × 3 area of the geoboard and ask them to produce the triangles on the computer screen using Logo. Challenge the children to see if they can use Logo to reproduce figures other children have constructed on the geoboard. Encourage children to attempt to make a variety of different polygons, first on the geoboard and then on the screen.

Objective: *to use the computer to design geometric figures.*

4. Continue to use Logo to strengthen students' abilities to define geometric figures and to develop procedures for complex designs and patterns. A **procedure** is a set of commands that may produce a simple figure or that may combine other procedures to form a more complex figure.

Ask children, as they work individually or in pairs, to develop a procedure for making a *box* with sides of 50 turtle steps. Next, have them make a *flag* using the box procedure. Then, have them make a *windmill* using the flag procedure. Finally, challenge the children to use the windmill procedure to make a *pinwheel*. The results of these four procedures are shown in Figures 12–19 a, b, c, and d. The procedures that may be used to draw these figures are as follows:

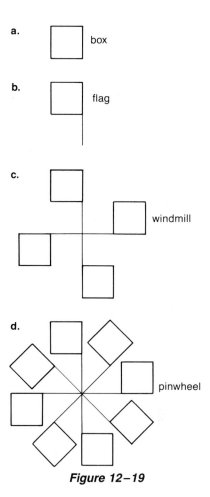

Figure 12–19

```
TO BOX
REPEAT 4 [FD 30 LT 90]
END
```

```
TO WINDMILL
REPEAT 4 [FLAG LT 90]
END
```

```
TO FLAG
FD 30
BOX
BK 30
END
```

```
TO PINWHEEL
REPEAT 2[WINDMILL LT 45]
END
```

The procedures presented above show how the repeat command can be used to replace a set of commands and streamline the procedures. This is a recognition that students should be encouraged to make after the repeat command has been introduced.

Ask children to develop procedures for producing a number of polygons of different sizes. This may involve using variables within the procedures. Variables present an added dimension to working with Logo. Challenge the students to use their skills to reproduce materials like the pattern blocks, the attribute blocks, or a "picture" drawn by students on squared paper. As students develop the ability to design figures, they learn valuable information about plane figures.

Besides using variables as they develop procedures, students will soon be able to use recursion, to employ coordinates to define locations, and to design complex figures. You will find resources for teaching Logo in the bibliography at the end of the chapter.

This group of activities for primary and intermediate children plays an important role in children's geometric learning. They help define plane figures and their properties in concrete and abstract terms. We now turn to the concepts of symmetry and motion.

Symmetry and Motion

The notions of symmetry and motion are exemplified by patterns in nature and in the art and architecture of human beings. **Symmetry** requires a line or lines about which a figure or design is balanced or a point about which a figure or design is rotated. There is something orderly and pleasant in balance. (**Balance** is the characteristic of a figure that suggests an equality of parts.) Children often generate symmetrical designs with building materials. Children who have been made aware of the symmetry around them have a fuller appreciation of their environment. Besides, many geometric figures contain fine examples of symmetry, having, in some cases, several lines of symmetry.

ACTIVITIES

Primary (K–3)

Objective: *to develop simple symmetrical patterns with objects.*

1. Provide the children with Cuisenaire rods, pattern blocks, or parquetry blocks. Encourage them to make designs that they think are pleasant. Compliment the students on their efforts and point out the unique characteristics of the designs. For example, point out those made of materials of the same color, those using pieces of the same shape, and those that have line symmetry. Discuss what it means for a figure to have balance, using one of the children's designs as an example. Have the children look around the room and point to shapes that appear to be the same on both sides.

   Ask the children to make a design with symmetry. You may tightly structure this activity by designating which pieces to use in making a design; for example, using the pattern blocks, have the children take two red pieces, four green pieces, and two orange pieces for their design. Ask the children to sketch the results or to glue colored paper cut into the shapes being used.

2. Provide mirrors with which the children may explore and develop symmetrical patterns. (Inexpensive mirrors are available through school supply catalogs that feature learning aids.) Using Cuisenaire rods, pattern blocks, or parquetry blocks and mirrors, have the children construct symmetrical designs and reaffirm their symmetry.

   Ask the children to make a design using three or four blocks or rods. Then have them place a mirror along one edge of the design, note the reflection, and copy the image in the reflection, placing the copy behind the mirror. Thus, the mirror is lying along the line of symmetry.

   Ask the children to remove the mirror and to discuss their symmetrical design. Say, for example, "What pattern do you see in your design? If your design were a picture, what would it show? See if you can take the reflected design away, mix up the pieces, and then put the design back the way it was before. Where do you think the line of symmetry is? Check it with the mirror. Can you make a new design and its reflection without using the mirror? Try it. Use your mirror to check to see if your design has symmetry."

   Finally, have the children sketch and color the pattern and its mirror image on a sheet of squared paper. Figure 12–20 illustrates this process.

3. Stretch a rubber band across a geoboard from edge to edge so there is ample space on each side of the rubber band. In the simplest example, the rubber band would be stretched across the center of the geoboard either horizontally or vertically. Construct a figure on one side of the rubber band and challenge the children to construct the image of the figure on the other side. In the beginning, have the children stand a mirror on its edge along the symmetry line and make the image while looking in the mirror. Figure 12–21 provides examples of this activity.

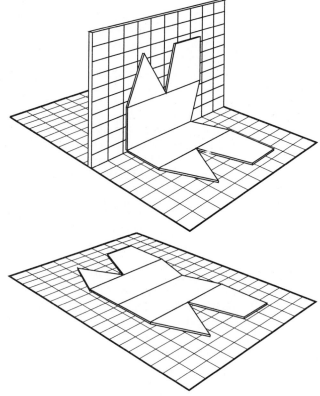

Figure 12–20

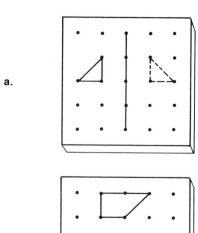

a.

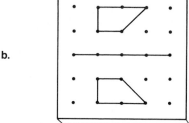

b.

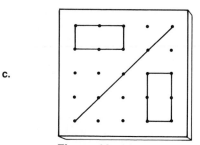

c.

Figure 12–21

Let the children make up figures and challenge the rest of the class to construct the mirror image of the figure on the geoboard across the line of symmetry, with or without a mirror. Provide dot paper on which the students may copy their symmetrical geoboard designs. As the students develop proficiency in recreating images, use diagonal lines as lines of symmetry.

Objective: to develop the ability to visualize symmetrical patterns.

4. Ask the children to fold a sheet of paper in half and to cut out some shape from the folded edge. Then have the children open the sheet and observe the symmetrical figure. Provide an opportunity for the children to share their designs.

   Next, challenge the class to plan shapes to cut out of folded sheets and to guess what the results will look like. The students may draw what they believe the figures will look like when the paper is unfolded. Then let them cut out the figures and check the results against their estimates.

   Another variation of this activity is to punch a hole through the folded sheet with a paper punch. Have the children guess how many holes there will be, then open the sheet to see. Try two holes, then three. Also, try folding the sheet of paper twice and then punching one or more holes through the paper. Add a challenge to this activity by having children guess where the holes will be as well as how many there will be. Display the children's work.

Objective: to identify symmetrical figures.

5. Have the children search through magazines for pictures that have symmetry. Have them cut out those pictures. On a bulletin board, put up the heading "These Pictures Have Symmetry" and the heading "These Pictures Don't Have Symmetry." Have the children classify the pictures they have cut out and place each of them under the appropriate heading.

   A variation of this activity is to go on a school or neighborhood walk to look for symmetry in the environment. As examples are found, have two or three children sketch the examples on squared paper. When the walk is over, have the students color the sketches and classify them on the bulletin board.

   Another variation of this activity is to provide each child with an object you have cut out from a magazine and then cut in half along its line of symmetry. For example, give children one side of a face, half of a flower in a pot, or half of an orange. Ask the children to paste the half picture onto a piece of drawing paper and to draw the other half of the object using crayons.

Thus far, we have been using line symmetry. With older children, introduce **rotational,** or **point, symmetry.** A figure has rotational symmetry if the figure can be rotated about a point in such a way that the resulting figure coincides with the original figure. Thus, the equilateral triangle in Figure 12–22 may be rotated clockwise about point P. In this case, the triangle will coincide with the original triangle three times during one full turn. Each of these positions is shown in Figure 12–22. The first activity that follows presents rotational symmetry.

position 1   position 2   position 3

Figure 12–22

ACTIVITIES

Intermediate (4–6)

Objective: to introduce the concept of rotational symmetry.

1. Construct a large equilateral triangular shape to serve as a model for rotational symmetry. On the floor, make a masking tape frame in which the triangle fits. Put a small hole through the model at its point of rotation and insert a pencil or a piece of doweling. Make some sort of mark in one corner of the shape to serve as a reference point when the figure is rotated. Put the shape in its frame, and have the class record its position on their paper.

   Invite students to carefully rotate the figure clockwise until it again fits the frame. Have the class record the new position.

   Have the students rotate the figure again until it once more fits the frame. Have the class record its new position.

   The next rotation will put the figure back in its starting position. Ask, "How many different positions are there when we rotate an equilateral triangle?" There are three positions. Continue, "We say this figure has rotational symmetry of order 3. What do you think will be the order for the rotational symmetry of a rectangle, a square, or a regular pentagon? Let's try these figures."

   You will need to investigate a variety of plane figures before the students will be entirely comfortable with rotational symmetry. As the students catch on, they will be able to think about and draw figures with a specified order of rotational symmetry.

Objective: to introduce the mira as a tool for exploring symmetry.

2. Activities involving the mira are particularly suited to a study of symmetry. A **mira** is a specially designed tool made of red transparent plexiglass that is used in place of a mirror for exploring line symmetry. The mira is superior to a mirror in several ways. In the first place, you can see through the mira, so images are easier to copy. Also, the mira stands by itself and does not need to be held.

   An extensive set of activities may be purchased to guide student work (see the reference at the end of the chapter).

   As with new manipulative aids, the initial activity with the mira should be a period of free play in which the students look for figures and pictures to be checked for symmetry. Provide materials such as pattern blocks, tiles, Cuisenaire rods, and magazine pictures. Encourage students to draw patterns on squared paper and to use the mira to investigate the patterns. Interesting discoveries will result.

3. Have the students consider the letters of the alphabet as shown in Figure 12–23. Ask the students to identify the

ABCDEFGHIJKLMNOPQRSTUVWXYZ
Figure 12–23

letters that have at least one line of symmetry and those with more than one line of symmetry. Have the students visually estimate, then have them write down the letters they believe have line symmetry.

Then, have the students use a mirror or a mira to check each letter for symmetry. It is appropriate at this time to see if the students are able to determine if any of the letters have rotational symmetry. That is, can the letter be rotated about a center point in such a way that the letter appears as it normally does before it has been rotated a full turn? For example, the letter *I* has rotational symmetry of order 2.

An extension of this activity involves finding words that have line or rotational symmetry. For example, both TOOT and CHOICE have line symmetry and NOON has rotational symmetry. Is there a word that has both?

4. Let the students explore various materials such as pattern blocks using two mirrors or miras. Suggest to the students that they tape the mirrors at right angles and place blocks at the intersection. Increase and decrease the angle of the mirrors to see what images result. Place the mirrors parallel to each other and observe the image of blocks placed between them.

Try using three mirrors, one lying flat and two at right angles on top. Have the students sketch the images they think will result. Examples of two mirror configurations are shown in Figure 12–24.

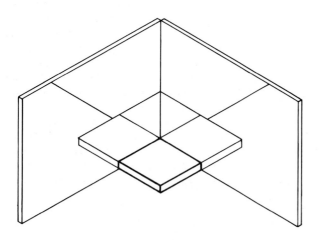

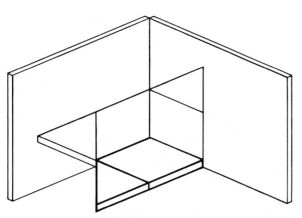

Figure 12–24

Another application of line symmetry and the images that result from using multiple mirrors can be found in computer software. For example, MacPaint, a dynamic program designed for the Apple Macintosh computer, allows the user to draw a figure and then, using commands, to produce the mirror or rotational image of the figure on the monitor. This exciting graphics feature allows students to instantly see the results of using mirrors or rotations.

Objective: *to construct symmetrical figures.*

5. Challenge the children by asking them to construct irregular figures on the geoboard. Provide a line of symmetry; this could be a vertical, horizontal, or diagonal line. Have the children construct the reflection of the figure on the opposite side of the line of symmetry. Then, let the children check their efforts with the mira or the mirror.

Let the students experiment with lines of symmetry other than those shown in Figure 12–21. See which, if any, other lines can be used to accurately construct reflected images.

Objective: *to challenge students with problems involving symmetry.*

6. Provide students with three green triangles and three blue diamonds from the set of pattern blocks. Have the students make triangles that measure three inches on a side and have (a) one line of symmetry and no rotational symmetry, (b) two lines of symmetry and no rotational symmetry (no solutions), (c) three lines of symmetry and rotational symmetry of order 3, (d) no lines of symmetry and rotational symmetry of order 3, and (e) no lines of symmetry and no rotational symmetry. Encourage the students to make up similar problems, creating other shapes using four to eight pattern blocks.

The activities above provide experiences with the symmetry found in various figures. The activities focus on line and rotational symmetry. These experiences help students not only learn the concept of symmetry but also develop the ability to visualize shapes in the mind's eye. We now turn to space figures.

Space Figures

Up to this point, the activities have dealt principally with plane figures—figures of two dimensions. Like all of us, children live in a three-dimensional world. All of their movements, explorations, and constructions have been in space. The exploration of space is the classic example of early mental growth.

As children continue their growth in geometry, activities with three-dimensional space figures are an important part of this learning. Whenever possible, tap children's environments—the classroom, home, and community. The activities that follow are designed to aid in the development of spatial concepts. Again, activities cannot by themselves teach. Augment them with reading, discussion, examples, and thought.

ACTIVITIES

Primary (K–3)

Objective: *to identify and draw two- and three-dimensional objects in the environment.*

1. Extend the playground or neighborhood walk mentioned earlier to include a search for three-dimensional figures. On a shape walk, ask students to sketch the shapes they observe. The shapes may be two or three dimensional. Students may draw the shapes of windows, doors, faces of bricks, or fences. Or they may draw the shapes of entire houses, individual bricks, garbage cans, or light posts. It is likely that you will need to discuss how to sketch three-dimensional figures. Have the children share with one another their own techniques. Descriptive stories by the class or individuals may help to conclude an investigation of shapes in the community.

2. Ask the children to bring empty containers from home to serve as a collection of commonly found space figures. Expect containers such as cereal boxes, cans with the tops and bottoms removed, plastic soap containers, and tubes from paper towels or toilet paper. Use these materials as a bulletin board or table display. Classify the various figures by overall shape or the shapes of various faces. Cut the container so it lies flat and the students can examine the pattern of the space figure. In how many different overall shapes are household items packaged?

Objective: *to copy and build space figures.*

3. Encourage children to use a variety of materials to build space figures. Large blocks and cardboard building bricks along with tiles, geoblocks, Unifix cubes, Cuisenaire rods, and pattern blocks are among those commonly found in primary classrooms. Make a construction and ask the children to copy it.

   A challenging series of work cards accompanies the set of geoblocks. Develop other, similar cards for use with the three-dimensional learning materials.

Objective: *to discover characteristics of polyhedra.*

4. Straws and pipe cleaners (or straws of two sizes) can be used to construct polyhedra. Initially, produce two-dimensional figures. As space figures are investigated, it should become apparent that the faces of all polyhedra are polygons. Thus, when a cube is constructed, an investigation of its faces yields squares. If a tetrahedron is constructed, an investigation of its faces yields triangles. Encourage children to construct various polyhedra. Several are shown in Figure 12–25.

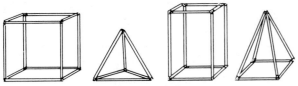

Figure 12–25

Have the children compare the space figures, noting the number and shapes of the faces, the number of vertices, and any interesting facts about their shapes. Have the children record these findings on a chart and prominently display it.

Allow the children to construct space figures. As children construct the figures, they develop a sense of how figures fit in space. As children begin to analyze space figures, they prepare the way for a more formal study of objects in space.

ACTIVITIES

Intermediate (4–8)

Objective: *to explore the characteristics of the regular polyhedra.*

1. Among the myriad space figures, there are only five regular polyhedra. A regular polyhedron is one in which all the faces are congruent, all the edges are the same length, and all the angles are the same size. The regular polyhedra are the tetrahedron (4 faces), hexahedron or cube (6 faces), octahedron (8 faces), dodecahedron (12 faces), and icosahedron (20 faces). They were illustrated in Figure 12–1i.

   Students explore these shapes most effectively when they can hold them, turn them, and note their characteristics. Provide materials and patterns so the students may construct their own set of regular polyhedra. (See Appendix B for patterns for the five figures.) The patterns may be copied onto heavy paper or oaktag. Have the students cut out the patterns, score the fold lines with a paper clip, make the folds, and glue the tabs.

   One systematic investigation of the regular polyhedra is discovering the relationship between the number of faces, the number of edges, and the number of vertices. A table, such as the table shown in Figure 12–26, can be used as an effective problem-solving tool to display the information gathered. The table provides a way to systematically organize the information as it is collected.

   Have the students handle the tetrahedron. Have them count the number of vertices, or corners, of the tetrahedron. There are four. Record that number in the table. Next, count the number of faces and record that information. Finally, count the number of edges and record that information. Continue counting vertices, edges, and faces for the other figures.

   Once the information has been recorded in the table, challenge the students to look for a relationship between the vertices, edges, and faces of a regular polyhedron. Have them look at the numbers for each of the regular polyhedra. Give the students time and support as they look for this relationship.

   A formula named after the mathematician Euler describes the relationship between the faces, edges, and vertices of

number of	tetrahedron	cube	octahedron	dodecahedron	icosahedron
vertices	4				
edges	6				
faces	4				

Figure 12–26

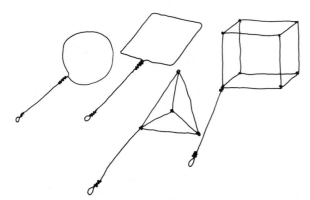

Figure 12-27

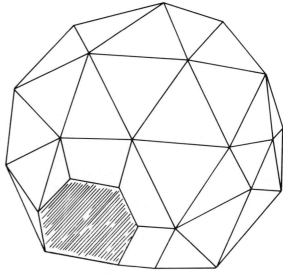

Figure 12-28

polyhedra. The formula states that $V + F - E = 2$, that is, the number of vertices plus the number of faces minus the number of edges equals 2. Many students are capable of finding this relationship.

To extend this activity, see if the students can determine if the relationship discovered for a regular polyhedron holds true for any pyramid or any prism.

Objective: *to explore space figures formed by soap films on wire frames.*

2. Provide the students with wire somewhat lighter than coat hanger wire; it should be easy to bend and cut the wire. The object is to construct shapes out of the wire that can be used with soapy water to produce various two- and three-dimensional figures. Figure 12–27 shows four possible wire shapes. Encourage the students to create wire shapes with tightly secured corners.

Have the students dip the two-dimensional shapes in a mixture of liquid soap and water (half and half) and record what happens. Let them trade their wire shapes and experiment some more. Possible explorations include blowing a bubble with a circular frame and then blowing a bubble with a triangular frame.

See what happens when the three-dimensional frames are dipped in soap and water. What happens when a diagonal is constructed inside a three-dimensional shape that is then dipped in soap and water? Construct shapes that are not polygons, then dip them and blow bubbles or just dip them.

Objective: *to combine imagination and knowledge of space figures to create a microworld.*

3. Projects using space figures offer motivation for creative learning experiences. One such project was initiated during an introductory class on space figures. As the children and the teacher looked at a set of geoblocks, one child noted that a particular piece looked like an Egyptian pyramid; another student thought that the word *prism* sounded like *prison*. Soon a boy in the class mentioned that it would be exciting to create a city full of shapes. The *geoworld* project was begun.

The geoworld was built on a platform of triwall construction board that measured four feet by eight feet. The very first piece of architecture that arose was *tetrahedra terrace*, a series of connected tetrahedra. Then came the *cuban embassy*, an idea sparked by surveying atlases for possibilities. The cuban embassy was a large cube. It was surrounded by several cubans, who were represented by smaller cubes with personal characteristics. Many other structures were added to geoworld; and when the project had been completed, ev-

ery member of the class felt a deep sense of pride in the creative work of their peers.

Objective: *to construct a geodesic dome.*

4. Another project is the construction of a large space figure. Thus, a cube that measures 1 or $1\frac{1}{2}$ meters on a side may be built and used as a quiet place or reading corner. A large dodecahedron may be constructed. Among the more interesting of all such figures is the geodesic dome, originally conceived of by the late Buckminster Fuller. The following steps result in a rather spectacular geodesic dome, whether it has a radius of 10 or 40 inches.

   a. Make the big decision. What size of dome do you want to build? Decide on the radius desired (half the width at the dome's widest point). Figure 12–28 illustrates what the finished dome will look like.

   b. Construct the dome using two different sizes of triangles. The size of each triangle is determined by the size of dome desired. One of the triangles, T1, is equilateral, with each side 0.6180 times the length of the dome radius. The other triangle, T2, has one side equal to the length of a T1 side and two shorter sides, each 0.5465 times the length of the dome radius. Thus, for a dome of radius 10 inches, the T1 triangle has sides 6.2 inches long, and the T2 triangle has one side 6.2 inches long and two sides 5.5 inches long.

   c. Make a pattern for each triangle. Figure 12–29 shows one such set of triangles. Note that there is a flap on each side. The flap is used to attach the triangles.

   d. Using the patterns, make fifteen T1 triangles and forty-five T2 triangles. For a 10-inch radius dome, oaktag is suitable material; for a 40-inch radius dome, cardboard appliance cartons are best. It is necessary to lightly score the fold lines on the flaps.

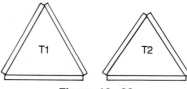

Figure 12-29

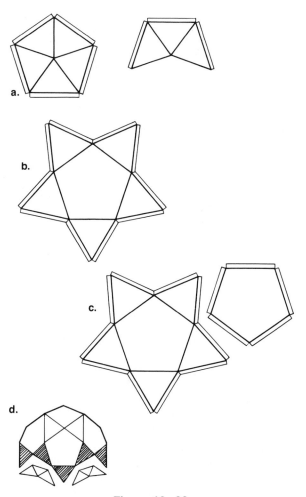

Figure 12–30

children grow in their abilities to grasp geometric concepts. Children must actively experience geometry. They must be guided in their explorations. Above all, geometry must be a part of the mathematics program. It must be extended far beyond the basal textbook and presented throughout the school year.

Developing and Practicing Geometric Skills

The skills of geometry involve readily identifying, copying, inventing, and constructing figures. Teach the skills in concert with teaching geometric concepts. Developing and practicing skills will, in most cases, follow conceptual development.

Identifying Shapes

The collection of shapes easily identified by primary children varies with the experience and maturity of the children. The most productive activities for shape identification are those in which the child is actively manipulating and discussing figures. A kindergartner or first grader may call a triangle a rectangle because the names are similar. A second or third grader who has used attribute blocks, pattern blocks, and geoboards and who has discussed the figures will seldom misname the triangle. A second or third grader may, however, misname a diamond or hexagon. Again, this difficulty can be alleviated through carefully designed experiences.

Primary students should be expected to develop geometric skills at a basic level. Thus, visual skills should include the ability to recognize different figures from a physical model or a picture. Verbal skills should include the ability to associate a name with a given figure. Graphical skills should include the ability to construct a given shape on a geoboard or to sketch the shape. Logical skills should include the ability to recognize similarities and differences among figures and to conserve the shape of a figure in various positions. Applied skills should include the ability to identify geometric shapes in the environment, in the classroom, and outside.

At the primary level, children develop skills as a result of extending activities used to develop the concepts. It is important that the teacher provide time, materials, and direction. Pay attention to developing visual, verbal, graphical, logical, and applied skills. Refer to the primary activities suggested earlier for developing geometric concepts.

Intermediate students should be expected to develop skills at a slightly higher level. Thus, visual skills should include the ability to recognize properties of figures, to identify a figure as a part of a larger figure, to recognize a two-dimensional pattern for a three-dimensional figure, to rotate two- and three-dimensional figures, and to orient oneself relative to various figures. Verbal skills should include the ability to describe various properties of a figure.

e. Begin construction. If you use oaktag, use white school glue to attach the triangles. It will take the cooperative effort of several students to put the final pieces in place and hold them while they dry. If you use cardboard, you will need a heavy-duty industrial stapler to attach the pieces. Follow these four steps: (1) Make six pentagons and five semipentagons from T2 triangles. See Figure 12–30a. (2) Add T1 triangles to the perimeter of one pentagon. See Figure 12–30b. (3) Fill the gaps between triangles with other pentagons. See Figure 12–30c. (4) Add T1 triangles between and below pentagons. Then, add semipentagons at the bottom. See Figure 12–30d.

As a final touch to the ball-shaped geodesic dome, fill the gaps around the base of the dome and attach the bottom flaps to a frame or to the floor to make the dome more rigid. It is helpful to cut windows and a door into geodesic domes large enough to enter.

During this project, students may wish to send away for a catalog from a company that prefabricates geodesic dome houses or to search for magazine articles about such homes. Some students may investigate some of Buckminster Fuller's other inventions.

This section has focused on how children learn geometric concepts and specific activities to reinforce this learning. The process of learning is developmental; that is,

Graphical skills should include the ability to draw a figure from given information and to use given properties of a figure to draw the figure. Logical skills should include the ability to classify figures into different types and to use properties to distinguish figures. Applied skills should include the ability to recognize geometric properties of physical objects and to draw or construct models representing shapes in the environment.

Like primary children, intermediate and middle grade children should learn geometry through physical activities with a variety of concrete materials. Again, extending the activities intended for conceptual development will provide skill activities. The teacher should facilitate activities and discussion during the learning process.

Copying and Inventing Shapes

Copying activities were mentioned earlier in conjunction with parquetry blocks and geoboards. For students at all levels, copying can be challenging and fun. The complexity of the figures to be copied should vary with the age and experience of the children. Inventing shapes is an outgrowth of copying the shapes formed by teachers and classmates. Asking primary children to find as many four-sided figures as possible challenges them to invent shapes.

Intermediate children can be challenged with the same problem. The results, however, are likely to be different. How many six- or eight-sided figures can be found? The geoboard is a helpful tool for investigating polygons. Rectangular and isometric dot paper are useful for both sketching and recording shapes. Figure 12–31 illustrates both dot patterns. Both rectangular and isometric dot paper can be found in Appendix B.

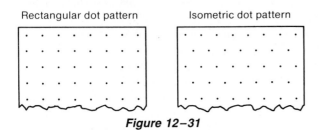

Figure 12–31

Another tool children may use to invent shapes is Logo. Figures may be designed on the computer and saved on a diskette for future access. Using Logo, the children can discover more than just what shapes are possible. They must consider the sizes of the exterior and interior angles and the length of each side of the figure. Once they invent a shape, have them make a sketch of it to serve as a challenge to other students and to you.

Constructing Shapes

A skill appropriate at the intermediate level is that of constructing simple geometric shapes using a compass and straightedge. At this level, the goal is to introduce stu-

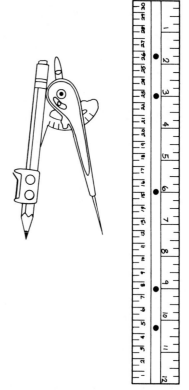

Figure 12–32

dents to techniques of constructing simple figures. The tools used in constructing figures are inexpensive and readily available. The compass uses the student's pencil as a marking tool, as shown in Figure 12–32. Also pictured is a straightedge; a standard school ruler works fine.

The initial activities should involve copying a given figure. Thus, copying a line segment, an angle, and a circle with a given radius are appropriate. It is expected that the students will have been exposed to terms such as *line, line segment, point, angle, arc, ray, bisector,* and *perpendicular.* Most of these terms will appear in the math book, although words such as *arc* and *bisector* may need to be explained. An **arc** is any part of a circle. A **bisector** is a line that divides an angle or line into two equal parts. **Perpendicular** means to be at a right angle with a line.

ACTIVITIES

Intermediate (4–8)

Objective: *to use a compass and straightedge to construct simple figures.*

1. Copy a line segment, AB, onto a line, m (see Figures 12–33 a, b, and c).
   a. Place the compass points on A and B.
   b. Mark the length of segment AB onto line m.
   c. Segment A'B' is the same length as segment AB.
2. Copy an angle, B, onto a given ray (see Figures 12–34 a,b,c,d, and e).

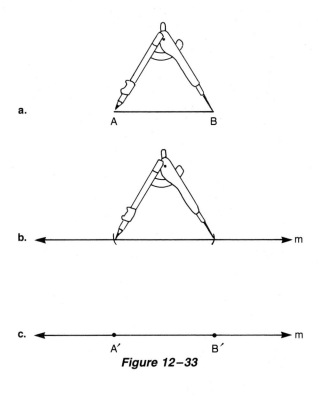

Figure 12–33

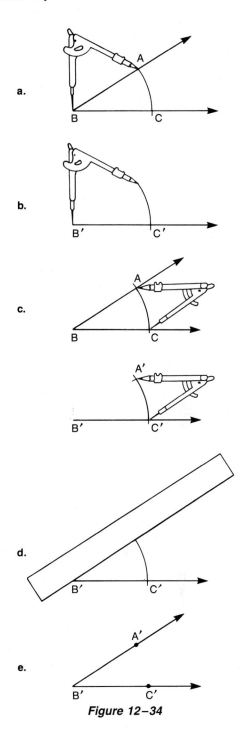

a. With B as the end point, make an arc crossing the rays at points C and A.

b. Using the same radius and B' as the end point, make an arc crossing the given ray at point C'.

c. Make A'C' the same length as AC.

d. Use the straightedge to draw ray B'A'.

e. Angle A'B'C' is the same size as angle ABC.

3. Construct a circle with a given radius, r (see Figures 12–35 a, b, and c).

   a. Spread the compass points to correspond to the length of the radius, r.

   b. Using the same radius, draw a circle.

   c. The completed circle has a radius equal to r.

The next three constructions require a somewhat higher level of skill. Instead of copying a given figure, they involve their own unique set of procedures. The first involves constructing the perpendicular bisector of a segment; the second, constructing a triangle from three given line segments; the third, constructing a hexagon inscribed in a circle.

ACTIVITIES

Intermediate (4–8)

Objective: *to use a compass and straightedge to construct a perpendicular bisector and a triangle.*

1. Construct the perpendicular bisector of a given line segment, AB (see Figures 12–36 a, b, c, and d).

   a. Using point A as the center, draw an arc.

   b. Using the same radius and point B as the center, draw another arc.

Figure 12–34

   c. Place the straightedge at the intersections of the two arcs, points X and Y. Draw segment XY.

   d. Segment XY is perpendicular to segment AB and bisects segment AB at point Z.

2. Construct a triangle with sides equal in length to three given line segments, AB, BC, and CA (see Figures 12–37 a, b, c, d, and e).

   a. Draw a line, m. On the line, copy segment AB.

   b. With point B as the center, draw an arc with a radius the same length as segment BC.

   c. With point A as the center, draw an arc with a radius the same length as segment CA.

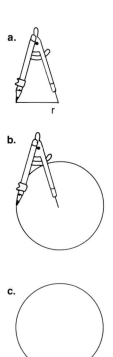

a.

b.

c.

Figure 12–35

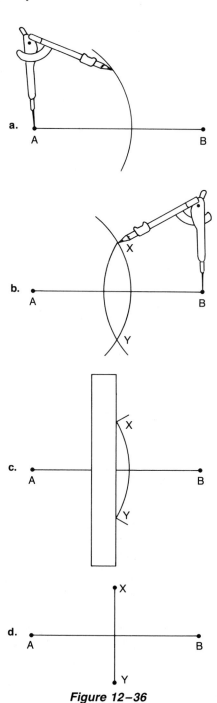

a.

b.

c.

d.

Figure 12–36

d. Use the straightedge to connect points A and B with the intersection of the two arcs at C.

e. Triangle ABC has sides equal in length to segments AB, BC, and CA.

3. Construct a hexagon inscribed in a circle (see Figures 12–38 a, b, c, and d).

   a. Construct a circle with a radius of your choice.

   b. Place the point of the compass at any location on the circumference of the circle and draw another circle.

   c. Place the point of the compass where the circumference of the second circle intersects with that of the original circle and draw a third circle. Continue around the circumference of the original circle using the points of intersection as centers until a total of seven circles have been drawn.

   d. Connect the points of the "star" to form a hexagon inscribed in the original circle. Note that the radius of the circle is also the length of each side of the hexagon. Can you find a simpler way to construct the hexagon?

The latter construction shows one of the attractive designs that can result from work with constructions. By coloring parts of the design, students can create attractive patterns that can serve as bulletin board or hallway displays. Do you see a way to connect another set of intersections on the figure to produce a second, larger hexagon?

These are but a small sampling of possible constructions using a compass and straightedge; there are many extensions of construction activities. For example, challenge students to use a mira to construct a perpendicular bisector of a given line or to call on skills learned previously to construct figures or designs. Seek out an old geometry textbook at a library or a book sale as a source of further activities.

Estimating

Throughout the activities presented in this chapter, we have suggested that you encourage children to estimate, asking, for example, "How many squares do you think can be constructed on a geoboard? How many turtle steps do you believe are necessary to hide the turtle under the shape? Which figures do you think have line symmetry?

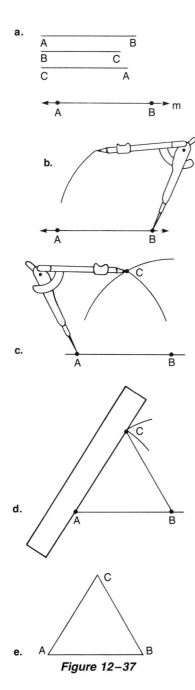

Figure 12–37

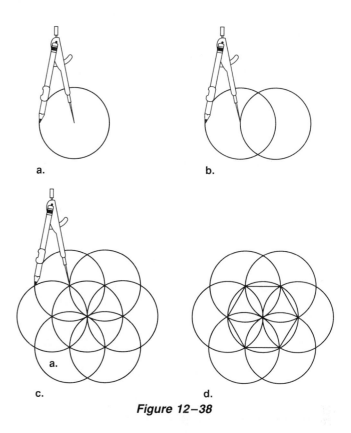

Figure 12–38

Can you tessellate with a pentagon?" All of these questions relate to estimating.

For those who actively pursue mathematical thinking, estimating is a valuable skill. As related to geometry, estimation involves the ability to reasonably guess how many, to visualize how figures will look before they are constructed, and to estimate the sizes of one-, two-, and three-dimensional figures.

Constantly challenge children to take a moment and estimate before they complete a project, activity, or exercise. After a while, estimating becomes a part of geometric thinking. The entire mathematics curriculum, then, provides students with practice in estimating. Following are several activities that reinforce estimation and relate to geometry.

ACTIVITIES

Primary (K–3)

Objective: *to estimate the sizes and shapes of various figures.*

1. On a sheet of paper, draw the outlines of five or six triangles. Use actual cutouts of the shapes to make the outlines. Then put the shapes on one table or counter and the outlines on another. Ask one child to pick up one of the triangles and then move to the edge of the table containing the outlines. Ask another child to look at the triangle being held by the first child and estimate which outline belongs to that shape. Have the child holding the shape put the triangle in the outline to see if it fits. Then ask another child to pick another of the triangular shapes. Continue the activity until all of the shapes have been fitted to outlines.

    Extend this activity by using different shapes. Use squares, rectangles, hexagons, and irregular quadrilaterals. To make the estimating more challenging, use twelve or fourteen outlines and two or three different shapes at the same time.

Objective: *to estimate and discover the number of noncongruent triangles that can be made on an isometric geoboard.*

2. Figure 12–39 shows an isometric geoboard. Begin by asking children to guess how many different triangles can be made using the first two rows of this geoboard. We find that there are twelve different triangles. (One example is shown on the geoboard in Figure 12–39). Then have the students construct as many triangles as they can.

    Later, have the students guess how many different triangles can be made using three rows of the geoboard. See how many of those triangles the children can construct. It is help-

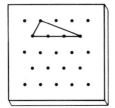

Figure 12–39

ful to provide isometric dot paper for the children to record their findings (see Appendix B). The results make a fine bulletin board display. This activity can eventually be extended to incorporate the entire geoboard.

Objective: to imagine and describe various space figures from their patterns.

3. Provide the children with patterns for various space figures. Include patterns for a cube, rectangular box, cylinder, cone, and tetrahedron. Have the children describe the figure they believe will result when the pattern is folded. Use dotted lines to indicate how the pattern will be folded. Encourage the children to sketch or find an example of the resulting space figure. Then have some children cut out and fold the figure. Compare the estimates with the final product.

   A variation of this activity is to show the children several household containers such as a cereal box, a paper towel tube, and a cracker box. Have the children sketch the pattern the container would make if it were cut apart and laid out flat. Cut the containers and compare them with the sketches.

It is helpful for children to have the opportunity to mentally visualize shapes. This allows them to gain experience in using the mind's eye as an aid in working with the visual aspects of geometry. We continue with activities for older students.

ACTIVITIES

Intermediate (4–8)

Objective: to visualize and construct a figure of a given size and shape.

1. Provide each student with one or more outlines of figures on oaktag or paper. These figures may be triangles, quadrilaterals, squares, rectangles, pentagons, or hexagons. Also provide construction paper.

   Have each student observe the outline of a figure and then using the construction paper, cut out the shape that will fill the outline. Encourage the students to devise ways to determine the appropriate size for the figure they are cutting out. When the figures have been cut out, have the students place them in the outlines and compare the results. Let the students then exchange outlines and try again.

   A variation of this activity is to put one outline on the chalkboard and provide students with construction paper. Have all the students cut out the shape that fits the outline on the board. Again, let students see how well their figures fit the outline.

Figure 12–40

Objective: to determine the results of a set of Logo commands.

2. Make a list of several Logo commands that will produce a geometric shape or design. Have the children read through the commands and attempt to draw what they believe the results will be. One set of design commands follows:

   REPEAT 2 [FD 40 RT 90 FD 60 RT 90]
   BACK 60
   END

   What do you think the results will be? (See Figure 12–40.) Have children act out the commands by walking around the room or on the playground.

   Invite individual children to suggest sets of commands and let the other children guess what the results will be. Try the commands on the computer. This particular activity helps children visualize geometric figures by mentally or physically acting out a sequential procedure.

Objective: to estimate and determine the number of squares that can be constructed on geoboards of varying sizes.

3. Have the students estimate how many squares they will be able to construct on a 5 × 5 rectangular geoboard without using any diagonal lines. After the students have guessed, encourage them to begin to systematically estimate and determine how many squares can be made on 2 × 2, 3 × 3, and 4 × 4 rectangular geoboards without using diagonals. The students should find one, five, and fourteen squares, respectively. See if they can use this information to discover how many squares can be made on the 5 × 5 geoboard.

   There is a number pattern involving the square numbers (1, 4, 9, 16, . . .) that will show that thirty squares can be made on the 5 × 5 geoboard without using diagonals. How many squares would you expect on a 6 × 6 geoboard? How many on a 10 × 10 geoboard?

   Extend this activity by including squares that involve diagonals. Encourage the students to break the problem into subproblems and then combine the results. Be sure to have the students estimate how many squares can be constructed.

Problem Creating and Solving

Just as estimating is an integral part of geometry, so is creating and solving problems. Once learned and practiced, skills in problem solving continue to serve the learner. Many of the activities discussed earlier were presented in a problem format. Following are other useful activities that provide problem-solving experiences.

ACTIVITIES

Primary (K–3)

Objective: to determine patterns for which clues have been given.

1. Make up pattern strips from railroad board, each having approximately ten 10 × 10 centimeter squares. Place objects on four to six of the squares so that a pattern is suggested. Ask the children to fill in or extend the pattern, depending on which squares have been left blank. For example, in Figure 12–41a the pattern is trapezoid, triangle, triangle, trapezoid, and so on.

    The pattern in Figure 12–41b is red triangle, red circle, red square, red diamond, then blue triangle, blue circle, and so on. In Figure 12–41c the pattern is two rectangles in a horizontal position, two rectangles in a vertical position, circle, two rectangles in a horizontal position, and so on. Finally, let's consider the pattern in Figure 12–41d.

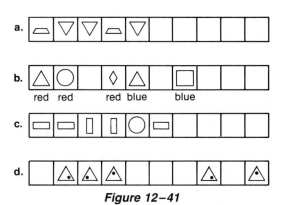

Figure 12–41

- *Understanding the problem.* What we need to do to solve this problem is to find shapes to put in the empty regions that fit the pattern already started. The figures that we can see are triangles with dots in them.
- *Devising a plan.* We will begin with the group of three triangles and look for likenesses and differences. If we find what we think is a pattern, we will move to the right along the row and see if the figures fit the pattern we have in mind (look for a pattern).
- *Carrying out the plan.* Because all of the triangles look alike, we look closely at the dots in the triangles. The first triangle has a dot in the lower right corner. In the next

triangle, the dot is in the lower left corner. In the next triangle, the dot is in the top corner.

It seems as though the dot is moving from corner to corner. If that is how the pattern works, the very first square should have a triangle with a dot in the top corner. The next three empty squares should have triangles with dots in the lower right, lower left, and top corners. The last empty region will have a triangle with a dot in the lower left corner.

We have found the pattern. It looks as if either the dots are moving around to the right inside the triangles or the triangles are rotating to the right.

- *Looking back.* When we put all of the triangles and dots in the empty regions, is the pattern of dots the same from the beginning to the end of the row? Yes, it is. The pattern must be correct.

There are many possibilities for patterns such as these. Invite children to make patterns for their classmates to complete. Children can be skillful problem creators.

Objective: to develop spatial visualization using tangram pieces.

2. Tangram pieces were used in Chapter 10 in activities related to fractions. Tangrams offer children the chance to solve puzzles and to engage in creative endeavors, as well. There are seven tangram pieces, as shown in Figure 12–42. All seven may be fitted together to make a square, as in Appendix B.

    Initial activities should include providing frames in which the children fit two or more of the tangram shapes. For example, using an *a* piece and a *d* piece, make the shape shown

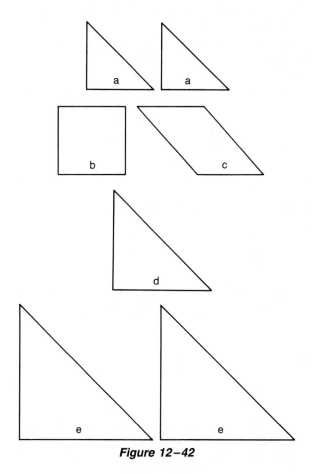

Figure 12–42

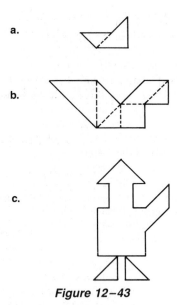

Figure 12-43

in Figure 12–43a. The children should be able to put the pieces together and achieve success. Later, use a greater number of pieces and make the shapes more difficult to complete. Ask experienced children to make a shape using all but one *e* piece, as in Figure 12–43b.

Another enjoyable tangram activity is to construct pictures of animals, people, objects, and houses using all or some of the tangram shapes. Children may fill in frames, construct their own pictures, or develop figures for other children to complete. The waving man in Figure 12–43c is an example of such a creation.

Objective: *To use clues to solve mystery shape problems on the geoboard.*

3. Provide the children with geoboards. Explain that they will be given clues to the mystery shapes. They should find at least one shape that matches each set of clues. Say, for example, "I am thinking of a shape that has 4 nails on its boundary and 1 nail inside. Can you find it?" Figure 12–44 shows three

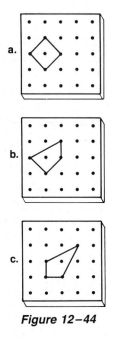

Figure 12-44

different shapes that fit the clues; there are others. Once the children have found one solution, encourage them to find others.

Here are additional clues that describe other shapes. I am thinking of a shape that has:

- 4 nails on its boundary and 0 nails inside
- 5 nails on its boundary and 0 nails inside
- 6 nails on its boundary and 0 nails inside
- 10 nails on its boundary and 2 nails inside

Once the children are able to find the mystery shapes, ask them to make up clues for shapes that other members of the class can find. Have them put the solutions on rectangular dot paper. Remind the children that often there is more than one shape that matches a set of clues.

Intermediate (4–8)

Objective: *to create dodecagons of the same size using a variety of shapes.*

1. Provide students with a set of pattern blocks and an example or two of dodecagons constructed using the blocks. Figure 12–45 shows two such figures. The challenge is to see how many different dodecagons of the same size students can make using the pattern blocks. There are more than sixty different dodecagons of the same size that can be constructed using the pattern blocks.

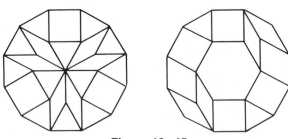

Figure 12-45

Provide outlines of dodecagons on which the students can sketch the pattern blocks used. Let the students color the sketches using the appropriate colors. Then place the figures on a bulletin board as a reference for others who are working on the project.

Objective: *to develop visual perception.*

2. Exploring pentominoes offers children the opportunity to test their perceptual and creative abilities while problem solving. A **pentomino** is a figure produced by combining five square shapes of the same size. There is one rule: Each square must share at least one complete side with another square in the figure. Three of twelve possible pentominoes appear in Figure 12–46.

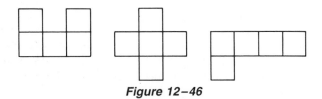

Figure 12-46

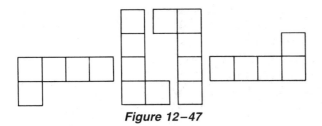

Figure 12-47

Pentominoes are two dimensional and two pentominoes are considered the same if one is a flip or a rotation of the other. For instance, the pentominoes in Figure 12-47 are considered the same.

Initially, give children numerous square shapes to explore. Squares of 3 centimeters on a side are ideal. Challenge students to find as many different pentominoes as they can. As they discover the figures, have them shade or color the patterns on a sheet of squared graph paper.

Extend this activity by having students select the pentominoes that they believe can be folded to make a box with an open top. Allow time for students to cut out the pentominoes and to attempt to fold them into boxes.

3. Another way to investigate pentominoes is to use the small milk containers commonly found in schools. Cut off the top of each container so that the bottom and the four sides are same-sized squares. Then ask the children to see how many of the twelve pentominoes they can make by cutting the cartons along the edges and without cutting any one side completely off. Figure 12-48a shows an example in which a cut was made along each of the four vertical edges and the sides were folded down. Figure 12-48b shows a different pentomino made by cutting along other edges of the milk carton.

4. Once children are comfortable with pentominoes, have them tessellate with various pentominoes. Using only one of the pentomino shapes, is it possible to cover a sheet of paper without leaving gaps? Figure 12-49 illustrates the beginnings of two tessellations.

5. Use pentominoes to further explore symmetry. Have students try to place a mirror or a mira on all or some of the pentominoes so that the reflection is the same as the part of the figure behind the mirror. In other words, do all pentominoes have line symmetry? Identify those that do and those that do not.

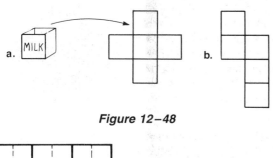

Figure 12-48

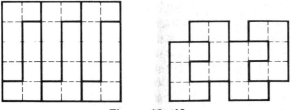

Figure 12-49

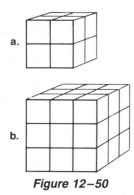

Figure 12-50

6. Have the children consider **hexominoes,** figures constructed using six square shapes. There are considerably more hexominoes than pentominoes. Each of the above activities, except the one using the milk container, can easily be done with hexominoes. Use Logo to visually construct pentominoes or hexominoes using turtle graphics.

Objective: *to analyze various cubes and determine color patterns.*

7. Make available twenty-seven small cubes with dimensions of 2 or 3 centimeters. Have the students construct a large $2 \times 2 \times 2$ cube using these smaller cubes (see Figure 12-50a). Then have the children imagine that the large cube has been painted blue. Encourage the children to make a table to record the number of faces of each smaller cube that are painted blue.

Then present the challenge. Have the students construct a large $3 \times 3 \times 3$ cube using the smaller cubes (see Figure 12-50b). Have them imagine that this cube is painted blue. Ask them to make a table to record the number of smaller cubes with (a) no faces painted blue, (b) exactly one face painted blue, (c) exactly two faces painted blue, and (d) exactly three faces painted blue. Extend the problem by asking the children to construct a $4 \times 4 \times 4$ cube and answer the same four questions regarding the faces painted blue. Here, the table will be especially useful. Then have the students try to construct a $5 \times 5 \times 5$ cube and answer the questions.

For a final, more difficult extension, see if anyone can find the various numbers of blue faces on a $10 \times 10 \times 10$ cube. This last problem may be a question of the week.

Objective: *to combine Logo procedures to generate other figures.*

8. When students have had an opportunity to work with Logo and can design certain simple shapes, such as a square, a triangle, and a circle, encourage them to solve problems using their skills. Have them construct a shape with each side a specified length in each corner of the computer monitor. Have them make the largest visible square or circle. Challenge the children to construct a large square with a circle inside it and a triangle inside the circle. See if they can construct three shapes side by side that just barely touch each other.

Here is another opportunity for students to develop problems to present to others in the class. When a design or figure has been developed, ask the inventor to sketch the design on a piece of squared paper and post it near the computer as a class challenge. Figure 12-51 shows one such student-generated problem. As well, Logo resources contain many problems for students.

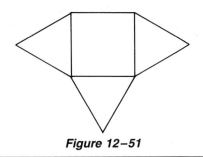

Figure 12–51

To supplement the many teacher-initiated problem-solving activities, a number of fine games, puzzles, and computer programs have been marketed commercially. Examples of these include the Rubik's Cube, Rubik's Magic, Soma Cube, The Super Egg, What's in the Square, Imagic, Mind's Eye, Perceptual Puzzle Blocks, and computer programs such as Factory, Superfactory, Gears, Building Perspective, Flip and Tip, and Right Turn.

Using a coordinate system such as that found on maps is a valuable application of geometry. The map can be an imaginary neighborhood or the neighborhood surrounding the school. The sample map shown in Figure 12–52 is the basis for the following activities and questions. Copy it for individual students. Five different, yet related, activities employ the map. These are briefly described below. Expand each activity to match the needs and abilities of the students.

ACTIVITIES

Intermediate (4–8)

Objective: *to use a rectangular coordinate system.*

1. *Where Is It?* Have the students study the map and answer the following questions:

   - Laura's Gas Station is at the corner of Second and Walnut. Where is Jack's Market?
   - Where is Fire House No. 46?
   - Where is Lincoln School? (Be careful!)
   - Where is Center City Park?

2. *How Far Is It?* Have the students use the map to follow the instructions and answer the questions below:

   - From Tom's Cafe to Fire House No. 32 is five blocks by the shortest route. See if you can draw the shortest route.
   - How many blocks is the shortest route from Alice's Place to Lincoln School?
   - Your bicycle had a flat tire on Sixth Street between Oak and Peach Streets. Give the address of the closest gas station.
   - Suppose you are standing at Third and Walnut and someone from out of town asks how to get to Jack's Market. Tell them how to get there.
   - You are a jogger and you want to jog around the outside of Center City Park for 2.5 kilometers. Every ten blocks equals 1 kilometer. Tell where you begin and finish your jog.

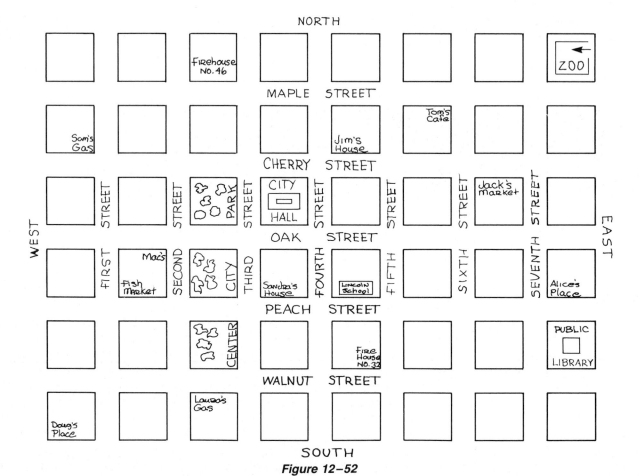

Figure 12–52

3. *Location Codes.* Have the students use the map to answer the following questions:

- Suppose you are part of a group at Doug's place. All of a sudden, one member of the group says, "I know a new way to tell where places are." He goes on, "Laura's Gas Station is (2,1)," and he writes it down. "Sam's Gas Station is (1,4)." Do you see the code?
- Using the code, where is Jim's house?
- Using the code, where is the Zoo entrance?
- What is at (7,1)?
- What is at (3,2)?

4. *Following Directions.* Have the students use the map to follow the directions below:
   a. Place an "A" at the corner of Fifth and Cherry Streets. The A will represent where you are.
   b. Walk two blocks east, three blocks south, two blocks west, and one block north. Place a "B" at the corner where you have stopped.
   c. Beginning at B, walk one block east, three blocks north, one block east and place a "C" at the corner where you have stopped.
   d. Beginning a C, start out walking south and zig-zag south and west, alternating one block at a time and walking five blocks in all. Place a "D" at the corner where you have stopped.
   e. Beginning at D, walk three blocks west, three blocks north, one block east and place an "E" at your final stopping point (Third and Maple Streets).

5. *A Trip to the Zoo.* Let students play the following game using the map: You and a friend decide to go to the Zoo. You both meet at Doug's place and agree to make the trip in an unusual way. You will need a pair of dice to give directions. (You may use dice like those in Figure 12–53 or regular dice.)

Each die has 3 "East" and 3 "North" faces.

Figure 12–53

The faces on the dice give you the directions east and north. East means to go one block east and north means to go one block north. (With regular dice, even numbers—2, 4, and 6—mean to go east and odd numbers—1, 3, 5—mean to go north).

You and your friend want to see who will get to the Zoo first by rolling the dice and following the directions. Begin now, and see who arrives at the Zoo entrance first. If you go directly past City Hall, you get an extra throw.

Children's awareness of geometry in the environment is heightened considerably as you focus attention on various applications of geometry. This awareness also strengthens students' appreciation for and understanding of geometry.

Grouping Students for Learning Geometry

When students engage in learning geometry, a variety of grouping patterns are appropriate. For example, when younger children are learning concepts such as *near, far, on, in,* etc., you may wish to have the children all together in a discussion corner or in another area of the room. This allows several children to participate simultaneously, allows the students to carry on a discussion, and allows the teacher to observe the work of the children. Other examples of whole-class activities include introductory work on the geoboard, introductory work with Logo, constructions of polyhedra models, and projects such as building a geodesic dome.

Cooperative learning groups are appropriate for activities in which materials may be shared and for those in which problems are presented. Examples of learning group activities include geoboard problems, tessellations, pentominoes, mirror symmetry, soap films on wire frames, constructing shapes with a compass and straightedge, and pattern block challenges.

Individual or pair learning may best take place when students are exploring the Logo environment or making line drawings or coloring patterns. Once children have learned how to make objects by paper folding, folding is done individually.

By and large the types of activities suggested in this chapter involved the active participation of the children. Such activities tend to be social activities; that is, they are effectively accomplished when children are working together and comparing and discussing their work. Even the skills of geometry are effectively learned as children work side by side informally.

Communicating in Learning Geometry

Description is an important part of communicating in learning geometry. As children observe shapes, discover their properties, develop definitions involving essential characteristics, and draw and construct shapes, their ability to communicate their thoughts is fundamental. For the students, part of the communication process is drawing sketches that illustrate their work. Written communication is used to describe their work and to put into words the shapes and forms with which they are working. Oral communication serves a similar purpose. For example, "In your group today, you are to describe the figure that has been provided. The reporter for your group will present the description orally to the rest of us and we will attempt to sketch the figure from the description. Tomorrow, we will do a similar activity but your group's recorder will write a description that you will share with other groups to see if the members of the other groups can sketch the figure that has been described."

Children enjoy writing and illustrating theme books, for example, a book featuring round objects with pictures and written descriptions of round things found at school and at home. Older students may keep journals of shapes with descriptions of the shapes and where they are found. These journals can spark ideas for creative stories about various shapes. The shape descriptions in Juster's *The Phantom Tollbooth* (1961) will surely inspire writing that includes shapes.

The oral language that children use to describe their movement as they walk around geometric figures made with tape or yarn on the floor becomes the basis for writing Logo procedures. The language of the children is translated into the language of Logo; likewise, children can read Logo procedures and describe them in their own words.

Cooperative and individual writing are appropriate activities for cooperative learning groups. As group members explore geometric concepts and develop skills, the group recorder provides a chronicle of the thinking process. This chronicle is then shared with other groups during the debriefing time at the end of the activity. Discussion is then invited. Children question one another and seek clarification of the ideas that have been put forth.

Evaluating Geometric Learning

In evaluating geometric learning, consider the objectives. When a school or a district adopts an elementary mathematics textbook series, it is, by and large, adopting a collection of objectives. The objectives are found throughout the teacher's manual for each level in the series. The geometry presented in a basal math series is reviewed and evaluated at the end of a chapter or section. Chapter tests help in evaluating geometric learning but do not tell the whole story.

Activities provide opportunities for observing the actual performance of the children and making additional evaluations. Observations are most helpful when teachers make anecdotal records at the time of or shortly after the observation. With the large number of useful geometry activities, there is ample opportunity for teachers to observe and note children's behavior as the children actively engage in learning geometry.

Further, we recommend that the teacher expand on the material presented in the math textbook. Thus, students have a greater opportunity to advance from the first and second van Hiele levels (recognizing and describing shapes) to the third level (establishing relationships between figures and their properties).

Teacher-made tests provide the surest evaluation of content not found in the math textbook. A test may re-

1. Sketch the following on the geoboards below.

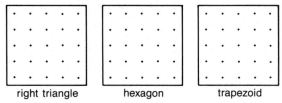

right triangle   hexagon   trapezoid

2. Circle each pentomino that has exactly <u>one</u> line of reflection.

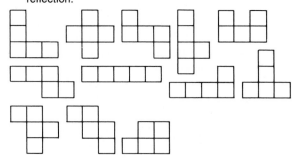

3. Make the following design with Logo commands. Use REPEAT and the procedure SQUARE.

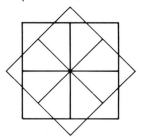

Figure 12–54

quire paper and pencil or it may be a task requested by the teacher. Figure 12–54 shows three sample paper-and-pencil test items.

The same material can be tested by asking students to construct a particular figure on their geoboards and then to hold the boards up for the teacher to see. Likewise, students can be given pentominoes and asked to use a mirror or a mira to find those figures with exactly one line of symmetry. Another task may be to produce a particular design or figure on the computer using Logo.

Evaluation is an ongoing task for the teacher. The more information you gather, the better able you will be to fit instruction to the learning styles of your students. Continually monitoring students as they work is among the most important tasks of the teacher.

Something for Everyone

Many of the activities in Chapters 12 and 13 require children to work in visual or in spatial learning modes. Of course, children use other learning modes as they learn the concepts and skills of geometry and measurement. To avoid repetition, the discussion of the learning modes associated with geometry and measurement is presented at the end of Chapter 13.

KEY IDEAS

Geometry abounds in our environment. From nature to architecture, we are continually surrounded by examples of geometry. This chapter presents ways that geometric concepts can be experienced by children. When children enter school, they are in the process of completing a transition from perceiving the world from a topological point of view to perceiving it from a Euclidean point of view. Exhibit geometric figures in many forms and allow children to see, touch, feel, and talk about them.

Textbooks play an important role in defining and illustrating basic geometric figures. Materials such as pattern blocks, geoblocks, tiles, geoboards, mirrors, and Logo also serve children as they work at the early van Hiele levels.

There are many fine projects in which classes of children can engage as they learn geometry. The skills of geometry include the ability to name figures and their various properties, the ability to copy figures, and the ability to construct figures. Children should practice visualizing figures. To help evaluate their progress in learning geometry, observe children as they work.

REFERENCES

BEARDEN, DONNA; MARTIN, KATHLEEN; AND MULLER, JAMES H. *The Turtle's Sourcebook*. Reston, Va.: Reston Publishing Co., 1983.

BENNETT, ALBERT; MAIER, EUGENE; AND NELSON, L. TED. *Math and the Mind's Eye: 1. Seeing Mathematical Relationships*. Salem, Or.: Math Learning Center, 1987.

_____ . *Math and the Mind's Eye: II. Visualizing Number Concepts*. Salem, Or.: Math Learning Center, 1987.

_____ . *Math and the Mind's Eye: III. Modeling Whole Numbers*. Salem, Or.: Math Learning Center, 1987.

_____ . *Math and the Mind's Eye: IV. Modeling Rationals*. Salem, Or.: Math Learning Center, 1987.

_____ . *Math and the Mind's Eye: V. Looking at Geometry*. Salem, Or.: Math Learning Center, 1987.

BILLSTEIN, RICK; LIBESKIND, SHLOMO; AND LOTT, JOHNNY W. *MIT Logo for the Apple*. Menlo Park, Ca.: The Benjamin/Cummings Publishing Co., 1985.

BURGER, WILLIAM F. "Geometry." *Arithmetic Teacher*. Vol. 32, No. 6 (February 1985), pp. 52–56.

BURGER, WILLIAM F., AND SHAUGHNESSY, J. MICHAEL. "Characterizing the van Hiele Levels of Development in Geometry." *Journal for Research in Mathematics Education*. Vol. 17, No. 1 (January 1986), pp. 31–48.

CLITHERO, DALE. "Learning with Logo 'Instantly'." *Arithmetic Teacher*. Vol. 34, No. 5 (January 1987), pp. 12–15.

COPELAND, RICHARD. *Diagnostic and Learning Activities in Mathematics for Children*. New York: Macmillan Co., 1974.

COPELAND, RICHARD W. *How Children Learn Mathematics*. New York: Macmillan Co., 1984.

COWAN, RICHARD A. "Pentominoes for Fun Learning." *The Arithmetic Teacher*. Vol. 24, No. 3 (March 1977), pp. 188–190.

CRUIKSHANK, DOUGLAS E., AND McGOVERN, JOHN. "Math Projects Build Skills." *Instructor*. Vol. 87, No. 3 (October 1977), pp. 194–198.

DIENES, Z. P., AND GOLDING, E. W. *Geometry Through Transformations: 1. Geometry of Distortion*. New York: Herder and Herder, 1967.

FOUKE, GEORGE R. *A First Book of Space Form Making*. San Francisco: Western Book Services, 1974.

FUYS, DAVID. "Van Hiele Levels of Thinking in Geometry." *Education and Urban Society*. Vol. 17, No. 4 (August 1985), pp. 447–462.

GIGANTI, JR., PAUL, AND CITTADINO, MARY JO. "The Art of Tessellation." *Arithmetic Teacher*. Vol. 37, No. 7 (March 1990), pp. 6–16.

JURASCHEK, WILLIAM. "Getting in Touch with Shape." *Arithmetic Teacher*. Vol. 37, No. 8 (April 1990), pp. 14–16.

KENNEY, MARGARET J., AND BEZUSKA, STANLEY J. *Tessellations Using Logo*. Palo Alto, Ca.: Dale Seymour Publications, 1987.

LAYCOCK, MARY. *Bucky for Beginners*. Hayward, Ca.: Activity Resources Company, Inc., 1984.

MANSFIELD, HELEN. "Projective Geometry in the Elementary School." *Arithmetic Teacher*. Vol. 32, No. 7 (March 1985), pp. 15–19.

McKIM, ROBERT H. *Thinking Visually*. Belmont, Ca.: Lifetime Learning Publications, 1980.

Mira Math for Elementary School. Palo Alto, Ca.: Creative Publications, 1973.

MOORE, MARGARET L. *LOGO Discoveries*. Palo Alto, Ca.: Creative Publications, 1984.

_____ . *Geometry Problems for LOGO Discoveries*. Palo Alto, Ca.: Creative Publications, 1984.

MORRIS, JANET P. "Investigating Symmetry in the Primary Grades." *The Arithmetic Teacher*. Vol. 24, No. 3 (March 1977), pp. 188–190.

NATIONAL COUNCIL OF TEACHERS OF MATHEMATICS. *Curriculum and Evaluation Standards for School Mathematics*. Reston, Va.: NCTM, 1989.

ONSLOW, BARRY. "Pentominoes Revisited." *Arithmetic Teacher*. Vol. 37, No. 9 (May 1990), pp. 5–9.

PIAGET, JEAN. "How Children Form Mathematical Concepts." *Scientific American*. Vol. 189, No. 5 (November 1953), pp. 74–78.

PIAGET, JEAN, AND INHELDER, BARBEL. *The Child's Conception of Space*. New York: W. W. Norton & Co., 1967.

"Play Dome." *Sunset*. Vol. 50, No. 1 (January 1973), pp. 51–53.

PRENTICE, GERARD. "Flexible Straws." *Arithmetic Teacher*. Vol. 37, No. 3 (November 1989), pp. 4–5.

RANUCCI, E. *Seeing Shapes*. Palo Alto, Ca.: Creative Publications, 1973.

SEYMOUR, DALE, AND BRITTON, JILL. *Introduction to Tessellations*. Palo Alto, Ca.: Dale Seymour Publications, 1989.

SEYMOUR, DALE, AND SCHADLER, REUBEN. *Creative Constructions*. Palo Alto, Ca.: Creative Publications, 1974.

STEVENS, PETER S. *Patterns in Nature*. Boston: Little, Brown & Co., 1974.

SUYDAM, MARILYN N. "Forming Geometric Concepts." *Arithmetic Teacher*. Vol. 33, No. 2 (October 1985), p. 26.

WENNINGER, MAGNUS J. *Polyhedron Models for the Classroom*. Reston, Va.: National Council of Teachers of Mathematics, 1975.

WIRSZUP, IZAAK. "Breakthrough in the Psychology of Learning and Teaching Geometry." In Martin, J. Larry, ed. *Space and Geometry: Papers from a Research Workshop*. Columbus, Oh.: ERIC Center for Science, Mathematics Environmental Education, 1976.

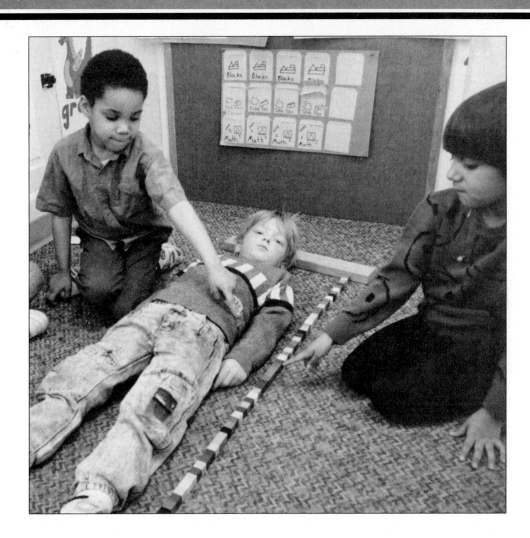

Measurement provides quantitative information about certain familiar aspects of our environment. These include length, distance, area, volume, capacity, weight, mass, temperature, time, and angle. Of course, this list is not complete. We know that certain specialized occupations require measures not commonly encountered. For example, surveyors, sailors, and pharmacists may use chain, nautical, and apothecaries' fluid measures, respectively.

In the child's world, it is enough to find out which is biggest, smallest, longest, shortest, fastest, slowest, warmest, or coolest. Children may ask how much or how long when they want to know about measured quantities. Later, youngsters are intrigued by the entries found in publications such as the *Guinness Book of World Records*. A majority of the records reported include some type of measurement.

A world record has little meaning if the units of length, time, or amount are not understood. The basis for establishing understanding is measuring. From measuring springs a sense of quantity relating to the item being mea-

sured. A sense of meters and seconds helps put the world record for the women's 1,500 meter run in perspective.

Where do these ideas begin to be established? Earlier we mentioned that children flailing about in their cribs explore space. They explore distance as well. Before children talk, they answer the question "How big are you?" by spreading their arms out, indicating to the pleasure of their parents that they are so big.

From these early beginnings, children compare objects and judge sizes. They order objects by size, weight, length, and duration. They develop an eye for size. "Her cookie is bigger than mine" expresses this visual comparison. In their early experiences, children develop the ability to measure by perception. By the time they enter first grade, they can begin simple measurement activities. Hiebert (1984, pp. 22–23) noted, "Effective instruction should take advantage of what children already know or are able to learn and then relate this knowledge to new concepts that may be more difficult to learn."

Most measurements require tools. Thus, the manipulative materials for teaching measurement include the standard measuring instruments. We recommend using rulers, meter sticks, tape measures, trundle wheels, graduated beakers, measuring cups, measuring spoons, pan balances, bathroom scales, thermometers, timers, and protractors. Other useful materials are geoboards, centimeter cubes, Cuisenaire rods, containers of various sizes, string, and Logo. Teachers should collect objects and materials to help in developing measuring concepts.

Helping students to learn measurement involves more than presenting a series of activities or textbook pages. Not only should measurement be taught, it should also be practiced throughout the school year whenever measuring is needed. You should be alert to measuring situations and encourage the students to measure when the opportunity arises.

Developing Measurement Concepts

The concepts that provide the foundation for measuring skills are those of length and distance, area, weight and mass, time, volume and capacity, temperature, and angle. Angle measure is a concept traditionally developed at age 10, but the utility and popularity of the computer language Logo suggests that an earlier introduction to angle measure is appropriate.

By and large, as measurement concepts are introduced, children should be provided with a sequence of activities: direct comparison, indirect comparison, arbitrary units, and standard units. Through these activities, children develop their concepts of measurement.

Direct comparison means that children take two objects and place them side by side or one on top of the other to discover if they are the same size. This requires that both objects be on surfaces of the same height or in containers of the same diameter.

Using **indirect comparison,** children determine if the sizes of two objects are the same when the objects cannot be directly compared. For example, indirect comparison would be used to find if a table would fit through a door if the table were not easy to move.

Arbitrary units of measure are used to strengthen a child's understanding of unit. The length of a drinking straw and the area of a floor tile may be used to measure a variety of objects. The transition to **standard units** of measure, those commonly accepted and used throughout the world, follows work with arbitrary units.

The two common standard sets of measuring units are the metric and the conventional systems. We use the metric system throughout this chapter. Most textbooks for children include sections on both metric and conventional units. The process of measuring is the same regardless of the specific unit being used. You should have little difficulty teaching either metric or conventional units.

As children move through the sequence just described, they begin by using **continuous measurement.** When a piece of string is stretched along the object being measured and then compared with another object or when two objects are directly compared, continuous measurement is used. The measuring tool (string) does not assign a number to the object; rather, it is used to compare lengths.

Later in their work, children begin to use **discrete measurement.** When a pencil is placed end to end to determine the length of an object or a meter stick is used, the type of measurement is discrete. The measuring tool is used repeatedly or it shows calibrations of a given unit such as centimeters. The transition from continuous to discrete measurement is seen in the activities of this chapter.

When teaching measurement, consider children's stages of readiness. Are children ready to learn measurement concepts and skills when they enter kindergarten or first grade? The work of Piaget (1960) suggests that until children have reached certain stages of intellectual development, they will have difficulty measuring successfully. For example, children who are unable to conserve length may believe that a measuring stick changes length as it is moved. Thus, measuring length should be held off until the child is able to conserve length.

More recently, Hiebert found that the absence of conservation did not seem to limit children in learning most measurement concepts. Hiebert (1984, p. 24) noted, " . . . it appears more productive to involve children in a variety of concrete measuring activities than to wait until they develop certain logical reasoning processes."

The teacher can provide experiences in informal measuring to serve as a foundation for later work. It is important during the primary years to provide activities to form the basis for measurement and to introduce measurement skills, thus allowing children to develop their personal understanding of measurement. Prior to involving children in organized activities, time should be provided for children to play with objects, containers, and water, rice, or sand. This play helps children establish the basis for early measurement activities. The activities that follow provide experiences in visual perception and direct comparison. Indirect comparison, arbitrary units, and standard units are presented in the next section, Developing Measurement Skills.

Length and Distance

Length refers to the measure of how long a thing is from end to end. **Distance** is the space between points or objects.

In introducing length, allow children to experience long and short distances. Permit the children to directly compare objects in order to determine which object is taller and which is shorter.

ACTIVITIES

Primary (K–3)

Objective: *to experience long and short distances.*

1. Ask the children to name something in the classroom that is long. A brief discussion about the meaning of long may be necessary. Children may suggest that these objects are long: a chalkboard, a work table, or the bar on the coat rack.

   Mention that something that is long has great length and ask which has the greatest length, the chalkboard or the work table. The children will use their visual perception to make this judgement. Next, ask the children to name something on the playground that is long and then something that is not found at school that is long. What is the longest thing they can think of, the object with the greatest length?

2. Mark off several long line segments (6 to 8 meters) and several short line segments (2 to 3 meters) on the classroom floor or on the playground with masking tape, yarn, or chalk. Challenge the students to walk along one short line segment, then another short line segment, then a long line segment, and so on, allowing them to find the various long and short line segments. Discuss the meaning of long and short segments as used here. As a variation, provide long and short curves on which the children may walk.

3. Begin by saying that when the students line up for lunch or recess, a line with four or fewer students is a short line and one with more than four is a long line. In the classroom or on the playground, mark seven or eight places at which to line up. Have the children get into long lines. This may result in one long line or several lines with at least five students in each one. Then have them get into short lines. Challenge them to get into one long line and three short lines or two long lines and two short lines. See if they can form the longest line possible or the most short lines possible.

4. Provide children with eight or ten classroom objects. Each object should be able to stand by itself. Objects may include a cottage cheese container, a can, a jar, a book, and a Cuisenaire rod (see Figure 13–1). Spread the objects out on a table.

   Let the children discuss which is the tallest object, then the shortest. Are two objects the same height? Find two objects taller than the cottage cheese container. If the children have some difficulty in visually determining the taller and shorter objects, challenge them to devise a way to determine these objects more easily. Putting the objects side by side is an answer you may expect.

   A variation of this activity is to provide objects that do not stand on end: pencil, spoon, eraser, stapler, and so on. To be directly compared, the ends of each pair of objects must be lined up. This task is slightly more difficult than the one above. Again, let the students discuss how the objects can be compared.

5. In each of two bags, place one Cuisenaire rod. The rods should be about the same length, such as dark green and black or brown. Let students take turns reaching into the bags and telling which rod is longer or shorter merely by touch. Reaching into both bags simultaneously may be the most effective procedure, but let the children experiment.

   A variation is to put an entire set of ten rods in one bag and display another set of ten rods on a table. Have a child reach into the bag, grab a rod, and then choose a rod from the table that is the same length.

Area

The activities above were intended to show how children can be introduced to perceptual and direct measurement of length. The following activities focus on area. **Area** is a measure of the size of a two-dimensional figure, such as a rectangle or a circle.

ACTIVITIES

Primary (K–3)

Objective: *to determine the shape with the largest area.*

1. Provide children with a collection of square shapes with sides ranging from 4 to 10 centimeters. Arrange the shapes in a random pattern as in Figure 13–2. Ask the children to tell which shape they believe is the largest or has the largest area. To show which one is largest, compare those selected by holding them up against each other. Order the shapes from smallest to largest by comparing them.

   Extend this activity by using several rectangles, circles, triangles, and diamonds. Initially, estimate and compare using the same shape. Later, estimate and compare area using different shapes.

2. Have the students draw around one of their feet onto a sheet of paper. Put several of the outlines in a cluster. Ask the children, "Whose foot do you think is the largest or has the largest area?" Then compare those selected by holding them up against each other in front of a window. Find out who has the smallest foot in the same way. Ask, "Are there two children who have the same foot size?"

   As a variation, repeat the activity but compare hands instead of feet. Then compare hands with feet and determine the largest area. It will be difficult to make the latter comparison; thus, the activity will help children to see a need for other ways to measure area.

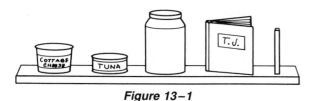

Figure 13–1

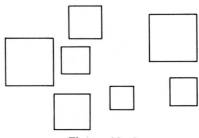

Figure 13–2

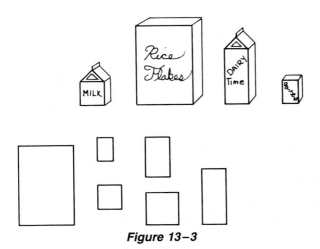

Figure 13–3

Objective: *to directly compare areas.*

3. Collect several boxlike containers such as a cereal box; half-pint, quart, and half-gallon milk containers; a raisin box; and a crayon box. Trace the faces of these containers onto tagboard and let the children cut them out as in Figure 13–3.

   Give each child several of the cutout faces. Have the children estimate to determine which of the containers their pieces belong to. Then have them see if the pieces actually fit these containers by holding the pieces against the faces of the boxes.

   A variation of this activity is to display the containers and ask the children to cut out a square or rectangular shape the same size as a face of a container. When the shapes have been cut out, have the children compare the cutouts by holding them up against the faces of the containers.

Objective: *to determine which of various shapes are larger or smaller than a given shape.*

4. Cut a piece of cardboard about the size of the top of a student desk. Hold the piece of cardboard up and ask the children, "Who can name a shape in the room that is bigger than this one? Look around and see if you can find one."

   After several shapes, such as the door, a window, and the floor, have been named, ask, "Who can see a shape in the room that is smaller than this shape? Let's name them." Suggestions may include a chair seat, a piece of tablet paper, a book, and a computer disk. As items are mentioned, compare them with the piece of cardboard to reinforce the notion of direct comparison. Periodically, provide a different-sized shape and repeat the process.

Weight and Mass

We turn our attention to activities of direct comparison that involve the weight and mass of objects. **Weight** is a measure of the force of gravity acting on an object. **Mass** is the amount of matter in an object.

ACTIVITIES

Primary (K–3)

Objective: *to determine which object is heavier.*

1. Have available several classroom objects such as a pencil, scissors, an eraser, crayons, a small box, an orange Cui-

Figure 13–4

senaire rod, a glue container, and a book. Have children compare the weights of various pairs of these objects and tell which object weighs the most by holding one object in each hand as in Figure 13–4.

   Perhaps some items are of equal weight. Have the children compare and recheck pairs until they are able to put the objects in order from lightest to heaviest. Among the objects used should be one or two large objects that are light, such as a paper cup, and small objects that are heavy, such as a rock. This is to help avoid confusing the properties of size and weight.

   To extend this activity, use a pan balance to compare the objects mentioned above to confirm or challenge the order determined by feel. Figure 13–5 illustrates one such balance. Do not be surprised to find discrepancies between the order established by holding objects and that established using a pan balance. Children enjoy experimenting with the balance and various objects found around the room.

Objective: *to practice estimating and weighing different materials.*

2. Fill several half-pint milk cartons with different materials such as rice, beans, split peas, clay, plaster of paris, and wooden

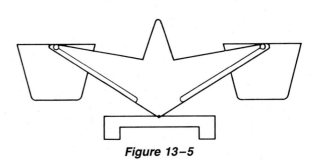

Figure 13–5

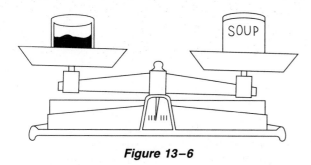

Figure 13–6

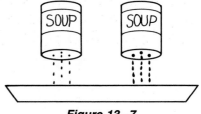

Figure 13–7

cubes. Seal the cartons and label them by color or letter. Tell the children what the materials are but do not identify the contents of a particular carton. Have the children guess how to order the cartons by weight according to what they contain. Then let the children order the cartons by weight, holding them in their hands and using the pan balance to check their estimates.

Objective: *to determine the amount of one material that weighs as much as a selected object.*

3. Collect several common objects such as a chalkboard eraser, a can of soup, a container of cleanser, a stapler, and a tape dispenser. Provide a pan balance and a material such as rice, water, clay, or sand.

   Have the children estimate how many small containers of sand it would take to balance the can of soup. Then see how many small containers must be used to balance the soup can, as in Figure 13–6. If the soup can does not balance, have the children increase or decrease the amount of sand in one of the containers. On the chalkboard, record both the estimates and the final results. If the final number of containers is not a whole number, the results should be recorded as, for example, "more than 5 containers but less than 6."

   Repeat this activity using other materials and objects. Be sure to have the children estimate each time they begin.

Time

Time is a measure of the period between two events or the period during which something happens. It is also a precise moment determined by a clock. Introduce time to children by having them directly compare the times of events, establishing whether one event takes more time or less time than another. Events may include twenty foot taps, a ball bouncing ten times, ice melting under different conditions, and water emptying out of cans with different-sized holes.

ACTIVITIES

Primary (K–3)

Objective: *to determine which event among several events takes the most or the least amount of time.*

1. Select several events that require a short period of time to complete. Events that may be used include tapping a foot twenty times; sitting in a chair, standing, and sitting again ten times; hopping on one foot from one side of the classroom to the other; reciting the words to "Row, Row, Row Your Boat"; moving twenty cubes one at a time from one container to another; bouncing a ball ten times; and pointing to and naming ten other students.

Ask the children, "Which of these events can be done the fastest? Let's write on the board those that we think can be done the fastest. Which will be the slowest events? Let's write those down."

Choose two of the events and begin them at the same time. Record the results. Repeat the process for each pair of the events and determine which event takes the least time and which event takes the most time.

Objective: *to investigate the amount of time it takes ice cubes to melt.*

2. Fill two equivalent containers with water, one with warm water and the other with cooler water. Leave a third container empty. Let the children see and feel the cups and the water. Explain, "We are going to put an ice cube in each container. Which ice cube do you think will melt first? Let's write our guesses on the board. How many believe the ice cube in this container will melt first?"

   After the guesses are recorded, put the ice cubes in the containers and observe the results. Variations of this activity include using salt water, using very cold and hot water, using varying amounts of water that are the same temperature, and dissolving sugar cubes instead of melting ice.

Objective: *to estimate and determine which of several containers empties in the least amount of time.*

3. Take several soup cans and puncture three holes in the bottom of each one, using a different-sized nail for each can. Have the children inspect the cans and estimate which they think will empty first, second, third, and so on.

   Using two cans at a time, pour equal amounts of water in each can and observe them (see Figure 13–7). Repeat this process until the cans have been ordered from fastest to slowest. Similarly, the cans may have different numbers of holes of the same size or different numbers of holes of varying sizes.

Volume and Capacity

The activities above focus on directly comparing events involving the passage of time. The following activities involve volume and capacity. **Volume** is the amount of space contained within a three-dimensional figure. **Capacity** is the amount of space that can be filled.

ACTIVITIES

Primary (K–3)

Objective: *to estimate and determine which of several containers holds the greatest amount.*

1. Provide the children with several small jars, cans, or plastic containers. For example, you may use a tuna can, a plastic

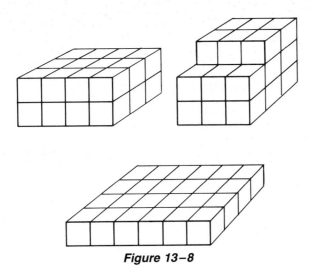

Figure 13–8

drinking cup, a baby food jar, a soup can, a paper cup, and a peanut butter jar. Have the children estimate which of the containers will hold the least amount, the next to the least amount, and so on, until they have found the container that will hold the greatest amount. Once the estimates have been made, have the children pour rice or water from container to container, checking to see which holds the most. Compare the actual order with the estimated order.

Objective: *to determine which containers will hold the same amount of material.*

2. Collect ten or twelve cups and cans of varying sizes. Point out one of the cups and fill it with rice. Have the children estimate which of the other cups will hold the same amount of rice as the first cup.

   Have the children pour the rice from the first cup into the cups they believe have the same volume. Then have them check the cups they thought were too large or too small by pouring the rice into them. In this way, the children can confirm or reject their guesses.

Objective: *to construct buildings of various shapes using identical numbers of cubes.*

3. Provide children with five or six groups of twenty-four colored cubes. Have them construct a building using the twenty-four cubes. When they have finished, challenge them to build different buildings using the twenty-four cubes, as in Figure 13–8. Talk to the children about the shapes of the buildings and whether the different buildings have the same amount of space. If it is unclear that they contain the same space, have the children count the cubes in each building.

Temperature

From activities involving direct comparisons of volume and capacity, we turn to activities dealing with temperature. **Temperature** is a measure of the hotness or coldness of a material.

ACTIVITIES

Primary (K–3)

Objective: *to determine relative temperature by feel.*

1. Partially fill five sour cream containers with water of varying temperature. Use unheated water, warm tap water, cold tap

water, water that has one ice cube in it, and water with five or six ice cubes in it. Label the containers with shapes or colors for identification. Have the children put the containers of water in order from warmest to coldest by putting their hands in the water. Have several small groups of children complete the ordering and compare results.

   After the containers of water have been in the classroom for an hour or so, have the children repeat the exercise. Discuss the results with the children.

2. Take the children to the playground and let them work in groups of three or four. Have the children search the play area for four things that feel cool to them and four things that feel warm. The children may find the slide and swing supports to be cool; they may find the asphalt and wooden play apparatus to be warm.

   Discuss the findings with the whole group. Ask which was the coolest single object and which was the warmest single object. Talk about what the objects are made of.

3. Arbitrarily group children into groups of six or eight. Have the children in each group shake hands and find the individual with the warmest hand and the individual with the coldest hand. Have those selected form a group and find the child with the warmest hand and the child with the coldest hand. Have those two children circulate among the other children and put their hands against the cheeks of the others.

   Ask the children if they know of ways to make their hands warmer. Suggestions may include putting their hands in warm water, rubbing their hands together or rubbing them against their clothing, and clapping their hands. Have the children warm their hands using one of the methods suggested and then have them shake hands again or put their hands against the cheeks of the others.

Angle

The activities above involve children comparing the temperatures of several materials by feeling them. The following activities introduce informally the idea of measuring angles. An **angle** is the space between two rays that share a common end point.

ACTIVITIES

Primary (K–3)

Objective: *to estimate and determine the sizes of angles.*

1. In Chapter 12, several activities are described in which children investigate geometry using Logo. It is necessary for children to experiment with various turns by using commands such as RT 90, RT 30, LT 50, LT 180, and RT 360. At this stage, the children are not measuring angles, but they are developing some initial understanding of angle size. Success with turtle geometry depends on the concept of angle size.

   Have children record and label angles of various sizes. Is the angle made with the commands FD 75, BK 75, RT 30, FD 75, and BK 75 the same size as the angle made with FD 75, BK 75, LT 30, FD 75, and BK 75? How are these two angles different?

   As children begin to construct simple figures, they will be using exterior angles. An **exterior angle** is the angle between the side of a polygon and the extension of the adjacent

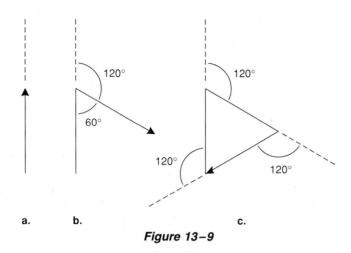

a.   b.   c.

Figure 13–9

side. For example, in order to construct an interior angle of 60° when drawing an equilateral triangle, the procedure makes the turtle turn an exterior angle of 120° as in Figure 13–9. Children develop skill with exterior angles by designing procedures for drawing a variety of polygons.

2. Trace angles of several sizes on overhead transparencies, putting one angle in the center of each transparency. Place a transparency on the monitor screen so the vertex of the angle is centered on the Logo turtle, with one ray in a vertical position as in Figure 13–10. To confirm the alignment of the angle, have the child move the turtle forward and then back to the original position. Realign the angle if necessary.

Have the child estimate the size of the turn necessary to follow the other ray of the angle. Let the child test the estimate by turning the turtle in the appropriate direction and moving forward and then back to produce a second ray. Have the child compare the angle on the transparency with the angle made using Logo.

Repeat the activity using various angle sizes. Variations include having the turtle turn left for some angles as well as right and having the turtle begin facing in a different direction rather than toward the top of the screen.

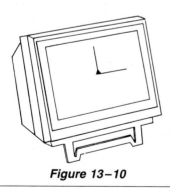

Figure 13–10

The activities in this section were presented to serve as examples to reinforce and supplement the math textbook and to provide a foundation on which to build the skills of measuring. Each activity involved direct comparison. You are encouraged to pay attention to the perceptions and observations of children as they begin measuring tasks.

Some of the perceptions will indicate that further direct measurement activities are needed.

Developing and Practicing Measurement Skills

The skills of measurement include using instruments or tools to determine how long, how much, what size, and what temperature. Specifically, the skills of measurement include measuring length, area, weight, time, volume and capacity, temperature, and angle. The skills follow the development of early measurement concepts. Then, consistent practice with the tools of measurement is necessary.

The activities presented in the previous section emphasized direct comparison, the first stage in a four-stage teaching sequence. The remaining stages are indirect comparison, using an arbitrary measuring unit, and using a standard measuring unit. The activities in the following section focus on the latter three stages for each type of measurement described.

Active experiences with measuring provide children with the chance to construct meaning relative to the measurement process. To build understanding of measurement is the focus of the children's active participation. This understanding leads to skill in measuring. Be cautious about telling children how to measure before they have had a chance to experiment.

Measuring Length

When primary children have had experiences using direct measurement of length for a variety of tasks, they should engage in other types of length measurement. Their experiences will provide a basis for constructing knowledge of length as they work through the measurement sections of their textbooks.

Intermediate level students should have worked with and used length measurement for some time. They will be engaged in estimating with accuracy and measuring with increased precision. The lengths they measure will be longer. They will easily use a variety of standard units of measure. The common metric units of length include centimeter, meter, and kilometer.

As you introduce various units of metric measurement, discuss the prefixes and their meanings. The mathematics textbook will also introduce and explain them. **Milli** means one thousandth, **centi** means one hundredth, **deci** means one tenth, and **kilo** means one thousand. There are several others, but the ones above are among the most commonly used.

In most instances of measuring with standard units, instruct children to place the ruler or meter stick along the item being measured, with the end of the ruler or zero lined up with one end of the item being measured, as in Figure 13–11a. Have children then read the spot on the

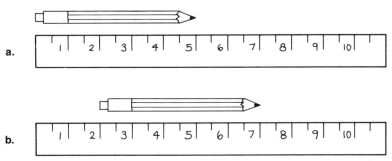

Figure 13–11

ruler where the other end of the item being measured falls, in this case, 5 centimeters.

If the starting point is not zero, as in Figure 13–11b, make children aware that the length is found by either counting the spaces along the item being measured or subtracting the starting point from the end point, $7 - 2$, or 5, centimeters in the case shown.

ACTIVITIES

Primary (K–3)

Objective: *to use indirect measuring to determine length.*

1. Cut a piece of adding machine tape approximately as long as the average height of a child in your class. Put the name of an imaginary child or a character from a story, such as Garfield, on the tape. Attach the adding machine tape to the wall so that one end touches the floor. Have the children estimate whether they are shorter than, taller than, or the same height as Garfield. Record the estimates.

   Then have the children cut strips of adding machine tape to represent their heights. Ask the children to compare their height with that of Garfield. Finally, have the children order their heights by comparing them and taping their paper strips along a classroom or hallway wall.

   Because they will have produced a pictorial representation of class heights, it is useful to describe and discuss their findings and to write down and post these observations beside the paper strips. The observations might include the following:

   - We found that nine of us are taller than Garfield. There are fifteen of us who are shorter than Garfield. Mark and Christine are just as tall as Garfield.
   - The shortest person in the class is Peter. The tallest person in the class is Dana.
   - Five of us thought we were taller than Garfield, but it turned out that nine of us actually were taller.

   A variation of this activity is to have students lie on pieces of newsprint or butcher paper and have others draw their outlines. Comparisons can then be made as before.

2. On one side of the classroom, arrange six to eight objects of varying lengths. Include items such as a piece of yarn, a piece of chalk, a meter stick, a ruler, a strip of adding machine tape, and the edge of a desk or table. Next, mark off a length on the chalkboard or bulletin board. Have the children estimate which of the objects are longer and which are shorter than the mark on the board. Have them check their estimates with the condition that they cannot move either the objects or the mark on the board.

A variation of this activity is to provide children with blocks or a similar material and invite one group of children to build a tower on the floor at the front of the room. Invite another group to build a tower on the floor at the back of the room. Challenge the children to estimate which tower is taller. When the estimates have been made, have the children see if they can devise ways to tell which tower is taller without moving the towers.

Objective: *to use arbitrary units of measure to determine the lengths of objects.*

3. Identify five or six lengths to be measured. For example, you might use the chalk tray, a student desk, a bookcase, a work table, a row of books, and a sink. Provide groups of children with sufficient numbers of different measuring units such as toothpicks, paper clips, straws, tongue depressors, and orange Cuisenaire rods. Have each group measure the lengths of the objects indicated using its particular unit, for example, straws as in Figure 13–12.

   Have each group record its results. Because it is unlikely that a measurement will be exactly a whole number of units, have the children record the results as, for example, "longer than 6 straws and shorter than 7 straws."

   When the measurements have been made, compare the results. Can any conclusions be drawn regarding the number of units used in measuring and the size of the unit? To illustrate the meaning of the above question, have each group put ten of their measuring units end to end and compare the resulting lengths.

4. Announce that a drawing will be held to find three new measuring units. Have all the children write their names on pieces of paper and put the pieces of paper into a container.

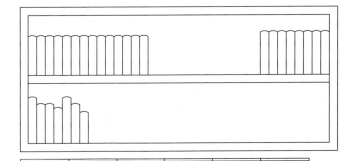

Figure 13–12

Draw the name of the child who will provide a *shoe unit*. Draw the name of another child, who will provide a *hand unit*. Draw the name of the third child, who will provide a *thumb unit*.

Have the first child place one foot, with the shoe on, on a piece of tagboard and draw around it. Using this as a pattern, have the children cut out a number of, say, *Margaret's shoes*. Repeat the process for *Sarah's hand* and *Bob's thumb*. Each of these will be a unit of measure.

For the next two weeks or so, have the children measure the lengths of many objects using these special units. To extend this activity, have the children use only one of the measuring devices and move it along the object being measured, counting the number of times it is used.

Objective: *to measure perimeter.*

5. Put several large shapes on the floor. The shapes may be drawn on butcher paper or made with masking tape, string, or yarn. Initially, have children walk around the figure and count the number of baby steps or walking steps they take.

Later, have children use other arbitrary units, such as a piece of dowel or an orange Cuisenaire rod. Eventually, have children use a meter stick or a trundle wheel. A **trundle wheel** is a plastic or wooden disk that is attached to a handle. The circumference of the disk is one meter. The trundle wheel is pushed along a line or boundary to determine its length.

Challenge the students to look for boundaries to measure. Items that lend themselves to boundary work include the outlines of a desk top, a bulletin board or chalkboard, the classroom, a work area, the gymnasium, and the playground.

6. A direct extension of walking around and measuring large floor figures is to use turtle steps in Logo. The side of a figure in Logo is defined in terms of a particular distance. By virtue of designing a figure, the length of its boundary is determined. Perimeter, then, comes to be viewed as an integral part of each shape.

7. Continued work on the geoboard helps to establish skill in measuring perimeter. Focus activities on counting the units around various figures. After children have gained experience in using geoboards, they are ready to use the diagonal distance between nails in their calculations. Initially, that distance may be known as a little more than one.

Objective: *to use standard units to measure various objects.*

8. Have available a large collection of centimeter cubes and eight or ten relatively small objects to measure. Objects may include a book, a pencil, a chalkboard eraser, a sheet of tablet paper, a crayon box, a sneaker, a stapler, a workbook, and a desk. Identify these items with letters, colors, or numerals. Have the children connect the interlocking centimeter cubes until they are as long as the item being measured, as in Figure 13–13.

Let the children know they are using a unit called a centimeter. Have them count the number of centimeters long each item is and record the length on a piece of paper. Because most items are not a whole number of centimeters long, have the children record the length as, for instance, "more than 7 and less than 8 centimeters."

9. Provide groups of three or four children with pieces of adding machine tape about 35 centimeters long and several

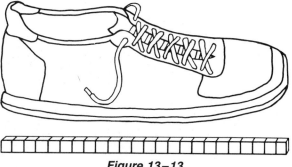

Figure 13–13

centimeter cubes. Have each group carefully mark along an edge of the paper strip using the centimeter cube as the unit and have them number each mark until they have marked off 30 centimeters. Make sure the children are aware that the first mark is labeled *1* and means the distance from the end of the paper strip.

Challenge the children to use their 30-centimeter ruler to find things in the classroom that measure 4, 10, 13, 19, 22, 27, and 30 centimeters in length. To extend this activity, have the children estimate and measure body parts using their ruler: width and length of hands, length of feet, width of a finger, distance around wrists, and length of smile. Have the students record the estimates and measurements.

10. In the gymnasium or on the playground, mark off six to eight lines using yarn, chalk, or masking tape and label each line with a color or letter. These lines should range from 3 to 10 meters in length and should be placed in a variety of directions.

Provide groups of two or three children with record sheets with a space to estimate the length of each line and a space to write in the actual measure. Have the children estimate and then measure each of the lines.

Later, discuss the range of estimates for each line and compare the measured lengths. It may be necessary to go back to the playground and remeasure. To extend this activity, have children find a length between 4 and 6 meters or measure various long objects in the classroom.

These activities have been illustrative of what primary children can do to strengthen their skill in measuring. The following activities for intermediate students begin with arbitrary or nonstandard measurement. By this age, the need to practice measuring with direct and indirect comparison has decreased. The following activities focus on the applications and the precision of linear measurements.

ACTIVITIES

Intermediate (4–6)

Objective: *to measure lengths using an arbitrary unit.*

1. Designate a *measuring unit of the day*, a unit that is easy to provide to students. Possible units might include spaghetti, unsharpened pencils, tongue depressors, and unmarked strips of oaktag. The unit selected may have marks to indicate one-fourth or one-half units.

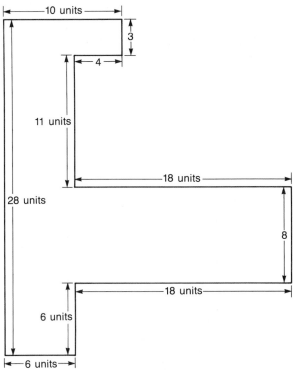

Figure 13–14

As a class, agree on a number of common classroom objects to measure, such as length, width, and height of desks; length and width of the room; length and height of the chalkboard and bulletin board; width and height of the door; and so on. In groups, have the students agree to measure certain of these items. Before measuring begins, have all children estimate the sizes of the objects using the selected unit. Then have the children measure the items to the nearest one-fourth unit and record the measures. To extend this activity, have the children draw or write a description of their classroom using, say, spaghetti units.

2. As a class, select an arbitrary unit with which to measure. For example, the width of a sneaker at its widest part may be the unit. Construct a measuring tape or string based on that unit. Then construct a design on butcher paper and carefully measure it using the sneaker width unit. Sketch the design on a sheet of notebook paper and indicate the measure of each of its lengths in terms of the new unit, as in Figure 13–14.

Send the sketch to another class in the school and tell them the measuring unit, the *widest part of a sneaker*. Ask the class to construct the design on their chalkboard or on butcher paper. Then compare the results by putting the two full-sized designs side by side. Discuss any discrepancy between the two designs.

It is likely that the sizes will differ because of the arbitrary nature of the unit. The need for a standard unit should be evident. Next, create another design using standard units of measure, have another class construct it, and compare the results.

Objective: to estimate and measure with standard units.

3. Begin this activity, entitled the shape of me, with a worksheet on which are listed several body parts (see Figure 13–15).

Ask the children to estimate in centimeters their head width and length, that is, the distance from one side to the other and from top to bottom. Next, have them estimate their shoulder width. Have them continue until they have estimated all parts. At this point, provide 2-meter lengths of newsprint, kraft paper, or butcher paper to individuals or pairs of children.

Ask the children to construct the body they have estimated using the paper provided. The students should use meter sticks, 30-centimeter rulers, or tapes to construct these figures. Have them complete the figures by dressing them using crayons or marking pens and cutting them out. Display the figures around the room. Finally, have the children measure their body parts and compare the results with their estimates.

4. Establish with the assistance of your class a competition called the metric olympics. This series of events will challenge the students to use their estimation and measuring skills. The metric olympics consists of six to eight events. These events may include the following:

- *Sponge throw.* Give students an opportunity to throw each of three sponges as far as they can from behind a line. Lightweight sponges can be thrown 3 to 5 meters. Have the officials mark and measure the length of each toss and record the greatest distance for each participant.
- *Length guess.* On a classroom or hallway wall, arrange a piece of yarn or string 12 to 15 meters long, as in Figure 13–16. Have each participant estimate the length of the yarn.
- *Cube toss.* Make two lines 3 meters apart. Have the students stand behind one line and estimate how close to the other line they will be able to toss a cube. They must toss the cube at least 1 meter. The winner of this event is the student whose estimate and performance are closest.
- *Paper plate sail.* Have students sail a paper plate as far as they can from behind a line. Measure the distance in meters using a trundle wheel.
- *Standing long step.* Have students predict how far they can step using one giant step. Have them measure from the starting line to where their heel lands. A variation is to estimate the distance covered in three giant steps. Either way, base the scoring on how close students are to their estimates.
- *String feel.* Seat students and ask them to hold their hands behind their backs. Give each one a piece of string, and ask each child to run both hands over the string behind his or her back and estimate the length of the string. Award students within 5, 10, and 15 centimeters of their estimates 3 points, 2 points, and 1 point, respectively.

Measuring Area

Measuring area involves determining how much space there is within a plane figure. At the primary level, children count units covering or placed inside a figure. At the intermediate level, students can discover the various formulas to calculate area. Both levels require continued manipulation of objects and tools. The common metric units of area include square centimeter, square meter, hectare (equivalent to a square with sides of 100 meters), and square kilometer.

THE SHAPE OF ME

Just for the record let's measure to find out what size we are. But first <u>guess</u> each item below. Then measure using the materials available. You might like to work with a friend.

		My Guess	The Real Me (Measure)
1. HEAD	Width	_____ cm	_____ cm
	Length	_____ cm	_____ cm
2. SMILE WIDTH		_____ cm	_____ cm
3. HEIGHT (top of head to floor)		_____ cm	_____ cm
4. ARMS (length of each arm from shoulder to fingertip)		_____ cm	_____ cm
5. HAND SPAN (width of palm at widest point when fingers are together)		_____ cm	_____ cm
6. SHOULDER WIDTH		_____ cm	_____ cm
7. LEGS (length from hip to ankle)		_____ cm	_____ cm
8. HIP WIDTH		_____ cm	_____ cm
9. FOOT LENGTH		_____ cm	_____ cm
10. WEIGHT		_____ kg	_____ kg

Figure 13–15

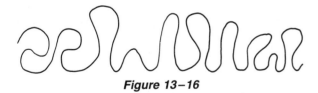

Figure 13–16

ACTIVITIES

Primary (K–3)

Objective: *to indirectly measure and compare the areas of objects.*

1. Draw several rectangles of various shapes and sizes on butcher paper or on the chalkboard. The largest dimension of any particular rectangle should be about 50 centimeters. Give each rectangle a letter or color for identification. Ask the children to estimate and then determine which rectangle is largest or smallest. Which rectangles are larger than the red one or larger than the red one and smaller than the yellow one?

Because the children cannot move the rectangles, they must invent a way to compare the sizes of the various shapes. They may cut a piece of paper the size of a given rectangle and compare the paper to other rectangles. To vary this activity, use shapes other than rectangles so children will have the chance to measure squares, triangles, circles, and so on.

Extend this activity by challenging children to determine the relative area of various classroom items. Ask, for example, "Which is larger, the top of your desk or a pane of glass from the window?" or "Which is smaller, the side of the filing cabinet or the top of the work table?" For each challenge, let the children cut a piece of butcher paper or newsprint the size of one of the items and compare the paper to the other item.

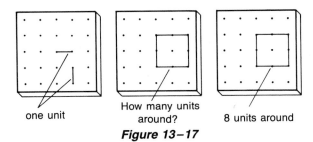

one unit   How many units   8 units around
                    around?

Figure 13-17

The shapes are likely to be different, so additional cutting and rearranging will be necessary.

Objective: *to use arbitrary units to measure area.*

2. Explore length and area on the geoboard by first designating the units of length and area. For example, designate the distance between two nails in any direction except diagonally as one unit. Then use this unit of length to determine the length of the boundary of a particular figure. Figure 13-17 illustrates this type of activity.

   Next, ask children to make a figure with a certain number of units in its boundary. Request, for example, "Make a rectangle with a boundary of eight units."

   To measure area, have children count the number of square units within a figure. Designate the smallest square on the geoboard as one square unit. Then inquire, "How many square units can we find in this figure?" Let the children count the number of square units contained in a given figure, as shown in Figure 13-18. As they gain experience, the children can construct squares or rectangles with a given number of square units, demonstrating an initial understanding of area.

   There are other ways to determine area, beginning with the area of a triangle. At this point, it becomes necessary to identify a half unit. A half unit results from constructing the smallest triangle possible on the geoboard, which is one half of a square unit. Now, the areas of many other geometric figures can be easily determined. Be sure to match the difficulty of the activity with the ability of the children.

3. On the floor of the classroom or gymnasium or on the playground, design several large (2-4 square meters) regular and irregular regions. Some may be squares, rectangles, and trapezoids. Others may be irregular curves, including quadrilaterals and shapes that are not polygons. Have the children estimate the areas of the various figures in terms of floor tiles or other square units designated by the students or teacher.

   Then place the tiles inside the figure and determine its area. Some judgements will be necessary, as not all figures will hold a whole number of square units. It is helpful to have some half units available to fit into the figure.

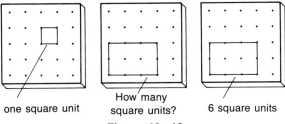

one square unit   How many   6 square units
                    square units?

Figure 13-18

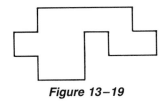

Figure 13-19

As a variation of this activity, determine how many students can stand within each region. Both of a child's feet must be within the region and all children should be standing comfortably upright. The children can then discover which figures are largest and which are smallest and how all of the figures can be ordered from largest to smallest.

Objective: *to use standard units to measure area.*

4. Provide groups of three or four children with centimeter cubes and various small square and rectangular regions (8-24 square centimeters). Have the children estimate the number of cubes that can be placed in each region. Then have the children fill the region with cubes and count them. Explain that each cube takes up 1 square centimeter of area in the region. The total number of cubes is equivalent to the number of square centimeters in the region. To extend this activity, include regions similar in shape to the region shown in Figure 13-19, irregular polygons, and other closed curves.

5. Have available transparencies of centimeter grids. Also, provide drawings of squares, rectangles, triangles, circles, and hexagons. Later, use drawings of closed curves that are not polygons or circles. Encourage the children to estimate the area of each figure in square centimeters.

   Then lay the centimeter grid over the top of each figure and count the numbers of whole and half square centimeters that fit in the figure. When the figures are less like polygons, it is more difficult to estimate and determine the areas. Introduce the various figures slowly over a period of several days.

Intermediate (4-6)

Objective: *to use arbitrary units to measure area.*

1. This activity compares the body surface areas of two students to see which student has the greatest body surface area. Ask for two volunteers to serve as patient subjects. Have the class estimate which of the two has the greatest body surface area. Use toilet paper to carefully wrap each of the students, barely overlapping the tissue. Wrap each leg, then the trunk, then each arm. Finally, lightly wrap the head. Then, even more carefully, unwrap each student and compare the amounts of tissue used. Either count the tissues or place them end to end and compare.

2. Procedures to introduce area using the geoboard continue to be valid for older elementary students. As well, it is helpful for children to be exposed to additional ways to find the areas of triangles and other polygons. For example, the area of a right triangle whose legs have lengths a and b is one half the area of a rectangle whose sides have lengths a and b. This is illustrated in Figure 13-20a.

   The area of any non-right triangle may be found by subdividing the triangle into right triangles by constructing a segment from the vertex to the base. Then, using the previous

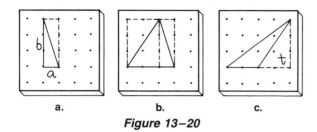

a.   b.   c.

Figure 13-20

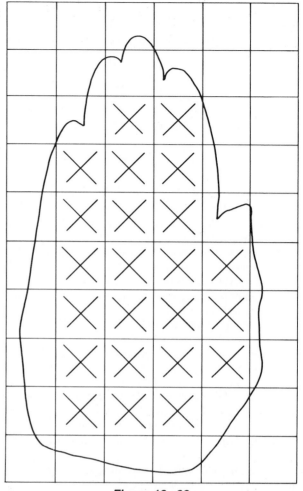

Figure 13-22

technique, find the areas of the right triangles and add them together. This is illustrated in Figure 13-20b.

Finally, if the triangle is similar to the one in Figure 13-20c, it may be stretched into a right triangle. The area of the original triangle is the area of the large right triangle minus the area of the right triangle resulting from the stretch, *t*.

Another method of finding the area of a triangle or the area of an irregular figure is to surround the figure with the smallest rectangle that contains it. Then, find the area of the rectangle and subtract the areas of the newly formed figures. In Figure 13-21, the irregular figure, *f*, has been surrounded by a square of nine square units. The areas of the newly formed figures outside *f* yet inside the square around *f* total six square units. Thus, *f* has an area of 9 − 6, or 3, square units. Challenge the students to form as many different shapes as they can on the geoboard with a total area of, say, 3 square units.

The areas of other polygons—parallelograms and trapezoids, for example—may be derived by means similar to those used to find the areas of triangles and irregular figures. This rather brief description is not meant to be complete in terms of the uses of geoboards. A number of fine references are available; some are listed at the end of this chapter.

3. Prepare sheets on which are drawn square grids measuring approximately 2 centimeters on a side or use the 1-inch graph paper from Appendix B. Encourage the children to take any objects that will fit on the grids and trace their outlines. Have the children count the total number of square regions that are partially inside the outline of the object plus the total number of squares that are completely inside the outline.

The hand outlined in Figure 13-22 completely contains 23 squares, plus it partially contains another 26 squares. We can say, then, that the area of the hand outline is greater than 23 square units but less than 49 square units. We can refine our calculation by finding the average of the two numbers, 23 and 49. Thus, the area of the hand is approximately 36 square units.

Objective: to measure the areas of objects using standard measuring units.

4. The procedure described in the activity above may be repeated using a grid with 1-centimeter squares. The results of

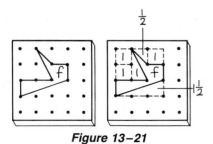

Figure 13-21

measuring should be reported in square centimeters. Similarly, an overhead transparency with a 1-centimeter grid drawn on it may be used. Place the transparency directly on the shape being measured.

The areas of two types of figures can be determined. First are squares and rectangles with dimensions that are whole numbers of centimeters. Second are other polygons with at least two dimensions that are whole numbers of centimeters. These include triangles, quadrilaterals, trapezoids, pentagons, and hexagons. Be sure to have students estimate the area before they find it.

5. This activity utilizes the knowledge students have gained through their use of geoboards. Having made many rectangular regions of varying sizes, students should be able to begin a table that shows the areas of these rectangles (Table 13-1).

Have the students examine the relationships between the lengths, the widths, the products of lengths and widths, and the areas. Once they have discovered that the length times the width equals the area, students are ready to apply that discovery to figures measured with standard measures. Thus, a rectangle with a length of 4 centimeters and a width of 3 centimeters will have an area of 4 × 3, or 12, square centimeters. This finding can be checked using the 1-centimeter transparency grid. The student is now able to

TABLE 13–1

Rectangles on the Geoboard

Length	Width	Length × Width	Number of Square Units
1	1	1	1
2	1	2	2
3	1	3	3
4	1	4	4
2	2	4	4
3	2	6	6
4	2	8	8

measure objects in centimeters or meters and to determine their areas with understanding.

To extend this activity, develop similar tables to show the areas of squares and triangles. From these examples, the formulas for the areas of squares and triangles can be developed.

6. Once students can use various formulas to determine the areas of simple polygons, they are ready to find the areas of large and small regions. Among the large regions are panes of window glass, the tops of tables or desks, the classroom floor or walls, doors, and chalkboards. Small regions include squares, rectangles, triangles, the faces of cereal boxes or milk containers, and the pages of books. In each case, children should use rulers or meter sticks to measure the dimensions of the region and apply the appropriate formula to determine the area.

To vary this activity, ask children to draw particular areas on paper, on the chalkboard, or on the playground. For example, ask them to draw a rectangle of area 6 square meters or a triangle of area 15 square centimeters.

The area formulas that children should be aware of include those for squares ($A = s \times s$, where s is the length of a side), rectangles ($A = l \times w$, where l is the length and w is the width), triangles ($A = \frac{1}{2} \times b \times h$, where b is the length of the base and h is the height), and circles ($A = \pi \times r^2$, where r is the radius and $\pi = 3.14$).

Measuring Weight and Mass

Remember that weight refers to the force of gravity acting on an object, and mass is the amount of matter in an object. For any given object on earth, its weight and mass are equal. On a simple balance scale, we simultaneously determine weight and mass when we find that a box balances 124 grams. On the moon or on another body in space, the weight of an object will be different from its weight on earth. The mass, however, remains constant.

We use the term weight in describing the activities in this section. You will see that some activities use the same procedures used to determine mass, namely, the pan balance.

In their work with weight thus far, students have made direct comparisons with objects to determine which are lighter and which are heavier. They have done some work with the pan balance. By the time children have completed the primary grades, they should have little difficulty using standard units of weight with either a pan balance or a spring scale. Intermediate grade students need activities that provide practice weighing a variety of objects. Common metric units of weight include the gram and the kilogram.

ACTIVITIES

Primary (K–3)

Objective: *to determine which of two objects is heavier using indirect measuring.*

1. Place an object, perhaps a full can of soup, on one side of the room and another object, perhaps a book, on the other side. Challenge the children to discover which object is heavier without moving either object to the other side of the room. Have available at least one pan balance. For younger children, suggest finding another object or a material such as clay to use to determine the weight of the can of soup. This intermediate object can then be used to check the weight of the book.

This activity may be extended by putting four or five objects on each side of the room and having the children order them by weight. All of the checking must be done with intermediate objects.

Objective: *to determine the weights of objects using arbitrary measuring units.*

2. Use wooden cubes as measuring units. Have available eight or ten objects ranging in weight from 10 to 300 grams. The objects may include a chalkboard eraser, a piece of chalk, a pencil, a box of crayons, and a pair of scissors. Ask the students to estimate the weight of each of these objects in terms of the number of wooden cubes it would take to balance the object. Let the children determine how many cubes it actually takes using the pan balance.

Most objects will not balance a whole number of cubes. In these cases, the weight should be reported as, for example, "more than 12 cubes and fewer than 13."

You may vary this activity by using different arbitrary units. Paper clips, washers, beans, and pennies will serve the purpose very well.

Extend this activity by determining the relationships of coin weights. Provide students with one quarter, five dimes, five pennies, and three nickels. The object is to determine the weight of each coin with respect to the other coins. Ask questions such as the following:

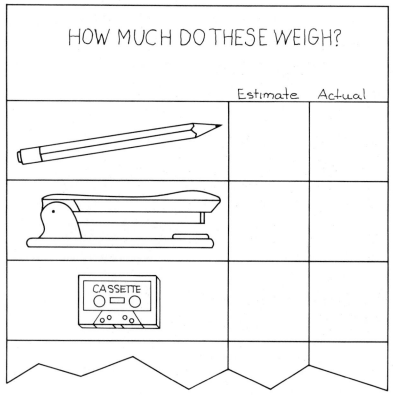

Figure 13-23

- A quarter weighs as much as how many pennies?
- A quarter weighs as much as how many dimes?
- A quarter weighs as much as how many nickels?
- A nickel weighs as much as how many pennies?
- A nickel weighs as much as how many dimes?
- A penny weighs as much as how many dimes?

Discuss how many pennies weigh as much as two or three quarters. Ask how many dimes weigh as much as three or four nickels, and so on.

Objective: *to use standard units to determine the weights of various items.*

3. Find eight or ten items with weights that range from 10 to 100 grams. Items may include a small school eraser, a pencil, a pair of scissors, a box of paper clips, a piece of chalk, and a cassette tape. Also provide the children with one hundred or so centimeter cubes, each of which weighs 1 gram. Let the children estimate the number of centimeter cubes it will take to balance a particular item. Then have them weigh each item using the cubes. Both the estimate and actual weight should be recorded. If it takes 45 centimeter cubes to balance an item, the children should know that the item weighs 45 grams.

Next, provide eight to ten objects that each weigh between 10 and 400 grams. These objects should include items found in the classroom or on the playground. This time, use the set of standard masses that accompany most pan balances. Again, ask the children to estimate the weight of the various objects before the masses are used. A worksheet picturing the objects with space for the estimate and the

actual weight may be helpful. Figure 13-23 illustrates part of one such worksheet.

4. Have the children work in groups of three or four. Each group should have access to a pan balance. If there is only one balance, let the groups take turns. Challenge each group to find three objects that each weigh 25 grams or less, three objects that each weigh 50 to 100 grams, three objects that each weigh 150 to 300 grams, and one object that weighs 1 kilogram. Hold the object search in the classroom. Have the students record the objects they find by drawing pictures of them or writing about them. Among the objects students could weigh are containers of water, rice, sand, or beans.

Intermediate (4-6)

Objective: *to recognize the need for uniformity when using units of measure.*

1. Provide a collection of various sizes of washers or stones. Have the students weigh five objects and record the results using the washers, as in Figure 13-24. Objects may include a crayon box, a ruler, a pair of scissors, a textbook, a chalkboard eraser—all objects that would be found in another classroom in the school.

The next step is to give another class the measuring units and the pan balance. Ask them to weigh the same objects that your class weighed and to report back to you on their findings.

Have the students compare the results of the two weighings and discuss why the results were not the same. It is important to realize that measuring units must be uniform, particularly if they are arbitrary units.

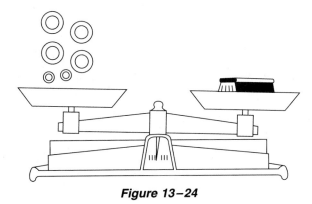

Figure 13–24

Objective: to use standard units to measure weight.

2. Collect four different materials, such as rice, beans, un-popped popcorn, and centimeter cubes. Put each of these materials in a separate bowl. Separate the class into groups of four. Each group will be a team for a prediction contest. Have each team designate a particular member to be a rice grabber, a bean grabber, and so on. Let the rice grabber estimate the weight of rice that she or he can grab using only one hand. The rice grabber should then grab the rice and put it into a plastic sandwich bag. The rest of the team should weigh and record the results. The other grabbers should do the same thing for their particular materials.

   Have the teams calculate the total estimated weight for the four items as well as the actual weight. Have them determine the difference between the estimated weight and the actual weight. The team with the lowest difference is the winner.

3. Display a canning or mayonnaise jar full of rice and an intermediate container such as a tuna can. First, have the students estimate the weight of the rice in the jar. Then announce, "We will be trying to find the weight of the rice. But there is a rule: The largest amount of rice that can be weighed is the amount that can be held by the tuna can." (See Figure 13–25.)

   Have groups of three or four students devise the most efficient plan they can to determine the weight of the rice in the jar. Let each group implement its plan and find the weight of the rice. When all groups have finished, discuss the procedures used and compare the various plans. Also compare the weights that each group derived.

   To extend this activity, use a jar of cubes and weigh to determine the number of cubes in the jar. This may involve weighing a sample of the cubes, counting them, and weighing the jar of cubes to find how many such samples are in the jar. Be sure to subtract the weight of the jar from the weight of the jar and the cubes.

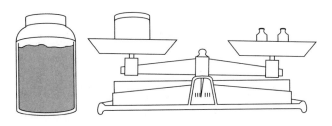

4. Provide each group of three or four students with an apple or orange. Have them find the weight of the apple in grams and record it. Have one member of the group take a bite out of the apple. Then weigh the remaining apple and record the weight. Calculate the weight of the bite and record it. Have the group member take another bite, weigh the remaining apple, record the weight, and calculate the weight of the second bite. Continue until the apple has been eaten.

   When the children are finished, have them answer questions about the apple: Are all bites the same size? What was the weight of the largest bite? What was the weight of the smallest bite? What was the weight of the average bite? What was the weight of the part of the apple that was eaten? Compare and discuss the results from each group.

Measuring Time

Children develop an understanding of time by experiencing events that last varying lengths of time. Younger children who believe yesterday was a long time ago later come to believe yesterday was not too long ago when compared with last week. The experiences that these children need include events of various durations. The time it takes the second hand to move from twelve to twelve, the time it takes the hour hand to move from nine to ten, the start of recess to the end, the start of the school day to the end, Monday to Friday, Monday to Monday, the month of February, and the school year are events that heighten the child's awareness of time.

In the primary grades, the skill of telling time depends on experiences children have had at home and at school. When children move into the intermediate grades, their abilities to tell time using a clock are more refined.

During the early years in school, students should measure time using various units. These include years, seasons, months, days, hours, minutes, and seconds. The tools used to measure time are the calendar and the clock.

Later, students tell time throughout the school day. Children should have practice with instruments such as a stop watch that can be used to measure time with greater precision than a sweep second hand.

ACTIVITIES

Primary (K–3)

Objective: to measure various events using arbitrary units.

1. Provide the children with eight soup cans, each of which has a hole punched in the bottom with a nail. Each hole should be a different size. Mark an "S" on one of the cans and use it as a standard. Have the children fill the standard can and one of the other cans, can A, with water while holding fingers over the holes. Let the cans drain together and compare can A with the standard can, telling whether it takes more time for can A to drain, less time for can A to drain, or the same amount of time for both to drain. Continue these comparisons until all cans have been compared with the standard can and rated.

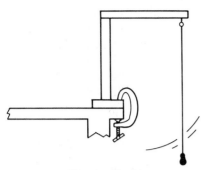

Figure 13–26

To extend this activity, compare each can with the others and order the entire collection of cans from those that take the least time to empty to those that take the most time.

2. Construct a pendulum using string and a 1- or 2-ounce lead fishing sinker, as in Figure 13–26. As the pendulum swings, have the children practice counting the number of times the sinker crosses a marker at the pendulum's base.

Once they can do this easily, fill one of the cans used as a water clock in the activity above. As the can empties, have the children count the number of pendulum swings. Record this number. Compare the cans based on the number of pendulum swings and again order the cans.

Extend this activity by timing various events by the number of pendulum swings. Using the pendulum, time events such as how long it takes for the children to take their seats and get ready to begin work after recess, to line up for lunch, to wash their hands before lunch, and to clean the floor before going home.

Objective: to use the calendar to measure days, weeks, and months.

3. Have a large, easily displayed calendar with ample regions representing each day. As an introductory activity each morning, discuss that day. Appropriate items for discussion include the date, the weather, important events, what day yesterday was, what date yesterday was, what day tomorrow will be, what date tomorrow will be, how many days it has been since the start of the month, how many days are left in the month, how many weeks it has been since the start of the month, and how many weeks are left in the month. Make notations in the region representing that day; a picture or a word or two will suffice. When a month has been completed, display it as the next month is begun.

Counting days and remembering what happened a day ago, counting weeks and remembering what happened a week ago, and counting months and remembering what happened a month ago will help children develop skill in using the calendar to tell time.

Objective: to use the clock to tell time.

4. Have a clock with hands that are easy to move. A large wooden or plastic clock with gears that allow the hands to move together is commercially available and can be very useful. It is also advantageous to have a real clock with a second hand to help show the passing of seconds and minutes. Each day on a regular schedule, spend 3 to 5 minutes setting the clock and having the children determine the time shown by the hands.

At first, show the time on the hour, such as 10 o'clock, 1 o'clock, and 7 o'clock. Soon, introduce time on the half hour,

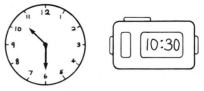

Figure 13–27

then on the quarter hour, five minutes, and one minute. Let the second hand move around and watch the minute hand move one minute.

Later, use a digital clock set at the same time to show the same time on both types of clocks (see Figure 13–27). It is possible that these young people will live in a world that contains only digital clocks and watches.

To extend this activity, challenge the children to play the minute game. Have the children face away from the clock. Then give a command to begin. When they believe a minute has elapsed, they should stand and face the clock. Facing the clock, they can immediately see if they stood before, after, or at the end of one minute. Use this activity for periods of 15, 30, and 45 seconds, as well.

Once the children begin to tell time, they should regularly practice using the clock to time events that take place during school. These events include the periods set aside for reading, mathematics, art, science, and social studies. There are regular times for outside-the-class activities: physical education, lunch, recess, and end of the day. Let children serve as designated timers for a morning or a day, with the responsibility to let the teacher or the class know when certain events are to take place.

Intermediate (4–6)

Objective: to develop timing devices with arbitrary and standard units.

1. There are several ways to develop timers. Using a funnel and sand is one way (see Figure 13–28a). Initially, fill the funnel with sand. The amount of time for the sand to empty out of the funnel becomes an arbitrary unit of time, named *funnel time* or whatever may have meaning.

Develop another timer by using a candle on which several equally spaced marks have been scratched, as in Figure 13–28b. The amount of time for the candle to burn from one mark to another becomes an arbitrary unit of time.

Construct a third timer by tapping or drilling holes in the bottom of a soup can, as in Figure 13–28c. The amount of

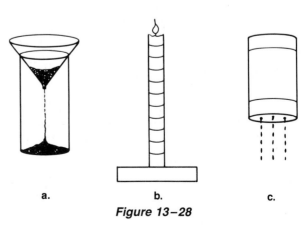

a.   b.   c.

Figure 13–28

time for the water to empty out of the can becomes an arbitrary unit.

Once the timers have been constructed, use them to time various events. Events may include how many blocks can be stacked before the sand timer empties, silently reading while the candle timer burns two intervals, and seeing if the entire class can wash and line up for lunch before the water escapes from the can.

Ask the children how they might make arbitrary units into standard units. Try some of the ideas. Calibrate each timer against a clock with a second hand or a stop watch. Perhaps the funnel and sand could be made into a one-minute timer. The candle could time intervals of five minutes. The soup can could be used as a three-minute timer. Each timer can be adjusted to measure time in standard units. Once completed, the timers should be used as a clock or watch would be used to time events.

An extension of this activity is for a class to construct one or two sundials. The sundial proved to be a reliable timepiece for centuries before the advent of mechanical clocks. Designing and building a horizontal sundial requires measuring a series of angles and compensating for the school's location on earth. Finding a book that describes the steps necessary to construct a sundial is essential. The finished product is very rewarding.

Objective: to use standard units to time events in the classroom.

2. As described earlier, the clock in the classroom should be used to time the events of the day. Fourth graders and some fifth grade youngsters may have some difficulty using the clock at first if they have not had regular experience telling time. Using both traditional and digital clocks on a regular basis will provide the experience needed by students.

When a period is set aside for sustained silent reading, have the students record the time they begin and the time they finish. Have a student time a recess and report to you when the time is up. Allow the students to earn minutes of special activity time for certain classroom behaviors.

To extend this activity, have the students survey the *Guinness Book of World Records* for some of the many timed events and prepare a collection of the most interesting. Let students collect or draw pictures illustrating these events.

3. Timing events with precision can begin by using the second hand on the clock to determine how much time it takes to clear desks and get ready to go home, or to wash and line up for lunch. To become more precise requires a stopwatch to time these same events to the nearest hundredth of a second. Team relay and individual events in physical education lend themselves to precise timing. Be careful to time events for the interest and motivation afforded by the timing, not to prove that one child is better than another.

A variation of this activity is to have three children, each with a stop watch, time the same event to the nearest hundredth of a second. Compare the results. Ask, "Why are there differences? Did some timers make a mistake? What is the best estimate of the event's time?" Perhaps an average of the three times will give the best estimate. Repeat this activity several times.

Measuring Volume and Capacity

Remember that volume refers to the amount of space contained in a three-dimensional object. Measures of volume include how much space is found in a box, a room, or a jar. Capacity refers to the amount of space that can be filled. Measures of capacity include how much water or rice will fit into a box or jar. Volume and capacity are closely related. A container's capacity is determined by its volume. At the primary level, we have a greater emphasis on capacity. At the intermediate level, we emphasize both volume and capacity.

Primary children expand their knowledge of volume and capacity through a variety of media. They build with blocks, large and small. They fill a space and find how much water, sand, or rice was used. They also find how many cubes it takes to fill a space. The cubes represent cubic units and provide the basis for the measurement of volume.

Middle grades children continue to use containers, filling them with a variety of materials. These experiences serve as a basis for developing formulas for calculating volume. The transition to abstract work depends upon the earlier manipulative activities. Common metric volume units include cubic centimeters, cubic decimeters, and cubic meters. Metric capacity units include the milliliter and the liter.

ACTIVITIES

Primary (K–4)

Objective: to determine the capacities of various containers using indirect measurement.

1. Provide the children with one particular container to use as a pouring container and several others to pour water (or sand) into. Ask the children to identify the containers that they believe can be completely filled when they start with the pouring container completely filled. Put those that the children believe can be filled together and those they believe cannot be filled together. Then have the children actually try to fill the containers to discover which ones can be filled.

As an extension, provide six clear containers of various shapes and capacities. Without the children present, pour exactly the same amount of rice into each container, as in Figure 13–29.

Have the children order the containers from those the children believe hold the least to those they believe hold the most. Then, together, pour the rice into a measuring container and let the children discover that all the containers held

Figure 13–29

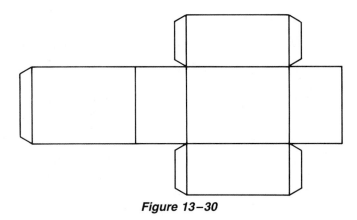

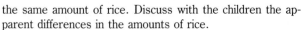

Figure 13–30

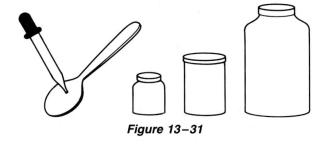

Figure 13–31

the same amount of rice. Discuss with the children the apparent differences in the amounts of rice.

2. This activity requires five jars with volumes that range from 500 milliliters to 2 liters and several infant formula bottles (or tuna fish cans) to use as measuring tools. Ask the children to estimate the number of formula bottles of rice it will take to fill each jar. Then have them find the amount of rice that fills the jars. If the amount is not a whole number, it should be noted as, for example, "more than 8 bottles and less than 9." Encourage the children to record the amount of their estimates and the actual capacities in terms of infant formula bottles. Finally, have the children order the containers by capacity based on the information they have collected.

Objective: to determine the volumes of several containers using standard measures.

3. Find or construct four or five boxes with metric dimensions. The boxes may be constructed by using a pattern similar to the one shown in Figure 13–30. The dimensions should be whole numbers of centimeters like the dimensions of the box shown; it measures $3 \times 4 \times 2$ centimeters.

Provide the children with centimeter cubes and the boxes. Have them estimate how many cubes they think it will take to fill each box. Then have them carefully stack the cubes in each box to find the actual volume. Explain to the children that the volume of each small cube is 1 cubic centimeter and ask for the volumes of the boxes in cubic centimeters.

Next, find two or three small boxes that do not have metric dimensions, and challenge the children to determine the volumes of the boxes using cubic centimeters. Here, the answers may be stated in terms of, for example, "more than 22 cubic centimeters."

4. This activity requires centimeter cubes and sets of measuring cups and spoons. The cups and spoons should be intended for use in the kitchen and should be calibrated in metric units. Provide the children with eight to ten common containers, the capacities of which must be determined using the materials at hand. The containers may include containers for shampoo, milk, syrup, peanut butter, ice cubes, sour cream, detergent, margarine, film, and greeting cards.

Ask the children which units, those that are stacked or those that are poured, are most appropriate for measuring the capacities of the containers. Have the children estimate the capacity of each container and then use the units to find the actual capacity.

Intermediate (4–6)

Objective: to use water displacement as a way to determine which of several objects has the greatest volume.

1. Construct six balls of clay of varying sizes. Also, find a glass jar with straight sides that will easily hold each ball of clay. A peanut butter jar or a quart canning jar should suffice. Put into the jar enough water to completely cover the largest ball of clay without flowing out of the jar.

Ask the students if they know how the volume of a ball of clay can be determined. Accept and discuss suggestions. Try reasonable suggestions.

If displacement is not suggested, have the children attach each ball of clay to the end of a piece of wire. Submerge the balls, one at a time, in the water while the students observe the water level. Discuss how the water level changes depending on the size of the clay. Have the students determine which ball of clay has the greatest volume.

Extend this activity by using three or four irregular rocks that will fit into a graduated cylinder. Have the students pour water into the cylinder to a level that is clearly marked. Then have them carefully immerse a rock in the water and note the new level of the water. Have them record the difference between the new level and the original level in cubic centimeters; this is the volume of the rock.

Objective: to use arbitrary units to determine the volumes of different containers.

2. For this activity, collect an eye dropper, a spoon of arbitrary size, and three containers of moderate size. The containers may be an infant formula jar, a peanut butter jar, and a mayonnaise jar (see Figure 13–31).

Ask the students how to determine how many eye droppers of water it takes to fill the large jar. They may suggest using the eye dropper to fill the large jar one drop at a time. Encourage the students to find a method that will take less time and effort. One suggestion may be to discover how many drops of water it takes to fill the spoon. Next, find how many spoons it takes to fill the formula jar. Then, find how many formula jars it takes to fill the peanut butter jar and how many peanut butter jars it takes to fill the mayonnaise jar. Finally, the volume of the mayonnaise jar can be stated in terms of drops of water.

3. Collect enough bottle caps of the same variety to fill a medium-sized paper bag. Asking for help from the children will speed up this collection. Provide the students with various containers: a soup can; a tuna can; half-pint, pint, and quart milk containers; a shoe box; a cereal box; and a drinking cup. Have the children estimate and then determine the volume of each container in terms of the number of bottle caps. Have them order the collection of containers based on their findings.

An extension of this activity is to select a second unit, such as small shell-shaped macaroni, and replicate the procedure. The results should prove to be the same. Do all students agree that the results will be the same before the second measure is made?

Objective: *to calculate the volumes of various containers using standard measurement.*

4. This is similar to Activity 2 above. Ask the students to use an eye dropper to find how many drops of water are in one milliliter. Next, have them determine the number of milliliters of water in a teaspoon and how many teaspoons fill a tablespoon. Then encourage the students to find how many tablespoons fill a paper cup. With the information they have collected, they should be able to calculate the number of drops of water it takes to fill a teaspoon, a tablespoon, and a paper cup.

   Finally, challenge the students to find the number of liters of water contained in 1 000 000 drops of water. A calculator will be very handy in this activity.

5. Begin by providing the students with small boxes constructed from oaktag. Make three or four boxes with measurements of (a) 2 × 2 × 2 centimeters, (b) 2 × 2 × 3 centimeters, (c) 2 × 3 × 3 centimeters, and (d) 3 × 3 × 3 centimeters. Also provide a large number of centimeter cubes. Have the children find how many cubes can be carefully stacked in each of the containers and record the number of cubes. Have them use a table like the following:

Box	Length	Width	Height	Length × Width × Height	Total No. of Cubes
a	2	2	2	8	8
b	2	2	3	12	12
c	2	3	3	18	18
d	3	3	3	27	27

Encourage the students to build "boxes" by stacking centimeter cubes using other dimensions and record their results in the table they have already started. After six or eight examples have been recorded, have the students examine the table for patterns.

Expect a variety of responses, but lead the students to notice the relationship between the product of length, width, and height and the number of cubes counted. The students will be discovering the formula for the volume of a box. The common formula for determining the volume of a rectangular container is $V = l \times w \times h$, where l is the length of the base, w is the width of the base, and h is the height.

Measuring Temperature

Much of what children learn and know about temperature comes from their interaction with their environments. When the air is cold, they bundle up. When it is hot, they wear fewer and lighter clothes. In the snow, they shiver. When they are ill, they may feel very warm or they may feel warm yet shiver.

Children's measures of temperature are generally from two sources, the weather and illness. Children hear the measure of the outside temperature daily as they watch television or listen to the radio. When they are ill, their body temperatures are recorded, but weather is by far the more common source of measures of temperature.

Primary children begin to become aware of temperature as they report the daily weather. Expressions like "It is warm today" and "It will be cooler tomorrow" and "It is frosty this morning" serve as indicators of temperature. Later, a daily temperature reading may be made and recorded as practice in reading a thermometer.

Middle grades students continue to read temperatures and to record and graph the results. There will be work in science that requires reading temperatures, in weather units and in other areas of study. The common metric temperature unit is degrees Celsius. As children are alerted to temperature, they should develop the ability to recognize if various temperatures, such as 20°C, are hot or cold. Also, if last night's low temperature was 4°C and today's high was 15°C, children should be able to determine that the difference in temperature from the low to the high temperature was 15°C − 4°C, or 11°C.

ACTIVITIES

Primary (K–3)

Objective: *to use indirect measurement to determine temperature.*

1. Provide students with thermometer bulbs and tubes from four inexpensive thermometers by removing their backings. About halfway up each thermometer tube, mark a spot with fingernail polish. This mark is a reference point with which to judge the movement of the mercury (see Figure 13–32).

   Ask the children to place the four thermometers in four different locations: (1) in a glass of cold tap water, (2) on a table in the classroom, (3) in a glass of warm tap water, and (4) outside the classroom window. Have the children guess which locations will be the warmest and coolest. Then, using their observations, let the children determine which locations are actually the warmest and coolest.

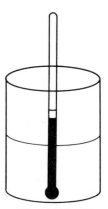

Figure 13–32

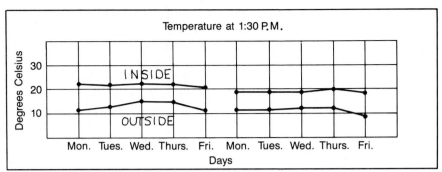

Figure 13-33

Objective: to use arbitrary units of temperature to measure various materials.

2. Use thermometer bulbs and tubes as in the preceding activity. On each tube, make five equally spaced marks. This may be done by laying the tubes on a piece of lined writing or notebook paper with the bottom of the bulb exactly on one of the lines and then making the five marks.

   Next, indicate six locations in which to check the temperature using the "new" thermometers. These locations may be on a countertop, grasped in the hand of a particular student, in a glass of water, in a sunny spot in the classroom or school, near a light source, and outside the window.

   Have the students record the temperatures they find as they test the various locations. The thermometers should be kept in a location for about five minutes to allow the mercury to stop moving. Because it is unlikely that the temperatures will be exact, the record should show the temperature in a location as, for example, "greater than 3 but less than 4."

   To extend this activity, have the children record the temperature inside the classroom and outside for each hour during a school day. Then let them construct a graph or table to show how the temperatures varied.

Objective: to use standard units to measure temperature.

3. Have the children observe a Celsius thermometer that has been placed in a container of boiling water. *Note: this activity requires you to take special precautions to avoid possible burns from spilled or splashed water.* Ask the children to read the temperature. Next, have the children read the temperature from a thermometer that has been placed in a glass containing chopped ice. (The thermometer should be at room temperature or cooler when it is placed in the ice, not still hot from the boiling water.) Then have the children read the thermometer at room temperature. Finally, check the temperature of the air outside. In each case, have the children write down the temperature they have read from the thermometer.

   It may be necessary to spend a little time explaining how the scale of a thermometer is constructed and how it is read. At first, the temperature may be read as, for example, "more than 95 and less than 100." As the students become more proficient at reading the scales, they will be able to be more precise.

Intermediate (4–6)

Objective: to use standard units to determine temperature.

1. Mount a Celsius thermometer outside the classroom window so that it may be read from inside the room. If this is impos-

sible, find a location outside where the thermometer may be placed. Mount another thermometer in the room. Have students record the temperature daily at a given time for two weeks. Keep a record on a table labeled for inside and outside temperature or graph the temperatures as in Figure 13–33.

2. Collect five different containers such as a paper cup, a heavy coffee mug, an aluminum cup, a glass measuring cup, and a thermos soup container. Into each one (or one at a time) pour the same amount of boiling water and insert a Celsius thermometer. After one minute, record the temperatures in the containers. Then, at five-minute intervals, record the temperatures in the containers. After 30 minutes, stop. To complete the activity, graph the results. Then, order the containers from the one that holds the temperature for the longest period of time to the one that holds it for the shortest amount of time.

3. This activity requires three oral Celsius thermometers. Ask for three volunteers. Have the volunteers place the thermometers in their mouths and keep them there for the prescribed amount of time (normally 3 to 5 minutes). Then have them read and record their temperatures.

   Discuss why the temperatures range from 36.5 to 37.5 degrees Celsius and when the children may expect the temperatures to be higher or lower. You may wish to check the temperatures of other students to collect more data. Be sure to thoroughly clean the thermometers before reusing them.

Measuring Angles

Experiences with Logo provide young children with the opportunity to use angle measure. With a little practice, children can direct the turtle to turn right or left a specified number. The number represents the degrees of a circle; thus, a complete rotation is 360 degrees. The more children work with turns, the greater their understanding of angle measure. As well, children will experience exterior angles as they write procedures for simple polygons. Apart from this exposure of primary children to angle measure, most angle measurement is introduced in the middle grades.

The protractor is used to measure angles. Once learned, the skill of measuring angles is fairly easy to maintain. Learning to use a protractor, however, requires careful teaching. Skill in measuring angles should begin with establishing an easily identifiable angle measure, 90 degrees.

ACTIVITIES

Intermediate (4–6)

Objective: *to use a model of 90 degrees to find right angles.*

1. The square corner or right angle is easily recognized by children and can be quickly constructed by folding a sheet of paper twice, as in Figure 13–34. Have the students use these models of 90 degrees to check the corners of books, the corners of desks, angles placed on a bulletin board, the corners of a classroom door, and the corners of the chalkboard.

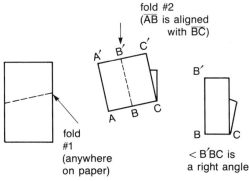

Figure 13–34

Next, have students determine if various angles are less than, equal to, or greater than a right angle. Provide the students with ten angles ranging in size from 70 to 110 degrees. Challenge students to estimate whether each angle is less than a right angle (acute), equal to a right angle (90 degrees), or greater than a right angle (obtuse). After the students estimate, have them use the folded right angle to check their estimates.

Objective: *to use a protractor to measure angle size.*

2. As you introduce the protractor, show students how to place it on the angle so that the vertex of the angle is aligned with the origin of the protractor and one side of the angle, called a ray, corresponds to the referent, or the 0/180 degrees line of the protractor. Figure 13–35 illustrates a properly placed protractor.

Determining the angle measure requires the student to read the degrees scale, which begins at zero along one of the rays of the angle. In the case of the angle in Figure 13–35, the measure is 60 degrees.

Providing students with explanations and assistance as they begin working with protractors is essential. Provide stu-

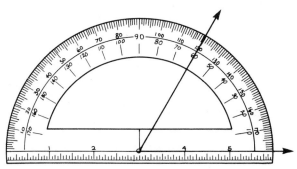

Figure 13–35

dents with angles to measure. The angles should have rays long enough to accommodate the size of the particular protractor being used. Otherwise, the measuring experience will be frustrating. Angles to measure may be provided on a worksheet or displayed on a bulletin board. If Logo is available, it is interesting to measure the angles generated by the turtle to check the accuracy of those angles.

3. In this activity, be sure students have protractors and paper available. Provide the students with five predrawn angles. First, have the students estimate the measure of each angle. Then, to provide an opportunity to construct angles using models, have the students measure the angles and construct copies of them adjacent to the originals.

Next, challenge the children to construct angles of given measures without using models. Have then construct the angles in any configuration on the paper to show that they know how to use both the left and the right scales on their protractors.

The activities presented in the skills section of this chapter give you some idea how extensive measurement is. There are many types and topics of measurement. Types of measurement include direct, indirect, arbitrary, and standard. Topics of measurement include length, area, weight, time, volume, temperature, and angle. Be prepared to use measurement whenever the opportunity arises in the classroom.

Estimating

In no other area of mathematics is estimating more prevalent than in measurement. Once children get in the habit of estimating before they measure, they are better able to determine if their measuring is accurate. Notice that nearly every activity in this chapter has suggested that students estimate before they measure; therefore, no further activities are presented.

As children gain experience in measurement, encourage them to develop a sense of the various units with which they are measuring. For example, when children are measuring length, they should know that a centimeter is about the width of a thumb, that a meter is about two average steps, and that a kilometer is about a 5- to 10-minute bicycle ride. When children are measuring volume or capacity, they should know that a cubic centimeter is about the size of a bean, that a cubic meter is a space in which they could put their desk and chair, and that a liter is an amount that would fill three or four drinking glasses. Having this sense of measuring units gives the children an important foundation for estimation and measurement.

Problem Creating and Solving

Many of the activities presented in this chapter have been presented in a problem-solving format. That is, students

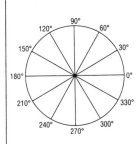

Over 2500 years ago, the Babylonians wrote numbers in a system based on the number 60. So they measured with units based on 60. Even today we use Babylonian ideas to measure time. That is why there are 60 minutes in an hour and 60 seconds in a minute. We also use Babylonian ideas in measuring angles.

The Babylonians divided a circle into 360 equally spaced units, which we call **degrees.** (The number $360 = 6 \times 60$ and is an estimate of the number of days in a year.)

An instrument that looks like half of the circle at left is the **protractor.** The protractor is the most common instrument used for measuring angles. Many protractors look like the one drawn below. Because protractors cover only half of a circle, the degree measures on the outside go only from 0° to 180°.

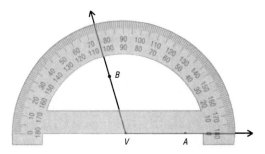

Every protractor has a segment connecting the 0° mark on one side to the 180° mark on the other. This segment is on the *base line* of the protractor. The middle point of this segment is called the *center* of the protractor. In the drawings above and on page 119 this point is named *V*. *V* is usually marked by a hole, an arrow, or a + sign. There are almost always two curved scales on the outside of the protractor. One goes from 0° to 180°. The other goes from 180° to 0°.

To measure an angle with a protractor:

(1) Put the center of the protractor on the vertex of the angle.

(2) Turn the protractor so that one side of the angle is on the base line and the other side of the angle is beneath the curved scales.

(3) The measure of the angle is one of the two numbers crossed by the other side of the angle. Which of the two numbers? The first side of the angle crossed one of the scales at 0°. Pick the number on that scale.

Figure 13–36

Vsiskin, Zalman, et al, Transition Math. Glenview, Ill.: Scott, Foresman and Company, 1991, p. 118. Reprinted by permission.

THE MATH BOOK

The textbook page shown in Figure 13-36 is from a transitional mathematics textbook. The textbook is designed to help students make the transition from a general elementary or middle school mathematics program to an algebra program. It is not labeled by grade level, but it is generally recommended for students in grades 6-9. Some districts use the program with highly gifted sixth graders who will be taking algebra in seventh grade, or with accelerated seventh graders who will be taking algebra in eighth grade, or with more traditional eighth or ninth graders who will be taking algebra the following year. It can also be used as a transition to a more integrated high school mathematics program.

In this lesson on using a protractor, you see a reference to the Babylonians. This textbook frequently makes reference to the history of mathematics to help students make the connection to the development of the mathematical concepts. Other lessons in the book discuss the contributions of famous mathematicians.

Geometry and measurement are woven throughout all the units of the book. Students are asked to use models and graphs to illustrate many of the numerical ideas. In this lesson, students use a protractor to learn to draw figures that are used later to illustrate other concepts. The recommendation of the NCTM *Standards* to make connections among a wide variety of mathematical topics is well-illustrated in this text.

have been challenged to use the skills of problem solving in seeking solutions. We hope you will encourage students to launch into solving a problem without fear of failure or frustration. This requires well-stated problems and considerable teacher support and encouragement. Following are a few examples of activities that provide problem-solving experiences.

ACTIVITIES

Primary (K–3)

Objective: *to use estimation and standard measures to solve problems.*

1. Present children with five events that require them to use the measurement skills they are learning. Following are examples of such events:

   - Find how much rice it takes to balance a soup can. Tell how much the rice weighs.
   - Here is a piece of string. Find an object that is as far from the doorknob as this string is long.
   - Here is a sheet of paper. Find a book in the room that is the same size as this piece of paper.
   - Find how many centimeter cubes it will take to fill a tuna can. Tell how much all the cubes weigh.
   - Use the Logo turtle to make a square that has sides of 55 turtle steps. Make two such squares on the screen.

   These events are intended for students working in groups of three or four, or for the entire class working with the teacher. The size of the class, the amount of help available, and the age and abilities of the youngsters play a role in the organization for this activity. An extension of this activity is presented below as an activity for middle grades children.

Objective: *to use the pan balance to solve a problem.*

2. Provide the children with eight balls of clay or play dough. They should weigh 5, 10, 15, 20, 25, 30, 35, and 40 grams. Challenge the children to divide the balls of clay among four children so that each child will have the same weight of clay. Have the children weigh various combinations of clay balls until they have found pairs that weigh 45 grams. They should match the 5 and 40 gram balls, the 10 and 35 gram balls, and so on. A pair of clay balls on one side of the scales should balance another pair on the other side.

   As an extension, ask the children to create four or six clay balls to give to other students to solve. Have the children create the balls without using the standard gram masses; let them use only the pan balance.

Intermediate (4–6)

Objective: *to estimate area and volume based on configurations of squares and cubes.*

1. On a table or shelf, make arrangements with various numbers of cubes. Challenge the students to find the number of cubes in a pattern without counting. Figure 13–37 shows two such patterns. Allow students to create patterns for the others, including you, to estimate the number of cubes.

   Then, move to estimating larger numbers using patterns shaded on squared paper. Perhaps the first estimates should

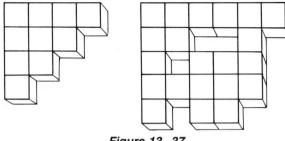

Figure 13–37

determine if the pattern has more or less than twenty-five squares. Next, have the students estimate more precisely how many squares are shown. Figure 13–38 illustrates two such square patterns.

   - *Understanding the problem.* I need to decide how many squares I believe are shaded in the figure on the left. I cannot count every square. I need to figure out a way to do this so I can do other problems like it.
   - *Devising a plan.* I am going to count some of the shaded regions. Then I'll try to fit the regions together in my mind to make a rectangular shape. I'll multiply the length and width to find the answer (make and use a drawing or model and look for a pattern).
   - *Carrying out the plan.* In the upper part of the shaded region on the left in Figure 13–38, I see a 3 × 7 rectangle (21 shaded squares). In the bottom part of that shape, I see what is more than another row but not two more rows. I believe there must be a little more than 4 × 7 shaded squares. Thus, I will say there are about 30 shaded squares in the shape. I am interested in knowing how close I am.
   - *Looking back.* When I counted the shaded regions, I found 29. My approach for finding the number of shaded regions in that shape was pretty good. I wonder if it will work for the next shape?

   An extension of this activity is to use stacks of cubes with the understanding that there are no holes within a figure. These stacks may actually be constructed or they may be pictured. See Figure 13–39a for a picture. As a variation, a top view of the stack may be shown with the number of cubes in each stack indicated, as in Figure 13–39b. Again, encourage the children to estimate how many cubes are contained in the figure.

Objective: *to determine when certain vegetables will be ready to harvest.*

2. Information for this activity may be easily gathered from garden or seed catalogs or from seed packets. Select two vegetables such as radishes and beans. Explain to the children

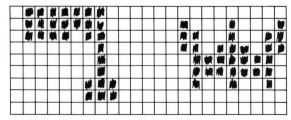

Figure 13–38

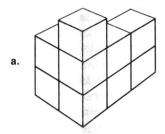

a.

b.

Figure 13–39

that three different kinds of radishes are ready to eat in 22, 25, and 28 days and that three different kinds of beans are ready to be picked in 50, 56, and 68 days.

If the seeds for the radishes are planted on April 20, when could you expect to have ripe radishes? If the bean seeds are planted on June 1, when could you expect to have ripe beans? If you wanted 56-day beans to be ripe on August 20, when should you plan to plant them? If you wanted 25-day radishes to be ripe on May 18, when should you plan to plant them?

This activity may be extended easily by selecting several kinds of vegetables for a garden that should all ripen on a particular day. When should each vegetable be planted? It is necessary to have calendars available for the children to use in seeking solutions to these problems.

Objective: to use estimation and standard measurement on a scavenger hunt.

3. Divide the class into teams of three or four students. Give each team a set of scavenger hunt challenges. Following are a sample of such challenges:

- Find how many square meters of floor space each person in our classroom has.
- If there are 100 students in the gym, how many cubic meters does each student have?
- How many square meters does the school playground have?
- Find the number of meters you must walk from our classroom to the principal's office and back.
- Find two pairs of students whose combined weights in kilograms are the same.
- How many turtle steps (Logo) would it take to be 1 meter long?
- Find the temperature of the oven in degrees Celsius needed to bake a cake.
- Write a problem similar to these for other students to solve.

It is helpful to have a metric tape measure, a trundle wheel, a metric bathroom scale, a computer with Logo, and a metric oven thermometer to assist in the above tasks. Give each team three tasks to complete. The winning team is the first to successfully complete its assigned tasks.

Grouping Students for Learning Measurement

Learning measurement is an active process. From the time children first experiment with measurement until they master measuring skills, their work should be active. Thus, much of the work will be done in individual and cooperative learning group settings. Young children will play individually or with a friend or two at the rice (or sand) table, pouring rice from container to container. At this stage, the experience is individual. Later, when a group of four children are given a particular task, such as ordering five containers at the rice table from smallest to largest, cooperative learning will be the focus.

Several types of instruction, such as introductory activities, developing skills in telling time, using the geoboard to find perimeter, and debriefing after a cooperative learning activity, may be best presented in the whole-class setting. These are times when it is either most efficient or most appropriate to use the whole-class setting for sharing and discussion to take place. By and large the whole-class setting is not used by teachers to tell children how to measure, except to summarize procedures that have come from the students' experiences.

Activities in which children work in small groups provide an opportunity for the children to experiment with measuring, and later to use standard measuring tools, and to discuss with one another their findings, to compare their procedures, and to get feedback from one another regarding the various techniques that they used. As the teacher, you will be busy moving from group to group, monitoring the progress of the students, offering occasional suggestions, and answering questions that all members of the group may have. It is important for the teacher to be a presence without interfering with the interaction of the children. An exception would be when you interrupt a group to encourage the children to return to the task at hand or to modify the behavior of disruptive students. As you read the activities presented in Chapter 13 think about ways in which they can be successfully used in the small group setting.

Communicating in Learning Measurement

As was the case in learning geometry, description in learning measurement is a focus that should be encouraged. Children should be encouraged to describe what they discovered and to see if others made the same discovery. They should compare and contrast procedures that they used during their activities. They should design instructions for other students to follow. These experiences may involve oral exchanges in which children can question one another and seek clarification. For example, the spokesperson for a small group may report how her group used a strip of adding machine tape to make a measuring tape and how they found objects in the room that were 10, 20, and 30 centimeters in length. Did other groups use the same procedures? Did they find the same or different objects? What made this task easy or hard?

Another type of important communication is describing specifically how a skill is performed. For example, a child may be asked to describe what he would expect a clock to look like if it showed that it is ten-thirty in the morning. This may be a writing task or an oral task. A picture may help the student formulate how he might describe the clock face. Many of the skills involved in measuring include specific procedures that can be the focus of descriptive communication.

A learning group journal can be used to record the work group members have done in measuring. Headings in such a journal might include "How We Made a Centimeter Ruler," "How to Measure Length," "Comparing Our Estimates with Our Actual Measures," "How to Tell Time," and "How We Solved the Volume Problem." The group recorder would be responsible for writing the ideas and descriptions of the group members. There should be an opportunity to share these entries with other class members.

Evaluating Measurement Learning

Evaluating measurement concepts and skills requires clearly stated objectives. Most likely the objectives will come from the section or chapter of the math textbook currently being used. The school district, school, or teacher may be responsible for determining whether the textbook objectives adequately represent measurement.

Throughout math textbooks, the teacher is provided with evaluative procedures. There are pre-tests for chapters about measurement. There are midchapter check-ups to see if the students are understanding the measurement skills presented. There is a review of the measurement chapter content, and there is a chapter test. These evaluations provide data regarding student proficiency in measurement. All of the evaluative data are keyed to the pages on which various aspects of measurement are presented. Both content and skills are evaluated.

It will be necessary to provide additional evaluative techniques for material not presented by the textbook. Many of the activities suggested in this chapter are nearly impossible to include in a basal math textbook. To evaluate the activities, observe how the children perform various tasks and jot notes about the children's performances. Anecdotal records are an important part of the evaluative process.

Evaluate measurement skills by asking children to make careful measurements. For example, have them cut a piece of string 17 centimeters long, make a rectangle with

sides of 6 and 9 centimeters, or construct a hexagon on the geoboard with an area of 7 square units. Ask children to make a clay ball weighing 35 grams. Have them put 250 milliliters in a container. Ask them to measure an angle of 55 degrees. Have them find how much time elapses between lunch and dismissal time. Ask the children to report the temperature inside and outside the classroom. Your first-hand observations as children perform these tasks will provide you with valuable information regarding the children's measurement abilities.

Something for Everyone

Geometry and measurement are good areas for children who may not excel in abstract numerical work in mathematics. These areas are perfect for students who have strengths in the visual and spatial areas. Such children often excel when they are asked to show something on a geoboard or with pattern blocks or to estimate a measurement. They may not be able to give a verbal explanation of what they did, but they can often perform even better than the teacher or the children who are the best students in computation.

Auditory learners may have difficulty in geometry and measurement. They will probably be good at defining geometric terms and stating metric conversions, but they may not be as good at spatial activities unless given a verbal explanation of what to do. They may have difficulty actually drawing figures or finding where to start measuring a given figure. They may be able to talk themselves through some spatial activities by reasoning aloud why two tangrams fit together in a certain way or how the faces of a polyhedron are shaped.

Tactile-kinesthetic learners excel when manipulating the geometry materials and the measuring tools. Most measuring requires some initial investigation with a ruler, a meter stick, a measuring cup, a clock, a thermometer, a protractor, or some other measuring tool. Geometry and measurement are topics that lend themselves to active learning and they are popular with students who perform well in active settings.

Students who exhibit gifted behaviors in geometry and measurement may not be the same students who exhibited gifted behaviors in other areas of math. This may be disconcerting for students who are used to being your star math students. They may be embarrassed to realize that they have no idea how tangrams fit together or where the line of symmetry is.

Students with different learning styles may learn from one another when they work in small groups. Try to develop a classroom atmosphere in which students feel free to ask each other for help. Encourage all types of learners to learn from one another. No one learns in strictly one mode, and everyone can improve skills in different areas.

Encourage students talented in geometry and measurement to go further in these areas. Let them combine geometry skills with skills in other subjects (such as art, mechanical drawing, woodworking, metal shop, or home economics) to create new applications for their learning. Students who are adept at tessellating, for example, may wish to study the artwork of Escher and create their own Escher-type drawings. The computer program Creativity Unlimited by Carlson and Kosel (from Sunburst) may help them in creating their designs. Students who are good at making scale drawings and at creating three-dimensional models from two-dimensional drawings may wish to make a scale drawing of the classroom and to create a model from a drawing of an ideal classroom.

Another good area for students talented in geometry is that of hypothesizing about geometric relationships. The computer program The Geometric Supposer by Schwartz and Yerushalmy (from Sunburst) is a good program for such capable students. It helps children explore geometric constructions and hypothesize about such things as the comparative lengths of diagonals of various quadrilaterals or the comparative areas of different types of triangles. In this way, students are encouraged to think like mathematicians—making hypotheses and trying to prove or disprove them. Children may do the same with a compass and straightedge as they learn to make their own constructions.

Another good computer program for talented students is The Factory by Kosel and Fish (from Sunburst). It helps students develop inductive thinking in a visual mode as they design geometric products on a simulated machine assembly line. Programming in Logo is also excellent for students displaying talents in geometric areas.

Students who have difficulty in geometry and measurement may need to be encouraged not to give up. Students may make comments such as, "Tangrams are stupid. I can never do one of those puzzles. Why do we need to do them anyway?" Let these students start with simple tangrams using only two or three pieces or with some of the outlines drawn in and then gradually move on to more difficult puzzles. Be sure to give them tasks at which they can succeed, and let them note their progress.

Discuss with the whole class some of the workers who must have spatial abilities, such as architects, artists, astronauts, construction workers, electricians, engineers, mapmakers, mechanics, plumbers, and surgeons. Above all, make geometry and measurement in the elementary school fun. Create experiences to help students move to at least the second or third van Hiele level in order to prepare them for later, more formal work in geometry.

KEY IDEAS

Measuring involves several concepts that need to be constructed by children through a variety of activities. Measuring also involves developing skill in a process that is widely used in our daily lives. Many of the activities presented in this chapter require children to use measurement in the same way they will use measurement outside of school. The teaching sequence suggested uses four types of measuring activities: direct comparison, indirect comparison, arbitrary units, and standard units. The topics of measurement include length, area, weight, time, volume, temperature, and angle. Objects used to measure include straws, cubes, and various containers for arbitrary units and rulers, balances, clocks, measuring cups, thermometers, and protractors for standard units.

Developing a sense of measurement units is emphasized. Estimation and problem solving play a prominent role in most of the measurement activities presented. Evaluating many of the activities requires you to observe and record how children go about measurement tasks.

REFERENCES

BARSON, ALAN. *Geoboard Activity Cards* (Intermediate). Fort Collins, Col.: Scott Scientific, 1971.

——— . *Geoboard Activity Cards* (Primary). Fort Collins, Col.: Scott Scientific, 1972.

BITTER, GARY G.; MIKESELL, JERALD L.; AND MAURDEFF, KATHRYN. *Activities Handbook for Teaching the Metric System.* Boston: Allyn & Bacon, 1976.

DOLAN, WINTHROP W. *A Choice of Sundials.* Brattleboro, Vt.: S. Greene Press, 1975.

HALLAMORE, ELISABETH. *The Metric Book . . . of Amusing Things to Do.* Woodbury, N.Y.: Barron's Educational Series, 1974.

HIEBERT, JAMES. "Why Do Some Children Have Trouble Learning Measurement Concepts?" *Arithmetic Teacher.* Vol. 31, No. 7 (March 1984), pp. 19–24.

KASTER, BERNICE. "The Role of Measurement Applications." *Arithmetic Teacher.* Vol. 36, No. 6 (February 1989), pp. 40–46.

LINDQUIST, MARY MONTGOMERY. "The Measurement Standards." *Arithmetic Teacher.* Vol. 36, No. 2 (October 1989), pp. 22–26.

McWHIRTER, ROSS. *The Guinness Book of World Records.* New York: Sterling Publishing Co., 1990.

NATIONAL COUNCIL OF TEACHERS OF MATHEMATICS. *Curriculum and Evaluation Standards for School Mathematics.* Reston, Va.: NCTM, 1989.

NELSON, DOYAL, AND REYS, ROBERT, EDS. *Measurement in School Mathematics.* Reston, Va.: National Council of Teachers of Mathematics, 1976.

PIAGET, JEAN; INHELDER, BARBEL; AND SZEMIUSKA, ALINA. *The Child's Conception of Geometry.* New York: Basic Books, 1960.

SHAW, JEAN. "Mathematical Scavenger Hunts." *Arithmetic Teacher.* Vol. 31, No. 7 (March 1984), pp. 9–12.

SIME, MARY. *A Child's Eye View.* New York: Harper & Row, 1973.

THOMPSON, CHARLES S., AND VAN deWALLE, JOHN. "Learning About Rules and Measuring." *Arithmetic Teacher.* Vol. 32, No. 8 (April 1985), pp. 8–12.

VARMA, VED P., AND WILLIAMS, PHILLIP, EDS. *Piaget, Psychology and Education.* Itasca, Ill.: F. E. Peacock Publishers, 1976.

One important aspect of teaching middle school mathematics is helping students bridge the gap between the mathematics of elementary school and the mathematics of high school. Frequently, this means helping students make connections between concrete and abstract notions. Algebra is an especially important part of this transition. Through an introduction to algebra, students begin to make the transition from the concrete world of arithmetic to the more formal world of algebra.

It is especially important that middle school students develop a very firm understanding of algebra that is related to concrete models and that they continue to develop the definition of mathematics in Chapter 1, namely, that mathematics is the science of pattern and order. The informal introduction to algebra in this chapter will continue to build upon the earlier concepts developed through the exploration of patterns. The focus will be on the development of algebraic concepts, with some discussion of practicing skills. The NCTM *Standards* (1989, p. 71) have called for a decreased emphasis on manipulating symbols, memorizing procedures, and drilling on equation solving when teaching algebra in the middle grades.

An understanding of algebra is increasingly important for success in our technological world, but algebra is frequently neglected in the mathematics curriculum in the United States. *The Underachieving Curriculum,* the report of the Second International Mathematics Study (McKnight, 1987), found that in the United States much of the middle school mathematics curriculum is simply a review of earlier concepts. This is not the case in the rest of the world. For example, nearly one-third of the eighth grade school year in the United States is spent on fractions, ratio, proportion, and percent, while the Japanese spend only 6% of their time on these topics. In the United States, teachers of eighth grade mathematics report spending about 20% of their time teaching algebra, while Japanese teachers spend nearly 37% of their time on this topic.

Many students in the United States never study algebra at all. Even in high school, students may take a general math course that is another review of elementary school arithmetic. The results are evident from tests such as the Fourth National Assessment of Educational Progress, which showed that only 6% of the seventeen-year-old students in the United States could solve multistep problems, especially if they involved geometry or algebra (Dossey, 1988, p. 16). Considering the large number of dropouts who left school before taking these tests, the average seventeen year old in the United States probably performs on an even lower level. The achievement of our pre-calculus students is in some cases the lowest in the industrialized world.

These results have led the committee that developed the NCTM *Standards* (1989) to call for an increased emphasis in grades 5–8 on developing an understanding of variables, expressions, and equations, on using a variety of methods to solve linear equations, and on informally investigating inequalities and nonlinear equations. This will also be the emphasis in this chapter. These ideas will be connected to earlier work in grades K–4.

Developing Algebraic Concepts

Integers and Their Operations

One topic that is crucial to the understanding of later work in algebra is the study of integers. Fremont (1969, p. 203) stated, "Everything that is done in any future work in algebra will involve using these new numbers and using them with ease. Our start in this area should be unhurried and as meaningful as we can make it." This is as true today as it was over 20 years ago. Students need to understand the meaning of negative numbers and why they function as they do, and not just memorize rules for their operations. It may take longer for students to develop a conceptual understanding of negative numbers than to just memorize the rules, but the goal of operations with integers is understanding, not rule memorization. If you recall the hierarchy of goals in Chapter 3, you will remember that the "doers," those who can remember rules but do not understand the mathematics, are toward the bottom of the diagram. You want students who understand the mathematics; the calculator knows the rules.

Even young children should be familiar with negative numbers from, for example, scores in games, temperatures, losing or borrowing money, being below sea level, and perhaps reports on the stock market. They may have encountered negative numbers on a number line when they explored what happens if you subtract $3 - 5$; the answer is found to the left of zero on the number line. Danny, a first grader, learned about negative numbers quite well when he borrowed a dollar to buy an action figure and then had to pay it back out of his allowance a week later.

ACTIVITIES

Intermediate Grades (2–4)

Objective: *to become acquainted with integers on a number line.*

1. Make a large number line on the floor with masking tape, placing zero in the center. Make the numerals to the right of the zero in blue and the numerals to the left of zero in red. Explain to the children that these numerals represent distances in blocks east (blue numerals) and blocks west (red numerals) from City Hall in Anytown, U.S.A. (see Figure 14–1). (You can use landmarks in your own town for this activity.) Place a picture or model of City Hall on zero and place pictures or models of other buildings or landmarks on the red and blue numerals. Pose questions to the students such as "How many blocks is it from school to the zoo?" and "If you start at the library and go five blocks east, where are you?" This activity can later be linked to integers by designating movement toward the east as adding, movement toward the west as subtracting, blue numerals as positive, and red numerals as negative. For example, the question "If the zoo is on the blue five and you go seven blocks west, where are you?" corresponds to the equation $5 - 7 = n$. You would land on the red 2 (-2), which may be the location of the drug store. Encourage the students to make up similar problems for each other. Younger children do not need to worry about writing equations for their situations, but you may wish to ask third or fourth graders to begin to use the integer notation. The important thing at this stage is for students to have a concrete understanding of integers.

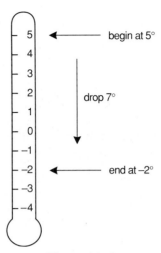

Figure 14–2

Objective: *to use a calculator to predict patterns involving integers.*

2. Give each child a calculator and let them explore using the constant feature. You might begin with addition. Start by pressing $3 + 4 = = = =$. Ask the children to explain what is happening. Do they notice that the calculator adds 4 each time they press the equal button? Encourage them to predict the next number before they press the equal button. After they are comfortable with addition, try $12 - 3 = = = = =$. Again, ask the children to explain what is occurring. Can they continue to predict the next number when the results are negative? Let the children work with a partner to explore other patterns. Third and fourth grade students should be asked to write down any generalizations they can make about adding or subtracting integers.

Middle Grades (5–8)

Objective: *to add and multiply integers on a number line.*

1. Make a large vertical number line (see Figure 14–2) that resembles a thermometer for the following number line problems. As with all new concepts, negative numbers should be introduced using concepts with which the children are already familiar. Temperature is an area in which students in many parts of the country have already experienced negative numbers, and therefore, it makes a good model for beginning experiences with integers. Ask the students to begin at 5 degrees and determine the temperature after a drop of 7 degrees ($5 + (-7) = -2$). Begin at -4 degrees and determine the temperature after an increase of 8 degrees ($-4 + 8 = 4$). A future time may be indicated by a positive number and a time in the past by a negative number. If the temperature drops steadily at a rate of 5 degrees an hour and the temperature is now 0 degrees, what will the temperature be in 3 hours?

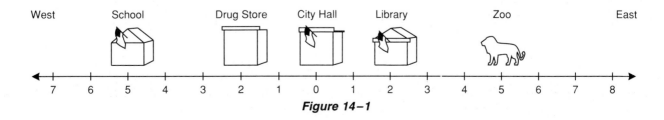

Figure 14–1

$(3 \times -5 = -15)$ What was the temperature 4 hours ago? $((-4) \times (-5) = 20)$ Ask the students to make up other questions for each other and to demonstrate solutions on the number line. After the students have worked several problems, ask them if they have discovered any patterns in their work. Keep a written record of the rules the children discover in a prominent place in the classroom. One of the first rules the children should discover is that of the **additive inverse.** That is, the sum of any number and its opposite are zero; these opposites are called additive inverses. For example, 1 and -1 are additive inverses because $1 + (-1) = 0$ and $(-1) + 1 = 0$. Rules for operating with integers will be needed all year (and for several years to come) as students work with signed numbers, not only for integers, but also for rational numbers and algebraic expressions.

As with all concepts, it is useful to experience algebraic concepts in a number of different contexts. Another real-life situation involving integers is Danny's experience borrowing money.

Objective: to add, subtract, multiply, and divide integers using discrete objects.

2. Each child will need a handful of bingo chips of two different colors for the following activities. This example uses blue and red chips. Set up a bank using blue chips for positive amounts (dollars) and red chips for negative amounts (IOUs). If you earn 2 dollars, you take two blue chips. This adds a positive amount to your account. If you owe 3 dollars, you take 3 red chips. This adds a negative amount to your account. If you buy a toy for 5 dollars, you take away 5 blue chips, thus subtracting a positive amount from your account. If you pay off a debt, you subtract IOUs or red chips, a negative amount, from your account. For example, Danny normally gets an allowance of $1 each week (adding a positive amount). The week he paid off his debt, instead of receiving $1 and adding it to his negative $1 for a net total of zero, his debt of $1 was taken away, again leaving him with a net total of zero. Therefore, subtracting a negative amount had the same outcome as adding a positive amount. Here are some problems to try with the chips.

   a. Julie has $5. She owes Sam $2. What is her net worth? The number sentence is $5 + (-2) = $ ___ . She has $5 and adds a $2 IOU. Take 5 blue chips for the positive $5 and 2 red chips for the IOU. Each red chip can be added to a blue chip, its additive inverse, to make $0. When the zeros are taken away, what is left? Three blue chips or a positive $3. (See Figure 14–3.)

   b. Jorge owes his dad $6. His father tells Jorge that he will take away $5 of the debt if Jorge rakes the yard. How much will Jorge have or owe after he rakes the yard? The number sentence is $-6 - (-5) = $ ___ . He has a $6

Figure 14–4

debt and his father will take away a $5 debt. Take 6 red chips for the -6. When his father takes away the $5 debt, he will take 5 of the red chips. One red chip is left, signifying a debt to his father of $1. (See Figure 14–4.)

   c. What if Jorge decides the following week to again rake the yard when he is only $1 in debt? If his father again agrees to take away a $5 debt, how much will he have or owe? The number sentence is $-1 - (-5) = $ ___ . He has a $1 debt and his father will take away a $5 debt. You start with one red chip to signify the $1 debt. Immediately you recognize that you have a problem; you do not have 5 red chips for Jorge's father to take away. You remember that a positive chip and a negative chip add to zero. You decide to add several zeros to the -1. You can add as many zeros as you want without changing the value of the chips, -1. Since you only need 5 red chips for Jorge's father to take away, you decide to add 4 zeros—4 blue chips and 4 red chips. You now have a total of 5 red chips and 4 blue chips. Verify that the chips still have a value of -1. Now you can take away 5 red chips. What is left? The 4 blue chips, signifying a positive $4. Jorge now has $4. (See Figure 14–5.)

   d. Marta has $4. She wants to buy a used video game for $9. How much will she owe if she buys the game? The number sentence is $4 - 9 = $ ___ . Start with 4 blue chips to signify the $4 she has to begin with. You immediately recognize that to spend $9 you need to take away 9 blue chips and you do not have enough blue chips. Again, you add zeros to your starting amount of 4 blue chips. This time you add 5 zeros—5 red chips and 5 blue chips—to your starting amount of 4 blue chips. At this point, you have 9 blue chips and 5 red chips. You take away the 9 blue chips to pay for the video game and you are left with 5 red chips, or a debt of $5.

Ask the students to make up similar problems of their own. After they have worked several problems, they will probably not want to use the chips any more. Encourage the students to just write the equation for the problem and then explain to each other (and you) how they would go about solving it. Remember that it is the method and not the answer that you are interested in here. After the students have worked several problems, ask them if they notice any patterns. Can they generate any rules? What happens if you subtract a negative number? Is the result always the same as that of adding a positive number? Students will find that this is true. They should also notice that subtracting a positive number has the same result as adding a negative number.

After working several problems involving addition and subtraction of integers, try some problems involving multiplica-

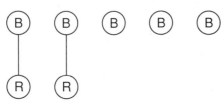

Figure 14–3

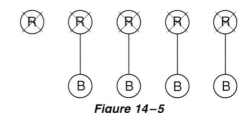

Figure 14–5

tion and division of integers. Again, a time in the future should be represented by a positive number and a time in the past by a negative number. Having a debt of $3 for five weeks in a row can be thought of as $5 \times (-3) =$ ___ . You would add 3 red chips every week for five weeks. Asking how your finances stood five weeks ago relative to today after this five weeks of debt would be $(-5) \times (-3) =$ ___ . You would have to start at some representation of zero and take away 3 red chips five times to find the answer.

Because division is the inverse operation of multiplication, the problem $15 \div (-3) =$ ___ can be thought of as the inverse of the problem above. When would you have $15 more than you have today if you add a debt of $3 each week? You know that if you continue to add the debt of $3 every week into the future, you will always have less money than you have today. You cannot go into the future that is, the number of weeks cannot be positive. To have more money when you add a debt every week, you must go into the past. Therefore, $15 \div (-3)$ must equal -5, representing 5 weeks ago. After working several multiplication and division problems with the students, ask them to note any patterns and to generate rules for multiplying and dividing integers. They should notice that multiplying or dividing two negative integers has the same result as multiplying or dividing two positive integers; when multiplying or dividing a positive integer by a negative integer or a negative integer by a positive integer, the result will always be negative.

Objective: *to use everyday situations to create integer word problems.*

3. Encourage the students to make up several problems of their own using integers. Be on the alert for other types of real-life situations involving positive and negative amounts. You might want to follow the ups and downs of the stock market for experience with positive and negative rational numbers. Looking at gains and losses of yardage on the football field makes for some interesting problems. What happens if a penalty of 5 yards is taken away (subtracting a negative number)? Does this have the same effect as a gain of 5 yards? Keep the students' problems in the learning center for other students to work on or put them on worksheets and use them instead of the examples in the textbook.

Functions, Variables, Equations, and Inequalities

A **function** is defined as a rule or process that sets up a correspondence between a first set (the **domain**) and a second set (the **range**) such that each element in the first set corresponds with *one and only one* element in the second set. Notice that this is not necessarily a one-to-one correspondence. A statement such as $y = x^2$ is a function of x, but it is not a function of y. For each x, there is one and only one y, but there are 2 x's for each y. For example, for $x = -3$, the only y is 9, but for $y = 9$, there are two possible x's, $x = 3$ and $x = -3$.

In Chapter 7, we used an input-output machine to demonstrate the concept of an operation. In algebra, we build upon that concept to further develop the concept of a function. We then relate using the function machine to

using tables, equations, and graphs so that students can broaden their concept of a function.

ACTIVITIES

Primary Grades (1–4)

Objective: *to develop the concept of a function through the use of an input-output machine.*

1. *Guess my rule* is played by pairs of students using the input-output machine described in earlier chapters and the attribute blocks. Ask one student to be in charge of placing blocks in the input of the machine and ask the second student to think of a secret rule, such as "change only the color of the block." After the first student places a block in the input of the machine, the second student finds a block that fits his or her rule and places this block in the output of the machine. Try this several times, until the first student can guess the secret rule. Discuss whether more than one output was possible for each input. Ask the students to trade roles and make up a new rule. Was more than one output possible this time?

   After playing the game with attribute blocks, play guess my rule with numbers and operations. Calculators are also useful in this activity. Students may place numeral cards in the input and output. The first student should choose a numeral card to place in the input while the second student decides on a secret rule. Begin with rules that use only one operation. The second student should place the numeral card that shows the result of the operation in the output. Do this several times, until the first student can guess the secret rule. Again, discuss whether more than one output was possible for each input. Ask the children to trade roles and choose another rule. Encourage them to discuss how they determined the rule each time.

Objective: *to explore functions on the computer.*

2. Use a computer game such as Jane Plus from Longman to give students the opportunity to beat the computer at a game of guess my rule. In Jane Plus, the computer uses stick figures of boys and girls on the screen to indicate a rule such as "add 5" or "multiply by 2." The boys always indicate addition and the girls indicate multiplication. You may use either one or two operations. The game is played in the same manner as guess my rule. Because the operations are limited to addition and multiplication, the answer will be a whole number if the input is a whole number. Therefore, this game can be played by younger students. You can also modify the calculator game to restrict the operations in the same way.

Middle Grades (5–8)

Objective: *to develop the concept of a function through the use of a function machine and a table.*

3. Draw a function machine and a table for recording inputs and outputs as in Figure 14–7 on an overhead transparency and put the transparency on the overhead projector for the students to use. Ask one student to write down a secret rule for the function machine on the transparency. Keep the rule hidden.

   You may wish to begin with a rule involving only one operation, such as "subtract 7," but the activity becomes much more challenging if you allow more than one operation per rule, such as "subtract 7 and then multiply by 2." The other

When you look at something through a magnifying lens, it usually has a bigger image. Some magnifying lenses can also invert the image. That means that the magnifying lens turns the image upside down.

Suppose we have 4 different magnifying lenses: one that magnifies twice, one that magnifies 3 times, one that magnifies twice and inverts, and one that magnifies 3 times and inverts. We'll call them (2), (3), (−2), and (−3).

Think about the following situations. What will the size and position of the images be? The pictures will help you.

Imagine looking at a small △ through the (2) and (3) lenses.

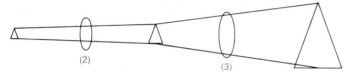

Imagine looking at the △ through the (2) and (−3) lenses.

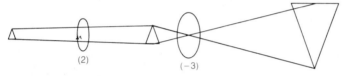

Imagine looking at the △ through the (−2) and (3) lenses.

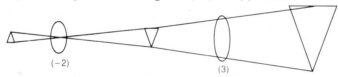

Imagine looking at the △ through the (−2) and (−3) lenses.

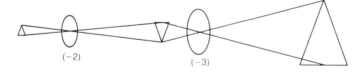

Figure 14–6
Willoughby, Stephen S., Carl Bereiter, Peter Hilton, and Joseph Rubinstein. Real Math. *Peru, Ill.: Open Court, 1991, p. 266. Reprinted by permission.*

THE MATH BOOK

The seventh grade textbook page shown in Figure 14–6 provides a semi-concrete illustration of the concept of multiplication of integers. This would be a very good activity to use in conjunction with the actual lenses so that students could construct the rules for multiplying integers. This makes it easy to see that when you have one lens that inverts images and one that does not, it does not matter if the lens for (-2) comes before or after the (3) lens. In both cases, the image is 6 times as large and inverted. For both the example with the non-inverting lenses (2) and (3) and the example with the inverting lenses (-2) and (-3), the final image is not inverted and is 6 times as large. After experimentation with several lenses, students should generalize that two non-inverting lenses together or two inverting lenses together always give you a non-inverted image. When the two lenses are different, however, one inverting and one non-inverting, the final image is always inverted. This corresponds to the mathematical rules for multiplying integers, "If both factors have the same sign, their product is positive. If their signs are different, their product is negative." These rules are introduced on the following page of this textbook.

The teacher's guide recommends reviewing addition and subtraction of integers orally in conjunction with this activity to give students more practice in mental arithmetic. To review, the teacher gives the students problems such as $2 - (-3)$ and they respond by putting their thumbs up if the answer is positive, thumbs down if the answer is negative, or fists together if the answer is zero. In this way, the teacher can check the whole class at once to see whether anyone needs to relearn the concept or skill. This same activity could be done with multiplication and division of integers once the concept is developed.

After the skills of addition, subtraction, multiplication, and division with integers are introduced and practiced, this textbook has directions for a dice game that reinforces skills with all the operations. The dice are included with the student's packet of materials for this series, and games are considered an essential part of the program. The program also includes a booklet for parents that gives tips on working with the students at home.

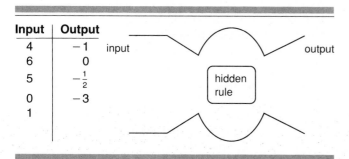

Input	Output
4	-1
6	0
5	$-\frac{1}{2}$
0	-3
1	

Figure 14–7

students will attempt to guess the rule by determining what happens when different numbers are put into the function machine. This is a good activity for using calculators. After the secret function has been written on the function machine and covered, the other students name an input, which is then written on the chart by the student with the secret rule. This student then puts the input into the calculator, performs the function, and records the output on the chart. This continues for several different inputs until the other students are confident that they know the secret rule. To avoid giving away the secret rule too early, when a student believes that he or she knows the rule, that student should predict the output for a given input, but not tell the rule. When most students can predict the output for a given input, ask the first student who predicted the output to tell the rule. If correct, this student is the next one to write a secret function. Figure 14–7 shows the inputs and outputs for the rule "divide by 2 and then subtract 3." Notice that if the game allows division and subtraction, you will frequently have outputs that are integers and other rational numbers, even if all the inputs are whole numbers.

The use of variables is another concept vital to the understanding of algebra. Algebra is often thought of by students as the area of mathematics that uses a lot of letters instead of numbers. These letters, however, are often not very well understood. Today, we generally use the term **variable** to mean a symbol for an element in a replacement set. This frequently does involve using a letter to stand for a number, but letters may also refer to points in geometry such as $\overline{AB} = \overline{BC}$, propositions in logic such as p $\wedge$ q, or a matrix in algebra. The symbol used may not even be a letter. For example, $3 + x = 7$, $3 + \_\_ = 7$, and $3 + ? = 7$ all have the same meaning, but the variable changes from x to $\_\_$ to ?. In this context, students have been studying variables since they began to work with symbols and operations. Primary and intermediate level teachers should begin to use n or x to acquaint students with the idea of a variable and to help make the bridge to algebra, which formalizes these concepts. It is important that students in the middle grades continue to have many informal experiences with variables.

A symbol or variable may stand for one or more numbers, which are called the values of the variable. The term *variable* may be used to mean a symbol that may have *two*

or more values during a particular discussion, such as *area* = $l \cdot w$, thus the concept of variance underlying the term variable. However, a variable may have only one value and students may have difficulty understanding why the c in $5c = 20$ is called a variable when the only amount it can stand for to make this a true statement is 4.

Another difficulty that students may have in working with variables is that they frequently think of a variable as a letter that stands for one particular number. If they see $2s + 5 = 9$ and $2c + 5 = 9$, they think that the s and the c must stand for different numbers. They may believe that it is all right to use l to stand for length and w to stand for width, but you cannot use t to stand for length. The variable t must stand for time, or turkeys, or tubas, or tambourines, but not length or width or dogs.

Students are introduced to variables when they learn simple computer programming languages such as BASIC, and they are sometimes confused. Variables may have different forms and they may be written in several different ways. For example, two versions of a program that accepts the length and width of a rectangle and returns the area of the rectangle are shown below. Two programs may appear different yet perform in the same way when run on a computer.

```
 10   PRINT "ENTER LENGTH"
 20   INPUT LENGTH
 30   PRINT "ENTER WIDTH"
 40   INPUT WIDTH
 50   PRINT "AREA = "; LENGTH*WIDTH
 60   PRINT "WANT TO TRY AGAIN? (Y OR N)"
 70   INPUT ANSWER$
 80   IF ANSWER$ = "N" THEN 100
 90   GOTO 10
100   END
```

```
 10   PRINT "ENTER LENGTH"
 20   INPUT L
 30   PRINT "ENTER WIDTH"
 40   INPUT W
 50   LET A = L*W
 60   PRINT "AREA = "; A
 70   PRINT "TRY AGAIN? (Y OR N)"
 80   INPUT R$
 90   IF R$ = "N" THEN 110
100   GOTO 10
110   END
```

Notice that the variables for length, width, and area are different in each program but do the same work. A variable may be represented by a single letter such as L or by a series of letters such as LENGTH. The variable for the answer, ANSWER$ or R$, is a different type of variable that accepts symbols rather than numbers.

These misconceptions about using variables emphasize the importance of giving students a number of different experiences with variables in the middle grades so that they can build their own concepts. Just as a toddler may not understand that potatoes may be mashed, baked, or french fried and still remain potatoes, middle grades students need to experience variables in many different contexts before they begin to understand all the ways variables can be used.

ACTIVITIES

Middle Grades (5–8)

Objective: to use variables with functions and graphs.

1. Repeat the activity described above with the function machine, using variables instead of the terms *input* and *output* on the chart. The most common variables used are x for the input and y for the output. These variables lead to graphing on a Cartesian coordinate system, which was introduced in Chapter 11 (Probability, Statistics, and Graphing). The function in the previous activity, where the rule was "divide by 2 and then subtract 3," would be written as $y = \frac{x}{2} - 3$. Ask the students to explore whether that is the same as the function $y = \frac{(x-3)}{2}$. How do the graphs differ? Encourage the students to graph several functions, and then ask them if the function graph was always a line. Can they think of a function whose graph would not be a line?

Objective: to explore order of operations on different calculators.

2. Use different calculators to determine if you always get the same answer when working out equations. Try equations such as $3 + 4 \times 2 =$ ___ and $2 + 3 \div 4 =$ ___ . Discuss with the students the difference between calculators that work the exercise in the order in which you put in the numbers and ones that use algebraic logic. If you have parentheses on your calculator, explore what happens when you use them. When using algebraic logic, you first compute the operations inside parentheses, beginning with the innermost parentheses, then you compute exponents, then you multiply and divide from left to right, and finally you add and subtract from left to right. This is called the **order of operations.**

The order of operations is important to students not only for work with rational numbers, but also for simplifying algebraic expressions and solving equations. An **algebraic expression** is a symbolic form involving constants, variables, operations, and grouping symbols such as parentheses. Two or more algebraic expressions joined by addition or subtraction are called **terms,** and two or more algebraic expressions joined by multiplication or division are called **factors.** If two or more algebraic expressions containing at least one variable are joined by an equal sign, it is called an **algebraic equation.**

The understanding of variables, expressions, and equations is one of the areas on which the NCTM *Standards* recommend an increased emphasis. Again, this understanding should be developed using concrete and pictorial materials. The graphs used above with functions can be expanded to further explore equations.

Objective: to explore the effects of changing amounts in equations using Cartesian graphs.

3. Begin with the equation $y = x$, and make a table of values such as the following:

x	y
2	2
5	5
−1	−1
0	0
$\frac{1}{2}$	

Ask the students to graph this equation using a Cartesian coordinate system. Now try the equation $y = 2x$. Make a table of values and then graph this equation on the same graph. Ask the students to make predictions for the graphs of $y = 3x$ and $y = \frac{1}{2}x$. Make a table of values for these equations and then graph them on the same graph. What do you notice? What happens to the slope as the coefficient of x changes? Encourage the students to ask other questions about the graph, such as what happens if the slope is negative? What if you add some value to the equation as in $y = x + 2$? Is the graph still a line? What has happened to the line?

Objective: to explore the graphs of linear equations using a graphics calculator.

4. Try the same activity as above, this time using a graphics calculator. You will not need to begin with the chart of values. Put the calculator in the graphics mode and ask for the graph of $y = x$. Then ask for the graphs of $y = 2x$, $y = 3x$, $y = -2x$, and $y = \frac{1}{2}x$ on the same graph. What do you notice? What is the effect of a negative coefficient of x? What happens as the coefficient gets larger? What happens if you add a constant to the equation? Can you make a graph that is nonlinear?

Algebra blocks are another material that can be used for the development of algebra concepts. They give the students a good picture of the concept, are related to earlier concepts learned with whole and rational numbers, and can be manipulated to determine answers to a variety of situations.

Algebra blocks can be purchased commercially, as Hands On Equations, Algebra Lab Gear, or Algebra Tiles, or constructed out of railroad board. A master for algebra blocks is included in Appendix B. This set may be used with the large square having dimensions of x by x for an area of x^2, the medium-sized rectangle having dimensions of x by 1 for an area of x, and the small square having dimensions of 1 by 1 for an area of 1. You may change the representations so that the small square has dimensions of y by y for an area of y^2, which would then give the rectangle dimensions of y by x for an area of xy. Changing the values of the squares helps the students realize the variable nature of algebra. As you do the activities in this

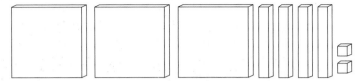

Figure 14–8

section, relate the work with the algebra blocks to the earlier work with base ten blocks. Earlier, the large square had dimensions of 10 by 10 for an area of 100, the rectangle had dimensions of 10 by 1 for an area of 10, and the small square had dimensions of 1 by 1 for an area of 1. When we moved to decimals, the large square represented 1, the rectangle 0.1, and the small square 0.01.

Figure 14–8 shows the representation for $3x^2 + 4x + 2$ if the large square is x^2, the rectangle x, and the small square 1. Using the representation in which the small square is y^2, Figure 14–8 represents $3x^2 + 4xy + 2y^2$.

ACTIVITIES

Middle Grades (5–8)

Objective: *to introduce concrete models for algebraic expressions.*

1. Ask the students to each construct a set of algebra blocks using the master from Appendix B. The blocks may be made of either railroad board or laminated construction paper. It is helpful to make one set out of different colored construction paper to represent positive numbers and one set out of red construction paper to represent negative numbers. The *Algebra Tiles for the Overhead Projector* have blue x^2s, green x's and yellow 1's. All of the negatives are red in that set.

Ask the students to use the blocks to show a variety of expressions. Begin by adding positive amounts. You might show $3x^2 + 2x + 5$, $4x^2 + 3x + 2$, etc. After the students

are comfortable with this activity, ask them to show negative amounts as well as positive amounts. Try $3x^2 + (-2x) + (-2)$. This might also be written as $3x^2 - 2x - 2$. After the students can represent amounts using the x's, tell them that they can use any letters to name the dimensions of the rectangles. Let them choose other letters for the dimensions and then tell you what the tiles represent. For example, Figure 14–9 shows $3z^2 - 4zw + 2w^2$ if the dimensions of the large square are z by z and those of the small square are w by w.

Objective: *to use algebra blocks to simplify expressions involving additive inverses.*

2. Remind students of the rules they developed for additive inverses with integers. Ask them to predict what $x + (-x)$ should be. Work this problem with the blocks. Use the blocks to show zero in a variety of ways. Figure 14–10 shows a few examples.

After the students are comfortable showing zero with the blocks, ask them to put down any combination of blocks, record the blocks they used, and then remove all the zeros. Figure 14–11 shows $3x^2 + (-2x^2) + 2x + (-3x) + 5 + (-2)$. After the zeros are removed, the figure shows $x^2 - x + 3$.

Ask the students to model a number of different algebraic expressions. Let them make up problems for each other that involve the addition and subtraction of expressions. Discuss with them the rules for adding integers and how these rules are related to the addition and subtraction of algebraic expressions or **polynomials**, algebraic expressions built up from constants and variables by adding, subtracting, or multiplying. In a polynomial, a variable cannot appear in a denominator, in an exponent, or within a radical sign.

Figure 14–9

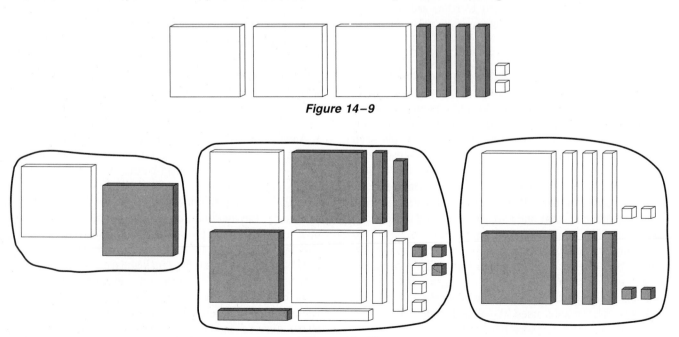

Figure 14–10

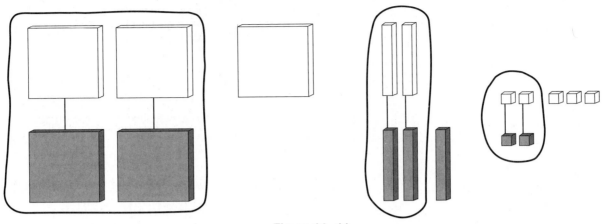

Figure 14–11

Objective: *to multiply and factor polynomials using algebra blocks.*

3. Review with the students the area model for multiplying whole numbers. In Figure 14–12, if the large square is 10 × 10 and the small square is 1 × 1, the picture represents 12 × 13. The area of the large square is 100, the area of the medium rectangle is 10, and the area of the small square is 1. When you add the values, you have $1(100) + 3(10) + 2(10) + 6 = 156$. If the large square is 1.0 × 1.0, the problem becomes 1.2 × 1.3 = 1.56. The same principles hold in algebra. If the large square is $x \times x$, the rectangle $x \times 1$, and the small square 1 × 1, Figure 14–12 now represents $(x + 2)(x + 3)$. The area is now $1(x^2) + 3x + 2x + 6$, or $x^2 + 5x + 6$.

Notice that this model relates to the FOIL method, which is frequently taught in algebra classes. FOIL is an acronym that stands for "first, outside, inside, and last." For $(x + 2)(x + 3)$, the FOIL method would involve multiplying the *first* terms in the parentheses $(x \cdot x)$, then the *outside* terms $(x \cdot 3)$, then the *inside* terms $(2 \cdot x)$, and finally the *last* terms $(2 \cdot 3)$. Locate each of these pairs of terms on the diagram. Try relating this to the multiplying of whole numbers, such as $(10 + 2)(10 + 3)$.

Have the students model other examples of multiplying polynomials, such as $(x + 3)(x + 4)$ and $(x + 1)(x + 3)$. Encourage them to make up problems for each other. Let one student make a model and ask another student to explain what is being shown.

After the students have modeled several problems involving the multiplication of two terms with positive amounts, ask them to reverse the operation and try factoring. Start with $x^2 + 5x + 4$ and ask the students to form a rectangle with the pieces. What are the dimensions of the rectangle? These

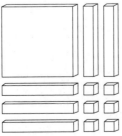

Figure 14–12

are the **factors** of the polynomial. Figure 14–13 shows the results, $(x + 4)(x + 1)$.

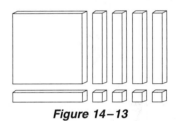

Figure 14–13

Solving Equations and Inequalities

One very important part of algebra is the solution of equations. To **solve an equation,** you must find replacements for the variables that make the left side of the equation equal to the right side. For example, if you wish to solve the equation $x + 3 = 5$, you must find a value for x that will make this a true statement. In this case, we know the answer is 2, simply because we recognize the addition fact. It is not always that easy to solve an equation, however. The following is an example of one way to introduce students to the idea of solving equations. Encourage the students to think of other examples of their own.

ACTIVITIES

Intermediate Grades (3–5)

Objective: *to use a balance to introduce the concept of solving equations.*

1. For this activity, you will need a pan balance, some inch or two-centimeter cubes, and a very lightweight paper or opaque plastic bag. (It should be light enough that its presence is not noticeable on the balance. If you cannot find a bag that is light enough, you can hold a screen in front of the balance to hide the secret number of cubes.) Have one student put a secret number of cubes in the bag. This may be named x cubes, or perhaps j cubes if Joanna put the cubes in the bag. Place the bag on the left pan along with two other

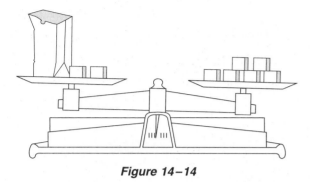

Figure 14–14

cubes that can be seen by the students. Now put enough cubes on the right pan to balance the left. Figure 14–14 shows a balance with the bag and 2 cubes on the left pan and 6 cubes on the right pan.

Ask the students to write an equation describing what they see. They might write $j + 2 = 6$. How can they tell how many cubes are in the bag? It would be easy if the bag were the only thing on the left pan. If they subtract two cubes from the left pan, what must they do to the right pan to keep the balance? Try it. Now you have only the bag on the left pan and four cubes on the right. This can be represented by $j = 4$. Look in the bag to see if you were right.

Ask other children to put different numbers of cubes in the bag and in plain sight on the left pan. Put cubes on the right pan to balance the left side and then determine the number of cubes in the bag. Do you always have to do the same thing to both pans to keep the balance? What if the bag is on the right side? Can you solve the equation $4 = x + 1$ in the same way that you solved the equation $x + 1 = 4$?

Try putting the same number of cubes in two bags along with another number of cubes in plain sight on one pan. What would you do if you saw $2c + 1 = 9$, as in Figure 14–15? If you begin by subtracting 1 cube from each pan, you then have $2c = 8$. Will the pans balance if you subtract 2 from each side? You cannot subtract 2 cubes from the side with the bags unless you open a bag. Can you take half the objects off each side? Yes, you can take off one of the bags and 4 of the cubes. What is left now? Is this the correct answer?

Objective: to use a balance to introduce the concept of inequalities.

2. A balance may also be used to introduce the concept of inequalities. Repeat the activity above, only this time put one too many cubes on one of the pans. Discuss with the children how you can tell by looking at the balance which side is heavier. Review the use of the greater than and less than signs. Does

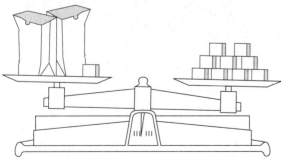

Figure 14–15

the balance change when you do the same thing to each side? Again, work the problem until you have only one bag on one of the pans. Ask the students to write the equation for each step. Leave the balance in the learning center so that students can pose different problems for each other.

On the computer, try Balancing Bear from Sunburst and Algebra Shop from Scholastic to give students practice with similar ideas.

Developing and Practicing Algebraic Skills

Most of the algebra concepts that are developed in the middle grades will not be fully mastered until high school. Some, such as manipulating symbols, memorizing procedures, and drilling on equation solving, are topics that no longer should receive much attention (NCTM, 1989, p. 71). Some topics, however, such as operations with integers, should be practiced and mastered in the middle grades.

ACTIVITIES

Middle Grades (5–8)

Objective: *to practice skills in operating with integers.*

1. Use a regular deck of playing cards without the jokers and face cards or make your own deck with 10 red cards (numbered 1–10) and 10 black cards (numbered 1–10). The black cards are worth positive amounts and the red cards are worth negative amounts. This is a game best played by four players, but it can be adjusted for other numbers of players by adding more cards. Begin the game by shuffling the cards and dealing five cards face down to each player. Each player then picks up his or her five cards, adds the numbers shown on the cards, and writes down the total number of points. This is the score for round one. On each turn, each player draws one card from the player on the left. If you take a red 5, you must add a negative five to your score. What happens to the player who lost the red five? That person will subtract a negative five. Does this have the same effect as adding a black five? Round two ends when each player has drawn one card from the player on the left. At this point, everyone again adds the points shown on his or her five cards. The score from round two is added to the score from round one. Play continues in the same manner until someone has a score greater in absolute value than 50 or until time is called. The player with the score greatest in absolute value at the end of the game wins.

    Note: An easy way to check that everyone has added correctly on each round is to check the total score for the four players. This should always be zero. Ask the students to explain why this is so. If you are using a different number of players and a different number of cards, determine the sum of all the cards before you begin.

Objective: to reinforce skills with graphing using the Cartesian coordinate system.

2. The computer program Green Globs and Graphing Equations, available from Sunburst Communications, contains several activities to develop skills related to graphing and writing equations. For example, Equation Plotter allows the student to enter an equation from the keyboard. Promptly, the graph of the equation appears on a coordinate grid on the screen. This provides students with an opportunity to explore how equations will appear in graphic form. Activity sheets provided with the software lead students to make various observations; as well, teachers and students have the flexibility to experiment on their own.

Another activity is Linear and Quadratic Graphs, in which graphs are displayed on a coordinate grid and students enter equations that match the displayed graphs. Participants may select lines, parabolas, circles, ellipses, and hyperbolas to be displayed. When the students believe they have discovered the appropriate equation, they enter it, and immediately the graph of the equation is plotted on the screen along with the target plot. They may continue entering equations until the graphs match.

Green Globs is an activity that presents thirteen "green globs" on the screen in a random fashion. The students attempt to explode all the globs by touching them with the graphs of equations that have been entered from the keyboard. The object is to use the fewest number of equations possible to explode the globs.

Estimating and Mental Calculating

Many algebraic equations can be solved mentally without writing out a lengthy series of equations. Encourage students to work these problems mentally whenever possible. A student should not need paper and pencil to find the solution for $x - 6 = 12$. If exact mental calculation is not possible, ask the students to estimate before working out an equation. Ask for justification for their estimates. For example, for the equation $4x + 920 = 256$, the students may have difficulty determining the exact answer mentally. One of the first questions they should ask themselves is whether x should be a positive or a negative number. How do you know? Is x greater or less than -100?

Problem Creating and Solving

Using algebra is an excellent way to explore real-life situations. It is not necessary to present these problems in categories, however, such as age problems, money problems, and rate and ratio problems. Many good examples of problems can be found in science books as well as in problem-solving books such as *Problem Solving: A Handbook for Teachers* by Krulik and Rudnick (1987). An example from their book of a problem that could be solved using algebra is the following (p. 103): "In a class election, the winning candidate received three times as many votes as her opponent. If 32 votes were cast, how many votes did the winner receive?"

Some students may try to do the problem using the strategy guess and check and may begin by guessing an amount such as 20. When they check whether 20 could be three times the number of votes received by the opponent, they find that this would not be possible. They would then guess to find a number that is a multiple of three and that also fits the other conditions of the problem. Students frequently find after working several problems of this nature using guess and check that using algebraic notation can be a much more efficient method. For this problem, a student might use a v to represent the number of votes the opponent received. What would represent the number of votes the winner received? Can this number be expressed in terms of v? Yes, it would be $3v$. The equation $v + 3v = 32$ would then represent the voting situation in the class. Students who have worked earlier problems solving equations would realize that this is $4v = 32$, or $v = 8$. Upon checking, the students should verify that if the loser received 8 votes and the winner 3 times as many, the winner received 24 votes. They find that $8 + 24$ does result in the total of 32 votes in the class. Therefore, the winner received 24 votes. Encourage the students to make up other problems for each other involving algebra. How many different ways can these problems be solved? Can they draw a picture or graph to illustrate the solution?

Use real situations to create other problems in the classroom. You might want to try looking at the stretchiness of a spring. Measure the length of a spring at rest and then measure the length again after hanging a heavy object such as a brick from it. What is the length of the spring now? Can you predict the length if you hang two bricks from it? What about three bricks? Make a graph of the results. What is the shape of the graph? Can you write an equation that would tell you the length of the spring for any given number of bricks? Encourage the students to ask other questions about the situation and explore the answers.

Other situations that are interesting to explore include looking at the distance a toy car travels beyond the end of a ramp compared to the angle on the ramp, the number of times a pendulum swings in 20 seconds compared to the length of the pendulum string, the time it takes water to empty from a paper cup with a hole in the bottom compared to the amount of water in the cup, and the length of time it takes a dropped object to hit the ground compared to the height of the drop. All of these lead to some very interesting physics concepts that can best be explored using very simple algebraic concepts. Encourage the students to carefully set up experiments in which they collect data to be organized in a table and then graphed. After the data are organized in this manner, ask the students to list all of the questions they can think of regarding the experiment. Which of these can be answered based on the data already collected and which need further exploration? Perhaps the science teacher would join you in this project.

Grouping Students for Algebraic Learning

The activities described above that involve experimenting with different physical concepts would best be done in small groups. Three or four students working together can set up the experiment, collect data, and ask and answer a variety of questions about their information. They can also work collaboratively to present their data in an interesting and understandable fashion.

Other activities, such as exploring the effects of changing the constants on the graphs shown on a graphics calculator, would probably best be accomplished by two people working cooperatively. It is difficult for more than two students to work with one graphics calculator, but more students might get together to discuss results.

Other problems might initially be presented to the whole class at once for individual exploration. After students have had an opportunity to explore on their own, they could get together in small groups or as a class to share results.

Communicating Learning of Algebraic Skills

Students working on projects such as exploring the effects of changing the length of a pendulum string or the angle of a toy car ramp should be encouraged to create different means of presenting their data. They might want to build a model, make a chart, draw a graph, search for equations to describe the graphs, and present questions and answers in written form. Remember that others are interested in the process that they used and not just the results. Several of these projects may make very good exhibits for a math or science fair.

It is interesting to look at results others have presented in such sources as newspapers and magazines. Do they use graphs to illustrate the results? Does the text seem to go along with the graphic information or are there misleading statements in the article? Students who have struggled with presenting their own results make very good critics of others' information.

Evaluating Algebraic Concepts and Skills

When evaluating students' skills in algebra, it is important to remember that the process is at least as important as the answer. Ask the students to explain their reasoning and not just write down numerical answers when working algebraic problems or exercises.

If you use group projects, let the students evaluate their own work and that of the others in the group. They can list both what they contributed to the project and what they learned from it. They can also list two or three of the most important contributions from the others in the group.

Following are some common areas of misunderstanding or difficulty in algebra. Be on the lookout for these misunderstandings. You might ask students to be on the lookout for them in their own work and in the work of others in the groups in which they are working.

1. A number and a variable placed next to each other should be multiplied, not added. For example, many students think $2a$ means $2 + a$ rather than $2 \times a$. This is understandable since 54 means $50 + 4$, not 50×4, and $4\frac{1}{2}$ means $4 + \frac{1}{2}$, not $4 \times \frac{1}{2}$.

2. Different variables may stand for the same number, and the same letter in a different equation may stand for a different number. If $2x + 3 = 7$ and $2y + 3 = 7$, both x and y stand for 2. If in one statement $x + 5 = 9$ and in another statement $x + 4 = 0$, x stands for 5 in the first statement and 4 in the second.

3. Translating from a word problem to an equation is also difficult. If there are 4 apples for every orange in the bowl, the situation is not represented by $4a = o$. If a stands for the number of apples and o stands for the number of oranges, ask the students which there are more of in the bowl. The statement tells us there are more apples, 4 times as many as oranges. In the equation $4a = o$, which is larger, a or o? The o must be larger. Does this fit the statement from the word problem? Be on the alert for other incorrect methods of translating a written statement into an equation.

4. It is often easier for students to interpret information in graphical form than in equation from. Be sure to use a geometric or graphical model whenever possible.

5. Tasks given in story form may be easier to understand than tasks given in a purely numerical form. For example, it may be much easier to understand why a negative times a negative is a positive if you use examples involving money and IOUs or temperature than if you simply state the mathematical rule.

6. Some students may believe that every relationship is a linear one. Let them try graphing $y = x^2$ on the calculator or the relationship of the acceleration of a falling object to the distance traveled to see that not all relationships can be graphed with a line.

7. The concept of a variable is a difficult one and students frequently struggle to correctly use symbols, variables, expressions, equations, and inequalities. Give them plenty of experience in defining and illustrating for you the meanings of these terms and symbols.

Something for Everyone

Algebra provides a challenge for students who excel in abstract mathematical work; on the other hand, when physical models are used to illustrate concepts and to reinforce skills, students who have strengths in visual and spatial areas are quite able to grasp the concepts and develop the skills. Because students are beginning their study of algebra in the middle grades, it is important to use a variety of instructional techniques.

Auditory learners will be able to describe and explain algebraic processes. They may, however, have some difficulty in displaying spatially the concepts that are being presented unless clear verbal explanations are provided. In cooperative group activities, students who are auditory learners may be very helpful in describing the problem situations. They will also be very helpful when it comes to writing the results of the experiments and the solutions to the problems.

Those who do well in tactile-kinesthetic situations will succeed when materials such as algebra blocks and balance scales are used to model algebraic equations. Having the chance to move the materials around and discuss the relationship between one configuration and another will help provide the link between the concrete and abstract aspects of algebra.

Students with special talent in mathematics will likely discover algebraic relationships and patterns quickly. Because the work in algebra is at the introductory level, there is plenty of room for these students to excel and move ahead. For example, when using Green Globs and Graphing Equations, students will be able to work at several levels and will be able to solve increasingly challenging problems. Talented students can select more difficult problems on which to work. Further, when working with manipulatives, talented students will be able to develop complex examples and solutions. These students should also be helpful when it comes to discussing and explaining concepts and skills in their cooperative work groups.

Those who have difficulty in learning algebra should be encouraged to work with the models that illustrate the concepts. Provide opportunities for them to be successful and to demonstrate to the whole class and to members of their cooperative work group those ideas that they do grasp. Allow these students more time as they begin to learn algebra. Be willing to seek out alternative methods and materials for reteaching concepts and skills that were not immediately learned.

KEY IDEAS

The informal introduction of algebra in the middle grades is assuming an important place in mathematics instruction. It is suggested that increased emphasis be placed on teaching variables, expressions, and equations along with using a variety of methods to solve linear equations and to informally investigate inequalities and nonlinear equations. The presentation of concepts involves using models and simple science activities and is tied to concepts learned in the primary grades. Properties are discovered and invented as students manipulate materials that model topics associated with algebra.

Emphasis is placed on using examples and problems from real-life situations. Thus, students have the opportunity to apply what they have learned to situations that are common in their experiences. Understanding is stressed and encouraged through both oral and written communication. Allowing for a diversity of thought is important at this stage in learning.

Students should have the chance to explore a variety of problem situations that require simple algebraic solutions and to present their own problems based on their experiences. Students are capable of challenging one another in ways that textbooks cannot. Teachers evaluate student progress by a variety of means, including listening to students as they describe their thinking processes and reading what they have written about their work. There are a number of common misunderstandings that surface as students learn algebra; making students aware of these misunderstandings can allow students to help themselves and to help one another.

REFERENCES

BERMAN, BARBARA, AND FRIEDERWITZER, FREDDA. "Algebra Can Be Elementary . . . When It's Concrete." *Arithmetic Teacher.* Vol. 36, No. 8 (April 1989), pp. 21–24.

COXFORD, A. F., ED. *The Ideas of Algebra, K–12.* Reston, Va.: National Council of Teachers of Mathematics, 1988.

DOSSEY, JOHN A.; MULLIS, INA V. S.; LINDQUIST, MARY M.; AND CHAMBERS, DONALD L. *The Mathematics Report Card: Are We Measuring Up? Trends and Achievement Based on the 1986 National Assessment.* Princeton, N.J.: Educational Testing Service, 1988.

FREMONT, HERBERT. *How to Teach Mathematics in Secondary Schools.* Philadelphia: W.B. Saunders Co., 1969.

HOWDEN, H. *Algebra Tiles for the Overhead Projector.* New Rochelle, N.Y.: Cuisenaire Co. of America, 1985.

KRULIK, STEPHEN, AND RUDNICK, JESSE A. *Problem Solving: A Handbook for Teachers* (2nd ed.). Boston: Allyn and Bacon, Inc., 1987.

McKNIGHT, C. C.; CROSSWHITE, F. J.; DOSSEY, J. A.; KIFER, E.; SWAFFORD, J. O.; TRAVERS, K. J.; AND COONEY, T. J. *The Underachieving Curriculum: Assessing U. S. School Mathematics from an International Perspective.* Champaign, Ill.: Stipes, 1987.

NATIONAL COUNCIL OF TEACHERS OF MATHEMATICS. *Curriculum and Evaluation Standards for School Mathematics.* Reston, Va.: NCTM, 1989.

PICCIOTTO, HENRY. *The Algebra Lab Middle School: Exploring Algebra Concepts with Manipulatives.* Sunnyvale, Ca.: Creative Publications, Inc., 1990.

SOBEL, M. A., AND MALETSKY, E. M. *Teaching Mathematics: A Sourcebook of Aids, Activities, and Strategies* (2nd ed.). Englewood Cliffs, N.J.: Prentice Hall, 1988.

WILLOUGHBY, STEPHEN S.; BEREITER, CARL; HILTON, PETER; AND RUBINSTEIN, JOSEPH H. *Real Math.* La Salle, Ill.: Open Court, 1991.

NCTM *Professional Standards*

Standard 4: Tools for Enhancing Discourse

The teacher of mathematics, in order to enhance discourse, should encourage and accept the use of—

- computers, calculators, and other technology;
- concrete materials used as models;
- pictures, diagrams, tables, and graphs;
- invented and conventional terms and symbols;
- metaphors, analogies, and stories;
- written hypotheses, explanations, and arguments;
- oral presentations and dramatizations.

Standard 5: Learning Environment

The teacher of mathematics should create a learning environment that fosters the development of each student's mathematical power by—

- providing and structuring the time necessary to explore sound mathematics and grapple with significant ideas and problems;
- using the physical space and materials in ways that facilitate students' learning of mathematics;
- providing a context that encourages the development of mathematical skill and proficiency;
- respecting and valuing students' ideas, ways of thinking, and mathematical dispositions;

and by consistently expecting and encouraging students to—

- work independently or collaboratively to make sense of mathematics;
- take intellectual risks by raising questions and formulating conjectures;
- display a sense of mathematical competence by validating and supporting ideas with mathematical argument.

Standard 6: Analysis of Teaching and Learning

The teacher of mathematics should engage in ongoing analysis of teaching and learning by—

- observing, listening to, and gathering other information about students to assess what they are learning;
- examining effects of the tasks, discourse, and learning environment on students' mathematical knowledge, skills, and dispositions;

in order to—

- ensure that every student is learning sound and significant mathematics and is developing a positive disposition toward mathematics;
- challenge and extend students' ideas;
- adapt or change activities while teaching;
- make plans, both short- and long-range;
- describe and comment on each student's learning to parents and administrators, as well as to the students themselves.

(NCTM, 1991, pp. 52, 57, 63. Reprinted by permission.)

Teaching mathematics to elementary school children requires you to be knowledgeable in both mathematics and psychology. Equally important is your ability to develop a smooth-running classroom and an environment that is stimulating and attractive. Organizational skills are required. Some individuals are naturally well organized; others need to spend extra time developing the ability to organize their classrooms.

The importance of organization in teaching is a major focus of current writings in mathematics education. The Mathematical Sciences Education Board challenges teachers by noting:

> In reality, no one can *teach* mathematics. Effective teachers are those who can stimulate students to *learn* mathematics. Educational research offers compelling evidence that students learn mathematics well only when they *construct* their own mathematical understanding. To understand what they learn, they must enact for themselves verbs that permeate the mathematics curriculum: "examine," "represent," "transform," "solve," "apply," "prove," "communicate." This happens most readily when students work in groups, engage in discussion, make presentations, and in other ways take charge of their own learning. (1989, pp. 58–59)

The NCTM *Professional Standards for Teaching Mathematics* (1991, p. 22) confirms this challenge by stating: "Because teaching mathematics well is a complex endeavor, it cannot be reduced to a recipe for helping students learn. Instead, good teaching depends on a host of considerations and understandings. Such teaching demands that teachers reason about pedagogy in professionally defensible ways within the particular contexts of their own work." Your work as a teacher will depend on your personal characteristics and preferences, the nature of the children with whom you work, the curriculum, and other factors unique to your teaching situation. These factors will affect how you organize your classroom for mathematics instruction.

In the early stages of your career, preparing to teach, developing timing, spur of the moment activities, and management strategies will consume much of your time. While we cannot offer *the* solution for you in your particular teaching situation, we can provide some guidelines. The more options and alternatives from which you have to choose, the more dynamic your classroom may become. Five aspects of classroom teaching are dealt with in the following pages:

1. fitting instruction to children's differing needs and styles
2. evaluating, recording, and reporting children's progress
3. children with special needs
4. arranging the classroom
5. providing learning aids

Fitting Instruction to Children's Differing Needs and Styles

Both children and teachers have distinct needs and styles of operation in the classroom. Each child's personal learning style is unique to that child. It determines how and when the child discovers relationships, learns to read, and develops the concepts of mathematics.

We have mentioned that children learn in different ways. Some learn best through tactile/kinesthetic experiences, that is, through touching objects and through physical movement. Auditory learners have the ability to listen to others and effectively understand. Those who are visual learners appreciate pictorial models. Children learn in all of these modes simultaneously, but tend to excel in one particular mode.

Then there are children who have particular difficulty learning; they take more time, have less interest, and seem unable to grasp concepts and skills. Likewise, some children are fast learners, enjoy intellectual challenges, and are ready to move ahead. One child may be ill-at-ease, self-conscious, and shy; another child may be at home in the classroom, comfortable with the surroundings and other children. As a consequence, one child may learn slowly and may be dependent upon an adult, whereas another child may learn quickly, barely assisted by the teacher.

When twenty-five or thirty children are brought together for instruction, the combinations of learning styles present a formidable challenge to a teacher. Likewise, your own personal teaching style affects how you teach. Your beliefs, experiences, education, and expectations of children's behavior will cause you to be a certain kind of teacher.

Learning and teaching must be carefully planned. Granted, some of the best learning is spontaneous or incidental, but for long-term, sequential learning to occur in an enriched environment, the teacher must lay the groundwork. You may wish to include children in the planning, to share the learning objectives with the children, and even to encourage the children to lead—all are a part of the learning process.

There are many alternative ways to teach. Although never complete, a listing of many teaching-learning strategies follows. Each particular strategy contains a major type of (1) student participation, (2) teacher behavior, and (3) student grouping.

When you are organizing for mathematics instruction based on sound principles of learning, the alternatives presented suggest a variety of different and effective teaching-learning strategies. Still other strategies emerge as thoughtful teachers reflect on how they teach and wish to teach. A particular strategy is a function of the teacher and may be successful only under the conditions experienced by that individual. If you have not yet taught, the following options demonstrate the range of available teaching-learning strategies. If you have taught, you will be reminded of the many possible teaching methods that exist.

Options for Student Participation

The type of student participation in a mathematics lesson depends on the objectives of the lesson. Then, in planning for teaching, you should consider how the students will be involved during the lesson. In some lessons, you may wish to have the students participating in several different ways at various times. For the sake of planning, it is important to identify the main types of student participation for a given lesson. Following are alternative types of student involvement and examples of each.

1. *Free exploring and discovering.* A group of children are left alone to explore and familiarize themselves with the geoboard. The children may talk with one another as they share their work.
2. *Directed exploring and discovering.* The class, divided into cooperative learning groups, is challenged to find all of the Cuisenaire rod trains that are equivalent in length to the yellow rod.
3. *Discussing.* Children and teacher discuss methods that may be used to determine how many grains of rice are in a peanut butter jar.
4. *Listening and looking.* Children listen and watch as the mother of a student demonstrates how the computer helps her in her job.
5. *Memorizing.* Three students engage in a card game that reinforces the basic multiplication and division facts.
6. *Evaluating.* The students complete an end-of-chapter test in their math textbooks to determine the level of skill attained.
7. *Writing.* Children record in their mathematics journals thoughts about a challenging problem.
8. *Presenting.* A student who has learned how to use Napier's rods explains the procedure to the rest of the class.
9. *Computing.* Two interested students work on the computer to develop a simple Logo program that will draw several flowers on the screen.
10. *Problem solving.* The class is provided with a 4 × 4 magic square and asked to find at least one solution.

Options for Teacher Behavior

The type of student participation planned dictates how the teacher will operate during a lesson. Of course, there are certain behaviors expected of all teachers, including maintaining a supportive learning environment and keeping order. The options for student participation mentioned above are used to illustrate how the teacher's behavior is directed.

1. When students are involved in freely exploring, teachers are observing and having some interaction with the children. This interaction is usually listening to children and reinforcing their efforts.
2. When students are involved in directed exploring and discovering, teachers are asking leading questions, giving few or no answers, answering questions with questions, and interpreting directions.
3. When students are involved in discussing, teachers are leading the discussion, asking questions, answering questions with questions, giving opinions, listening to students, encouraging students to interact with each other, and mediating.
4. When students are listening and looking, teachers are lecturing, explaining, describing, demonstrating, using educational media, and participating with children.
5. When students are memorizing, teachers are timing, encouraging, coaching, working with individuals, and observing.
6. When students are evaluating, teachers are examining, supplying data, discussing, negotiating, observing, reading, and recording.
7. When students are writing, teachers are assigning worksheets and journal topics, dictating, assisting individuals, reading and responding, and observing.
8. When students are presenting, teachers are listening, discussing, and responding.
9. When students are computing, teachers are assigning, guiding, encouraging, making suggestions, and debugging.
10. When students are problem solving, teachers are questioning, providing information, suggesting skills to be used, challenging, and reinforcing.

Options for Student Grouping

How a particular class of children is organized for learning depends on what is being taught, the nature of the students, and the mood the teacher wishes to create. The more mathematical experiences children have, the greater the diversity of their interests and talents.

It is helpful for teachers to know the alternatives available for grouping children of varying backgrounds. For example, when only one set of attribute blocks is available for a class of twenty-eight, eight students may be working with attribute blocks, while the other twenty students may be in small groups using worksheets or participating in a language activity.

A listing of alternative types of student grouping follows. It is quite possible to use two or more types of grouping simultaneously.

1. *Small group, 2 to 6 students.* Four children are working with two dice to determine the probability of rolling a 7.
2. *Large group, 7 to 20 students.* A group of seventeen students is participating in a mental calculation activity.

3. *Individual, a single student.* The teacher is reviewing the addition of mixed numerals with a single student.
4. *Whole class, one classroom of children.* The entire class views an animated western about equivalent fractions.
5. *Extra class, more than one classroom of children.* Two fifth grade classes compete in a metric olympics.

As you ponder developing a teaching-learning strategy, you may identify one alternative from options for student participation, an accompanying teacher behavior, and a type of student grouping. Variety in teaching approaches results from selecting from the numerous combinations that are possible. As you gain further experience, other possibilities will emerge. Be willing to try them.

Planning for Teaching

Excellent teaching occurs, in part, because it is well planned. Teachers are seldom able to spontaneously lead children day after day without pondering, reflecting, deciding, anticipating, and researching. You must be every bit the learner you would expect one of your students to be.

Planning for teaching takes numerous forms. Newcomers to teaching must carefully plan their lessons often writing a detailed lesson plan so that the teaching situation can be controlled and later analyzed by both the teacher and the cooperating teacher or supervisor. (See Appendix B for a lesson plan outline to assist in planning lessons.) When writing a plan, the student teacher gains skill and confidence in his or her ability to think through the teaching act. Such a plan helps teachers develop a thinking process that eventually frees them from having to write out detailed lesson plans.

Guidelines for designing a lesson plan are presented below. It is assumed the mathematical topic (concept or skill) to be taught has been selected from a math textbook or school district curriculum guide.

Designing a Lesson Plan

I. Research the topic as follows:
   A. Discover various introductory techniques appropriate to the topic.
   B. Select teaching aids and materials that will best illustrate the topic.
   C. Identify various reintroductory methods to aid in mastery of the topic.
II. Consider how you will make this topic (concept or skill) interesting and worthwhile for both you and the children.
III. Plan each individual lesson using the following:
   A. State specific instructional objectives for each lesson in terms of what the children should be able to do when the lesson is completed.
   B. List the mathematical terms that may need to be reviewed or learned by the children.
   C. List your learning aids and materials and explain the part they will play in the presentation of the lesson.
   D. Outline your teaching strategy as follows:
      1. Introduction—Exactly what device or technique will you use to motivate this particular lesson? How will you ensure that the children will be interested?
      2. Procedure—How do you plan to succeed in achieving the instructional objectives? Be descriptive. What will the students be doing? How will they be organized? Is this different from the way you taught them last time?
      3. Follow-up—Have you planned a game, problem-solving activity, braintwister, discussion, or worksheet? What have you planned and what is its purpose?
   E. State how you will determine if the child has achieved the behavior stated in the instructional objectives.
IV. Evaluate your teaching performance in light of how the children responded, how you responded, and the relative success of the lesson. What might you do differently the next time you teach this lesson or group of lessons?

By the time you begin full-time teaching, you should be able to plan a week at a time by jotting down topics and key ideas you wish to teach. You will not need to detail every activity but may write out activities that need special planning. Much of the planning, including objectives, activities, grouping, and room arrangement, can be done mentally, but it is sound practice to have the week's plans written out in global terms.

Experienced teachers also must plan. Experience has taught them what to expect, how to react, how to time a lesson, and ways to interest children. Most of the planning done by experienced teachers is done mentally. They must refine and prepare their teaching to fit the particular group of children with whom they are working. Experienced teachers are aware of their personal teaching styles and adjust their styles to fit the learning styles of their children. Experienced teachers can be a bit more spontaneous and less tied to a fixed lesson plan.

There are times when specific instructional strategies are more appropriate. For example, when a manipulative material is first introduced to children, free exploring and discovering are called for. The accompanying teacher behavior is observing and interacting with the children. The grouping pattern will likely be whole class or small group. Experience has shown that children must freely interact with the manipulative materials before they work with them in a structured situation; otherwise, the children are distracted by the material.

When teachers are concerned about children's skill achievement, they use more direct instruction. This includes having the students discussing, listening and look-ing, memorizing, and writing, while the teacher is leading discussions, questioning, lecturing, explaining, describing, demonstrating, coaching, and assigning. Rosenshine and Stevens (1986) reported that certain teacher behaviors are associated with increased academic achievement on the part of their students. These include beginning a les-son with a short review of previous learning and statement of goals; presenting new material in small steps, with prac-tice after each step; giving clear and detailed instructions and explanations; providing active practice for all students; asking many questions, checking for understanding, and having all students respond; guiding students during initial practice; providing systematic feedback; and providing ex-plicit instruction and practice for seatwork exercises (p. 377).

Throughout the process of planning, teachers must be active learners. They must be current in what is happen-ing in the world because the children will voice their concerns, and teachers should be ready to use children's concerns when they arise and in future lessons. Teachers must also be current in the various subjects for which they are responsible.

To be current in the teaching of mathematics, the teacher may enroll in graduate or in-service courses of-fered by colleges or school districts. Teachers may attend local, regional, or national meetings held by the National Council of Teachers of Mathematics and other organiza-tions whose purposes include improving mathematics in-struction. Another way for teachers to keep current is to become active in state mathematics organizations. Teach-ers can attend workshops and later help provide work-shops for local teachers, students, and parents. Journals such as the *Arithmetic Teacher, Learning,* and *Instructor,* along with many of the fine journals published by state mathematics organizations, are rich sources of ideas for teaching mathematics.

Cooperative Learning

Competition. After visiting classrooms to observe, participate with, or teach children, you may begin to de-velop various concerns. Typical among comments is the following: "I saw the children playing a computation game and the boys were against the girls. The girls always seemed to win, and the teacher constantly praised the girls for being so fast. Competition is useful, isn't it? After all, life outside of school is competitive and children need to learn to compete early to survive." Typical among com-mercial advertisements is one telling consumers that "Be-ing the best is not everything; it is the only thing." Seldom can one be "the best," but competition is heightened. In the classroom, emphasis on being the best, fastest, or brightest can prove destructive. For every "best" child in a group of twenty, nineteen feel less able, weaker, or insignificant. Diminishing the worth of children through

competition can diminish the worth of the entire group. Employing competition to improve motivation or quality is often ineffective. The children who can win compete, but the others ignore the competition. A firm distinction must be maintained between being the best and doing one's best. This is not to say that all competitive situations are harmful. Friendly competition sometimes increases friend-ship and common appreciation among competing individu-als or groups.

Cooperating and having a common cause foster personal growth and identification. Mathematics learning should be cooperative. Cooperation involves other people in construc-tive roles. Children who receive support from other people can see the worth of working together as opposed to work-ing against one another. More productive personal growth results through cooperation than through competition. But cooperation does not just happen; it is learned and requires a teacher who is a model of the cooperative spirit.

Grouping. The nature of young children does not lend itself to extensive group work. Six year olds have not fully developed the capacity to work in groups. They do enjoy games involving several other children, but these games should allow them considerable individual freedom. This is the time to introduce whole-class and small-group activi-ties and to carefully guide children through the activities. When whole-class projects are undertaken, the teacher is the catalyst and leader. Likewise, opportunities arise to teach sharing and cooperation by developing small-group projects, such as dramatic play or mathematics activities with a short-term, definite purpose.

Eight year olds have developed sufficiently to work in small groups with considerable adult leadership. They are able to understand tasks and to work through them. They are able to respond to the teacher's questions and guidance. They are learning to cooperate without constantly grappling for attention. They are able to assume some leadership in small groups but can rarely lead the entire class. Take care to teach skills of democratic living and to provide time for these skills to be practiced. This is the beginning of a life-time learning process in group interaction.

Ten year olds can effectively work in groups with little adult supervision. At this age, the group tasks and indi-vidual responsibilities must be well defined. Those who work well together will be pleased with the group product. Individual students can assume leadership roles and are comfortable leading the whole class.

Grouping children in cooperative learning groups as an alternative to competitive and individualistic learning has emerged in the writings of Johnson and Johnson and oth-ers. **Cooperative learning** is an organizational pattern in which children work together in small groups to accom-plish academic and collaborative tasks. The proponents of cooperative learning point to the advantages and benefits to students of working together for a common purpose. Johnson and Johnson discuss these advantages:

Achievement will be higher when learning situations are structured cooperatively rather than competitively or individualistically. Cooperative learning experiences, furthermore, promote greater competencies in critical thinking, more positive attitudes toward the subject areas studied, greater competencies in working collaboratively with others, greater psychological health, and stronger perceptions of the grading system's fairness. The implications of these results for teachers are as follows:

1. Cooperative learning procedures may be used successfully with any type of academic task, although they are most successful when conceptual learning is required.
2. Whenever possible, cooperative groups should be structured so that controversy and academic disagreements among group members are possible and are managed constructively.
3. Students should be encouraged to keep each other on task and to discuss the assigned material in ways that ensure elaborative rehearsal and the use of higher level learning strategies.
4. Students should be encouraged to support each other's efforts to achieve; to regulate each other's task-related efforts; to provide each other with feedback; and to ensure that all group members are verbally involved in the learning process.
5. As a rule, cooperative groups should contain low-, medium-, and high-ability students to help promote discussion, peer teaching, and justification of answers.
6. Positive relationships among group members should be encouraged. (1987, p. 40)

There are reference books that provide detailed descriptions of cooperative learning and how to implement cooperative learning in the classroom. There are college-level and in-service courses that help teachers perfect the skills needed to use cooperative learning. But the best way for you to develop the ability to successfully use cooperative learning is to practice it in your own classroom, to evaluate your own progress, and to talk with others who are also using cooperative learning. You are encouraged to use cooperative procedures as one of your options for teaching.

Evaluating, Recording, and Reporting Children's Progress

Evaluation is a multifaceted process. It involves determining the amount and quality of children's growth, development, and achievement. It involves knowing the objectives of the mathematics program. It requires you to know and understand children. Evaluation also includes diagnosing, recording, and reporting children's progress.

Evaluation is not an adjunct to the mathematics program. It is an integral part of daily instruction, for you cannot proceed without knowing how the children are performing each day. Evaluation begins whenever you first come into contact with children and parents. It occurs every day throughout the school year.

The NCTM *Standards* (1989) includes a listing of fourteen evaluation standards and a complete discussion of each standard. In its initial discussion, the working group on evaluation notes:

> The main purpose of evaluation, as described in these standards, is to help teachers better understand what students know and make meaningful instructional decisions. The focus is on what happens in the classroom as students and teachers interact. Therefore, these evaluation standards call for changes beyond the mere modification of tests. (p. 189)

Just as the curriculum will change as a result of the standards mentioned throughout this textbook, so will the process of gathering information to be used for student evaluation. The major thrust of the evaluation standards is fourfold, that

1. student assessment be integral to instruction;
2. multiple means of assessment methods be used;
3. all aspects of mathematical knowledge and its connections be assessed; and
4. instruction and curriculum be considered equally in judging the quality of a program. (p. 190)

Thus, by becoming a dynamic part of mathematics instruction, evaluation will provide you with more appropriate ways to assess progress and a broader base of information on which to judge children's mathematical ability.

Evaluating is gathering information using many available techniques. Recording is a way of keeping the information once gathered. Reporting is a way of sharing the data with those most concerned with the children's progress—the children themselves, parents, teachers, and administrators. These three aspects of the evaluative process are discussed below.

Evaluating Children's Progress

Traditionally, standardized tests and textbook tests have been used as the primary means by which to evaluate children's mathematical progress. This procedure is limited because standardized and textbook tests provide limited information about children's thinking processes. Children's thought processes are dynamic; they should be assessed accordingly. Thus, it is important to know about and practice alternative assessment techniques, including observation and questioning, performance-based assessment, diagnostic interviews, teacher-made tests, writing activities, and group problem solving. Each is discussed in the following paragraphs.

Standardized Tests. Standardized achievement tests are among the most controversial of evaluative procedures. They must be used with understanding and discrimination. Standardized, or norm-referenced, tests are typically used for comparing the work of an individual or group with norms established for children in similar age groups or at similar grade levels.

Sometimes, standardized tests help survey children's skill and knowledge to provide a basis for evaluating a school district's curriculum. Standardized achievement tests measure knowledge, skills, speed and accuracy, the ability to solve one- or two-step word problems, and vocabulary. Children's scores are reported as percentiles or grade-level equivalents based on national norms.

Teachers should avoid relying heavily on standardized tests as a basis for evaluating their children's specific mathematical abilities. Standardized achievement tests tend to be merely approximations of how well particular children in a particular section of the country perform. Such tests give little evidence of children's resourcefulness or confidence in attacking new problems.

Comparative scores are based on the assumption that all teachers present the same material in the same sequence during the school year. There is the assumption that all children worked equally diligently when answering each question and that they did not randomly guess at answers. Further, it is assumed that all teachers administer the tests in the same way, carefully following the written instructions provided with the tests.

Standardized achievement tests do provide a general statement about the progress that groups of children are making. They are revised periodically to reflect current trends in teaching mathematics. Standardized tests are prepared by experts in test construction and leaders in mathematics education. The most up-to-date tests for children can provide information difficult to obtain by other means.

Textbook Tests. Textbook tests or supplemental tests provided with textbook series supply necessary information. They diagnose children's readiness to begin a topic. They provide a basis for grouping or individual work. They let you know if the children have mastered the facts and skills presented in each chapter, partial chapter, or section of the textbook. A sample test from a first grade textbook is shown and discussed in the math book insert for this chapter.

Throughout this book, we have encouraged you to use textbook tests to help judge how well content is being learned. Be aware that a paper and pencil test often cannot tell you what children think and understand. However, textbook tests are convenient to use. Most textbooks or teacher's manuals let you know exactly what pages are being tested by each section of the test. You are also advised what to do if students do not measure up to the standards you or your district have established.

Tests are usually provided in each student's textbook. Parallel forms of the tests are provided in a format that is easy to reproduce for classroom use. Once the test results are compiled, you can begin to reteach topics or provide remedial instruction to individuals or small groups. The teacher's edition of the textbook will guide you to resources available to help children who need additional as-

sistance. You will probably find that conferencing with or focusing observations on individuals having difficulty will provide you with valuable information about how to help them.

Observation and Questioning. Teacher observation and questioning are among the most valuable methods for assessing the progress of children. Observation should focus on the individual child and the specific mathematical effort in which that child is engaged.

Many aspects of children's behavior are not evaluated by tests. These behaviors include attitudes and interests, creative tendencies, thinking processes, and children's abilities to explain their own and other children's solutions. During discussions, work periods, play time, and instructional time, teachers observing mentally collect information about individual children. This information should then be acted upon or written down in the children's mathematics folders, on an annotated class list, or wherever such information is kept. Many of the activities presented in this text are best evaluated by teacher observation.

To illustrate evaluative observation, we present the following description. Mr. Edwin notices that Daron and Rob have been making numerous similar errors as they begin subtracting mixed numerals. One example of the type of error they are making is shown below:

$$6\frac{3}{4} = \quad 6\frac{6}{8}$$
$$-3\frac{7}{8} = \quad -3\frac{7}{8}$$
$$\overline{\qquad\qquad 3\frac{1}{8}}$$

It is apparent that the boys are subtracting the top fraction from the bottom fraction, rather than the bottom from the top. When he has a chance, Mr. Edwin takes the boys aside and asks them to answer this question, "What is $6 - 7$?" He writes the question down as shown below:

$$\begin{array}{r} 6 \\ -7 \\ \hline \end{array}$$

Both boys say the answer is 1. Then Mr. Edwin writes $7 - 6$ as follows and asks for the answer:

$$\begin{array}{r} 7 \\ -6 \\ \hline \end{array}$$

Again, the boys say the answer is 1.

Mr. Edwin asks, "Are both problems the same?"

The boys indicate they think they are.

Continuing, Mr. Edwin puts 6 cubes on the table and asks Rob to remove 7 cubes from those on the table.

Immediately, Rob says he can't do that and begins to understand why 6 − 7 and 7 − 6 don't mean the same thing. Even though the boys had *learned* subtraction several years before, some of its meaning had slipped away. Observation and questioning provide Mr. Edwin a chance to determine where the difficulty is.

The above example of observing and questioning is specific to a computational procedure. Throughout the mathematics program, there are opportunities to question children on every aspect of mathematics. In its booklet, *Assessment Alternatives in Mathematics*, EQUALS and the California Mathematics Council suggest a number of questions that teachers may use to assess children's progress in several aspects of problem solving. For example, in questioning students about problem comprehension, these questions are suggested:

- What is the problem about? What can you tell me about it?
- How would you interpret that?
- Would you please explain that in your own words?
- What do you know about this part?
- Do you need to define or set limits for the problem?
- Is there something that can be eliminated or that is missing?
- What assumptions do you have to make? (p. 24)

Many other questions are presented in this booklet and the reader is encouraged to build a collection of good questions to use in the classroom. Clearly, the types of questions you ask must be appropriate for the age and developmental level of the children you teach.

The observing and questioning teacher will gain information about all aspects of children's cognitive growth—language and communication, social awareness, curiosity—not merely mathematical growth. This information will help guide the teacher in planning classroom experiences in a range of subjects. Because teachers make observations every day they are with children, teachers should accept this technique as providing the most consistent and abundant source of information about children.

While accepting observation and questioning as a key evaluation technique, you also must accept the need to record the important observations and to collect samples of children's work to serve as written evidence of the observations. Then you will be able to track children's progress, use the information to prepare individual instruction, and accurately report the progress to children and their parents.

Finding time to make and record observations will be a challenge with your already busy schedule. Remembering your observations about a student's mathematics work from among all your daily observations will take practice and concentration, as will remembering to record the information.

Performance-Based Assessment.   Some states, such as Kentucky, have opted to substitute performance-based assessment for standardized testing. In this type of assessment, students are asked to perform a task, either individually or in small groups, which requires problem solving and higher-level thinking. Frequently, the task involves using a variety of concrete materials, is aimed at solving a "real-life" problem, and cuts across several different subject areas. Tasks are scored according to a scoring rubric that describes superior, satisfactory, and unsatisfactory behaviors.

Diagnostic Interviews.   Diagnostic interviews take place when the teacher sits with individual children and asks them to perform simple tasks or to comment on tasks that are performed for them. Piaget devised many such tasks for use with young children to determine the children's readiness to learn various mathematical topics. Topics for which tasks have been devised include early counting procedures, the four basic arithmetic operations, measurement, logical thought, space, time, and fractions. Often, the most helpful sets of questions are devised by teachers in their daily interaction with youngsters. You should find out how children are thinking. The best way to do this is to ask them.

Piagetian and similar tasks are carefully described by Copeland and others. Baratta-Lorton provides extensive diagnostic procedures in her program, entitled *Mathematics Their Way*. References may be found in the chapter bibliography.

Teacher-Made Tests.   Teacher-constructed tests are useful when specific information cannot be gathered by other means. The main advantage in constructing your own test is the flexibility to choose the format and the test items. Computer programs have been devised to help teachers construct tests. The chief disadvantage to constructing your own test is the time that it takes to develop a good test.

Students may be invited to write test items based on the work they have been doing. The quality of their questions will improve if you instruct students on how to write test items.

Writing Activities.   Writing activities can help in assessing children's mathematical progress. As they write, students reflect on a variety of topics and feelings and provide both themselves and the teacher with the chance to review their work and their thinking processes. The writing children do may be completed in several contexts. One opportunity for writing occurs when teachers provide questionnaires regarding recent mathematical work. For example, following several days of working with probability events, students were asked (1) to describe the activity that they thought best helped them understand how to determine the probability of an event occurring; (2) to mention one activity that surprised them; and (3) to explain to another person why some games are fair and others are not fair.

Journal writing gives students the chance to express attitudes and feelings about mathematics. Journal entries may be made each day or even once a week, but they should be made regularly. It is often helpful for you to provide a topic about which students can write. Journal "starters" might include the following: (1) Describe what you think about doing math in school. (2) If you could be a number, what number would you be and why? (3) In math, I sometimes wonder about. . . . You will gain insights about the children as they write down their thoughts about mathematics. You will be better able to serve the needs of individuals as you discover how they are thinking and what they are thinking about.

Writing activities may include describing how a particular problem was solved, how to perform a new algorithm, or how to explain an event. These writing activities focus on describing a specific skill that requires careful thinking and explanation. For example, students are asked to explain to a classmate why certain figures, such as equilateral triangles, cannot be constructed on a rectangular geoboard.

Another useful writing activity emerges from cooperative learning group work. As a group works on an assignment, problem, or project, the group recorder might be asked to describe how the group decided to complete the assignment or in what ways members of the group praised one another as the work progressed. The written record from group work will have been read to the group and will likely reflect some group editing.

Group Problem Solving.   One way to evaluate mathematical ability is to assess group work. We have discussed cooperative learning as an organizational procedure that encourages academic and collaborative work, typically surrounding a problem-solving task. To assess group work, you will need to determine how well the assigned problem or task was dealt with. The time following a cooperative activity may be spent as a whole class *debriefing,* that is, discussing the procedures used by each group to solve the problem. The group reporter may orally describe the work or the reporter may read the summary that was part of a writing assignment.

Sometimes a group will be assigned a project that may take several days to complete. The outcome of the project may be a display, a skit, a video, or a written paper. The outcome of the project represents the collective work of the group. Because of the interdependence of group members in completing the project, the product is seen as the collective work of all group members. By using the techniques discussed above, the value of the project is assessed and insight into the mathematical learning is provided.

Another important aspect of group assessment is to determine how well group members worked with one another to complete the task. Did each group member participate? Were ideas readily accepted and discussed? Were members praised for their contributions? Did each group member perform the role assigned to him or her? Were disagreements discussed and resolved in a constructive manner? Did group members feel good about their contribution and the group product? Did group members enjoy working with one another? You may find the answers to these questions by observing groups as they work, listening to the debriefing reports, talking with individuals, and reading journal entries.

Using broad-based assessment does not mean that all assessment techniques are used all of the time. Rather, you have available to you alternative techniques that will provide sound evidence of children's mathematical growth, thought processes, and abilities. You must select the most appropriate technique for the type of task being evaluated.

Teachers have a major obligation once data have been gathered by any of the above means. Namely, they must decide what the data mean. The teacher's interpretations of the children's work are crucial in laying the groundwork for further instruction.

One useful function of interpretation is diagnosis. For example, children may be grouped for instruction based on concept or skill deficiencies that show up during evaluative procedures. If testing reveals that three children are unable to solve word problems involving multiplication, it is appropriate to design instruction to assist them.

Tallying right and wrong answers is not nearly so important as determining how the child is thinking. It has been said, "Children are not wrong; they merely respond to the stimulus according to their knowledge and development at the particular time when we are checking their progress" (Biggs and MacLean, 1969, p. 191). Teachers should keep this in mind to provide children with the encouragement they need to maximize their potential. Most children naturally learn from failure and mistakes. All children should realize that evaluative procedures represent another natural step in the learning process.

Recording Children's Progress

During the entire evaluation process, organizing the information and work samples gathered is essential. The test scores and other numerical data should be recorded in a gradebook, a computer database, or a similar repository for quantitative information. The work samples, including qualitative and quantitative material, may be kept in a student **portfolio.** The mathematics portfolio is a collection of materials that demonstrate the child's ability to think about, use, and apply mathematics. It is material that transcends the purely quantitative assessment data (test scores and worksheet or workbook scores). The portfolio will show a profile of the student's abilities in mathematics in a variety of mediums. The portfolio should show the student's best work, including revisions of early work.

Kenney, in his work with the State of Vermont to develop a three-part approach to assessment, commented specifically on using portfolios for instruction and assessment. He noted that items in a portfolio could include any of the following:

- A solution to a problem assigned as homework or given on a test or quiz. The solution should show originality or deviation from the usual procedure, not just a neat set of figures. Different solutions to the same problem would constitute one entry.
- A problem made up by the student, with or without a solution, depending on the complexity.
- A paper done for another subject that contained some mathematics, such as an analysis of data presented in a graph, particularly if the data was collected by the student.
- A report of some group activity or project, with comments as to the individual's contribution. These could include surveys, such as of adults and their use of mathematics in their work and daily lives, reviews of the use of mathematics in the media in both good and bad ways, as well as other projects involving the collection and analysis of data.
- A picture or sketch made by the student (or the teacher) of a student's work with manipulatives or two or three dimensional figures as a solution to a problem or a description of a mathematical concept or situation.
- Art work done by the student involving mathematics such as string designs, coordinate pictures, scale drawings or maps, and other applications.
- A description by the teacher of some student activity that displayed an understanding of a mathematical concept or relation.
- A videotape of a student or a group of students giving a presentation involving mathematics.
- A report on the history and/or application of some mathematical concept.
- An entry or entries from a journal or a log. (1990, p. 8)

Throughout the year, the teacher and the children can review the children's growth and the progress they make. When parents confer with the teacher, the mathematics portfolio can provide a dynamic source of information to reinforce the teacher's specific evaluative comments.

Reporting Children's Progress

Parents and teachers should communicate about children's growth and progress. The interchange between parents and teachers helps the parents determine their child's progress and helps the teacher determine the nature of the child's home environment and the child's perception of the teacher and school. The key to two-way communication is your willingness to share any information you have and any observations you have made.

You should welcome parents to school and, if the occasion arises, feel comfortable in the children's homes. The common interest of parents and teacher in the child's growth, development, and school progress should draw them together. The teacher's assessment and the parents' observations should blend together to strengthen the parent-teacher partnership.

Parents of elementary children are particularly interested in hearing about their children. Besides quarterly or mid-year conferences, parents may be invited to the classroom to observe the daily routine or specific activities such as mathematics.

Send notes to parents pointing out a significant event or accomplishment of individuals or small groups of children. A note home need not have a negative connotation. A summary of the past week's activities or of events to come may be distributed to keep parents informed about classroom life. Solicit the help of parents when the parents are known to have particular skills or experiences that can enrich a class. Invite parents to participate.

When reporting the mathematical progress of children, written descriptions along with samples of the children's work are helpful. Sometimes a check sheet of concepts and skills may substitute for the written description. When children have been involved in self-assessment, their views and perceptions add to the sum of information about mathematical progress. At least twice a year, and hopefully more often, teachers and parents should meet face-to-face to discuss the mathematical work and growth of the children. Above all, be open and straightforward in discussing children. Your concerns and those of the parents should be coordinated for the benefit of the children.

Children with Special Needs

In every chapter of this book, we have mentioned children's different learning styles. We also have mentioned children who learn more slowly or more quickly than others. Your own observation of children will confirm these situations.

In addition, there may be children in your classroom with physical or mental handicaps and children qualified for assistance under federal or state guidelines. Resource rooms may be available for special instruction. Because support services are invaluable in meeting the needs of all children, use them to benefit every child in your classroom.

Children with Special Problems

Terms such as *low achievers, emotionally disturbed,* and *culturally deprived* are labels used to describe children having difficulties in school. These labels often result in a child's negative performance, both socially and academically. Avoid labeling children. Consider children for both their strengths and weaknesses, as full members of the human family. Children who happen to learn more slowly than others deserve special attention to help them develop mathematical concepts and skills.

Slow learners may be identified by intelligence quotient, achievement, teacher observation, reading ability, readiness level, or other means. They generally fall below an average in one or more of these areas. Slow learners may have many things in common, but each child has a unique set of strengths and weaknesses. Characteristics frequently possessed by slow learners are listed below.

- *Negative self-concept.* Children may come to believe at a very young age that they are stupid. Failure is too easily learned. Many children will not even attempt a task, particularly a new task, because they are afraid of failure.

  Make an effort to ensure success and to look at failure as an acceptable route toward learning. If children view themselves as worthwhile, they are more apt to approach a problem with confidence and have a greater chance of success.
- *Short attention span.* Slow learners often have short attention spans. This may be because problems are too difficult, too long, or uninteresting.

  Children will work for relatively long periods on interesting problems suited to their ability level. They will play in class much longer than they will work. Ensure that children are positively motivated toward appropriate tasks.
- *Specific mathematical disability.* Terms such as *dyscalculia, number blindness,* and *specific minimal brain damage* are used to describe children with specific problems in learning quantitative concepts. There may also be perceptual problems that affect learning spatial concepts. These include difficulties in forming concepts such as position in space (near, far, up, down, left, right), in distinguishing a figure from the surrounding background, and in eye-hand coordination.

  Children are easily distracted by extraneous stimuli. Too many problems or pictures on a workbook page, as well as too many objects or people in the classroom, can be distracting. Workbook pages and classroom environments should be relatively plain and simple for children with perceptual problems. Be alert for children with specific mathematical disabilities, and seek professional help when problems demand it.
- *Poor self-control.* Some children are explosive, hyperactive, or erratic. They always seem to be in motion. They rarely sit still and often wander aimlessly about the room.

  Some research indicates that poor self-control may be triggered or aggravated by diet. Much research still needs to be done on how diet affects children.

Hyperactive children require a structured environment with few extraneous distractions.

- *Language problems.* Children who have difficulty learning mathematics often have language difficulties. In addition, the language of mathematics is often abstract and complex. Children may not understand such common mathematical vocabulary as *up, down, in, out, two,* and *plus.* Even more difficult are phrases such as *divided by* and *divided into.* Children may be unable to read simple directions, equations, or mathematical symbols. They may be unable to communicate concepts they do understand.

  Remember to keep conversations with children having language problems as simple as possible. Be alert for any misunderstanding of terms. Develop concepts through physical manipulation and language.

Memory and Application.   Studies show that slow and retarded children are capable of learning complex motor and verbal skills. Their retention may be similar to that of younger children of the same mental age. Overlearning may be required to ensure retention.

Allow for practice, drill, and repetition only after a concrete understanding of concepts has been developed. Transfer of learning is difficult for slow and retarded children but may be accomplished if it is incorporated into the lesson. Such children can retain and apply skills when they have transferred and practiced them.

Complex problem solving may be too difficult for slow and retarded children, but they can memorize simple, rote, factual material and they can learn and apply basic facts. Teaching skills such as how to use a calculator, how to be an effective consumer, and other life skills is appropriate.

General Principles of Good Teaching.   General principles of good teaching have been mentioned throughout the book and pertain to all children. They are reiterated here because of their significance for children with special problems.

1. All children are ready to learn something. You must determine the level of readiness.
2. Success is important. Carefully sequence learning to ensure several levels of success. Immediate positive feedback is helpful.
3. Self-concept affects success and vice versa. Children must be worthwhile in their own eyes and in the eyes of their peers.
4. Practice is important and should follow the concrete development of concepts. It should be applied in practical situations and should contain provisions for transfer.
5. Be prepared with a variety of teaching strategies. Do not present them all at once, but if one method fails, try another. The child may be capable of mastering the concept, but not in the context in which it was

first presented. Several methods of presentation that involve the senses may be needed to meet each child's learning style.

6. Begin work on a concrete level. Move to work on an abstract level only after children understand concepts concretely.

7. Analyze children's mistakes carefully. Children periodically make careless mistakes, but there is often another reason for an error. Look for patterns in children's errors, and discuss their thinking processes to correct mistaken concepts.

8. Learning is different for each child. This book mentions several ways to diagnose and evaluate individual children. Lessons should be planned according to the diagnoses.

Children with Special Mathematical Talents

Just as children having problems in mathematics need special consideration, so do children with special talents in mathematics. The regular math curriculum may be as unsuited for gifted children as it is for slow children. The right of children to an education suited to their individual needs has prompted some states to legislate special programs for gifted children. Whether gifted children are in a regular classroom or a special program, there are several things to consider.

Gifted children, of course, are not all alike. Although they are commonly identified by high intelligence quotients, such as a score of 130 or better, there are several other characteristics to consider. Some of these characteristics are listed below.

- *Creativity.* Many children, from preschool on, have special creative, mathematical talents. Encourage creative children to explore, to manipulate, to suggest a variety of solutions to a problem, and to suggest new problems for exploration. Minimize rote memory. Stress flexible thinking.

- *Awareness.* Gifted students are often sensitive to and aware of their surroundings. They perceive problems readily and can see patterns and relationships easily. They do not need to have every step of a problem spelled out for them.

- *Nonmathematical abilities.* Although some children have special talents only in mathematics, many children who are mathematically talented are also mentally and emotionally mature. Many gifted children are highly verbal and can express their thought processes.

- *Abstract reasoning abilities.* Talented children may be able to reason at a higher level of abstraction than their age-mates. They may work symbolically with quantitative ideas but may still benefit from manipulation of concrete materials.

- *Transfer ability.* Gifted children can transfer skills learned in one situation to novel, untaught situations. They may apply learning in social situations, when working in other subject areas, at home, and so on. Encourage gifted children to generalize rules and principles and to test the generalizations in new settings.

- *Good memory.* Talented children often have the ability to remember and retain what they have learned. They do not need as much drill as other children.

- *Curiosity.* Mathematically gifted children often display intellectual curiosity. They are reluctant to believe something just because the teacher says it is true. They want to know why it works. They are interested in a wide range of ideas and often explore topics through independent reading. They ask many questions, enjoy solving puzzles, and delight in discovering winning strategies in games. Make available challenging articles, books, puzzles, and games for their use.

Teaching talented and creative children is challenging. It is even threatening to some teachers. Such teachers are sometimes concerned that they will not be able to answer all the children's questions or that the children may be smarter than they are. This may be true but should not necessarily present a problem. Some suggested techniques that are especially appropriate for teaching gifted children follow:

1. Gifted children need challenging problems. Many of the problem-solving activities suggested in Chapters 4 through 14 are especially appropriate for gifted children. Encourage gifted children to create original problems for others to solve. Gifted children enjoy strategy games and complex problems.

2. Gifted children do not need busywork. Because they often finish assignments early, they may be asked to do additional problems of the same type. If they can do the initial problems, it is likely that they do not need more practice. Let them move on to more challenging tasks.

3. Encourage independent research. Gifted children are often capable of independent study and research at a young age. They may even be able to set their own goals and evaluate their progress. This, of course, does not mean they should always be left alone.

4. Set up a buddy system. Let children work with peers of similar ability, or let them help others who may be having problems. Children can often communicate with and learn from peers better than from adults.

5. Encourage creative and critical thinking. Avoid forcing children to memorize one method to the exclusion of others. Accept any correct method, and lead children to discover why some methods do not

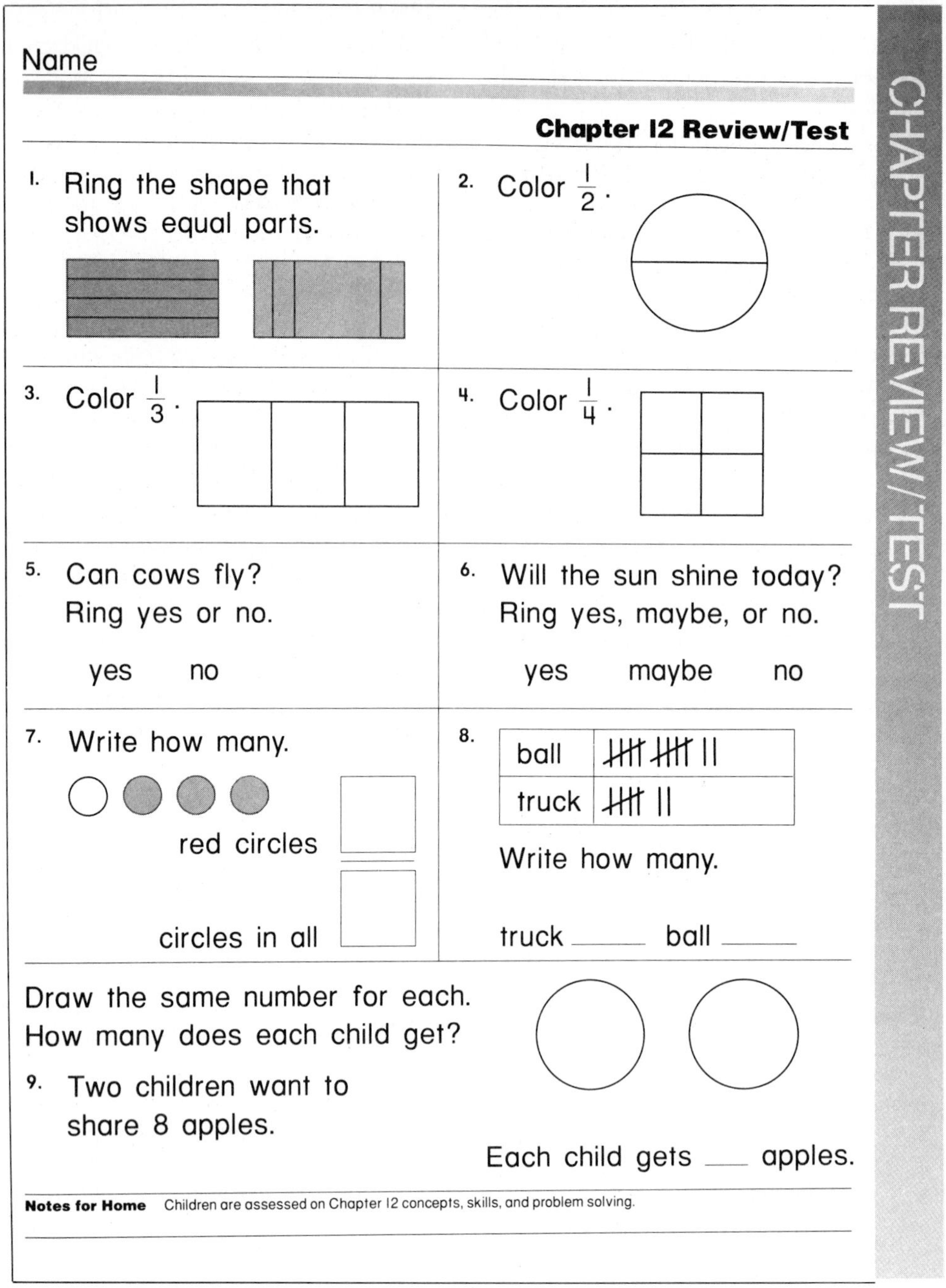

Name _____

1. Ring the shape that shows equal parts.

2. Color $\frac{1}{2}$.

3. Color $\frac{1}{3}$.

4. Color $\frac{1}{4}$.

5. Can cows fly? Ring yes or no.

   yes   no

6. Will the sun shine today? Ring yes, maybe, or no.

   yes   maybe   no

7. Write how many.

   _____ red circles

   _____ circles in all

8.

| ball | IIII IIII II |
| truck | IIII II |

   Write how many.

   truck _____   ball _____

Draw the same number for each. How many does each child get?

9. Two children want to share 8 apples.

   Each child gets ___ apples.

Notes for Home   Children are assessed on Chapter 12 concepts, skills, and problem solving.

Figure 15–1
Bolster, Carey, et al, Exploring Mathematics. *Glenview, Ill.: Scott, Foresman and Company, 1991, p. 395. Reprinted with permission.*

THE MATH BOOK

This first grade page shown in Figure 15–1 is an end-of-chapter review or test on the concepts of fractions, probability, using charts, and an introduction to division. This particular series contains other versions of this test, including two that are in a multiple-choice, standardized-test format. These tests are all coded to coincide with the objectives of the chapter, and the pages on which the concepts are covered are noted for the teacher. These tests are also included in a computer management system.

In addition to end-of-chapter tests such as this one, the series also contains an inventory test to use at the beginning of the year, four cumulative tests to use during the year, and a mid-chapter check-up in each chapter. Suggestions for reteaching, remediating, and enriching are included in the teacher's manual and may be used after the teacher has analyzed the test to discover the strengths and weaknesses of each child.

Because this test is a paper and pencil type of assessment, it has some limitations, which are acknowledged in the teacher's manual. Suggestions in the manual include ideas for alternate assessment using manipulative materials such as paper-folding and using paper models of fractional pieces while answering questions about fractions. Other methods of alternate assessment in this series include recording forms for assessments involving student writing, observations, interviews, self-assessment, surveys, attitude assessment, and holistic scoring for problem solving. Teachers should use as many different types of assessment as are feasible in a given classroom to provide the most appropriate instruction for each individual child.

always work. Have the children evaluate solutions for appropriateness, ease, originality, and the like.

Perhaps you recognized the suggested approaches for teaching gifted children. The same approaches should be used with *all* children. We mentioned them to alert you to sound practices and to remind you that extra planning is needed for teaching mathematics to talented students.

Educators often debate whether programs for gifted children should be enriched or accelerated. Some states have mandated one type of program. Currently, the trend seems to be toward enrichment or a combination of enrichment and acceleration. **Enrichment** includes introducing topics not normally considered part of the curriculum and exploring the traditional topics in more depth. Thus, enrichment requires the availability of materials other than textbooks and workbooks. Construct games, task cards, and other materials for enrichment, and make full use of commercial manipulative materials and their suggested challenging activities.

A wide range of resource books should be available either in the classroom or in the library. For the teacher, a professional library, the National Council of Teachers of Mathematics, journals, and activity books are rich sources of good ideas.

Arranging the Classroom

Classrooms are for children. Classrooms should provide the best physical environment possible for learning. It is possible to have a classroom that is too stimulating, with so many bright pictures and objects that children have difficulty finding a single object or an area in which to work that is not distracting. It is also possible to have a classroom so barren and unchanging that children have little or nothing in which to become interested. It is possible to have a classroom so informal that children do not know what to do or where to go or to have a classroom so formal that children become regimented and repressed. It is a challenge to provide a balanced, comfortable, flexible classroom, so that no matter what activity has been planned, there is space available that lends itself to the activity. Also to be considered are the different learning and teaching styles.

Several basic tenets should be kept in mind in providing the physical learning environment.

1. The physical learning environment provides a support system for the educative process. It is important that the physical environment stimulate learning to complement textbooks, teachers, games, and visual aids.
2. As learning environments improve, so does the learning. That is, there is improvement in mathematical concept and skill development as well as communication.
3. Full use of the learning environment will assist you in teaching every child more fully, so that children receive help from both you and the environment.
4. Classrooms for children should mirror the decisions and interests of the children. Children should have the opportunity to help design parts of their own learning environments. Teachers should also help design the learning space.
5. Children's behavior is affected by their learning environment. Anything that affects behavior also affects learning. The quality of the physical environment must be maintained.
6. Be aware of the physical environment as part of the learning process. An awareness of alternative ways to use space is fundamental.
7. The utility of a classroom depends on the awareness of those who spend time in that classroom. Children and teachers do not need a new building or a new classroom to have a rich learning environment.
8. Consider the human and physical needs of children in designing learning environments. Taylor and Vlastos underscore this belief: "If the child's experiences are a starting point for education, then, by definition, his own language and the culture of his home, neighborhood, and community should be utilized in the educational environment. The classroom and outdoor area should play a vital role in reflecting a child's cultural background and his interests" (1975, p. 24).

These tenets provide the key to developing more effective learning environments: adapting existing classroom space. Aspects of the children's world and their learning of math as described in Chapter 3 are that children (1) have many number experiences, (2) are active, (3) observe relationships, (4) learn mathematics in concert with other subjects, and (5) have feelings that should be considered. These characteristics of children provide the starting point in determining how a specific classroom should be structured.

How can the learning environment be enhanced? Provide bulletin boards that invite a response from children, like that shown in Figure 15–2. Provide bulletin boards that inform, such as one that illustrates a variety of historic mathematical tools. Construct large yarn shapes such as squares, triangles, and pentagons to decorate walls and

Figure 15–2

ceilings. Let the children construct space figures by cutting and folding construction paper. Have tetrahedrons, cubes, prisms, octahedrons, dodecahedrons, icosahedrons, and so forth, hanging from the ceiling or along a wall. Build a geodesic dome that can be used by three or four children as a quiet reading place. Rearrange the student desks and learning centers every so often. Display number lines in various locations and configurations around the room. Most of all, be willing to change the decorations occasionally to provide variety and interest.

Whatever classroom you inherit, you can do much to make it a rich learning environment. Evertson and others offer sound suggestions for setting up the physical environment at the beginning of the school year (1989, pp. 3–14). A complete transformation is not expected immediately. There are limits imposed by the classroom structure, time, money, and other school personnel, but an awareness of the existing potential of each classroom and a sensitivity to the children who will spend so many hours in the classroom should serve you in developing, maintaining, and changing the learning environment.

Providing Learning Aids

Although some classes and school districts do not use math textbooks, the predominant learning aid in most classrooms is and will probably continue to be the math textbook. Whether a hardback book or an expendable workbook, the text will provide a curricular framework for your mathematics program, but as you strive to become a more effective teacher, you will find it necessary to move beyond textbook pages and worksheets. When concepts are to be learned, manipulative materials are needed. When skills are to be sharpened or facts are to be memorized, repetitive games or activities are needed. When independent work is prescribed, activity cards, learning centers, and computer programs are needed. Providing quality learning resources to support a sound program of mathematics instruction is basic to effective teaching.

Acquiring learning aids depends on three factors: (1) the financial resources of the school or district, (2) the energy and creativity of the teachers, and (3) the priorities established by the school or district and its intent in providing a strong mathematics program. The last factor is influenced by the other two.

Knowledgeable teachers and curriculum specialists should serve on advisory committees for developing school- or district-wide guidelines for mathematical instruction. They should make recommendations for the wise expenditure of financial resources to provide the school and classrooms with useful learning aids.

Commercial Materials

Excellent commercial materials are available for use in teaching mathematics in the elementary and middle schools. They include textbooks and workbooks, kits, games, structural materials, activity cards, and computer software. Each is briefly discussed.

Textbooks and Workbooks.   For most mathematics programs from kindergarten through high school, textbooks and workbooks provide the foundation on which the program rests. These textual materials are carefully prepared by recognized authorities in mathematics education. A textbook or workbook presents a unified sequence of concepts and topics, which are reintroduced at appropriate intervals throughout the book. Textbooks and workbooks tend to be attractive, colorful, and appealing.

Selecting the mathematics textbook or workbook series is an important task, usually performed by a school district or building textbook committee. Most often, teachers have a number of series from which to choose. This list of acceptable textbooks may be compiled by the state department of education. Several general criteria for selecting textual materials should be considered. One listing of criteria is presented below. A final list of criteria is the responsibility of those who actually choose the mathematics series for a district or school. It is these individuals who have considered the goals of the mathematics program, local priorities, budgetary limitations, and teacher resources.

Here are selected general criteria for choosing mathematics textbooks or workbooks. Does the textbook or workbook

1. encourage active student involvement and investigation and discovery of mathematical ideas?
2. present task-oriented problems at the student's level of understanding and encourage higher levels of thinking?
3. suggest the use of manipulative materials?
4. use correct vocabulary yet avoid wordiness and undue difficulty that may interfere with the student's learning?
5. provide adequate exercises to assist in introducing mathematical concepts and skills?
6. spiral the mathematical ideas so students confront an idea several times in the elementary years, each time at a slightly more advanced level, but avoiding needless rote repetition?
7. have an accompanying teacher's edition with valuable suggestions for introducing, reinforcing, diagnosing, and reintroducing mathematical concepts and skills?
8. relate mathematical concepts that have common parts, for example, ordering relations with objects (is taller than), numbers (is greater than), and measurement (is longer than)?
9. build mathematical concepts and skills on previously understood concepts and skills?
10. support the learning of basic addition, subtraction, multiplication, and division facts?

11. allow for students to progress at different rates, reflecting individual differences among children?

12. interest children because it is attractive, colorful, and motivating, page after page?

Kits.   Commercial kits for use in mathematics learning are typically of three different types. **Kits of diagnostic and learning materials** may contain manipulative materials used to test youngsters relative to their development of prenumber and early number skills, for example, classifying, ordering, corresponding, and counting. As well, kits are available that provide materials useful in developing an initial understanding of classifying, relationships, number, fractions, and measurement.

Throughout this text, we have spoken of materials available in kits. Unifix and Cuisenaire materials are two such types of manipulatives. It is common to find structural materials in kits. *Explorations* and *Mathematics Their Way*, popular primary-level math programs, employ a wide variety of manipulatives in kits.

Kits designed to accompany textbook or workbook series provide manipulative materials illustrated or suggested by the teacher's edition of the series. These kits may contain attribute materials, colored rods, counters, or measuring apparatus. This type of kit is a valuable resource, since it allows children to work with concrete materials as they learn mathematical ideas. Similarly, manipulative kits can be specially prepared by distributors for school districts or to meet the recommendations of state departments of education.

The third type of kit is a **skills kit.** The skills kit provides audio cassettes, cards, or games for basic skills practice. These skills may be computation with whole numbers, fractions, geometry, problem solving, or other math topics. The advantage of such materials is that they allow children to work independently or in small groups at their own level, freeing the teacher to work with other children.

The best way to find out what sorts of kits are available is to peruse catalogs from commercial distributors of educational materials. (See Appendix A for a list of suppliers of educational materials.) Exhibits of kits and other materials are found at professional meetings, such as those sponsored by the National Council of Teachers of Mathematics.

When choosing kits for classroom use, carefully review the goals of the mathematics program. The amount of time devoted to kit materials should be determined by the nature of the mathematics program and the particular kits being considered. The convenience of a kit must be weighed against the amount of interest the kit will generate and maintain. If the kit can provide a function not provided by other components of a mathematics program or by classroom materials, it should be considered for possible purchase.

Games.   There is a wide selection of games available to reinforce mathematical skills. Card games, race board games, tile games, target games, dice games, table games, word and picture games, and games of probability may help develop skills in recognition of characteristics, counting, recognition of numerals and number patterns, probability, matching, developing strategies, and problem solving. Children are usually motivated by commercial games.

Often, children are unaware of the mathematical value of games. At times, it may be appropriate for you to explain the relationship of a game to a particular skill the children are learning in a nongame context. For example, the popular card game old maid helps develop recognition and matching. Once children have played this game, you may help develop the transfer of matching playing cards to matching attribute blocks, parquetry blocks, or tessellation patterns. Be careful, however, not to destroy the children's enjoyment of the game and their motivation for the sake of the mathematics.

Games for classrooms are available from distributors of educational materials, department stores, and toy stores. Select games that fit clearly within the context of the total mathematics program. Sometimes the expense of games may prohibit their purchase. In such cases, teachers are encouraged to construct their own games using materials available in the school. A discussion of constructing games is presented shortly.

Structural Materials.   Structural learning aids are usually designed to help teach particular mathematical concepts and clearly illustrate the concepts for which they were developed. Examples of structural materials are Cuisenaire rods, multibase arithmetic blocks, and attribute blocks. Structural materials help develop concepts of number, properties, place value, and sorting, as well as logical thinking. Sources of structural learning aids appear in Appendix A.

Although sets of structural materials tend to be expensive, they are often the most useful purchases teachers can make to support a sound mathematics program. Shop around for structural materials and purchase materials only when you know the advantages, disadvantages, and uses of the materials. Sales representatives or educational consultants sometimes offer workshops in using certain materials. Take advantage of workshops to learn the full value of materials.

Activity Cards.   Activity cards are designed to lead children through developmental sequences that provide guided discovery. Use of activity cards is often associated with classroom learning centers. Activity cards have the advantage of encouraging independent work by children, thus freeing the teacher to work with other children. There are cards for use with attribute blocks, geoboards, pattern blocks, geoblocks, color cubes, and other manipulative materials.

Pictures and diagrams sometimes replace words to direct children to specific activities. The key in considering activity cards for children is the clarity of the instructions. A set of activity cards that children are unable to understand represents a waste of valuable funds.

When considering purchasing activity cards, consider certain general characteristics that the cards should possess. Characteristics of good activity cards are listed below:

1. The questions or activities on an activity card should tend to be open-ended; that is, they should encourage children to give a number of responses.
2. The objectives of an activity card should be clear to the child and the teacher.
3. The wording or directions should be concise and should be presented at the children's reading level.
4. Activity cards should provide some way for children to record their answers or responses.
5. Activity cards should allow for higher levels of thinking than just memorization.
6. Activity cards should be attractive.
7. The cards should make use of the environment—the classroom or outside.
8. The cards should encourage active exploration and manipulation of materials.

Computer Software.   Computer software abounds for the educational market. It serves many functions. It provides direct instruction, skill practice, follow-up instruction for work with manipulatives such as attribute blocks and Fraction Bars, interaction between student and computer, evaluation, and record keeping.

Some of the software available is of high quality in both the mathematics presented and the clever graphics that accompany the presentation. The future will see a greater number of software programs that are even more sophisticated and useful. New generations of computers introduced into the classroom will allow children more creative applications.

Not all software is equally useful or of the same high quality. You must be a discriminating consumer. Before you make final selections of software for your school or classroom, take the opportunity to examine the content of each program carefully. Use it on your computer. If possible, let children work with it. Test your reactions and those of your children. See if the objectives of the software match the objectives of the mathematics program. Read reviews of the software in journals devoted to computer education. Make sure the software is compatible with your computer. If color plays an important role in a program, the program will be of little use unless you have a color monitor available.

Computer software is available through general educational material catalogs and software distributors and their catalogs. (See Appendix A for a list of suppliers of software.) Once on a mailing list, you will have a source for a great variety of software. Attend workshop sessions at professional meetings to find out more about computer

offerings. Check with curriculum specialists to see what they recommend. And ask your colleagues in a school. Those who consistently use software will be able to suggest specific programs and sources.

Teacher-Made Learning Aids

Teachers may wish to construct their own learning aids. By doing so, they save money and at the same time tailor activities to fit their students. All that is required is an interest in such a task and the willingness to devote the time and energy needed to do a quality job. Effective teachers are renowned for these characteristics. Following are some examples of games or activities intended to assist children in practicing and remembering basic facts associated with addition, subtraction, multiplication, division, and fractions. The significance of the games described rests in the adaptability of a single game idea to many useful games that employ the same strategy.

Adapting the Rules or the Action of an Existing Game. Most games that teachers invent are adaptations of existing games they have seen or played. The simplest way to invent games is to modify the rules or the action of one already known. For example, addition *top it,* which we introduced in Chapter 7, is an adaptation of a popular children's card game called *war* and may be considered merely an addition game. A little more reflection reveals many other potential practice activities. Multiplication *top it* was presented in Chapter 8. Then we began to find other uses for *top it.*

1. If addition is to be stressed, the cards can be rewritten with series of three or perhaps four addends. Thus, cards may appear in the format in Figure 15–3a.
2. When children are ready for introductory multiplication work, the addition cards may have three, four, or five addends with the same value, as shown in Figure 15–3b.
3. The original instructions did not mention subtraction. Since addition and subtraction facts may be learned in concert, subtraction *top it* is appropriate. Sample cards are shown in Figure 15–3c.
4. Cards with mixed addition and subtraction expressions can be used as practice for both operations. The cards might look like those in Figure 15–3d.
5. If younger children are involved, they can play a similar game using dot patterns instead of addition or subtraction expressions. Figure 15–3e illustrates sample cards.
6. Another way to vary the game is to use numerals, as shown in Figure 15–3f.

The essential action has been retained in each variation shown in Figure 15–3; that is, two cards are drawn, and the larger number takes the smaller. In the case of a tie, another card is picked by both of the players who tied.

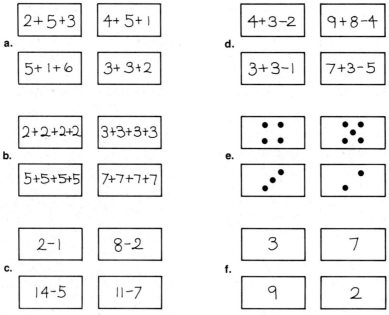

Figure 15–3

The second way to modify a game is to change the rules. One general example will suffice to illustrate changing the rules. Each game discussed above could be constructed so that each numeral or figure is one of two colors, blue or orange. The rules could be altered, so that when both players draw the same color numeral or figure, the larger takes the smaller, but when the two players draw different colors, the smaller takes the larger. Thus, with one rule change, the games assume a different character, although they maintain the goal of providing practice in basic mathematical skills.

Take care to prevent boredom by spacing the uses of variations of the same game and by using only variations appropriate for the practice needed. The multitude of examples listed show how games are invented by modifying the action or the rules of existing games.

Adapting the Rules and the Action.   There are times when simultaneously adapting the rules and the action of a game produces an activity quite different from the original. For example, consider three activities involving the creative use of dominoes. First, some description of the dominoes and their construction is appropriate. Although it is quite all right to use commercially produced dominoes for these activities, children are particularly attracted by larger sets, which are easily constructed from railroad board. A set that measures 10 × 20 centimeters is a good size. The complete set of 28 double-six dominoes is shown in Figure 15–4.

Domino Activity 1: Sum

Objective: *to score as many points as possible in a single round by adding the numbers of dots on five dominoes.*

Materials: *dominoes, paper, pencil.*
Players: *2 to 5.*
Play:

1. Spread out the dominoes face down.
2. Each player picks one domino, turns it over, and adds the numbers of dots. The player with the highest total plays first.
3. Each player then chooses five of the unexposed dominoes but does not look at them.
4. The first player turns over the five dominoes one at a time. When a domino is turned over, the player adds the numbers of dots, then announces and records the total. The first player continues to turn over the dominoes, announcing and recording each sum.
5. When the first player has finished with all five dominoes, he or she determines the score for the round by adding the five numbers recorded.
6. The next player continues in the same manner, hoping to find a larger total.
7. The player with the highest score wins that particular game. Play continues as before, with different children having the opportunity to win the game. Sample play for one player in one round is shown in Figure 15–5.

Domino Activity 2: Going Down

Objective: *to get as close as possible to zero without going below it.*
Materials: *dominoes, paper, pencil.*
Players: *2 to 5.*

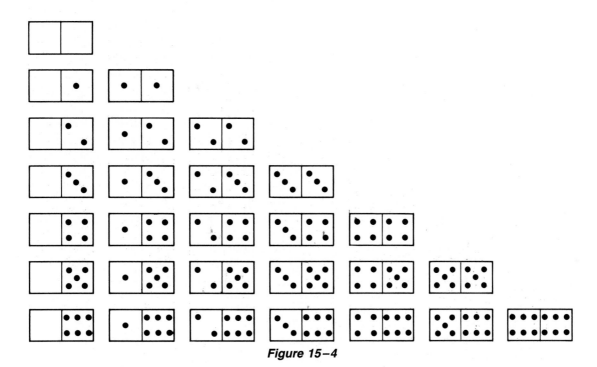

Figure 15–4

dominoes drawn	domino score	announced value
	3 + 4	7
	6 + 0	6
	1 + 3	4
	5 + 2	7
	1 + 1	2
	total for round:	26

Figure 15–5

3. Each player then chooses five of the unexposed dominoes but does not look at them. The player must use exactly four of the five dominoes during the round.
4. At the start of play, each player has ten points.
5. The first player turns over the five dominoes one at a time. When a domino is turned over, the player finds the difference between the numbers on the two halves. The player then subtracts the difference from the starting score of 10 points and records the new score. Because the player may use only four of the five dominoes, he or she must decide which one to discard. The decision whether or not to discard a domino must be made when the difference is determined. Once a domino is used, it may not later be discarded. Likewise, once a domino has been discarded, all of the remaining dominoes must be played.
6. The first player continues until four dominoes have been used, or until the score goes below zero, in which case the player loses the round. Play continues until all players have played.
7. The player closest to zero without going below it wins the round. Sample play for one player in one round is shown in Figure 15–7.

Domino Activity 3: Match

Objective: *to collect as many dominoes as possible.*
Materials: *dominoes.*
Players: *2.*

Play:

1. Spread out the dominoes face down.
2. Each player picks one domino, turns it over, and finds the difference between the numbers of dots on the two halves. The player with the highest difference plays first. (See Figure 15–6. Player 2 would play first.)

player 1 draws
difference is
6 − 4 = 2

player 2 draws
difference is
3 − 0 = 3

player 3 draws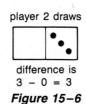
difference is
5 − 5 = 0

Figure 15–6

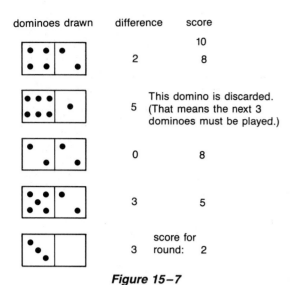

dominoes drawn	difference	score
		10
	2	8
	5	This domino is discarded. (That means the next 3 dominoes must be played.)
	0	8
	3	5
	3	score for round: 2

Figure 15–7

Play:

1. Shuffle the dominoes and stack them face down in the center of the playing area.
2. Each player picks one domino, turns it over, and adds the numbers of dots. The player with the highest total plays first.
3. The first player declares "Match," picks a domino from the stack, and turns it face up. The second player then picks a domino from the stack and turns it face up.
4. If there is a match, the player declaring match picks up and keeps both dominoes. If no match occurs, the second player picks up both dominoes. A match occurs when the dots on either end of one domino match the dots on either end of the other domino or when the sum of dots on one domino matches the sum of dots on the other domino.
5. Play continues with the second player declaring match. Both players then turn over dominoes to see who will keep the dominoes.
6. Play continues with players alternately declaring match until all the dominoes have been taken.
7. The player with the greatest number of dominoes wins the game.

As you can see, these domino activities are different from the game of dominoes. They do, however, represent the creative use of a common material. With additional thought, you should be able to develop some domino activities of your own. How might the dominoes be used with multiplication, fractions, or decimals?

Children-Made Learning Aids

Children are capable of developing and constructing learning aids. We have encouraged you to involve children in inventing problem-solving activities. We also encourage you to challenge children to develop fun games and activities that may be used to reinforce skills. Their first at-

tempts to develop *new* games will most likely involve changing the rules of an existing game. Children may wish to alter games in the ways mentioned earlier for teachers.

After a little practice with games, children will be interested in trying to invent games of their own. One way we have found to generate enthusiasm for inventing games is to announce a games contest. A games contest can be effective if it is preceded by discussion and analysis of the components of various games. Kohl (1974) described six such components in detail: theme, playing board, pieces, decision devices, goals, and teaching the game. Once children know about the structure of games, they will be more effective inventors of games. Cruikshank and Martin (1981) describe how one successful games contest was organized and operated. Whether participating in a games contest, suggesting changes in how games are played, or experimenting with making games, children prove to be imaginative and creative.

Constructing Learning Aids

Once you decide on a sound, usable learning aid, begin construction. Game boards and cards should be attractive and colorful. Children enjoy bright, cheery materials with which to play. These materials can be made with brightly colored railroad board or with colorful lettering and drawings on white railroad board. Pictures cut from magazines, ready-made stickers, or children's drawings sometimes add an extra touch of color that attracts children.

Materials should be durable. Teachers willing to spend the time and effort to produce a long-lasting, durable aid will be rewarded by time saved repairing or remaking the activity. Most teacher-made materials may be protected by covering the board or cards with either plastic laminate or clear contact paper. Both protective materials are readily available at art, hardware, and variety stores. Once covered, front and back, a learning aid will last for months or even years.

The quality of an aid is improved if care is taken when lettering or attaching pictures to the material. Letters and numerals should be easy to read. They may be affixed by hand, such as rub-on letters, or stenciled. Gummed letters or numerals also may be used. Using a lettering machine such as an Ellison to cut out letters, numerals, and teaching aids such as tangrams, pentominoes, fraction pieces, attribute blocks, pattern blocks and base ten blocks can also enhance the quality of your learning aids.

Some activities should have written instructions or should be explained by the teacher. Because an activity with poorly explained rules is of little value, the instructions should be carefully worded to be clear and concise. Pictures or drawings are often useful in explaining the action of a game.

Besides motivating children, learning aids should help reinforce or teach a concept or skill. There should be no question in your mind about what concept or skill is being presented by a learning aid. As well, for nearly every learning aid, consider the potential for adapting it to make another learning aid for teaching mathematics, communication skills, social studies, science, and so on.

Concentration games are among the most versatile. Sixteen to forty-eight cards may reinforce shapes, number patterns, numeral-number recognition, addition, subtraction, multiplication, division, words with the same beginning sound, homonyms, names and faces, animals and habitats, and so forth. For younger children, fewer cards (eight to sixteen) should be used and the cards should be laid out in two rows. The simplicity of concentration games makes them practical learning aids.

When you are assigned your first classroom, take the opportunity to go into the room alone and look it over. If it is during the summer, there may be no furniture or the furniture may be in disarray. Imagine how you would like the classroom to look. Imagine children in the room and instruction taking place. Begin to sense how all of the space can be useful. When you leave, let your mind work on how that learning environment will be molded in the months to come into an exciting, alive, dynamic environment for all the children. What a wonderful place it will be in which to learn mathematics!

KEY IDEAS

Fitting mathematics instruction to children's differing needs requires careful planning. Give consideration to both the children's and your own needs and styles. A teaching-learning strategy contains student participation, teacher behavior, and student grouping. A number of options for each component of a strategy result in many combinations of ways to teach. You are encouraged to teach using a variety of approaches.

Evaluating mathematical learning involves collecting data through textbook tests, teacher observation and questioning, standardized tests, writing activities, diagnostic interviews, teacher-prepared examinations, and group problem solving. Once collected, evaluative data should be used as a basis for further mathematics instruction and discussions with children and parents.

There are a number of children with special needs in most classrooms. Some children need additional help and time as they learn mathematics. Continued work with manipulatives helps to lay the groundwork for success. Other children have special talent when it comes to mathematics. They, too, need additional time and direction. Exposure to rich resources benefits the gifted.

The physical environment for learning mathematics should be inviting. It is a special place for children and should reflect their interests as well as those of the teacher. As a part of the learning environment, learning aids provide the framework for the mathematics curriculum. Textbooks, kits, games, structural materials, activity cards, and computers are commonly found in well-established classrooms.

REFERENCES

BARATTA-LORTON, MARY. *Mathematics Their Way.* Menlo Park, Ca.: Addison-Wesley Publishing Co., 1976.

BARSON, ALAN. "Task Cards." *Arithmetic Teacher.* Vol. 26, No. 2 (October 1978), pp. 53–54.

BIGGS, EDITH E., AND MACLEAN, JAMES R. *Freedom to Learn.* Reading, Ma.: Addison-Wesley Publishing Co., 1969.

BURNS, MARILYN. "Groups of Four: Solving the Management Problem." *Learning.* Vol. 10, No. 2 (September 1981), pp. 46–51.

CHARLES, RANDALL; LESTER, FRANK; AND O'DAFFER, PHARES. *How to Evaluate Progress in Problem Solving.* Reston, Va.: National Council of Teachers of Mathematics, 1987.

CHILDS, LEIGH, AND ADAMS, NANCY. *Math Sponges.* San Diego, Ca.: National Institute for Curriculum Enrichment, 1979.

COOMBS, BETTY, AND HARCOURT, LALIE. *Explorations 2.* Don Mills, Ontario: Addison-Wesley Publishing Co., 1986.

COPELAND, RICHARD W. *How Children Learn Mathematics.* New York: Macmillan Co., 1984.

CRUIKSHANK, DOUGLAS E., AND MARTIN, JOHN A., JR. "The Mathematical Game Contest." *Arithmetic Teacher.* Vol. 28, No. 5 (January 1981), pp. 42–45.

DESIGN GROUP. *The Way to Play.* New York: Paddington Press, 1975.

Enrichment Mathematics for the Grades. National Council of Teachers of Mathematics, 27th Yearbook. Washington, D.C.: NCTM, 1963.

EQUALS AND CALIFORNIA MATHEMATICS COUNCIL ASSESSMENT COMMITTEE. *Assessment Alternatives in Mathematics: An Overview of Assessment Techniques That Promote Learning.* Berkeley, Ca.: EQUALS, Lawrence Hall of Science, University of California, 1989.

EVERTSON, CAROLYN M., ET AL. *Classroom Management for Elementary Teachers.* Englewood Cliffs, N.J.: Prentice-Hall, 1989.

JOHNSON, DAVID W., AND JOHNSON, ROGER T. *Learning Together and Alone: Cooperative, Competitive, and Individualistic Learning.* Englewood Cliffs, N.J.: Prentice-Hall, Inc., 1987.

JOHNSON, DAVID W.; JOHNSON, ROGER T.; AND HOLUBEC, EDYTHE JOHNSON. *Circles of Learning: Cooperation in the Classroom.* Edina, Minn.: Interaction Book Co., 1986.

KENNEY, ROBERT. "Vermont's Assessment System—How Will It Work?" Unpublished paper presented to the Association of State Supervisors of Mathematics, 1990.

KOHL, HERBERT R. *Math, Writing and Games.* New York: The New York Book Review, 1974.

LANE COUNTY MATHEMATICS PROJECT. *Problem Solving in Mathematics.* Palo Alto, Ca.: Dale Seymour Publications, 1984.

MOOMAW, VERA, ET AL. *Expanded Mathematics Grades 4–5–6.* Eugene, Or.: School District No. 4, Instruction Department, 1967.

NATIONAL COUNCIL OF TEACHERS OF MATHEMATICS. *Professional Standards for Teaching Mathematics.* Reston, Va.: NCTM, 1991.

PECK, DONALD M.; JENCKS, STANLEY M.; AND CONNELL, MICHAEL L. "Improving Instruction Through Brief Interviews." *Arithmetic Teacher.* Vol. 37, No. 3 (November 1989), pp. 15–17.

PETERSON, DANIEL. *Functional Mathematics for the Mentally Retarded.* Columbus, Oh.: Charles E. Merrill, 1973.

ROSENSHINE, BARAK, AND STEVENS, ROBERT. "Teaching Functions." In Wittrock, Merlin C., ed. *Handbook of Research on Teaching.* New York: Macmillan Co., 1986.

SCHOEN, HAROLD L., AND ZWENG, MARILYN J., EDS. *Estimation and Mental Computation.* National Council of Teachers of Mathematics, 1986 Yearbook. Reston, Va.: NCTM, 1986.

The Slow Learner in Mathematics. National Council of Teachers of Mathematics, 35th Yearbook. Washington, D.C.: NCTM, 1972.

TAYLOR, ANN P., AND VLASTOS, GEORGE. *School Zone: Learning Environments for Children.* New York: Van Nostrand Reinhold Co., 1975.

Appendixes

Appendix A

Suppliers of Manipulative Materials and Computer Software

Appendix B

Blackline Masters

Attribute Shapes
Base Ten Patterns (Decimal Patterns)
Hundreds Chart
Table for Addition or Multiplication
Centimeter Graph Paper
Inch Graph Paper
Circular Fraction Patterns
Rectangular Fraction Patterns
Fraction Strips
How to Make a Geoboard
Rectangular Dot Paper
Isometric Dot Paper
Tangram Pattern
Regular Polyhedra
    Octahedron
    Tetrahedron
    Cube
Regular Polyhedra
    Icosahedron
Regular Polyhedra
    Dodecahedron
Algebra Blocks
Lesson Plan Outline

Appendix A   Suppliers of Manipulative Materials and Computer Software

Activity Resources Company, Inc.
P.O. Box 4875
Hayward, CA 94545

Addison-Wesley Publishing Co., Inc.
2725 Sand Hill Road
Menlo Park, CA 94025

AIMS Education Foundation
P.O. Box 8120
Fresno, CA 93747

Apple Computer, Inc.
20525 Mariani Avenue
Cupertino, CA 95014

Broderbund Software
P.O. Box 12947
San Rafael, CA 94913-2947

Conduit
Oakdale Campus
Iowa City, IA 52242

Creative Publications
5005 West 110th Street
Oak Lawn, IL 60453

Cuisenaire Co. of America, Inc.
12 Church Street, Box D
New Rochelle, NY 10802

Dale Seymour Publications
P.O. Box 10888
Palo Alto, CA 94303

Delta Education
Box M, Math Department
Nashua, NH 03061-6012

Didax Educational Resources, Inc.
One Centennial Drive
Peabody, MA 01960

DLM Teaching Resources
P.O. Box 4000
One DLM Park
Allen, TX 75002

Educational Resources
1550 Executive Drive
Elgin, IL 60123

Educational Teaching Aids
199 Carpenter Avenue
Wheeling, IL 60090

Hartley Courseware, Inc.
P.O. Box 419
Dimondale, MI 48821

International Business Machines Corp.
1133 Westchester Avenue
White Plains, NY 10601

Lawrence Erlbaum Associates, Inc.
365 Broadway
Hillsdale, NJ 07642

Lawrence Hall of Science
University of California
Berkeley, CA 94720

The Learning Company
545 Middlefield Road
Menlo Park, CA 94025

Math Learning Center
P.O. Box 3226
Salem, OR 97302

Media Materials
2936 Remington Avenue
Baltimore, MD 21211

Midwest Publications
P.O. Box 448
Pacific Grove, CA 93950

Minnesota Educational
   Computing Consortium (MECC)
3490 Lexington Avenue, North
St. Paul, MN 55126

National Council of Teachers of Mathematics
1906 Association Drive
Reston, VA 22091

Scholastic Book Services
730 Broadway
New York, NY 10003

Scott Resources, Inc.
1900 East Lincoln, Box 2121
Fort Collins, CO 80522

Selective Educational Equipment, Inc.
3 Bridge Street
Newton, MA 02159

Summit Learning
P.O. Box 493
Fort Collins, CO 80522

Sunburst Communications
39 Washington Avenue
Pleasantville, NY 10570-9971

Weekly Reader Family Software
245 Long Hill
Middletown, CT 06457

Wings for Learning, Inc.
1600 Green Hills Road
P.O. Box 660002
Scotts Valley, CA 95067-0002

Attribute Shapes

Base Ten Patterns (Decimal Patterns)

Base Ten Patterns (Decimal Patterns)

Base Ten Patterns (Decimal Patterns)

Hundreds Chart

1	2	3	4	5	6	7	8	9	10
11	12	13	14	15	16	17	18	19	20
21	22	23	24	25	26	27	28	29	30
31	32	33	34	35	36	37	38	39	40
41	42	43	44	45	46	47	48	49	50
51	52	53	54	55	56	57	58	59	60
61	62	63	64	65	66	67	68	69	70
71	72	73	74	75	76	77	78	79	80
81	82	83	84	85	86	87	88	89	90
91	92	93	94	95	96	97	98	99	100

Table for Addition or Multiplication

	0	1	2	3	4	5	6	7	8	9
0										
1										
2										
3										
4										
5										
6										
7										
8										
9										

Centimeter Graph Paper

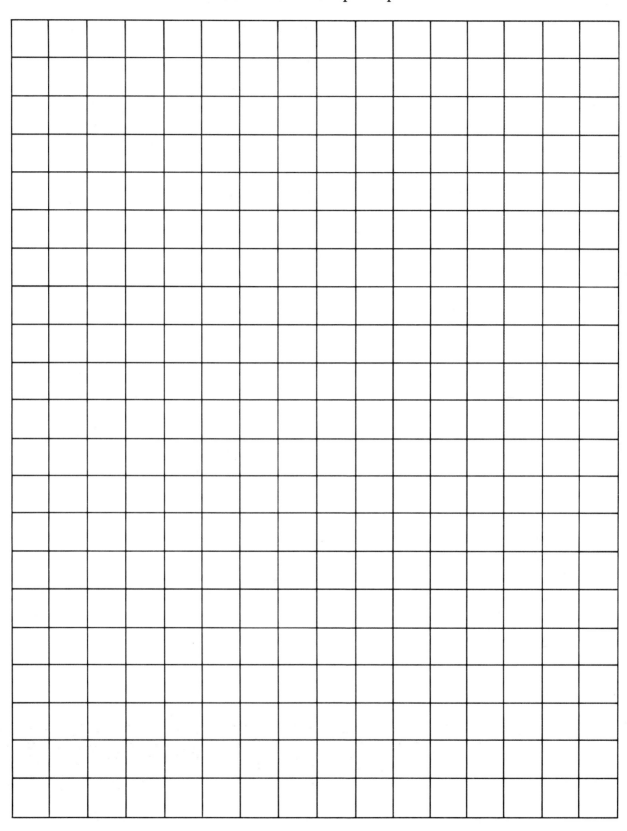

Inch Graph Paper

Circular Fraction Patterns

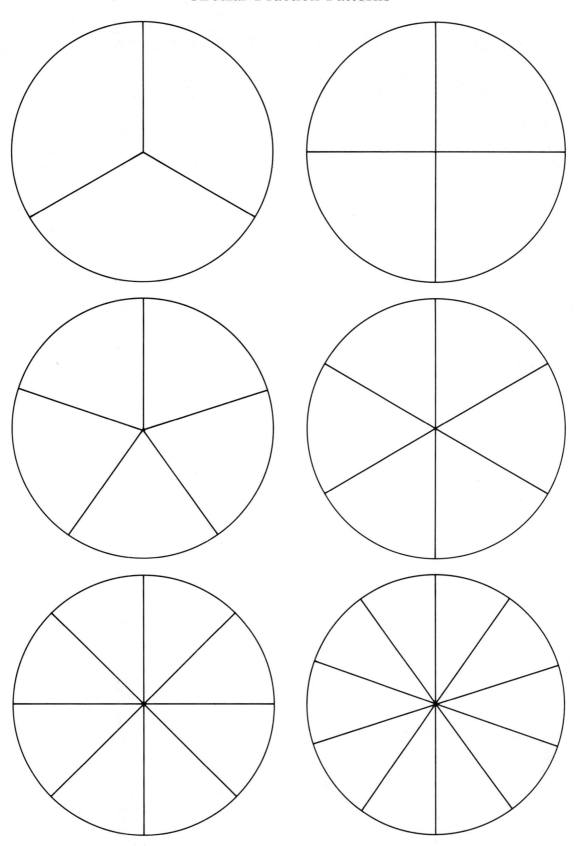

Rectangular Fraction Patterns

one unit

Fraction Strips

1

$\frac{1}{2}$	$\frac{1}{2}$

$\frac{1}{3}$	$\frac{1}{3}$	$\frac{1}{3}$

$\frac{1}{4}$	$\frac{1}{4}$	$\frac{1}{4}$	$\frac{1}{4}$

$\frac{1}{6}$	$\frac{1}{6}$	$\frac{1}{6}$	$\frac{1}{6}$	$\frac{1}{6}$	$\frac{1}{6}$

$\frac{1}{8}$	$\frac{1}{8}$	$\frac{1}{8}$	$\frac{1}{8}$	$\frac{1}{8}$	$\frac{1}{8}$	$\frac{1}{8}$	$\frac{1}{8}$

$\frac{1}{12}$	$\frac{1}{12}$	$\frac{1}{12}$	$\frac{1}{12}$	$\frac{1}{12}$	$\frac{1}{12}$	$\frac{1}{12}$	$\frac{1}{12}$	$\frac{1}{12}$	$\frac{1}{12}$	$\frac{1}{12}$	$\frac{1}{12}$

$\frac{1}{24}$	$\frac{1}{24}$	$\frac{1}{24}$	$\frac{1}{24}$	$\frac{1}{24}$	$\frac{1}{24}$	$\frac{1}{24}$	$\frac{1}{24}$	$\frac{1}{24}$	$\frac{1}{24}$	$\frac{1}{24}$	$\frac{1}{24}$	$\frac{1}{24}$	$\frac{1}{24}$	$\frac{1}{24}$	$\frac{1}{24}$	$\frac{1}{24}$	$\frac{1}{24}$	$\frac{1}{24}$	$\frac{1}{24}$	$\frac{1}{24}$	$\frac{1}{24}$	$\frac{1}{24}$	$\frac{1}{24}$

How to Make a Geoboard

You need: One 8″ × 8″ piece of $\frac{1}{2}$″ plywood

Twenty-five $\frac{5}{8}$″ Escutcheon Pins (#16 or #18)

One straight and accurate ruler

One pencil

One hammer

Some fine sandpaper

Step 1: Use the sandpaper to smooth the edges of your board.

Step 2: Carefully measure and lightly mark on your board five evenly spaced rows and columns (see Figure A–1). Where each intersection occurs, drive a nail. Each row and column is $1\frac{3}{4}$″ from the next row and column. The border is $\frac{1}{2}$″.

Step 3: Pound the nails into the board at the intersections of the rows and columns. Be sure the nails are in the straightest rows possible. Constantly line them up by sighting along the rows of escutcheon pins.

Figure A–1

Rectangular Dot Paper

Name:

Isometric Dot Paper

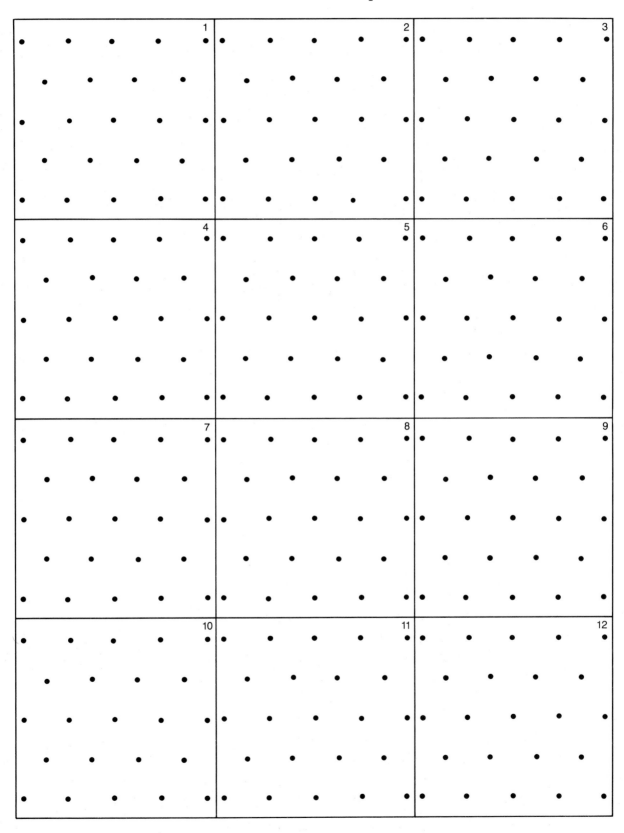

Tangram Pattern

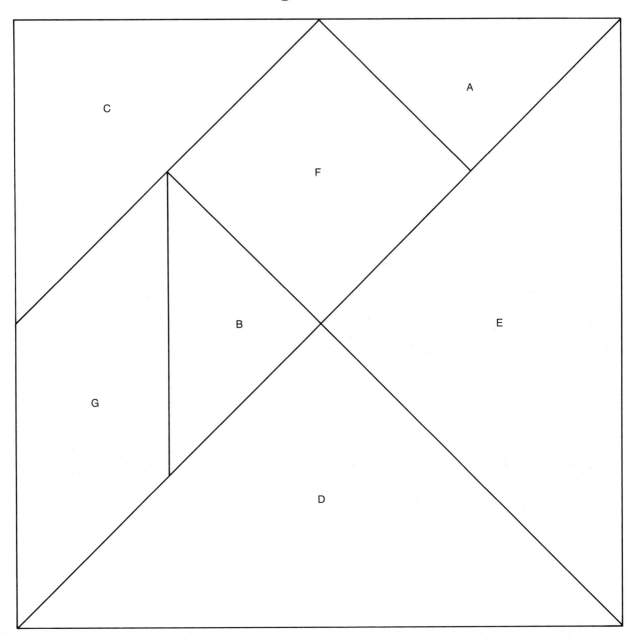

Regular Polyhedra

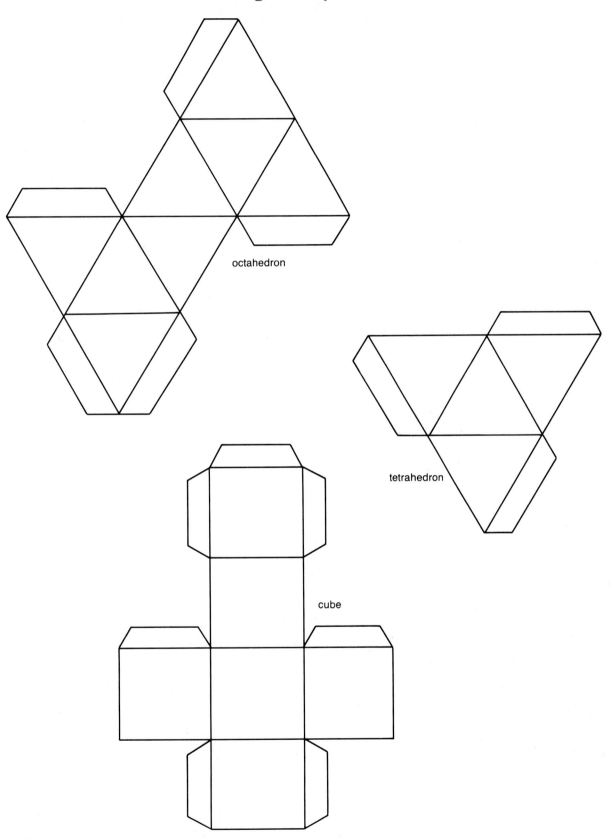

octahedron

tetrahedron

cube

Regular Polyhedra

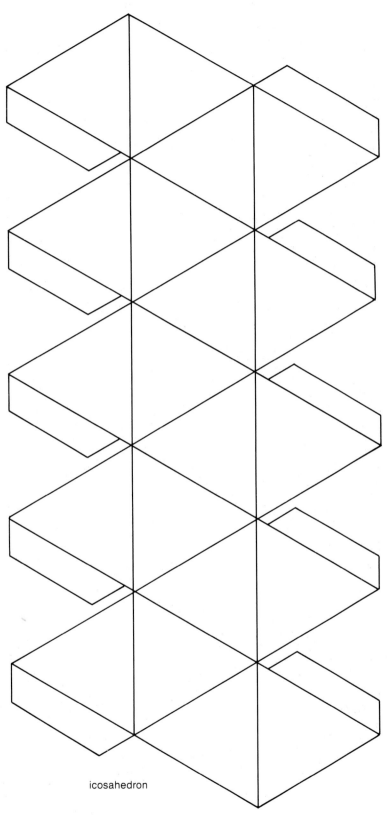

icosahedron

Regular Polyhedra

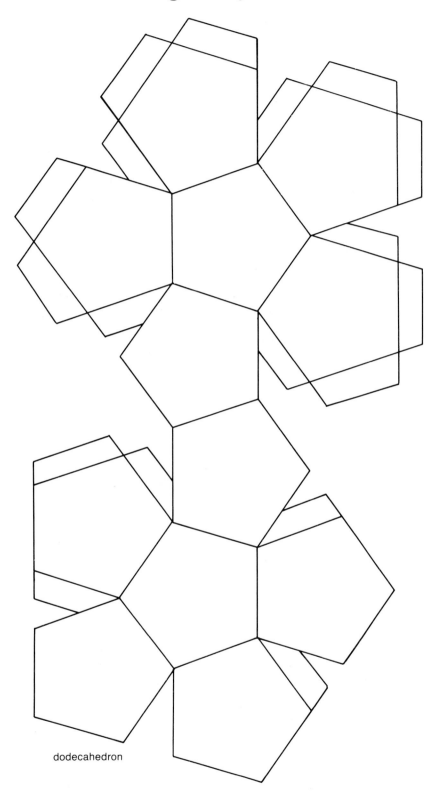

dodecahedron

Algebra Blocks

Lesson Plan Outline

Name: _____

Subject: _____ Date: _____ Time: _____

INSTRUCTIONAL OBJECTIVES:

TERMS or VOCABULARY:

LEARNING MATERIALS/AIDS:

TEACHING STRATEGY:
    *Introductory Activity (Anticipatory Set):

    *Meeting the Objectives (Instructional Input, Modeling, Monitoring, and Adjusting):

    *Practice (Guided and Independent):

    *Closure:

ASSESSMENT (Determine If Instructional Objectives Were Met):

Index